Lecture Notes in Mathematics

Edited by A. Dold and B. Eckmann

1357

S. Hildebrandt R. Leis (Eds.)

Partial Differential Equations and Calculus of Variations

Springer-Verlag
Berlin Heidelberg New York London Paris Tokyo

Editors

Stefan Hildebrandt
Mathematisches Institut, Universität Bonn
Wegelerstr. 10, 5300 Bonn, Federal Republic of Germany

Rolf Leis
Institut für Angewandte Mathematik, Universität Bonn
Wegelerstr. 10, 5300 Bonn, Federal Republic of Germany

Mathematics Subject Classification (1980): 35A15 & A30, 35B, 35G, 35J, 35K, 35L, 49F, 53C

ISBN 3-540-50508-3 Springer-Verlag Berlin Heidelberg New York
ISBN 0-387-50508-3 Springer-Verlag New York Berlin Heidelberg

Printed in Germany

Printing and binding: Druckhaus Beltz, Hemsbach/Bergstr.
2146/3140-543210

PREFACE

This volume contains 18 papers by members and guests of the former Sonderforschungsbereich 72 (SFB 72) "*Approximation und Mathematische Optimierung in einer anwendungsbezogenen Mathematik*", who have, over the years, collaborated in the research group "*Lösung partieller Differentialgleichungen und Variationsrechnung*". It is an account of 15 years of research.

The SFB 72 was in existence from 1971 to 1986 as an institute at the University of Bonn. It was supported by the Deutsche Forschungsgemeinschaft, by the University, and by the State of Nordrhein-Westfalen. Members of the special research group were mathematicians holding chairs at Bonn University as well as younger scientists with temporary positions. They collaborated with colleagues from neighbouring universities and with numerous guests from abroad who visited the SFB 72.

In 1986 the SFB 72 reached its maximum life span of 15 years. On this occasion we invited our colleagues and our former guests to contribute a research paper or a survey of a part of our common field of interest, the theory of partial differential equations and the calculus of variations. We hope to have attained a good balance of surveys and of original contributions. The emphasis is on existence and regularity results for variational problems and for nonlinear differential equations, on special equations of mathematical physics, and on problems of scattering theory.

We would like to take this opportunity to thank all those who have over the years supported the work of the Sonderforschungsbereich 72. Our special gratitude goes to the Deutsche Forschungsgemeinschaft and its referees. We are also deeply indebted to the State of Nordrhein-Westfalen, to our University, to our former colleagues and coworkers, and to our secretaries. We particularly thank Mrs. Anke Vogt who, as always, has overcome all difficulties in preparing this volume.

Last but not least we thank the series editors and the publisher for including our celebration issue in the series *Lecture Notes in Mathematics* of Springer-Verlag.

Bonn, 15th April 1988

Stefan Hildebrandt
Rolf Leis

TABLE OF CONTENTS

ON THE EXISTENCE IN THE LARGE OF SOLUTIONS TO THE ONE-DIMENSIONAL, ISENTROPIC HYDRODYNAMIC EQUATIONS IN A BOUNDED DOMAIN

H.D. Alber
Mathematisches Institut A der Universität Stuttgart
Pfaffenwaldring 57, 7000 Stuttgart 80

1. Introduction

We study the initial value problem given by the equations

$$\begin{aligned} &\rho_t + (\rho v)_x = 0, \\ &(\rho v)_t + (\rho v^2)_x + p_x = 0, \end{aligned} \qquad a < x < b,\ t \geq 0 \tag{1.1}$$

and the boundary and initial conditions

$$v(a,t) = v(b,t) = 0\ ,\ t > 0 \tag{1.2}$$

$$v(x,0) = V(x)\ ,\ \rho(x,0) = P(x)\ ,\ a \leq x \leq b. \tag{1.3}$$

This system of equations describes the one-dimensional, compressible flow of a gas enclosed in the interval (a,b), with the density $\rho(x,t)$, the flow velocity $v(x,t)$, and the pressure

$$p = A\rho^\gamma, \tag{1.4}$$

which only depends on the density. Here $A > 0$ and $\gamma > 1$ are constants. We prove the following result.

Theorem 1.1: *Let constants $0 < P_0 < P_1$ be given, and for the initial values (P,V) let*

$$P_0 \leq P(x) \leq P_1 \tag{1.5}$$

for all $x \in [a,b]$ and

$$V(a) = V(b) = 0\ ,\ TV((P,V)) \leq C_0. \tag{1.6}$$

If the constant C_0 is sufficiently small, then there exists a weak solution $u = (\rho,v)$ of (1.1) - (1.4) in $(a,b) \times \mathbb{R}^+$, which satisfies the entropy condition and the inequalities

$$\left|\rho(x,t) - (b-a)^{-1}\int_a^b P(x)\,dx\right| \leq C_1\ TV((P,V)) \tag{1.7}$$

$$TV(u(\cdot,t)) \leq C_2\ TV((P,V)). \tag{1.8}$$

Here

$$TV((P,V)) = TV(P;[a,b]) + TV(V;[a,b]), \tag{1.9}$$

and TV is the total variation. From (1.7) it follows that the solution does not contain cavities if C_0 is sufficiently small. We only consider solutions without cavities, since the equations (1.1) become singular at $\rho = 0$, which makes the methods of this paper inapplicable. A function $u = (\rho,v) : (a,b) \times \mathbb{R}^+ \longrightarrow \mathbb{R}^+ \times \mathbb{R}$ is said to be a weak solution of (1.1) - (1.4), if it belongs to $\mathcal{L}_\infty((a,b) \times \mathbb{R}^+)$ and satisfies the equations

$$(1.10) \qquad \int_0^\infty \int_a^b [\rho\phi_t + \rho v\phi_x]\, dx\, dt + \int_a^b P(x)\phi(x,0)\, dx = 0$$

$$(1.11) \qquad \int_0^\infty \int_a^b [\rho v\Psi_t + (\rho v^2+p)\Psi_x]\, dx\, dt + \int_a^b P(x)V(x)\Psi(x,0)\, dx = 0$$

for all infinitely differentiable functions $\phi,\psi : [a,b] \times [0,\infty] \longrightarrow \mathbb{R}$, which vanish for all sufficiently large t and satisfy $\psi(a,t) = \psi(b,t) = 0$ for all t. Equation (1.10) contains the boundary condition

$$(\rho v)(a,t) = (\rho v)(b,t) = 0.$$

The boundary condition (1.2) follows from this condition and from $\rho > 0$.

A function $\eta : \mathbb{R}^+ \times \mathbb{R} \longrightarrow \mathbb{R}$ is called entropy, if another function $q : \mathbb{R}^+ \times \mathbb{R} \longrightarrow \mathbb{R}$ exists such that for each continuously differentiable solution $u = (\rho,v)$ of (1.1) the additional conservation law

$$\eta(\rho(x,t),(\rho v)(x,t))_t + q(\rho(x,t),(\rho v)(x,t))_x = 0$$

holds. By definition, a weak solution u of (1.1) satisfies the entropy condition if for each convex entropy

$$\eta_t + q_x \leq 0$$

holds in the distribution sense. Regarding these definitions confer [8].

In [4] DiPerna showed with the method of compensated compactness, that the equations of the compressible isentropic flow defined on the whole real axis have a global solution for all initial values $(P,V) \in \mathcal{L}_\infty(\mathbb{R})$. However, his proof cannot be transfered to boundary-value problems in bounded domains, since difficulties appear with boundary terms of the entropies. The proof of Glimm [5] for the existence of weak solutions defined on all of $\mathbb{R} \times \mathbb{R}^+$ can be transfered to regions of the form $(a,\infty) \times \mathbb{R}^+$, but not to regions of the form considered here. However, we guess that the method used by Glimm & Lax in [6] to show global existence for an initial-value problem in case of periodic initial data can be transfered to the initial-boundary value problem (1.1) - (1.4).

The idea of proof in the present paper is based upon the fact that the total strength of the shock waves in the solution is not increased by interactions of waves, if the strength of shock waves is measured in an appropriate coordinate system in the ρv-plane. This idea was used by Bakhvarov [2]. DiPerna [3] applied this method to show global existence of the solution for the hydrodynamic equations in Lagrangian representation in $\mathbb{R} \times \mathbb{R}_0^+$.

This coordinate system is introduced in section 2. The basic idea of the proof is contained in section 3, where the interaction of discontinuities is studied. Theorem 1.1 is proved in section 4. For the proof we use the sequence of approximating solutions constructed in [1] and show how a-priori estimates for the total variation of these approximating solutions can be derived by means of the results from section 3.

2. Discontinuities of weak solutions

A piecewise constant weak solution of (1.1) satisfies the Rankine-Hugoniot conditions

$$(2.1)\qquad \begin{aligned} s(\rho-\rho_0) &= \rho v - \rho_0 v_0, \\ s(\rho v-\rho_0 v_0) &= \rho v^2 - \rho_0 v_0^2 + p - p_0, \end{aligned}$$

where (ρ,v) and (ρ_0,v_0) are the constant values of the solution on either side of the curve $x = x(t)$, along which the solution jumps, and where $s = dx/dt$ is the speed of the discontinuity. From the Rankine-Hugoniot conditions one obtains by some calculations that

$$v = v_0 \pm \left[(p-p_0)\left(\frac{1}{\rho_0} - \frac{1}{\rho}\right)\right]^{1/2}$$

and

$$(2.2)\qquad s = \frac{\rho v-\rho_0 v_0}{\rho-\rho_0} = v_0 \pm \left(\frac{\rho}{\rho_0}\right)^{\frac{1}{2}}\left[\frac{p-p_0}{\rho-\rho_0}\right]^{\frac{1}{2}} = v_0 \pm \left(\frac{\rho}{\rho_0}\right)^{1/2}(A\gamma \rho^{*\gamma-1})^{1/2},$$

with a suitable number ρ^* between ρ and ρ_0. The last equation on the right hand side of (2.2) follows from (1.4) and from the mean value theorem.

Let the functions $c_{1(\rho_0,v_0)}, c_{2(\rho_0,v_0)} : \mathbb{R}^+ \longrightarrow \mathbb{R}$ be given by

$$(2.3)\qquad c_{1(\rho_0,v_0)}(\rho) = \begin{cases} v_0 + \left[(p-p_0)\left(\frac{1}{\rho_0} - \frac{1}{\rho}\right)\right]^{1/2}, & \rho \geq \rho_0 \\[2ex] v_0 - \left[(p-p_0)\left(\frac{1}{\rho_0} - \frac{1}{\rho}\right)\right]^{1/2}, & \rho \leq \rho_0 \end{cases}$$

$$(2.4)\qquad c_{2(\rho_0,v_0)}(\rho) = \begin{cases} v_0 - \left[(p-p_0)\left(\frac{1}{\rho_0} - \frac{1}{\rho}\right)\right]^{1/2}, & \rho \geq \rho_0 \\ v_0 + \left[(p-p_0)\left(\frac{1}{\rho_0} - \frac{1}{\rho}\right)\right]^{1/2}, & \rho \leq \rho_0 \end{cases} .$$

(ρ_0,v_0) and (ρ,v) are the states on either side of a discontinuity of a weak solution if and only if (ρ,v) either lies on the graph of $c_{1(\rho_0,v_0)}$ or on the graph of $c_{2(\rho_0,v_0)}$. The discontinuity is called an i-wave if (ρ,v) is a point of the graph of $c_{i(\rho_0,v_0)}$. From (2.2) - (2.4) it follows that the velocity of a 1-wave is larger than v_0 and that the velocity of a 2-wave is less than v_0. This holds no matter whether (ρ_0,v_0) is the state to the left or to the right of the discontinuity. If (ρ,v) is the state to the left of a 1-wave or to the right of a 2-wave, then the discontinuity is said to be a shock wave if $\rho > \rho_0$, otherwise it is said to be a rarefaction wave.

It is our next goal to define various weak solutions $u = (\rho,v)$ of the Riemann initial-value problem of (1.1) to initial data given by

$$(2.5)\qquad u(x,o) = \begin{cases} u^l, & x < 0 \\ u^r, & x > 0 . \end{cases}$$

Let $w = (w_1,w_2) \in \mathbb{R}^+ \times \mathbb{R}$ and $i \in \{1,2\}$. To begin with, we define for each non-negative integer n a function $c^n_{iw} : \mathbb{R}^+ \longrightarrow \mathbb{R}$ and for each positive integer n a sequence $\{w^m(w,n,i)\}_{m=0}^{\infty} \subseteq \mathbb{R}^+ \times \mathbb{R}$. Let

$$(2.6)\qquad c^0_{iw} = c_{iw},$$

with the function c_{iw} defined in (2.3), (2.4). For $n > 0$ let

$$w^0(w,n,i) = w$$

and

$$(2.7)\qquad c^n_{iw}(\tau) = c_{iw}(\tau),$$

for all $\tau \geq w_1$. If $w^m(w,n,i)$ is defined, let $w^{m+1}(w,n,i) = (w_1^{m+1},w_2^{m+1})$ be the point on the graph of c_{iw^m} with $w_1^{m+1} < w_1^m$, chosen such that the arclength of this graph between w^{m+1} and w^m is equal to $1/n$. We set

$$(2.8)\qquad c^n_{iw}(\tau) = c_{iw^m}(\tau)$$

for all τ with $w_1^{m+1} \leq \tau \leq w_1^m$. This way c^n_{iw} is defined for all $\tau > 0$. Note that the sequence $\{w^m(w,n,i)\}_{m\geq 0}$ consists of points, which all are located on the graph of c^n_{iw}. The part of this graph belonging to $\tau \le w_1$ is divided by these points into pieces of length $1/n$. $c^n_{iw}(\tau)$ is independent of n for $\tau \ge w_1$, but for $\tau \le w_1$ it converges towards $\epsilon_{iw}(\tau)$ as $n \to \infty$, where the function ϵ_{iw} is defined by

$$\epsilon_{iw}(\tau) = w_2 - (-1)^i \frac{2(A\gamma)^{1/2}}{\gamma-1} (\gamma^{(\gamma-1)/2} - w_1^{(\gamma-1)/2}). \tag{2.9}$$

More precisely: Let τ be a point between w_1^{m+1} and w_1^m. Then from Taylor's formula we conclude that

$$c^{(n)}_{iw}(\tau) - \epsilon_{iw}(\tau) = \sum_{j=1}^{m} \frac{1}{3!} \partial^3[c_{iw^j} - \epsilon_{iw}](\tau^{*j})(w_1^j - w_1^{j-1})^3 \tag{2.10}$$

$$+ \frac{1}{3!} \partial^3[c_{iw^m} - \epsilon_{iw}](\tau^*)(\tau - w_1^m)^3,$$

where τ^{*j} is an appropriate point between w_1^j and w_1^{j-1}, and where τ^* lies between τ and w_1^m. Equation (2.10) follows from the facts that the graph of ϵ_{iw} is a level curve of a Riemann invariant of the system (1.1), that the graph of c_{iw} describes the family of states, which can be connected to w by discontinuities of solutions of (1.1), and that between these curves the inequalities

$$C_1|\tau - w_1|^3 \le |c_{iw}(\tau) - \epsilon_{iw}(\tau)| \le C_2|\tau - w_1|^3$$

hold with $C_1, C_2 > 0$, see [7]. A more precise investigation of (2.3), (2.4), and (2.9) yields

$$(-1)^{i-1} \partial^3[c_{iw} - \epsilon_{iw}] \ge 0.$$

From (2.10) and (2.11) we thus obtain

$$c^n_{iw}(\tau) - \epsilon_{iw}(\tau) = H_i(\tau, w, n)(\tau - w_1) \tag{2.11}$$

for all $\tau > 0$ with $H_1 \ge 0$, $H_2 \le 0$, and with

$$C_1|\tau - w_1|^2 \le |H_i(\tau, w, n)| \le C_2|\tau - w_1|^2$$

for all t and w_1 from a compact subset K of $(0,\infty)$, where the constants C_1, $C_2 > 0$ depend upon K. Moreover, from (2.10) it follows that for all $0 < c_1 < c_2$ there is a constant C with

$$|H_i(\tau, w)| \le C/n^2,$$

thus

$$|c^n_{i\omega}(\tau) - \epsilon_{i\omega}(\tau)| \le C/n^2, \tag{2.12}$$

for all $c_1 \le \tau \le w_1 \le c_2$.

Now we can define the solutions to (1.1) and (2.5). Let n and k be positive integers, and let $u^l = (\rho^l, v^l)$, $u^r = (\rho^r, v^r)$ be the points from $\mathbb{R}^+ \times \mathbb{R}$ given in (2.5). The graphs of $c^k_{1u^r}$ and $c^n_{2u^l}$ intersect at exactly one point u^+. We define points $u^{(0)},\ldots,u^{(\upsilon)}$ on this graph as follows. Let $\{w^i(u^l,n,2)\}_{i=1}^{m-1}$ be the set of points on the graph of $c^n_{2u^l}$ between u^l and u^+, and let $\{w^i(u^r,k,1)\}_{i=1}^{j-1}$ be the set of points on the graph of $c^k_{1u^r}$ between u^r and u^+. Depending upon the location of u^+, it is of course possible that one or both of these sets are empty, in which case we set m = 1 or j = 1. We obtain $u^{(0)},\ldots,u^{(\upsilon)}$ with $\upsilon = m + j$ by arranging the set

$$\{u^l,u^+,u^r\} \cup \{w^i(u^l,n,2)\}_{i=1}^{m-1} \cup \{w^i(u^r,k,1)\}_{i=1}^{j-1}$$

in such a way that the arclength of (graph $c^n_{2u^l}$) $\cup$ (graph $c^k_{1u^r}$) between u^l and $u^{(\mu+1)}$ is larger than the arclength between u^l and $u^{(\mu)}$. Thus, $u^{(0)} = u^l$, $u^{(m)} = u^+$, and $u^{(\upsilon)} = u^r$. Now let the solution of (1.1) to the initial values (2.5) be defined by

$$(2.13)\qquad u(x,t) = \begin{cases} u^l & , \ x \le s_1 t \\ u^{(i)} & , \ s_i t < x \le s_{i+1} t \ , \ 1 \le i \le \upsilon-1 \\ u^r & , \ s_\upsilon t < x, \end{cases}$$

where

$$(2.14)\qquad s^{(i)} = \frac{\rho^{(i)}v^{(i)} - \rho^{(i-1)}v^{(i-1)}}{\rho^{(i)} - \rho^{(i-1)}} \ , \quad 1 \le i \le \upsilon,$$

with $u^{(i)} = (\rho^{(i)},v^{(i)})$. Note that by construction $u^{(i+1)}$ belongs to the graph of $c_{1u^{(i)}}$ or $c_{2u^{(i)}}$. From (2.2), (2.14), and from the choice of the points $u^{(i)}$ it thus results that $s^{(i)} < s^{(i+1)}$ and that the function defined in (2.13) is a weak solution of (1.1) and (2.5).

Definition 2.1: We call the weak solution u of (1.1) and (2.5) defined in (2.13) decomposition of the discontinuity $[u^l,u^r]$ of type (n,k).

Next we define weak solutions $u = (\rho,v)$ of (1.1) in the domain $[a,\infty) \times \mathbb{R}^+$, which satisfy the boundary condition

$$(2.15)\qquad v(a,t) = 0 \quad , \quad t > 0$$

and the initial condition

$$(2.16)\qquad u(x,0) = u^r \quad , \quad x > a$$

with $u^r = \text{const} = (\rho^r,v^r)$, where $\rho^r > 0$. By definition of $c^n_{1u^r}$ in (2.6) - (2.8) this function coincides piecewise with functions c_{iw}. Therefore $c^n_{1u^r}$ is monotonically increasing and satisfies

$$c^n_{1u^r}(\tau) \longrightarrow -\infty \text{ as } \tau \longrightarrow 0; \; c^n_{1u^r}(\tau) \longrightarrow \infty \text{ as } \tau \longrightarrow \infty.$$

Consequently, the graph of $c^n_{1u^r}$ intersects the axis $v = 0$ in exactly one point $u^+ = (\rho^+,0)$ with $\rho^+ > 0$.

Definition 2.2: *The weak solution of (1.1), (2.15), and (2.16) defined in $[a,\infty) \times \mathbb{R}^+$ and obtained by decomposition of the discontinuity $[u^+,u^r]$ of type $(0,n)$ is called "adaptation of the state u^r to the left boundary of type n".*

Since u^+ belongs to the graph of $c^n_{1u^r}$, the decomposition of the discontinuity $[u^+,u^r]$ only consists out of 1-waves, and since $v^+ = 0$, it follows from (2.2) that the speed of these 1-waves is positive. For the solution u given in definition 2.2 therefore the equation $u(a,t) = u^+ = (\rho^+,0)$ holds for all $t > 0$. Consequently, (2.15) is satisfied in the classical sense for all $t > 0$ and thus also in the weak sense of (1.10). In the same way we define the adaptation of the state u^l to the right boundary of type n.

We also use the following terminology. Let u be a piecewise constant weak solution of (1.1), let λ be a real function defined on an interval, let u be discontinuous along the curve $x = \lambda(t)$, and assume that the discontinuity is an i-wave. Then we say that λ is an i-wave.

Definition 2.3: *Let λ be an i-wave, and let u, w be the constant states to the right and to the left of the discontinuity. Then the strength $|\lambda|$ of λ is defined by*

$$|\lambda| = |f_i(\epsilon_{ju}(0)) - f_i(\epsilon_{jw}(0))|, \tag{2.17}$$

where $j \in \{1,2\}$ with $i \neq j$, and

$$f_i(\tau) = \exp((-1)^{i+1}c\tau). \tag{2.18}$$

The positive constant c will be chosen later. The functions ϵ_{jw} are defined in (2.9). Note that the mapping

$$w \longrightarrow (f_1(\epsilon_{2w}(0)),f_2(\epsilon_{1w}(0))) : \mathbb{R}^+ \times \mathbb{R} \longrightarrow \mathbb{R}^+ \times \mathbb{R}^+$$

defines a new coordinate system on $\mathbb{R}^+ \times \mathbb{R}$.

3. Interaction of discontinuities

In this section we formulate and prove the fundamental estimates for the interaction of discontinuities. The basic idea is to show that the total strength of the shock waves emerging from an interaction is not larger than the strength of the shock waves before the interaction, if the strength of shock waves defined in (2.17) is used. It then can be shown that the total variation of the solution stays bounded, uniformly with respect to time. However, a difficulty arises from the fact that this is not true for the approximating solutions that we use to approximate the exact solution, since in these approximating solutions smooth rarefaction waves are approximated by a "fan" of "rarefaction discontinuities".

In the approximating solutions only two kinds of interaction occur. In the following two lemmas these interactions are examined.

Lemma 3.1: Let λ_1^- be a 1-wave and λ_2^- be a 2-wave, which interact. Let the weak solution of (1.1) be continued across the point of intersection of λ_1^- and λ_2^- by decomposition of the discontinuity at the point of intersection of type (0,0). In this case exactly one 1-wave λ_1 and one 2-wave λ_2 are issued from the point of intersection, and we speak of an interaction of type 1. For this interaction we have:
(i) The discontinuity along λ_i is a rarefaction wave if and only if the discontinuity along λ_i^- is a rarefaction wave.
(ii) Let λ_i^- be a shock wave. Then

$$|\lambda_j| \le |\lambda_j^-|$$

for $j \neq i$, if the constant c in (2.18) is chosen sufficiently large.
(iii) Let λ_i^- be a rarefaction wave, and let u^l, u^-, u^r be the constant states of the solution to the left of λ_1^-, between λ_1^- and λ_2^-, and to the right of λ_2^-. Then to each compact subset K of $\mathbb{R}^+ \times \mathbb{R}$ there is a constant C_1 with

$$|\lambda_j| \le (1 + C_1 |\lambda_i^-|^3) |\lambda_j^-|$$

for $j \neq i$, if all the states u^l, u^-, u^r are contained in K.

Lemma 3.2: Let λ_1^-, λ_2^- be interacting i-waves, and let the solution be extended across the point of intersection by decomposition of the discontinuity at the point of intersection of type (n,0) if i = 1, and by decomposition of this discontinuity of type (0,n) if i = 2. Then at most one i-wave λ is issued from the point of intersection. Perhaps though, several j-waves $\lambda_1, \ldots, \lambda_m$ with $j \neq i$ are issued from the point of intersection. We speak of an interaction of type 2. We have:
(i) The discontinuities along λ_1^- and λ_2^- cannot both be rarefaction waves. λ is a shock wave. If λ_1^- and λ_2^- both are shock waves, then

$\lambda_1,\dots,\lambda_m$ *all are rarefaction waves. If one of the two waves* λ_1^- *and* λ_2^- *is a rarefaction wave, then* $m = 1$ *and* λ_1 *is a shock wave.*

(ii) To each compact subset K of $\mathbb{R}^+ \times \mathbb{R}$ *there is a constant* C_2 *with*

$$\sum_{l=1}^{m} |\lambda_l| \le C_2 \ |\lambda_1^-| \ |\lambda_2^-| \ (|\lambda_1^-| + |\lambda_2^-|),$$

if $u^l, u^-, u^r \in K$, *where* u^l, u^-, u^r *are defined in Lemma 3.1.*

(iii) If λ_1^- *and* λ_2^- *are shock waves then to each compact subset K of* $\mathbb{R}^+ \times \mathbb{R}$ *there are constants* C_3, C^* *with*

$$|\lambda| \le |\lambda_1^-| + |\lambda_2^-| + \frac{C_3}{n^2} |\lambda_1^-| \ |\lambda_2^-|,$$

$$|\lambda_k| \le C^*/n \ , \ k = 1,\dots,m,$$

if $u^l, u^-, u^r \in K$.

(iv) Let λ_2^- *be a rarefaction wave. Then to each compact subset K of* $\mathbb{R}^+ \times \mathbb{R}$ *there exist constants* C_4, C_5 *with*

$$|\lambda| \le |\lambda_1^-| - C_4 \ |\lambda_2^-|,$$

$$|\lambda_1| \le C_5 \ |\lambda_2^-|,$$

if $u^l, u^-, u^r \in K$. *Moreover, we have* $C_5 < C_4$ *if the diameter of K is less than a suitable constant. This constant depends on the distance between K and the axis* $\{0\} \times \mathbb{R}$.

In the next lemma the reflection of waves at the boundary is examined.

Lemma 3.3: *Let* λ^- *be an i-wave which intersects the boundary of the region* $(a,b) \times (0,\infty)$, *and let the solution be extended across the point of intersection by adaptation of the state w to the boundary of type 0, where w ist the constant state to the left of* λ^- *if* $i = 1$, *and to the right of* λ^- *if* $i = 2$. *Then exactly one j-wave* λ *is issued from the point of intersection, where* $j \ne i$. *We say that* λ *is the wave obtained from* λ^- *by reflection at the boundary. We have:*

(i) λ *is a rarefaction wave if and only if* λ^- *is a rarefaction wave.*

(ii) Let λ^- *be a shock wave. Then*

$$|\lambda| \le |\lambda^-|,$$

if the constant c in (2.18) is chosen sufficiently large.

(iii) Let λ^- *be a rarefaction wave, let* $w^- = (w_1^-, 0)$ *be the constant state between the boundary and* λ^-, *and let w be the constant state on the other side of* λ^-. *Then for each compact subset K of* $\mathbb{R}^+ \times \mathbb{R}$

$$|\lambda| \le (1 + C_1 \ |\lambda^-|^3) \ |\lambda^-|,$$

if $w, w^- \in K$, *where* C_1 *is the constant of Lemma 3.1(iii).*

<u>Proof of Lemma 3.1:</u> This proof contains the basic idea of this paper. Let $u^{\ell} = (\rho^{\ell},v^{\ell})$, $u^{-} = (\rho^{-},v^{-})$, $u^{r} = (\rho^{r},v^{r})$, and $u^{+} = (\rho^{+},v^{+})$ be the constant states of the solution to the left of λ_1^-, between λ_1^- and λ_2^-, to the right of λ_2^-, and between λ_1 and λ_2. We consider the case i = 1.

(i) λ_1 is a rarefaction wave if and only if $\rho^+ < \rho^r$, and λ_1^- is a rarefaction wave if and only if $\rho^{\ell} < \rho^-$. The fact that $\rho^+ < \rho^r$ and $\rho^{\ell} < \rho^-$ are equivalent follows from the inequality $c'_{1w}(\rho) > 0$, which holds for all ρ and w, and from the fact that the graphs of c_{2u^-} and $c_{2u^{\ell}}$ do not intersect if u^- is different from u^{ℓ}. We prove that these graphs do not intersect and leave the remaining parts of the proof to the reader. Assume that there is ρ with $c_{2u^-}(\rho) = v = c_{2u^1}(\rho)$, and let $u = (\rho,v)$. From (2.4) it then follows that u^- and u^{ℓ} both lie on the graph of c_{2u}. On the other hand, also u^- and u^{ℓ} are both contained in the graph of c_{1u^-}, since u^- and u^{ℓ} are the constant states to the right and to the left of the discontinuity λ_1^-. However, because of $c'_{1u^-} > 0$, $c'_{2u} < 0$, the graphs of c_{1u^-} and c_{2u} intersect in exactly one point, which implies $u^- = u^{\ell}$.

(ii) We consider the case i = 1 and show that $|\lambda_2^-| - |\lambda_2| \geq 0$. The graphs of the functions occurring in the proof are depicted in figure 1 at the end of this article. (2.17) and (2.18) imply

$$|\lambda_2^-| - |\lambda_2| = |f_2(\epsilon_{1u^r}(0))-f_2(\epsilon_{1u^-}(0))|-|f_2(\epsilon_{1u^+}(0))-f_2(\epsilon_{1u^{\ell}}(0))|$$

$$= |\int_{\epsilon_{1u^-}(0)}^{\epsilon_{1u^r}(0)} f_2'(\tau)d\tau| - |\int_{\epsilon_{1u^{\ell}}(0)}^{\epsilon_{1u^+}(0)} f_2'(\tau)d\tau| = |\int_{\epsilon_{1u^-}(0)}^{\epsilon_{1u^r}(0)} f_2'(\tau)d\tau|$$

$$(3.1)\qquad - |\int_{\epsilon_{1u^-}(0)}^{\epsilon_{1u^r}(0)} f_2'\left(\frac{\epsilon_{1u^+}(0) - \epsilon_{1u^{\ell}}(0)}{\epsilon_{1u^r}(0) - \epsilon_{1u^-}(0)} [\tau-\epsilon_{1u^-}(0)] + \epsilon_{1u^{\ell}}(0) \right) \cdot \frac{\epsilon_{1u^+}(0) - \epsilon_{1u^{\ell}}(0)}{\epsilon_{1u^r}(0) - \epsilon_{1u^-}(0)} d\tau| .$$

Since $u^+ = (\rho^+,v^+)$ is the only point of intersection of the graphs of c_{1u^r} and $c_{2u^{\ell}}$, it follows that ρ^+ is an infinitely differentiable function of $|u^r-u^-|$ and of $|u^\ell-u^-|$, if u^- is kept fixed. From (2.9) it thus follows that

$$\text{(3.2)}\quad \epsilon_{1u^+}(0) - \epsilon_{1u^r}(0) = \epsilon_{1u^+}(\rho^+) - \epsilon_{1u^r}(\rho^+) = v^+ - \epsilon_{1u^r}(\rho^+)$$
$$= c_{1u^r}(\rho^+) - \epsilon_{1u^r}(\rho^+) = F(|u^r-u^-|,|u^\ell-u^-|),$$

where $F(\sigma,\tau)$ is infinitely differentiable for $\sigma,\tau > 0$ and satisfies the relations

$$\text{(3.3)}\quad \partial_\tau^m F(\sigma,\tau)_{|\tau=0} = 0 \quad , \quad m = 0,1,2$$

$$\text{(3.4)}\quad F(|u^r-u^-|,|u^\ell-u^-|) = H_1(\rho^+,u^r,0)(\rho^+-\rho^r)$$
$$\geq C_1(\rho^+-\rho^r)^3 \geq C|u^\ell-u^-|^3$$

with an appropriate constant $C > 0$. These relations result from (2.11), from the properties of H_i stated after (2.11), and from $\rho^+ > \rho^r$. The last inequality holds since the assumption and the assertion (i) of lemma (3.1) imply that λ_1 is a shock wave. Furthermore,

$$\text{(3.5)}\quad F(0,|u^\ell-u^-|) = \epsilon_{1u^\ell}(0) - \epsilon_{1u^-}(0),$$

since $u^r = u^-$ implies $u^\ell = u^+$. From (3.2), (3.3), (3.5), and from Taylor's theorem we obtain

$$\text{(3.6)}\quad |(\epsilon_{1u^+}(0) - \epsilon_{1u^\ell}(0)) - (\epsilon_{1u^r}(0) - \epsilon_{1u^-}(0))|$$
$$= |(\epsilon_{1u^+}(0) - \epsilon_{1u^r}(0)) - (\epsilon_{1u^\ell}(0) - \epsilon_{1u^-}(0))|$$
$$= |\frac{1}{2!}\int_0^{|u^r-u^-|}\int_0^{|u^\ell-u^-|} \partial_\sigma\, \partial_\tau^3 F(\sigma,\tau)(|u^\ell-u^-|-\tau)^2\, d\tau\, d\sigma|$$
$$\leq C|u^\ell-u^-|^3\, |u^r-u^-|.$$

Since λ_2 is a rarefaction wave if and only if λ_2^- is a rarefaction wave, it follows that $\epsilon_{1u^+}(0) - \epsilon_{1u^\ell}(0)$ and $\epsilon_{1u^r}(0) - \epsilon_{1u^-}(0)$ have the same sign. (3.6) thus yields

$$\text{(3.7)}\quad 0 < \frac{\epsilon_{1u^+}(0) - \epsilon_{1u^\ell}(0)}{\epsilon_{1u^r}(0) - \epsilon_{1u^-}(0)} \leq 1 + \frac{C|u^\ell-u^-|^3|u^r-u^-|}{|\epsilon_{1u^r}(0) - \epsilon_{1u^-}(0)|}$$
$$\leq 1 + C_1\, |u^\ell-u^-|^3,$$

where we used $|\epsilon_{1u^r}(0) - \epsilon_{1u^-}(0)| \geq C_2|u^r-u^-|$. This inequality follows from the fact that the mapping

$$w \longrightarrow (\epsilon_{1w}(0),\epsilon_{2w}(0)) : \mathbb{R}^+ \times \mathbb{R} \longrightarrow \mathbb{R}^2$$

defines a coordinate system on $\mathbb{R}^+ \times \mathbb{R}$. Furthermore,

$$(3.8)\qquad \frac{\epsilon_{1u^+}(0) - \epsilon_{1u^\ell}(0)}{\epsilon_{1u^r}(0) - \epsilon_{1u^-}(0)}\left[\tau - \epsilon_{1u^-}(0)\right] + \epsilon_{1u^\ell}(0)$$
$$\geq \tau + \min\left[\epsilon_{1u^+}(0) - \epsilon_{1u^r}(0)\ ,\ \epsilon_{1u^\ell}(0) - \epsilon_{1u^-}(0)\right]$$
$$\geq \tau + C|u^\ell - u^-|^3$$

for all τ between $\epsilon_{1u^-}(0)$ and $\epsilon_{1u^r}(0)$, with $C > 0$. The first inequality in (3.8) holds since the terms on both sides of the inequality are linear functions of τ, and since the inequality is satisfied for $\tau = \epsilon_{1u^-}(0)$ and for $\tau = \epsilon_{1u^r}(0)$. The second inequality in (3.8) results from (3.2), (3.4), and (3.5).

Now choose the constant c in (2.18) such that

$$c = C_1/C,$$

where C_1 and C are the constants in (3.7), (3.8). For all $|u^\ell - u^-|^3$ we then get

$$(3.9)\qquad (1 + C_1|u^\ell-u^-|^3)\exp\{-c(\tau + C|u^\ell-u^-|^3)\}$$
$$\leq \exp\{C_1|u^\ell-u^-|^3\}\exp\{-c(\tau + C|u^\ell-u^-|^3)\} = \exp(-c\tau).$$

From (3.1), (3.7) - (3.9) and from (2.18) we thus obtain

$$|\lambda_2^-| - |\lambda_2| = \left|\int_{\epsilon_{1u^-}(0)}^{\epsilon_{1u^r}(0)} \left[-c\exp(-c\tau) + c\,\frac{\epsilon_{1u^+}(0) - \epsilon_{1u^\ell}(0)}{\epsilon_{1u^r}(0) - \epsilon_{1u^-}(0)}\right.\right.$$
$$\left.\left.\cdot\exp\left\{-c\left[\frac{\epsilon_{1u^+}(0) - \epsilon_{1u^\ell}(0)}{\epsilon_{1u^r}(0) - \epsilon_{1u^-}(0)}\left[\tau - \epsilon_{1u^-}(0)\right] + \epsilon_{1u^\ell}(0)\right]\right\}\right]d\tau\right|$$
$$\geq \left|\int_{\epsilon_{1u^-}(0)}^{\epsilon_{1u^r}(0)} [-c\exp(-c\tau) + c\exp(-c\tau)]d\tau\right| = 0.$$

This proves (ii) in case i = 1. In the other case the proof proceeds in the same way.

(iii) Note that

$$(3.10)\qquad \big||\lambda_2^-| - |\lambda_2|\big| = \big||f_2(\epsilon_{1u^r}(0)) - f_2(\epsilon_{1u^-}(0))|$$
$$- |f_2(\epsilon_{1u^+}(0)) - f_2(\epsilon_{1u^\ell}(0))|\big|$$
$$= \left|\left[f_2(\epsilon_{1u^r}(0)) - f_2(\epsilon_{1u^-}(0))\right] - \left[f_2(\epsilon_{1u^+}(0)) - f_2(\epsilon_{1u^\ell}(0))\right]\right|$$
$$= \left|\left[f_2(\epsilon_{1u^r}(0)) - f_2(\epsilon_{1u^+}(0))\right] - \left[f_2(\epsilon_{1u^-}(0)) - f_2(\epsilon_{1u^\ell}(0))\right]\right|.$$

For, since λ_2 is a rarefaction wave if and only if λ_2^- is a rarefaction wave, it follows from some simple geometrical arguments that $f_2(\epsilon_{1u^r}(0)) - f_2(\epsilon_{1u^-}(0))$ and $f_2(\epsilon_{1u^+}(0)) - f_2(\epsilon_{1u^\ell}(0))$ have the same sign. Exactly as in (3.2) we obtain that

$$f_2(\epsilon_{1u^r}(0)) - f_2(\epsilon_{1u^+}(0)) = G(|u^r-u^-|,|u^\ell-u^-|),$$

where the function G has properties analogous to those of F in (3.3) and (3.5). Therefore we can show exactly as in (3.6) that the expression on the right hand side of (3.10) is bounded by

$$C|u^\ell-u^-|^3\ |u^r-u^-|. \tag{3.11}$$

Again by geometrical considerations it follows from the definition of ϵ_{iw} in (2.9), from the fact that u^ℓ, u^- both lie on the graph of c_{1u^-}, and from the fact that u^r, u^- both lie on the graph of c_{2u^-} that

$$|u^\ell-u^-| \le C|f_1(\epsilon_{2u^\ell}(0)) - f_1(\epsilon_{2u^-}(0))| = C|\lambda_1^-|$$

$$|u^r-u^-| \le C|f_2(\epsilon_{1u^r}(0)) - f_2(\epsilon_{1u^-}(0))| = C|\lambda_2^-|.$$

These inequalities and (3.10), (3.11) together yield

$$||\lambda_2^-| - |\lambda_2|| \le C_1|\lambda_1^-|^3\ |\lambda_2^-|.$$

Thus (iii) is shown for i = 1. In case i = 2 the proof proceeds along the same lines. Lemma 3.1 is proved.

Sketch of the proof of Lemma 3.2: We assume that i = 1 and that λ_2^- lies to the left of λ_1^-. This means that

$$\partial_t\ \lambda_2^- > \partial_t\ \lambda_1^-. \tag{3.12}$$

(i) It is well known that two rarefaction waves of the same family cannot intersect. For, if λ_1^- and λ_2^- both are rarefaction waves, then it follows from (2.2) by some considerations that the speed $\partial_t\lambda_2^-$ of λ_2^- is smaller than the speed $\partial_t\lambda_1^-$ of λ_1^-. This contradicts (3.12). For similar reasons it follows that the discontinuity along λ must be a shock wave. The proof is similar to the proof of Lemma A4 in [1], and hence we skip it here. The other assertions of (i) follow immediately from the geometrical properties of the functions c_{1w}^n and c_{2w}^n.

Also the estimates in (ii) - (iv) follow from the geometrical properties of the functions c_{iw}^n defined in (2.3), (2.4) and (2.6) - (2.8). Mostly the proofs are similar to the proofs of the well-known estimates for intersections of discontinuities, see for instance [5, 9]. To prove (iii) one needs (2.11) and (2.12). The inequality $|\lambda_k| \le c^*/n$ in (iii) follows from definition 2.1 and from the fact that $|\lambda_k|$

is bounded by some multiple of the arclength of c_{2w^ℓ} between the points w^ℓ and w^r, where w^ℓ and w^r are the constant states to the left and right of λ_k.

<u>Proof of Lemma 3.3:</u> Again we consider the case $i = 1$. Let $u^\ell = (\rho^\ell, v^\ell)$, u^-, and u^+ be the constant states of the solution to the left of λ^-, between λ^- and the right hand boundary, and between λ and the right hand boundary. The proof of lemma 3.3 can be completely reduced to the proof of Lemma 3.1, if one bears in mind that the type 1 interaction of the 1-wave λ^- with the 2-wave λ_2^-, which we define to be the discontinuity between the constant state u^- and the constant state $u^r = (\rho^\ell, -v^\ell)$, leads to the same state u^+ and to the same 2-wave λ as the reflection of λ^- at the right hand boundary. This follows from the symmetry of the functions c_{1w}^n and c_{2w}^n defined in (2.3), (2.4), and in (2.6) - (2.8). Because of $|\lambda_2^-| = |\lambda^-|$, all the assertions and estimates of Lemma 3.3 result from the corresponding assertions and estimates of Lemma 3.1.

4. Proof of Theorem 1.1

In order to prove Theorem 1.1 we construct a sequence $\{u^{(n)}\}$ of approximating solutions with $u^{(n)}$ defined on the set $[a,b] \times [0,T_n)$. Then we derive an estimate for the total variation of the functions $u^{(n)}$, and conclude from this estimate that $T_n \longrightarrow \infty$ and that there is some subsequence of $\{u^{(n)}\}$, which converges in $\mathcal{L}_1^{loc}$ to a weak solution of (1.1) - (1.4). For the definition of the sequence $\{u^{(n)}\}$ we use the construction from section 2 in [1], and we also use the terminology introduced at the end of section 2 in [1]. Since this construction and the definitions are rather long, we do not describe them here but refer the reader to [1].

In order to construct $\{u^{(n)}\}$ extend the initial values $U = (P,V)$ given on $[a,b]$ to a function $\tilde{U} = (\tilde{P},\tilde{V})$ defined on $\mathbb{R}$ such that

$$(4.1) \qquad \begin{aligned} (\tilde{P}(a+x),\tilde{V}(a+x)) &= (\tilde{P}(a-x),-\tilde{V}(a-x)) \\ (\tilde{P}(b+x),\tilde{V}(b+x)) &= (\tilde{P}(b-x),-\tilde{V}(b-x)) \end{aligned}$$

holds for all $x \in \mathbb{R}$. These two equations define $(\tilde{P},\tilde{V})$ uniquely on all of $\mathbb{R}$. Let $\{\tilde{u}^{(n)}\}$ be the sequence of approximating solutions to the initial values $(\tilde{P},\tilde{V})$ constructed as in section 2 of [1]. In constructing $\tilde{u}_n$ there is a certain freedom with respect to the choice of the positions of the "interaction points". We choose these points in such a way that for the function $\tilde{u}^{(n)} = (\tilde{\rho}^{(n)},\tilde{v}^{(n)})$ the equations

$$(4.2) \qquad \begin{aligned}(\tilde{\rho}^{(n)}(a+x),\tilde{v}^{(n)}(a+x)) &= (\tilde{\rho}^{(n)}(a-x),-\tilde{v}^{(n)}(a-x))\\ (\tilde{\rho}^{(n)}(b+x),\tilde{v}^{(n)}(b+x)) &= (\tilde{\rho}^{(n)}(b-x),-\tilde{v}^{(n)}(b-x))\end{aligned}$$

hold for all $t \geq 0$ and all $x \in \mathbb{R}$. This is possible because of the choice of $(\tilde{P},\tilde{V})$ in (4.1). The functions $\tilde{u}^{(n)}$ are defined on the sets $\mathbb{R} \times [0,T_n)$ with $T_n > 0$. Let $u^{(n)}$ be the restriction of $\tilde{u}^{(n)}$ to $[a,b] \times [0,T_n)$.

The function $u^{(n)}$ defined this way satisfies the boundary conditions (1.2) for almost every $t \geq 0$. For, from the construction of $\tilde{u}^{(n)}$ and from (4.2) it follows that if λ_1 and λ_2 are two "discontinuities" of $\tilde{u}^{(n)}$ with interaction point located on the straight line $x = a$ or on the straight line $x = b$, then the interaction between λ_1 and λ_2 is of type 1. If λ_1^+ and λ_2^+ are the discontinuities arising from this interaction, then λ_i^+ coincides with the wave obtained from λ_j, $i \neq j$, by reflection at the boundary. This has already been used in the proof of lemma 3.3. From this fact it follows that the second component of $\tilde{u}^{(n)}$ vanishes on the straight lines $x = a$ and $x = b$ at all points except for the countable set of interaction points. Therefore (1.2) is satisfied for almost every t.

Lemma 4.1: *Let K be a compact subset of $\mathbb{R}^+ \times \mathbb{R}$ with $u^{(n)}([a,b] \times [0,T_n)) \subseteq K$ for all n. Then*

(i) There is a constant C with

$$TV(u^{(n)}(\cdot,t)) \leq C \sum_{\lambda\in I_t} |\lambda|$$

for all n and for almost every $t \geq 0$. Here I_t is the set of all shock waves $\lambda \in L$, which intersect the line $[a,b] \times \{t\}$.

(ii) Let $u^{(n)}(x,t) = (\rho^{(n)}(x,t),v^{(n)}(x,t))$. Then there are constants $C_1,C_2 > 0$ with

$$|\rho^{(n)}(x,t) - (b-a)^{-1}\int_a^b \rho^{(n)}(y,0)dy| \leq \frac{C_1}{n} + C_2 \sum_{\lambda\in I_t} |\lambda|$$

for all n and for all $t \geq 0$.

The set L is the set of the discontinuities of $\tilde{u}^{(n)}$ and is defined in section 2 of [1].

Proof: (i) The function $u^{(n)}$ is piecewise constant and jumps along the discontinuities $\lambda \in L$. Let λ be a discontinuity which intersects the line $[a,b] \times \{t\}$, and let $u^l = (\rho^l,v^l)$, $u^r = (\rho^r,v^r)$ be the values of

$u^{(n)}$ to the left and right of λ. From (2.3) and (2.4) it follows that $v^r > v^l$ if λ is a rarefaction wave and that $v^r < v^l$ if λ is a shock wave, irrespective of whether λ is a 1- or 2-wave. Since $v^{(n)}(a,t) = v^{(n)}(b,t) = 0$ for almost every t, as shown above, it follows that

$$\text{(4.3)} \qquad TV(v^{(n)}(\cdot,t)) \leq 2\,\text{var}^-(v^{(n)}(\cdot,t)),$$

where $\text{var}^-(v^{(n)}(\cdot,t))$ is the negative variation of $v^{(n)}$. Since by assumption the values of $u^{(n)}$ are contained in a compact subset of $\mathbb{R}^+ \times \mathbb{R}$, it follows from the definition of $|\lambda|$ in (2.17), from (2.3), (2.4), and from (2.9) by geometrical considerations that there is some constant $C_1 > 0$ with

$$|v^l - v^r| \leq C_1|\lambda|,$$

where $|v^l-v^r|$ is the absolute value. From this estimate and from the remarks above we conclude that the negative variation of $v^{(n)}$ can be estimated by $\sum_{\lambda\in I_t} |\lambda|$. Together with (4.3) it follows that the variation of $v^{(n)}$ can be estimated by $\sum_{\lambda\in I_t} |\lambda|$. Since from (2.3), (2.4) we also obtain $|u^l-u^r| \leq C_2|v^l-v^r|$, we infer that the variation of $u^{(n)}$ can be estimated by the variation of $v^{(n)}$. Together we obtain statement (i).

(ii) By construction the functions $u^{(n)}$ are approximate solutions of (1.1) - (1.4). The equation (1.10) does not hold exactly. We rather obtain as in the proof of corollary 3.2 in [1] that

$$\text{(4.4)} \qquad \left|\int_0^\infty\int_a^b [\rho^{(n)}\phi_t + \rho^{(n)}v^{(n)}\phi_x]\,dx\,dt + \int_a^b \rho^{(n)}(x,0)\phi(x,0)\,dx\right| \leq \frac{1}{n}\max|\phi(x,t)|.$$

We now choose a sequence $\{\phi_m\}$ of test functions, which approximate the characteristic function of the set $[a,b] \times [0,\tau]$ and obtain from $v^{(n)}(a,t) = v^{(n)}(b,t) = 0$ and from (4.4) that

$$\text{(4.5)} \qquad \left|\int_a^b \rho^{(n)}(x,\tau)dx - \int_a^b \rho^{(n)}(x,0)dx\right| \leq C/n$$

for all $\tau \geq 0$ and all n. From (4.5) we infer that for all τ and n there is a number $x_0 \in [a,b]$ with

$$\text{(4.6)} \qquad \left|\rho^{(n)}(x_0,\tau) - (b-a)^{-1}\int_a^b \rho^{(n)}(x,0)\,dx\right| \leq \frac{C}{n(b-a)}.$$

Because of $|\rho^{(n)}(x,\tau) - \rho^{(n)}(x_0,\tau)| \leq TV(u^{(n)}(\cdot,\tau))$, assertion (ii) results from (4.6) and from the assertion (i) of this lemma. This completes the proof of Lemma 4.1.

Now we use the local existence result of [1]. This result is proved for a system of equations describing elastic oscillations of a string and for initial values with bounded variation and compact support. But since the properties of the interaction of discontinuities stated in Lemma 3.1 and Lemma 3.2 coincide with the properties used in [1], and since the result in [1] is a local result with respect to the x-variable because of the finite propagation speed, we can show for the present system (1.1) exactly as in [1] that a subsequence of the sequence $\{\tilde{u}^{(n)}\}$ converges in a strip $\mathbb{R} \times [0,T)$ to a weak solution $\tilde{u}$ of (1.1) to the initial values defined in (4.1). Examination of the proof of Lemma 3.4 in [1] shows that the positive constant δ occuring in this lemma, whose size is determined in the proof of this lemma, only depends on the total variation of the initial values $(\tilde{P},\tilde{V})$ in the interval $(x_0-\delta, x_0+\delta)$. More precisely, there is a constant $C > 0$ such that for all $x_0 \in \mathbb{R}$ the number δ in Lemma 3.4 of [1] can be chosen larger than $3/2(b-a)$ and such that the estimates (3.9) - (3.12) in this lemma hold if

$$TV[\tilde{u}^{(n)}(\cdot,0);\ (x_0 - \tfrac{3}{2}(b-a), x_0 + \tfrac{3}{2}(b-a))] < C.$$

Because of (4.2),

$$TV[\tilde{u}^{(n)}(\cdot,0);\ (x_0 - \tfrac{3}{2}(b-a), x_0 + \tfrac{3}{2}(b-a))]$$
$$\leq 3\ TV(u^{(n)}(\cdot,0);\ [a,b]) \leq C \sum_{\lambda\in I_0} |\lambda|.$$

Here we used Lemma 4.1(i). Therefore we can restate the Lemma 3.4 in [1] as follows.

<u>*Lemma 4.2:*</u> *Let $0 < P_0 < P_1$, and for $x \in \mathbb{R}$ let $\Omega = \Omega(x) = (x-3/2(b-a), x+3/2(b-a))$. Assume that the approximate solution $\tilde{u}^{(n)}(x,t) = (\tilde{\rho}^{(n)}(x,t), \tilde{v}^{(n)}(x,t))$ satisfies the inequalities*

$$P_0 \leq \rho^{(n)}(x,0) \leq P_1 \tag{4.7}$$

for all $x \in [a,b]$ and

$$\sum_{\lambda\in I_0} |\lambda| \leq V_1. \tag{4.8}$$

Then the following holds: If V_1 is sufficiently small, then there are constants V_2, V_3, and to all $x \in \mathbb{R}$ a function $\omega : \Lambda(\Omega(x)) \longrightarrow \mathbb{R}_0^+$ with

$$|\lambda| \leq \sum_{\substack{\alpha\in\Gamma(\Omega) \\ \text{with } \lambda\in\alpha}} \omega(\alpha,\lambda) \quad \text{for all } \lambda \in L(\Omega), \tag{4.9 a}$$

and with

$$\sum_{\alpha\in\Gamma(\Omega)} \sup_{\lambda\in\alpha} \omega(\alpha,\lambda) \leq V_2. \tag{4.9 b}$$

If $\lambda \in L(\Omega)$ is a rarefaction wave, then

(4.10) $$|\lambda| \leq V_3/n.$$

Exactly as in [1] we conclude from this lemma that there exists a constant $T > 0$ such that $\tilde{u}^{(n)}$ exists in the strip $\mathbb{R} \times [0,T)$ when $\tilde{u}^{(n)}$ satisfies the inequalities (4.7) and (4.8), where in the present case T only depends upon the constants P_0, P_1, and V_1 in (4.7) and (4.8). By construction of $\tilde{u}^{(n)}$ in [1],

$$|\tilde{u}^{(n)}(x,0) - (\tilde{P}(x),\tilde{V}(x))|_\infty \longrightarrow 0$$

$$TV(\tilde{u}^{(n)}(\cdot,0)\ ;\ [a,b]) \longrightarrow TV((P,V)\ ;\ [a,b]).$$

If $TV((P,V)) < V_1/2$ and $P_0 + \epsilon \leq P(x) \leq P_1 - \epsilon$ hold for all $x \in [a,b]$, then it follows from these relations that the inequalities (4.7) and (4.8) are satisfied for all n sufficiently large. Therefore $\tilde{u}^{(n)}$ exists for all n in the strip $\mathbb{R} \times [0,T)$, and exactly as in [1] it follows that $\{u^{(n)}\}$ has a subsequence, which converges to a weak solution u of (1.1) - (1.4) defined on $[a,b] \times [0,T)$. We now show

<u>Lemma 4.3:</u> *There is a constant C with the following property: If $u^{(n)}$ satisfies the inequalities (4.7) and (4.8), then*

(4.11) $$\sum_{\lambda \in I_t} |\lambda| \leq exp(Cn^{-2}) \sum_{\lambda \in I_0} |\lambda|$$

for all $0 \leq t \leq T/2$.

From this lemma we obtain that the solution u of (1.1) - (1.4) exists globally as follows. Assume that the function $\tilde{u}^{(n)}$ satisfies

(4.12) $$P_0 + \epsilon \leq \rho^{(n)}(x,0) \leq P_1 - \epsilon,$$

(4.13) $$\sum_{\lambda \in I_0} |\lambda| \leq V_1/2,$$

where P_0, P_1, V_1 are the constants from (4.7) and (4.8), and where $\epsilon > 0$. Then $\tilde{u}^{(n)}$ exists in the strip $\mathbb{R} \times [0,T)$ and satisfies (4.11). If n is sufficiently large, it follows from (4.11) and (4.13) that

$$\sum_{\lambda \in I_{T/2}} |\lambda| \leq \exp(C/n^2) \sum_{\lambda \in I_0} |\lambda| \leq V_1,$$

and from lemma 4.1(ii) and (4.12) we obtain

$$P_0 \leq \rho^{(n)}(x,T/2) \leq P_1,$$

if n is sufficiently large and V_1 is sufficiently small. Hence $\tilde{u}^{(n)}$ satisfies the inequalities (4.7) and (4.8) also on the straight line $t = T/2$, and we can apply lemma 4.2 anew with $\tilde{u}^{(n)}(x,T/2)$ as initial values. It follows that $\tilde{u}^{(n)}$ exists on the strip $\mathbb{R} \times [0,3T/2)$. From lemma 4.3 we then conclude that

$$\sum_{\lambda\in I_t} |\lambda| \le (\exp(C/n^2))^2 \sum_{\lambda\in I_0} |\lambda| \le \exp(2C/n^2) \sum_{\lambda\in I_0} |\lambda|$$

holds for all $0 \le t \le T$, and in case that n is sufficiently large the procedure can be continued. Thus $\tilde{u}^{(n)}$ is defined on the strip $\mathbb{R} \times [0,T_n)$ with $T_n = n^2T(2C)^{-1} \ln(2)$. By selection of a diagonal sequence we therefore can show that the weak solution u of (1.1) - (1.4) exists on all of $[a,b] \times [0,\infty)$ and has the properties stated in Theorem 1.1. It thus remains to prove Lemma 4.3.

Proof of Lemma 4.3: We use the terminology and notation introduced at the end of section 2 in [1]. In addition, we define the following set. Let Γ^* be the set of all subsets $\alpha = \{\alpha_m\}_{m\in I} \subseteq L$ with the following properties:

a) I is a subset of the non-negative integers with $0 \in I$.
b) The discontinuities α_m are all contained in $[a,b] \times [0,\infty)$.
c) $\alpha_0 \in L_i$ is a shock wave and either issues from the line $[a,b] \times \{0\}$, or there are intersecting $\lambda_1,\lambda_2 \in L_j$ with $j \ne i$ such that α_0 issues from the point of intersection.
d) Let α_m be defined on $[t_k,t_{k+1}]$ and either let $\alpha_m(t_{k+1}) = a$ or $\alpha_m(t_{k+1}) = b$. Then α_{m+1} is the discontinuity obtained from α_m by reflection at the boundary.
e) Let $\alpha_m \in L_i$ be defined on $[t_k,t_{k+1}]$ and let $\alpha_m(t_{k+1})$ be different from a and b. Our construction of $\tilde{u}_n$ implies that in this case there is exactly one discontinuity $\lambda \in L_i$ defined on $[t_{k+1},t_{k+2}]$ with $\lambda(t_{k+1}) = \alpha_m(t_{k+1})$. Then $\alpha_{m+1} = \lambda$.

If $\alpha = \{\alpha_m\}_{m\in I} \in \Gamma^*$, then α_0 is a shock wave, by property c), and thus also all of the α_m's, by the properties of the interaction and reflection stated in the Lemmas 3.1, 3.2, and 3.3. Hence, Γ^* can be considered to be the set of shock waves reflected at the boundary. From these lemmas it also follows that if λ is a shock wave, then there is at least one $\alpha \in \Gamma^*$ with $\lambda \in \alpha$.

Let now $\lambda \in I_t$ with $t \le T/2$, where I_t is the set of shock waves defined in Lemma 4.1. There are finitely many $\alpha^{(1)},\dots,\alpha^{(\ell)} \in \Gamma^*$ with $\lambda \in \alpha^{(m)}$. We use the properties of the interaction of discontinuities stated in the Lemmas 3.1 and 3.2, gather from Lemma 3.3 that the strength of shock waves does not increase when reflection takes place, and obtain

$$(4.14)\qquad |\lambda| \le$$

$$\le \sum_{m=1}^{\ell} \left[\prod_{\mu\in J(m,1)} (1+C_1|\mu|^3)\right]\left[\prod_{\mu\in J(m,2)} \left(1+\frac{C_3}{n^2}|\mu|\right)\right] |\alpha_0^{(m)}| - \sum_{\mu\in J(\lambda)} C_4|\mu|$$

$$\le \sum_{m=1}^{\ell} \exp\left\{C_1 \sum_{\mu\in J(m,1)} |\mu|^3 + \frac{C_3}{n^2} \sum_{\mu\in J(m,2)} |\mu|\right\} |\alpha_0^{(m)}| - \sum_{\mu\in J(\lambda)} C_4|\mu|.$$

Here $J(m,1)$ is the set of all rarefaction waves $\mu \in L$ lying in $[a,b] \times [0,T/2]$ which interact with $\alpha^{(m)}$ of type 1, $J(m,2)$ is the set of all shock waves lying in $[a,b] \times [0,T/2]$, which interact with $\alpha^{(m)}$ of type 2, and $J(\lambda)$ is the set of all rarefaction waves, which interact with one of the $\alpha^{(1)},\ldots,\alpha^{(m)}$ of type 2 and lie in $[a,b] \times [0,T/2]$. C_1 is the constant from Lemma 3.1(iii), C_3 is the constant from Lemma 3.2(iii), and C_4 is the constant from Lemma 3.2(iv). If we choose $T > 0$ sufficiently small, then all discontinuities lying in $[a,b] \times [0,T/2]$ are contained in $L(\Omega)$ with $\Omega = (a-(b-a),b+(b-a))$. This follows from the considerations after Corollary 3.2 in [1]. Therefore we can apply the inequalities (4.8) - (4.10). If we further note that each $\alpha \in \Gamma^*$ can interact with each $\beta \in \Gamma$ twice, at most, which follows from the definitions of Γ and Γ^*, we consequently obtain that

$$C_1 \sum_{\mu\in J(m,1)} |\mu|^3 + \frac{C_3}{n^2} \sum_{\mu\in J(m,2)} |\mu|$$

$$\le C_1 \frac{V_3^2}{n^2} \sum_{\mu\in J(m,1)} |\mu| + \frac{C_3}{n^2} \sum_{\mu\in J(m,2)} |\mu|$$

$$\le \left[C_1 \frac{V_3^2}{n^2} + \frac{C_3}{n^2}\right] \left[\sum_{\mu\in J(m,1)} |\mu| + \sum_{\mu\in J(m,2)} |\mu|\right]$$

$$\le (C_1V_3^2 + C_3)n^{-2} \left[\sum_{\mu\in J(m,1)\cup J(m,2)} \Big(\sum_{\substack{\alpha\in\Gamma(\Omega)\\ \text{with } \mu\in\alpha}} \omega(\alpha,\mu) \Big)\right]$$

$$\le (C_1V_3^2 + C_3)n^{-2} \left[\sum_{\mu\in J(m,1)\cup J(m,2)} \Big(\sum_{\substack{\alpha\in\Gamma(\Omega)\\ \text{with } \mu\in\alpha}} \sup_{\nu\in\alpha} \omega(\alpha,\nu) \Big)\right]$$

$$\le 2(C_1V_3^2 + C_3)n^{-2} \sum_{\alpha\in\Gamma(\Omega)} \sup_{\nu\in\alpha} \omega(\alpha,\nu) \le C\, n^{-2},$$

with $C = 2V_2(C_1V_3^2 + C_3)$. This inequality and (4.14) imply

$$(4.15)\qquad |\lambda| \le \exp(C\, n^{-2}) \sum_{m=1}^{\ell} |\alpha_0^{(m)}| - C_4 \sum_{\mu\in J(\lambda)} |\mu|.$$

From the definition of Γ^* it follows that if $\lambda_1,\lambda_2 \in I_t$ with $\lambda_1 \ne \lambda_2$ and if $\alpha,\beta \in \Gamma^*$ with $\lambda_1 \in \alpha$, $\lambda_2 \in \beta$, then $\alpha \ne \beta$ and, in particular, $\alpha_0 \ne \beta_0$. From (4.15) we thus obtain

$$(4.16)\qquad \sum_{\lambda\in I_t} |\lambda| \le \exp(Cn^{-2}) \left[\sum_{\alpha_0\in K_1} |\alpha_0| + \sum_{\alpha_0\in K_2} |\alpha_0| \right] - C_4 \sum_{\mu\in K_3} |\mu|.$$

Here $K_1 \subset L(\Omega)$ is the set of shock waves emanating from $[a,b]$, hence $K_1 = I_0$. $K_2 \subset L(\Omega)$ is the set of shock waves λ with the following properties: λ is issued from an interaction of type 2, and if $\lambda \in L_i(\Omega)$ then the interacting shock waves λ_1, λ_2 belong to $L_j(\Omega)$ with $i \ne j$ and lie in $[a,b] \times [0,T/2]$. From Lemma 3.2 it follows that one of the two disconinuities λ_1, λ_2 must be a rarefaction wave, the other one a shock wave. Exactly one shock wave $\lambda \in L_i(\Omega)$ is issued from such an interaction, and therefore K_2 is isomorphic to the set of all interaction points of a shock wave λ_1 and a rarefaction wave λ_2 which both lie in $[a,b] \times [0,T/2]$ and interact of type 2. Finally, $K_3 \subset L(\Omega)$ is the set of all rarefaction waves in $[a,b] \times [0,T/2]$, which interact with a shock wave of type 2. Since K_3 is isomorphic to the corresponding set of interaction points, also K_2 and K_3 are isomorphic. From Lemma 3.2(iv) we therefore obtain

$$\sum_{\alpha_0\in K_2} |\alpha_0| \le \sum_{\mu\in K_3} C_5|\mu|.$$

Choose n_0 such that $\exp(Cn_0^{-2})C_5 \le C_4$, which is possible because of $C_5 < C_4$. (4.16) now yields

$$\sum_{\lambda\in I_t} |\lambda| \le \exp(Cn^{-2}) \sum_{\mu\in I_0} |\mu| + (\exp(Cn^{-2})C_5 - C_4) \sum_{\mu\in K_3} |\mu|$$

$$\le \exp(Cn^{-2}) \sum_{\mu\in I_0} |\mu|$$

if $n \ge n_0$. This proves Lemma 4.3.

Bibliography

[1] H.D. Alber: Local existence of weak solutions to the quasi-linear wave equation for large initial values. Math. Z. 190 (1985), 249-276.

[2] N. Bakhvarov: On the existence of regular solutions in the large for quasilinear hyperbolic systems. Zhur. Vychisl. Mat. i. Mathemat. Fiz. 10 (1970), 969-980.

[3] R. DiPerna: Existence in the large for quasilinear hyperbolic conservation laws. Arch. Rat. Mech. Anal. 52 (1973), 244- 257.

[4] R. DiPerna: Convergence of the viscosity method for isentropic gas dynamics. Comm. in Math. Phys. 91 (1983), 1-30.

[5] J. Glimm: Solutions in the large for nonlinear hyperbolic systems of equations. Comm. Pure Appl. Math. 18 (1965), 697-715.

[6] J. Glimm, P.D. Lax: Decay of solutions of systems of nonlinear hyperbolic conservation laws. Mem. Am. Math. Soc. 101 (1970).

[7] P.D. Lax: Hyperbolic systems of conservation laws II. Comm. Pure Appl. Math. 10 (1957), 537-566.

[8] P.D. Lax: Shock waves and entropy. In: Contributions to nonlinear functional analysis. Zarantonello, E.A. (ed.) New York: Academic Press, 603-634 (1971).

[9] J. Smoller: Shock waves and reaction-diffusion equations. Grundlehren der mathematischen Wissenschaften, Bd. 258, Berlin: Springer-Verlag (1983).

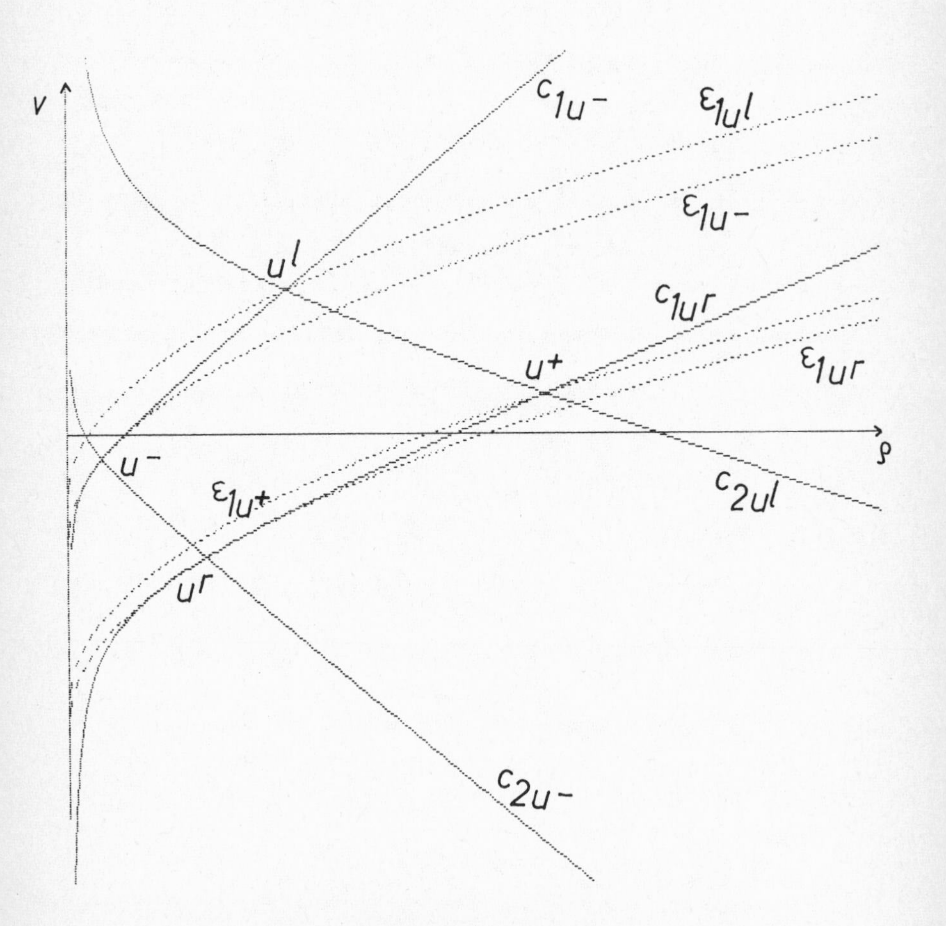

Figure 1

INITIAL-BOUNDARY VALUE AND SCATTERING PROBLEMS IN MATHEMATICAL PHYSICS

H.D. Alber
Mathematisches Institut A der Universität Stuttgart
Pfaffenwaldring 57, D-7000 Stuttgart 80

R. Leis
Institut für Angewandte Mathematik der Universität Bonn
Wegelerstraße 10, D-5300 Bonn 1

Initial-boundary value problems are often found in mathematical physics. For example, one can write the wave equation, the plate equation or the system of equations of linear elasticity in the form

$$u_{tt} + Au = 0 \quad \text{with} \quad u(0) = u^0 \quad \text{and} \quad u(0) = u^1.$$

Here, A is a linear differential operator in a domain $G \subset \mathbb{R}^3$.

$$u_t + iAu = 0 \quad \text{with} \quad u(0) = u^0$$

represents the Schrödinger equation, the system of the Maxwell equations or the system of Linear Acoustics, and by

$$u_t + Au = 0 \quad \text{with} \quad u(0) = u^0$$

heat conduction processes like initial-boundary value problems in Linear Thermoelasticity are described. In the first examples, A is selfadjoint.

For the treatment of such problems the equations are to be solved above all. In particular the notion of a solution has to be made precise. Subsequently one is interested in special properties of the solutions obtained, for instance one inquires about their regularity or about their asymptotic behavior for large times and proves the existence of wave and scattering operators. Problems of inverse scattering theory (from the signals reflected data like initial values, the boundary or the medium should be regained) are of great mathematical and practical interest. For the treatment of such problems one has to know, for instance, how the scattering operator is determined by the boundary, and this mapping then must be inverted. One obtains simple formulae in the limiting case of high frequencies.

The problem how acoustic or electromagnetic waves behave at high frequencies belongs to the important area of determining asymptotic properties of solutions of partial differential equations depending upon a parameter. From experience one knows that high-frequency waves spread according to the laws of geometrical optics. The statement that in the limit of high frequencies geometrical optics asymptotically describes the propagation of waves correctly is one of the most interesting propositions of this type of questions. The investigation of the transition from quantum mechanics to classical mechanics also belongs into this area. The corresponding differential equation is the Schrödinger equation, and the parameter tending towards zero in this case is Planck's constant h. Time harmonic wave propagation is usually described by an elliptic equation and the non-stationary time dependent wave propagation by a corresponding hyperbolic equation. Since one can obtain the solution of the hyperbolic equation by superposition of solutions of the elliptic equation, the asymptotic behavior of the solutions of the elliptic equation at high frequencies determines the singularities of the solutions of the hyperbolic equation. Therefore a further central point of this type of problems is the determination of the position and, especially, the form of singularity appearing in the solutions of the hyperbolic equations. This reveals the connection with microlocal analysis, whose main objective is to determine the position of singularities appearing within solutions of hyperbolic equations.

These questions were of fundamental importance for the development of mathematics and physics within the last 300 years. We only remind of the scientific dispute about the question whether the wave propagation theory set up by Huygens at the end of the 17th century or the corpuscle theory put up by Newton describe the propagation of light correctly. It was only in our century that this question could be clarified satisfactorily within the framework of quantum mechanics. The works of W.R. Hamilton at the beginning of the 19th century revealed connections with variational analysis.

In the sequel a survey of results with emphasis on those obtained at the "Sonderforschungsbereich 72" shall be given. We are only concerned with equations of classical physics and do not treat quantum scattering. In the first section initial-boundary value problems for an exterior domain in $\mathbb{R}^3$ are formulated and solved by looking at the model of the linear system of acoustics. Subsequently the corresponding free space problem with homogeneous isotropic medium is gone through. It will be used for reference purpose later on. In the third section the general case is taken up again and the spectrum of the underlying operator A is looked at more closely; boundary value problems for exterior domains are solved and it will be proved that the continuous spectrum is absolutely continuous. Hence it follows in the fourth section that the wave operators and the scattering operator exist. Thereby the behavior of the solutions for large times is described and in the fifth section an explicit representation of the scattering operator is derived by means of the radiation pattern. Subsequently the asymptotic behavior of solutions of the Helmholtz equation at high frequencies is studied. The sixth section presents a short introduction into this type of problems and describes the results obtainable with a more direct approach (though also here results about highly intricate questions like the exponential decay of energy are used); results requiring refined analysis will be discussed in the following sections. In Section 7 the asymptotic behavior of the solution on tangential rays is studied, in Section 8 the asymptotic behavior of the scattering amplitude is presented, and in Section 9 it is finally shown, how one can determine the shape of the obstacle from the knowledge of the scattering amplitude at high frequencies.

1. An initial-boundary value problem in Linear Acoustics

In order to have a specific problem in mind, we shall start formulating an initial-boundary value problem from Linear Acoustics. Corresponding problems for the other equations mentioned in the Introduction are being treated in Leis (1986). Thus, let Ω be an exterior domain in $\mathbb{R}^3$ (an open and connected set with bounded complement). We assume that Ω has the segment property, so that - locally - the Rellich Selection Theorem is applicable. Let κ and $\rho_{ik} = \rho_{ki}$ $(i,k = 1,2,3)$ be real-valued, bounded, and measurable functions defined on Ω with

$$\exists\, \kappa_1 > 0 \qquad \forall\, x \in \Omega \qquad \kappa(x) \geq \kappa_1$$

$$\exists\, \rho_1 > 0 \quad \forall\, \xi \in \mathbb{R}^3 \quad \forall\, x \in \Omega \qquad \xi_i \rho_{ik}(x) \xi_k \geq \rho_1 |\xi|^2$$

$$\exists\, r_a, \kappa_0, \rho_0 \qquad \forall\, x \in \Omega_a \quad k(x) = k_0 \text{ and } \rho_{ik}(x) = \rho_0 \delta_{ik}.$$

Here, $\Omega_a := \{x \in \Omega \mid |x| > r_a\}$. Thus, sufficiently far outside, the medium shall be homogeneous and isotropic.

Now let

$$v : \Omega \to \mathbb{R}^3 \text{ and } p : \Omega \to \mathbb{R}$$

be the velocity and the pressure, respectively. Then the underlying equations read

$$\rho v_t + \operatorname{grad} p = 0$$

$$\kappa p_t + \operatorname{div} v = 0.$$

In addition we have the initial conditions

$$v(0) = v^0 \text{ and } p(0) = p^0$$

and a boundary condition.

In order to formulate a Hilbert space approach, we choose

$$\mathcal{H} := (L^2(\Omega))^3 \times L^2(\Omega),$$

the weight matrix (with four rows)

$$M := \begin{bmatrix} \rho_{ik} & 0 \\ 0 & \kappa \end{bmatrix}$$

and for the scalar product

$$(U,V)_{\mathscr{H}} := (U,MV)_{L^2}.$$

Here, U stands for $U := (v,p)'$ and $U^0 := (v^0,p^0)'$.

From now on we consider the Neumann boundary value problem. In order to formulate it we define

$$\mathscr{D}(\Omega) := \{(C_1^*(\Omega))^3;\ \|\cdot\|_{\mathscr{D}}\}^{\sim},$$

where $\|\cdot\|_{\mathscr{D}}^2 = \|\cdot\|_{L^2}^2 + \|\mathrm{div}\,\cdot\|_{L^2}^2$ and $C_1^* := \{E \in C_1 \mid \|\cdot\|_1 < \infty\}$, as well as

$$\mathscr{D}^{\mathrm{o}}(\Omega) := \{C_\infty^{\mathrm{o}}(\Omega))^3;\ \|\cdot\|_{\mathscr{D}}\}^{\sim}.$$

$$\mathscr{D}^{\mathrm{o}}(\Omega) = \{E \in \mathscr{D}(\Omega) \mid \forall\ f \in H_1(\Omega)\ (U,\mathrm{grad}\ f) = -\ (\mathrm{div}\ U,f)\}$$

holds ("strong equals weak"), and $H_1(\Omega) := \{f \in C_1^*(\Omega);\ \|\cdot\|_1\}^{\sim}$ is the standard Sobolev space.

Now let

$$A := \mathscr{D}^{\mathrm{o}} \times H_1 \subset \mathscr{H} \to \mathscr{H}$$

$$AU := -iM^{-1} \begin{bmatrix} 0 & \mathrm{grad} \\ \mathrm{div} & 0 \end{bmatrix} U.$$

Then we are looking for a weak solution $U \in C(\mathbb{R}_0^+,\mathscr{H})$ of

$$U_t + iAU = 0 \quad \text{with} \quad U(0) = U^0 \in \mathscr{H}$$

that is of

$$\int_{\mathbb{R}^+\times\Omega} MU\ \overline{(\phi_t + iA\phi)} + (U^0,\phi(0))_{\mathscr{H}} = 0$$

for all $\phi \in C_0^{\mathrm{o}}(\mathbb{R},\mathscr{D}(A)) \cap C_1(\mathbb{R},\mathscr{H})$.

There are various notions of solution (e.g. classical, strict, weak or distribution solution). The notion of solution chosen here is physically appropriate, since it yields "solutions with finite energy", i.e. with

$$E(t) := \|U(t)\|_{\mathscr{H}}^2 < \infty.$$

E ist the energy of the process.

The inital-boundary value problem just formulated is uniquely solvable. This follows relatively easily, if one uses some spectral or semi-group theory. A is a symmetric operator, and from the definition of A^* the selfadjointness of A easily follows by using "strong equals weak". But then from

$$U(t) := e^{-iAt}U^0 = \int_{-\infty}^{\infty} e^{-i\lambda t}\, dP(\lambda)U^0$$

one obtains the solution of the problem using the Spectral Theorem. The energy $E(t) = \|U^0\|^2$ is constant. Details can be found in Leis (1986), p. 129 f.

From the Poincaré lemma one obtains for the null space and the range of A

$$\mathcal{N}(A) = \mathfrak{D}_0^{\circ} \times \mathcal{O} \quad \text{with} \quad \mathfrak{D}_0^{\circ} := \{E \in \mathfrak{D}^{\circ} \mid \operatorname{div} E = 0\}$$

$$\mathcal{R}(A) = \rho^{-1} \operatorname{grad} H_1 \times \kappa^{-1} \operatorname{div} \mathfrak{D}^{\circ},$$

respectively, and it is $\mathcal{H} = \mathcal{N}(A) \oplus \overline{\mathcal{R}(A)}$ orthogonally.

Finally we note the following local compactness criterion, which follows from the Rellich Selection Theorem

Theorem 1.1: *Let* $U^n \in \mathfrak{D}(A)$, $r' < r$ *and*

$$\exists\, k(r) > 0 \quad \forall\, n \in \mathbb{N} \quad \|U^n\|_A(\Omega_r) \le k(r).$$

Then $\{PU^n\}$ *contains a subsequence converging in* $\mathcal{H}(\Omega_{r'})$.

Here, $\|U\|_A^2 := \|U\|^2 + \|AU\|^2$ is the operator norm, $P : \mathcal{H} \to \overline{\mathcal{R}(A)}$ is the projector onto $\overline{\mathcal{R}(A)}$ and $\Omega_r = \Omega \cap B(0,r)$. It should be remarked that when treating the other equations mentioned in the Introduction it may be much more difficult to prove the analogue of Theorem 1.1. This especially happens when we are dealing with Maxwell's equations. Solutions of Maxwell's boundary value problems need not belong to H_1 so that the Rellich Selection Theorem is not applicable directly. Theorem 1.1 has been proved for the Maxwell operator by Weck (1974) and Weber (1980), cf. also Leis (1986), p. 165 f.

2. The free space problem

In this section we are treating the "free space problem", i.e. we let $\Omega = \mathbb{R}^3$, $\mathcal{H}_0 := (L^2(\mathbb{R}^3))^3 \times L^2(\mathbb{R}^3)$, $\kappa = \kappa_0$, $\rho_{ik} = \rho_0\delta_{ik}$ and A_0 be the corresponding operator. The free space problem will be used for reference purpose later on. It can be gone through completely by means of Fourier transformation with respect to the space coordinates. So let $\mathfrak{D}(A_0) := \mathfrak{D} \times H_1 \subset \mathcal{H}_0$ and consider

$$U_t + iA_0U = 0 \quad \text{with} \quad U(0) = U^0 \in \mathcal{H}_0.$$

Furthermore let $\hat{U} := FU$ be the Fourier transform of U and $\mathfrak{D}(\hat{A}_0) :=$

$\{\hat{U} \in \mathcal{H}_0 \mid U \in \mathcal{D}(A_0)\} \subset \mathcal{H}_0$. Then

$$(A_0U)^\wedge(p) := M^{-1} \begin{bmatrix} 0 & p \\ p' & 0 \end{bmatrix} \hat{U}(p) =: \hat{A}_0(p)\hat{U}(p),$$

and we obtain

$$\hat{U}_t + i\hat{A}_0\hat{U} = 0 \quad \text{with} \quad \hat{U}(0) = \hat{U}^0 \in \mathcal{H}_0.$$

The spectral family of $\hat{A}_0$ can be explicitly given. By means of the Heaviside function

$$H(s) := \begin{cases} 1 & \text{for } s \geq 0 \\ 0 & \text{for } s < 0, \end{cases}$$

and the projectors onto the eigenspaces of $\hat{A}_0(p)$

$$Q_\pm(p) = \frac{1}{2} \begin{bmatrix} p_0p_0' & \pm \sqrt{\kappa_0/\rho_0}\, p_0 \\ \pm \sqrt{\rho_0/\kappa_0}\, p_0' & 1 \end{bmatrix}$$

$$Q_0(p) = \begin{bmatrix} -p_0\times p_0\times & 0 \\ 0 & 0 \end{bmatrix}$$

where $p_0 = p/|p|$, one finds (cf. Leis (1986), p. 133 f.)

$$\hat{P}_0(\lambda) = H(\lambda + \frac{|\cdot|}{\sqrt{\kappa_0\rho_0}})Q_- + H(\lambda)Q_0 + H(\lambda - \frac{|\cdot|}{\sqrt{\kappa_0\rho_0}})Q_+.$$

From this one obtains the spectral family of A_0 in the form

$$P_0(\lambda) = F^*\hat{P}_0(\lambda)F =: \Pi(\lambda) + \Pi_0H(\lambda).$$

Here, $\Pi_0 := F^*Q_0F$ is the projector onto the null space $\mathcal{N}(A_0) = \mathcal{D}_0 \times \mathcal{O}$.

The solution of the initial-boundary value problem of the free space problem hence can be represented in the form

$$U(t) = \int_{-\infty}^{\infty} e^{-i\lambda t}\, dP_0(\lambda)U^0$$

$$= \Pi_0U^0 + \int_{-\infty}^{\infty} e^{-i\lambda t}\, d\Pi(\lambda)U^0 =: U_0 + U_1(t).$$

$U_0 := \Pi_0U^0$ is the stationary part.

Now we want to discuss $U(t)$ for large values of t. With $\tau := t/\sqrt{\kappa_0\rho_0}$ one obtains from the Fourier representation

$$U_1(t,x) = (\frac{1}{2\pi})^{3/2} \int_{\mathbb{R}^3} e^{ixp} \{e^{i|p|\tau} Q_-(p) + e^{-i|p|\tau} Q_+(p)\}\, \hat{U}^0(p)dp$$

$$= \left(\frac{1}{2\pi}\right)^{3/2} \int_0^\infty \{e^{ir\tau}\, V_-(x,r) + e^{-ir\tau}\, V_+(x,r)\}\, r^2 dr$$

with

$$V_\pm(x,r) = \int_{S^2} e^{ir(xz)}\, Q_\pm(z)\, \hat{U}^0(rz)\, dz.$$

The asymptotic behavior of such functions has been stated by Wilcox (1975) for the d'Alembert equation. One first discusses $V_\pm(x,r)$ for large values of $|x|$ by means of the Stationary Phase Method and for $U_1(t)$ then obtains

$$\lim_{t\to+\infty} \|U_1(t) - U^+(t)\| = 0, \quad \lim_{t\to-\infty} \|U^+(t)\| = 0,$$

$$\lim_{t\to-\infty} \|U_1(t) - U^-(t)\| = 0, \quad \lim_{t\to+\infty} \|U^-(t)\| = 0,$$

with

$$U^\pm(t,x) := \frac{Q_\pm(x_0)}{\sqrt{2\pi}\; i|x|} \int_{-\infty}^{\infty} e^{ir(|x|\pm\tau)}\, \hat{U}^0(rx_0)\, r dr.$$

To get this result one starts choosing C_∞-initial data $\hat{U}^0$ compactly supported in $\mathbb{R}^3\setminus\{0\}$ and obtains the result by completion, in particular also $U^\pm(t,\cdot) \in \mathcal{H}_0$.

Thus

$$(2.1)\qquad \begin{aligned} &\lim_{t\to+\infty} \|Q_+U_1(t) - U^+(t)\| = 0, \quad \lim_{t\to-\infty} \|Q_+U_1(t)\| = 0 \\ &\lim_{t\to-\infty} \|Q_-U_1(t) - U^-(t)\| = 0, \quad \lim_{t\to+\infty} \|Q_-U_1(t)\| = 0 \end{aligned}$$

as well as

$$(2.2)\qquad \lim_{t\to\pm\infty} \|Q_0U_1(t)\| = 0.$$

hold. So $Q_+U_1(t)$ asymptotically behaves like an outgoing wave, $Q_-U_1(t)$ like an incoming one, while $Q_0U_1(t)$ vanishes.

By means of the Fourier transformation one also easily obtains a fundamental solution for $A_0 - \lambda$. It will be needed in the next section. So let

$$(A_0 - \lambda)G_\lambda = \delta\ \mathrm{id}$$

or

$$(\hat{A}_0 - \lambda)\hat{G}_\lambda = \frac{\mathrm{id}}{(2\pi)^{3/2}}.$$

It then follows for $\lambda \in \mathbb{C}\setminus\mathbb{R}$ by simple checking

$$\hat{G}_\lambda = \frac{\kappa_0\rho_0}{(2\pi)^{3/2}}\left(\hat{A}_0 + \frac{1}{\lambda}\hat{A}_0^2\right)\frac{1}{|\cdot|^2 - \lambda^2\kappa_0\rho_0} - \frac{\mathrm{id}}{\lambda(2\pi)^{3/2}}$$

or

$$G_\lambda = \kappa_0\rho_0(A_0 + \frac{1}{\lambda} A_0^2)g_\lambda - \frac{\delta}{\lambda}\mathrm{id}$$

with

$$g_\lambda(x) = \frac{1}{4\pi|x|}\begin{cases} e^{i\lambda\sqrt{\kappa_0\rho_0}|x|} & \text{for } \mathrm{Im}\,\lambda > 0 \\ e^{-i\lambda\sqrt{\kappa_0\rho_0}|x|} & \text{for } \mathrm{Im}\,\lambda < 0. \end{cases}$$

For $F \in \overline{\mathcal{R}(A_0)}$ this simplifies to

$$G_\lambda * F = \kappa_0\rho_0(A_0 + \lambda)g_\lambda * F.$$

From the representation of G_λ one finds for large values of $|x|$ uniformly in $\{\lambda \mid |\lambda| > \epsilon > 0\}$

$$(Q_0G_\lambda)(x) = \mathcal{O}(|x|^{-2})$$

(2.3) $$(Q_-G_\lambda)(x) = \mathcal{O}(|x|^{-2}) \quad \text{for } \mathrm{Im}\,\lambda > 0$$

$$(Q_+G_\lambda)(x) = \mathcal{O}(|x|^{-2}) \quad \text{for } \mathrm{Im}\,\lambda < 0.$$

One can use this in order to formulate the Sommerfeld radiation condition for outgoing or incoming waves, respectively, as it is done for the Helmholtz equation. One obtains an outgoing solution of $(A_0 - \lambda)U = F$ (for real λ) by passing to the limit $\mathrm{Im}\,\lambda \downarrow 0$. In Linear Acoustics the outgoing radiation condition thus reads (compare with Equation (2.1))

(2.4) $$Q_-U \in \mathcal{H}$$

or for $|x| \to \infty$ because of $Q_+(x_0) - Q_-(x_0) = \sqrt{\kappa_0\rho_0}\,\hat{A}_0(x_0)$

$$(\sqrt{\kappa_0\rho_0}\,\hat{A}_0(x_0) - \mathrm{id})U(x) = \mathcal{O}(|x|^{-2}).$$

It suffices to impose this condition for one component only, thus e.g.

$$\sqrt{\kappa_0\rho_0}\,x_0\,U_2(x) - U_1(x) = \mathcal{O}(|x|^{-2}).$$

Correspondingly the incoming radiation condition reads

(2.5) $$Q_+U \in \mathcal{H}$$

or

$$(\sqrt{\kappa_0\rho_0}\,\hat{A}_0(x_0) + \mathrm{id})U(x) = \mathcal{O}(|x|^{-2}).$$

3. On the spectrum of A

In this section we want to show that the continuous spectrum of A is absolutely continuous. To this end we start mentioning that the value $\lambda = 0$ is the sole eigenvalue of A. For differentiable medium (which shall be presumed in the sequel) this follows from the Rellich estimate for the solutions of the Helmholtz equation (Rellich (1943)),

$A^2U = \lambda^2 U$ and the Unique Continuation Principle as it was first proved by Müller (1954), cf. Leis (1986), p. 64 and 144. As in the case of the Helmholtz equation one also obtains that $\mathbb{R}\backslash\{0\}$ belongs to the continuous spectrum.

With $\mathcal{H}_p := \mathcal{N}(A)$ and $\mathcal{H}_c := \overline{\mathcal{R}(A)}$ thus

$$\mathcal{H} = \mathcal{H}_p \oplus \mathcal{H}_c,$$

and for all $F \in \mathcal{H}_c$ $(P(\lambda)F,F)$ is continuous. Our aim is to prove that $(P(\lambda)F,F)$ is even absolutely continuous for these F, i.e. that

$$\mathcal{H}_c = \mathcal{H}_{ac}$$

holds.

For that purpose we use Stone's formula and for $F \in \mathcal{H}_c$ obtain the representation

$$(P(\lambda)F,F) = (P(\lambda_1)F,F) + \frac{1}{2\pi i} \lim_{\epsilon \downarrow 0} \int_{\lambda_1}^{\lambda} (R(\mu+i\epsilon)F - R(\mu-i\epsilon)F,F)\, d\mu$$

with $R(\lambda)F = (A - \lambda)^{-1}F$. We want to show, that on the right hand side passing to the limit can be interchanged with integration for fields F with bounded support ($F \in \mathcal{H}^f$), and that therefore $\mathcal{H}_c \cap \mathcal{H}^f \subset \mathcal{H}_{ac}$. $\mathcal{H}_c = \mathcal{H}_{ac}$ then follows from the fact that $\mathcal{H}_c \cap \mathcal{H}^f$ is dense in $\mathcal{H}_c$ and $\mathcal{H}_{ac}$ is closed in $\mathcal{H}_c$.

For the proof we choose $\lambda_1, \lambda_2 \in \mathbb{R}$, some $\tau > 0$ and set

$$Q^+ := (\lambda_1,\lambda_2) \times (0,\tau) \subset \mathbb{C}$$
$$Q^- := (\lambda_1,\lambda_2) \times (-\tau,0) \subset \mathbb{C}.$$

Furthermore let $\sigma(x) := 1 + |x|$, $\alpha := \sigma^{-1}$, $\|U\|_\alpha := \|\alpha U\|_{\mathcal{H}}$, $B(x_0) := \sqrt{\kappa_0 \rho_0}\, \hat{A}_0(x_0)$,

$$(U,V)_\alpha^\pm := (U,V)_\alpha + ((B \mp id)U,(B \mp id)V)$$

and $\mathcal{H}_\alpha^\pm := \{\mathcal{H};\ \|\cdot\|_\alpha^\pm\}^\sim$. In this way we have embodied the outgoing or incoming radiation condition in the inner product. The assertion then follows from the "Limiting Absorption Principle". Dealing with boundary value problems this principle was first proved by Eidus (1962) for the Helmholtz equation. We formulate it as

Theorem 3.1: *Let* $F \in \mathcal{H}_c \cap \mathcal{H}^f$. *Then the mappings*

$$R(\cdot)F : Q^\pm \to \mathcal{H}_\alpha^\pm$$

are uniformly continuous.

The principle states that $R(\mu \pm i\epsilon)F$ converges toward an outgoing or incoming solution of the exterior boundary value problem $(A - \mu)U = F$, respectively. Both are uniquely determined. In order to indicate the proof of Theorem 3.1 we note that since $U(\lambda) := R(\lambda)F \in \mathcal{H}_c$ there is a $u \in J_1$ with $U_1 = \rho^{-1}\nabla u$. Here, $J_1 := \{C_1^{*f}(\Omega); \|\nabla \cdot \|\}^{\sim}$. We choose a $\phi \in C_\infty$ with $\phi|B(0,r_a) = 0$, $\phi|\{x \mid |x| > r_a + 1\} = 1$ and set

$$V := \begin{bmatrix} \rho^{-1} \nabla\phi u \\ \\ \phi U_2 \end{bmatrix} \in \mathcal{H}_c.$$

V then is a solution in all of $\mathbb{R}^3$ and can be estimated by means of the fundamental solution. Hence the a priori estimate

$$\exists\, k,K \quad \forall\, \lambda,\ |\lambda| \le c, \quad \forall\, U \quad \|U\|_\alpha \le k\,\{\|U\|(K) + \|F\|_\sigma\}$$

with a compact $K \subset\subset \mathbb{R}^3$ follows. From this a priori estimate and the uniqueness of the solutions of the exterior boundary value problem one obtains

$$\exists\, \gamma > 0 \quad \forall\, \lambda \in Q^\pm \quad \forall\, F \in \mathcal{H}_c \cap \mathcal{H}^f \quad \|R(\lambda)F\|_\alpha^\pm \le \gamma\, \|F\|_\sigma,$$

and from that the Limiting Absorption Principle follows. For details we refer to Leis (1986), p. 145 and p. 184 f.

For the Helmholtz equation exterior boundary value problems have been treated by several authors. Uniqueness was first shown by Kupradze (1934), Freudenthal (1938) and Rellich (1943). Existence theorems were given by Vekua (1943), Weyl (1952), Müller (1952) and Leis (1958) using integral equation methods. It is interesting to remark that using integral equation methods it often happens that the equation obtained possesses eigenfunctions whereas the original problem does not. Thus it is especially important to find a suitable ansatz which avoids this and really transforms the problem in a one-to-one way. For the Helmholtz equation Werner (1962) was the first who achieved this using a volume layer; a surface layer ansatz was given by Leis (1964). Variable coefficients and nonsmooth boundaries were treated by Werner (1960-1963), Eidus (1962), Leis (1962, 1965), Jäger (1967) and others.

4. The existence of wave operators

These preparations being done we are able to describe the asymptotic behavior of the solutions $U(t)$ of $U_t + iAU = 0$ with $U(0) = U^0$. From now on we assume $U^0 \in \mathcal{H}_c$ since we are not interested in stationary solutions. We start noticing that from $\mathcal{H}_c = \mathcal{H}_{ac}$ the "Principle of Local Energy Decay" follows, i.e.

$$\forall\ r > 0 \quad \lim_{t\to\pm\infty} \|U(t)\|(\Omega_r) = 0$$

holds (again $\Omega_r := \Omega \cap B(0,r)$). The principle follows from the representation

$$U(t) = \int_{-\infty}^{\infty} e^{-i\lambda t}\, dP(\lambda) U^0$$

of the solution and the Riemann-Lebesgue Lemma (cf. Leis (1986), p. 113).

The Principle of Local Energy Decay states that $U(t)$ vanishes in every neighborhood of the boundary $\partial\Omega$ as $t \to \pm\infty$. Thus one expects that $U(t)$ behaves like a free space solution for large values of t. To specify that, we choose an extension operator J and a cut off operator J_0

$$J \ : \mathcal{H} \to \mathcal{H}_0$$

$$J_0 : \mathcal{H}_0 \to \mathcal{H}$$

respectively, with

$$(Jg)(x) := \begin{cases} j(x)g(x) & \text{for } x \in \Omega \\ 0 & \text{otherwise} \end{cases}$$

$$(J_0 g)(x) := j(x)g(x) \qquad \text{for } x \in \Omega$$

where $j \in C_\infty(\mathbb{R}^3)$ with $j|B(0,r_a) = 0$, $j|\{|x| > r_a + 1\} = 1$ and $0 \le j \le 1$. Furthermore let P and P_0 be the projectors onto $\mathcal{H}_{ac} = \overline{\mathcal{R}(A)}$ and $\mathcal{H}_{0,ac} = \overline{\mathcal{R}(A_0)}$, respectively.

Then we are looking for $V_0^+, V_0^- \in \mathcal{H}_{0,ac}$ with

$$\lim_{t\to\pm\infty} \|JU(t) - e^{-iA_0 t} V_0^\pm\| = 0$$

or

$$\lim_{t\to\pm\infty} \|e^{iA_0 t} J e^{-iAt} U^0 - V_0^\pm\| = 0.$$

In the sequel we will show the existence of the "wave operators"

$$W^\pm : \mathcal{H} \to \mathcal{H}_0$$

$$W^\pm := \text{s-}\lim_{t\to\pm\infty} e^{iA_0 t} J e^{-iAt} P.$$

The operators $W^\pm$ do not depend upon the particular choice of J and are unitary mappings from $\mathcal{H}_{ac}$ onto $\mathcal{H}_{0,ac}$.

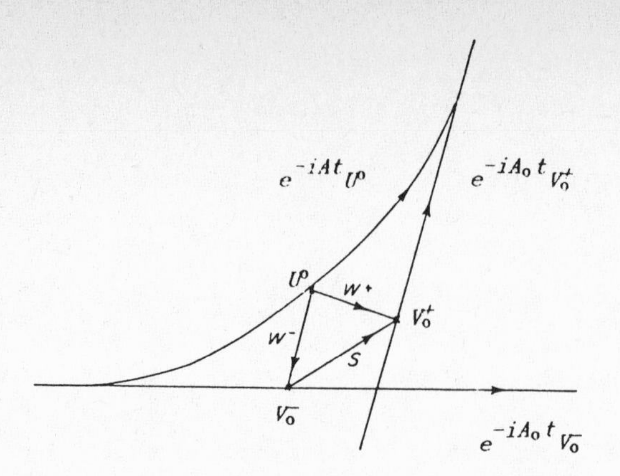

Thus for $t \to \pm\infty$ $U(t)$ behaves like the free space solutions

$$e^{-iA_0t}V_0^{\pm} \quad \text{with} \quad V_0^{\pm} := W^{\pm}U^0,$$

which we have discussed in the second section. The mapping

$$S := W^+(W^-)^* \quad : \quad \mathcal{H}_{0,ac} \to \mathcal{H}_{0,ac}$$

is called the "scattering operator".

On $\mathcal{H}_{ac}$

$$P(\lambda) = (W^{\pm})^* P_0(\lambda) W^{\pm}$$

also holds, i.e. A and A_0 are unitarily equivalent on the subspaces $\mathcal{H}_{ac}$ and $\mathcal{H}_{0,ac}$, repectively, $A = (W^{\pm})^* A_0 W^{\pm}$.

To prove the existence of the wave operators we formulate the following theorem which is proved in Leis (1986), p. 116 f. and which goes back to Kato (1976), Belopolskii & Birman (1968), Pearson (1978) and Picard & Seidler (1984).

Theorem 4.1: *Let* $\{I_m\}$ *be a family of disjoint open intervals with*

$$\mathbb{R} = \bigcup_{m=1}^{\infty} I_m + N$$

where N *is a Lebesgue null set. Let* H, I *be bounded intervals and*

$$M := \bigcup_{m=1}^{\infty} I_m \cap I.$$

Let furthermore

(i) $\quad J_0\mathcal{D}(A_0) \subset \mathcal{D}(A)$ *and* $J\mathcal{D}(A) \subset \mathcal{D}(A_0)$

(ii) $\quad \forall\, I \in \mathbb{R} \quad (JJ_0 - id)P_0(M) \in \mathcal{B}_\infty(\mathcal{H}_0,\mathcal{H}_0)$

(iii) $\quad \forall\, I \in \mathbb{R} \quad (J_0J - id)P(M) \in \mathcal{B}_\infty(\mathcal{H},\mathcal{H})$

(iv) $\quad \forall\, I \in \mathbb{R} \quad (AJ_0 - J_0A_0)P_0(M) \in \mathcal{B}_1(\mathcal{H}_0,\mathcal{H})$

(v) $\quad \forall\, H, I \in \mathbb{R} \quad P_0(H)(JA - A_0J)P(M) \in \mathcal{B}_\infty(\mathcal{H},\mathcal{H}_0)$

hold. Then the wave operators

$$W^{\pm} : \mathcal{H} \to \mathcal{H}_0, \quad W^{\pm} := \operatorname*{s-lim}_{t\to\pm\infty} e^{iA_0t} J e^{-iAt} P$$

$$W_0^{\pm} : \mathcal{H}_0 \to \mathcal{H}, \quad W_0^{\pm} := \operatorname*{s-lim}_{t\to\pm\infty} e^{iAt} J_0 e^{-iA_0t} P_0$$

exist and are partially isometric with

$$(W^{\pm})^* W^{\pm} = P, \quad (W_0^{\pm})^* W_0^{\pm} = P_0, \quad \text{and} \quad (W^{\pm})^* = W_0^{\pm}.$$

Here $\mathcal{B}_\infty$ are the compact and $\mathcal{B}_1$ the nuclear operators.

In order to apply Theorem 4.1 to our case we choose the I_m such that the origin is contained in none of them. Therefore the point spectrum is not contained in M but may be in H. We sketch the proof of the assumptions (i) - (v) for our case.

(i) is clearly fulfilled. (ii) and (iii) are similar; we show (iii). Let $\{U^n\}$ be bounded in $\mathcal{H}$. Then $V^n := P(M)U^n \in \mathcal{H}_{ac} = \overline{\mathcal{R}(A)}$,

$$\|AV^n\|^2 = \int_M \lambda^2 \, d\|P(\lambda)V^n\|^2 \leq \text{const},$$

and from the compactness result at the end of the first section (Theorem 1.1) the existence of a convergent subsequence of $\{(J_0J - id)P(M)U^n\}$ follows.

To prove (iv) we note that formally

$$(AJ_0 - J_0A_0)U = DU$$

holds where D is a bounded multiplication operator whose support is contained in $K := \{x \mid r_a < |x| < r_a + 1\}$. The proof of $DP_0(M) \in \mathcal{B}_1(\mathcal{H}_0,\mathcal{H})$ then either follows from explicit representation of $P_0(M)U$ by means of Fourier transformation (cf. Leis (1986), p. 189), or by exploiting the fact that for all n $V := P_0(M)U$ belongs to $\mathcal{D}((A_0)^n)$ and that

$$\|(A_0)^n V\| \leq c(n) \|U\|$$

holds. Using standard regularity theorems (cf. Agmon (1965), p. 129) it then follows that $V \in (H_4(K))^4$ and $DV \in (\overset{\circ}{H}_4(K))^4$. Now the inclusion

$$i : (\overset{\circ}{H}_4(K))^4 \to (L^2(K))^4$$

is nuclear (cf. Yosida (1974), p. 279). So,

$$(AJ_0 - J_0A)P_0(M) \quad : \quad \mathcal{H}_0 \to \mathcal{H}$$

$$(AJ_0 - J_0A)P_0(M) = iDP_0(M) \in \mathcal{B}_1(\mathcal{H}_0,\mathcal{H}).$$

Property (v) finally follows analogously to (iii) from

$$(JA - A_0J)P(M) = - DP(M)$$

and the boundedness of $P_0(H)$. This proves the existence of the wave operators.

5. The scattering matrix

In this section we want to give an explicit representation of the scattering operator. To that end we proceed as we did in the free space case and correspondingly have to generalize the Fourier transformation to functions defined in an exterior domain.

In order to simplify our presentation we confine ourselves from now on to the wave equation and a homogeneous medium. This equation will be treated in more detail in the next section. One can formulate and solve initial-boundary value problems for

$$(\partial_t^2 - \Delta)u(t,x) = 0$$
$$u(t,\cdot)|\partial\Omega = 0$$
$$u(0,x) = u^0(x) \quad \text{and} \quad u_t(0,x) = u^1(x)$$

similarly as was done in Section 1 (cf. Leis (1986). p. 33 f.). If one makes the separation ansatz

$$u(t,x) = e^{-ikt}v(x,k)$$

with $k \in \mathbb{R}^+$, then v is the solution of a boundary value problem for the Helmholtz equation, namely

$$(5.1) \qquad (\Delta + k^2)v(\cdot,k) = 0$$

$$(5.2) \qquad v(\cdot,k)|\partial\Omega = 0.$$

Exterior boundary value problems for the Helmholtz equation can be uniquely solved using the Limiting Absorption Principle if one adds the radiation condition, either

$$(5.3) \qquad (x_0\nabla - ik)v(\cdot,k) \in L^2(\Omega)$$

for outgoing waves or

$$(5.4) \qquad (x_0\nabla + ik)v(\cdot,k) \in L^2(\Omega)$$

for incoming waves.

The Fourier transformation $F : L^2(\mathbb{R}^3) \to L^2(\mathbb{R}^3)$

$$(Fu)(p) := \hat{u}(p) := \lim_{r\to\infty} \int_{|x|<r} u(x)\,\overline{w_0(x,p)}\,dx =: (u,w_0(\cdot,p))$$

where

$$w_0(x,p) = (2\pi)^{-3/2}e^{ixp}$$

is a unitary mapping with

$$(F^*u)(x) = (u,\overline{w_0(x,\cdot)})$$

as is generally known. The function

$$u(t,x,p) := e^{-i|p|t}w_0(x,p)$$

solves the wave equation and represents a "plane wave" which proceeds in p-direction. The planes $xp = \text{const}$ are wave fronts. With the help of these plane waves we may derive a representation of the solution of our initial-boundary value problem for the wave equation in the free space similarly to the representation we got in Section 2.

In order to generalize this procedure to arbitrary exterior domains Ω we now define two unitary mappings

$$F^{\pm} : L^2(\Omega) \to L^2(\mathbb{R}^3)$$

by using so-called "distorted plane waves" instead of plane waves $w_0(x,p)$ (cf. Wilcox (1975), p. 84 f, and Ikebe (1960)), namely

$$w^{\pm}(x,p) = w_0(x,p) + v^{\pm}(x,p) \tag{5.5}$$

with

$$(\Delta + |p|^2)v^{\pm}(\cdot,p) = 0 \tag{5.6}$$

$$(w_0(\cdot,p) + v^{\pm}(\cdot,p))|\partial\Omega = 0 \tag{5.7}$$

$$(x_0\nabla \mp i|p|)v^{\pm}(\cdot,p) \in L^2(\Omega). \tag{5.8}$$

The functions $v^{\pm}$ are uniquely determined; one therefore obtains two generalizations of the Fourier transformation, namely an "outgoing" and an "incoming" one.

$$(F^{\pm}f)(p) = (f,w^{\pm}(\cdot,p)),$$

$$((F^{\pm})^*f)(p) = (f,\overline{w^{\pm}(\cdot,p)})$$

and

$$v^-(x,p) = \overline{v^+(x,-p)}.$$

hold.

In this way one obtains the representations

$$(F^{\pm}(Af))(p) = |p|^2(F^{\pm}f)(p),$$

for the underlying operator A (the Δ-operator in Ω with Dirichlet boundary condition); the spectrum of A is absolutely continuous, $\sigma(A) = \mathbb{R}_0^+)$ and one speeks of an expansion in terms of "generalized eigenfunctions".

Now let $h := u^0 + iA^{-1/2}u^1$ with real-valued u^0, u^1 and

$$v(t) := e^{-itA^{1/2}}h = (F^-)^*e^{-it|\cdot|}F^-h.$$

Then $\operatorname{Re} v(t)$ solves the initial-boundary value problem for the wave equation and we may decompose v to become

$$v(t) = v_0(t) + v_1(t)$$

with

$$v_0(t) := F^*e^{-it|\cdot|}F^-h.$$

Because of Equation (5.14) (cf. Wilcox (1975), p. 126)

$$\lim_{t\to+\infty} \|v_1(t)\| = 0.$$

Let A_0 be the Δ-operator in $\mathbb{R}^3$. It then follows from the above and from the Principle of Local Energy Decay

$$W^+h = \text{s-}\lim_{t\to\infty} e^{itA_0^{1/2}} J e^{-itA^{1/2}} h = F^*F^-h$$

and analogously $W^- = F^*F^+$. Thereby we obtain the corresponding scattering operator $S = W^+W^{-*}$ in the form

(5.9) $\quad S = F^*\hat{S}F \quad \text{with} \quad \hat{S} : L^2(\mathbb{R}^3) \to L^2(\mathbb{R}^3), \quad \hat{S} := F^-(F^+)^*.$

The operator $\hat{S}$ is not defined on $(L^2(\mathbb{R}^3))^2$ as one should expect but only on $L^2(\mathbb{R}^3)$, since we have chosen the inital values to be real-valued and have put them into a complex h. $\hat{S}$ is unitary.

In the sequel we want to explicitly represent $\hat{S}$. The equations id $= (F^{\pm})^*F^{\pm}$ and id $= F^{\pm}(F^{\pm})^*$ in distribution notation read

(5.10)
$$(w^{\pm}(x,\cdot),w^{\pm}(y,\cdot)) = \delta(x - y)$$
$$(w^{\pm}(\cdot,p),w^{\pm}(\cdot,q)) = \delta(p - q),$$

respectively. Hence it follows immediately

(5.11) $$(\hat{S}\ \overline{w^+(x,\cdot\ )})(p) = \overline{w^-(x,p)},$$

and one again recognizes how the scattering operator $\hat{S}$ maps.

In order to find a representation for $\hat{S}$, we first introduce the notations of "radiation pattern" and "scattering amplitude", respectively. To that end we start proving the follwing lemma (cf. Müller (1957), p. 114).

Lemma 5.1: *Let* v *be a solution of equations (5.1) and (5.3) and let* $x = r|x_0|$. *then as* r *tends to* ∞

$$v(x,k) = \frac{e^{ikr}}{r}\,\phi(x_0k) + o(\tfrac{1}{r})$$

holds, uniformly with respect to $x_0 \in S^2$ *and* $k \in K \subset\subset \mathbb{R}_0^+$. *Here*

$$\phi(x_0,k) := \frac{-1}{4\pi} \int_{|y|=c} e^{-ik(x_0y)} \{ik(x_0n(y))v(y,k) + \frac{\partial}{\partial n}v(y,k)\}\, dS_y,$$

and the constant c *may be chosen arbitrarily as long as* $\mathbb{R}^3\setminus\Omega \subset B(0,c)$.

In Lemma 5.1 we assume n to be the outward normal vector to $\partial B(0,c)$ and $S^2 := \{x \in \mathbb{R}^3 \mid |x| = 1\}$.

To prove Lemma 5.1 one uses the fundamental solution

$$g(x,y,k) := \frac{1}{4\pi}\,\frac{e^{ik|x-y|}}{|x-y|},$$

and for v one obtains the representation

(5.12) $$v(x,k) = -\int_{|y|=c} \{g(x,y,k)\,\frac{\partial}{\partial n}v(y,k) - v(y,k)\,\frac{\partial}{\partial n}g(x,y,k)\}\, dS_y.$$

From this the assertion follows by expanding the fundamental solution for large values of r. ϕ is the radiation pattern of v; ϕ corresponds to v in a one-to-one manner. To see the latter let ϕ be a radiation pattern of both v_1 and v_2. Then $v_1 - v_2 = o(1/r)$ which together with Rellich's estimate implies $v_1 = v_2$.

Now let $v(x,p)$ be an outgoing solution of Equations (5.6) - (5.8). Then consequently there is an $F(x_0,p)$ with

$$v(x,p) = \frac{e^{i|p|r}}{r} F(x_0,p) + o(\tfrac{1}{r}),$$

and from

$$0 = \int_{|x|=r} \{w(x,p_1) \frac{\partial}{\partial n} w(x,p_2) - w(x,p_2) \frac{\partial}{\partial n} w(x,p_1)\} \, dS_x$$

one obtains with $p := k\omega$ (cf. Ramm (1986), p. 53)

$$F(x_0,k\omega) = F(-\omega,-kx_0). \tag{5.13}$$

With

$$f(x_0,\omega,k) := (2\pi)^{3/2} F(x_0,k\omega)$$

we can hence define

Definition 5.1: *Let* $\mathbf{v(rs,k\omega)}$ *be the outgoing solution of Equations* (5.6) - (5.8), *and* $f : S^2 \times S^2 \times \mathbb{R}^+ \to \mathbb{C}$

$$f(s,\omega,k) := \lim_{r\to\infty} (2\pi)^{3/2} r e^{-ikr} v(rs,k\omega).$$

Then f *is called scattering amplitude of* v.

Analogously one obtains the scattering amplitude of incoming solutions of the Helmholtz equation. From $v^-(x,p) = \overline{v^+(x,-p)}$ it follows

$$v^-(x,p) = \frac{e^{-i|p|r}}{r} \overline{F(x_0,-p)} + o(\tfrac{1}{r}). \tag{5.14}$$

We are now able to show the main result of this section which in this form goes back to Lax & Phillips (1967), p. 170, namely

Theorem 5.1: *Let* f *be the scattering amplitude just defined. Then*

$$\hat{S} = id + K$$

holds with (a.e. with respect to k *and* s*)*

$$(Kj)(k,s) := \frac{ik}{2\pi} \int_{|\omega|=1} j(k\omega) \, f(s,\omega,k) \, dS_\omega.$$

Here we have set $p := ks$, $s \in S^2$, and $j(p) =: j(k,s)$. The operator K is the scattering matrix.

To prove this we "test" the assertion with functions F^+h and, since $\hat{S} = F^-(F^+)^*$ holds, we have to show

$$F^-h(p) = F^+h(p) + \frac{ik}{2\pi} \int_\Omega h(x) \int_{|\omega|=1} f(s,\omega,k)\, \overline{w^+(x,k\omega)}\, dS_\omega\, dx,$$

or

$$0 = -(h,w^-) + (h,w^+) + \frac{ik}{2\pi}\, \overline{(h, \int_{|\omega|=1} \ldots)} = (h,u(\cdot,p))$$

with

$$u(x,p) := -w^-(x,p) + w^+(x,p) - i\sqrt{2\pi}k \int_{|\omega|=1} \overline{F(s,k\omega)}\, w^+(x,k\omega)\, dS_\omega.$$

The assertion thus follows from $u = 0$. In order to show this, we first note that u satisfies Equations (5.6) and (5.7). From the uniqueness theorem for solutions of the Helmholtz equation hence $u = 0$ follows if we can also prove that u satisfies the radiation condition for outgoing waves. In order to do that, we first omit all terms in u which one immediately sees to fulfill this condition, and discuss

$$u_1(x,p) := -v^-(x,p) - i\sqrt{2\pi}k \int_{|\omega|=1} \overline{F(s,k\omega)}\, w_0(x,k\omega)\, dS_\omega$$

$$= -\frac{e^{-ikr}}{r} \overline{F(x_0,-p)} - \frac{ik}{2\pi} \int_{|\omega|=1} e^{ikx\omega}\, \overline{F(s,k\omega)}\, dS_\omega + o(\tfrac{1}{r})$$

as $r := |x| \to \infty$. Using the Stationary Phase Method one obtains for the last integral (cf. Leis (1986), p. 247) for fixed k as $r \to \infty$

$$\int_{|\omega|=1} \ldots = \frac{2\pi}{ikr} \{e^{ikr}\, \overline{F(s,kx_0)} - e^{-ikr}\, \overline{F(s,-kx_0)}\} + o(\tfrac{1}{r}).$$

Together with Equation (5.13) it thus follows as $r \to \infty$

$$u_1(x,p) = -\frac{e^{ikr}}{r}\, \overline{F(s,kx_0)} + o(\tfrac{1}{r}).$$

Therefore u_1 also satisfies the radiation condition for outgoing waves and Theorem 5.1 is proved.

6. The WKBJ-ansatz in high frequency asymptotics

So far it has been shown how the solutions of initial-boundary value problems can be approximated in the limiting case of large times by the more easily constructable solutions of free space problems. There are other limiting cases in which approximations by more simple approximating functions are possible. To them the limiting case of high frequencies belongs, which will be discussed now.

We first discuss in this section the results which can be achieved by the WKBJ-ansatz. This method is named after G. Wentzel, H.A. Kramers, L. Brillouin, and H. Jeffreys, who treated problems from quantum mechanics and from the theory of partial differential equations with this ansatz. The ansatz though has already been known earlier and traces back to P. Debye, A. Sommerfeld, and I. Runge. Further details regarding the historical background can be found in Alber (1982a).

This section is also meant as an introduction to this type of problems and therefore contains more details. In the following sections we discuss a selection of the results obtained at the Sonderforschungsbereich on this subject, which relate to the scattering amplitude and to the asymptotic behavior in the neighborhood of tangential rays.

We do not discuss the results obtained in Alber (1980, 1984) for inhomogeneous media, and for the scattering kernel. Many other problems have been treated in the vast literature regarding this type of problems. Examples where interesting questions appear are the theory of analytic singularities (cf. for example Sjöstrand (1981), Hörmander (1983), 1985)) which is connected with the asymptotic behavior of Green's function in the shadow (cf. for example Babich (1965), Levy & Keller (1959), Popov (1987), Ursell (1957)), Maxwell's equations (cf. for example Born (1933), Kline (1965), Ludwig (1961)), and the system of equations of elasticity theory, especially in the case of wave propagation in anisotropic media. Applications to underwater acoustics can be found for example in Keller (1977). In addition to the papers on applications in scattering theory we mention here the papers of Petkov (1980) and of Jensen & Kato (1978).

As an example we again choose the propagation of acoustic waves. Although the following considerations can also be made for systems of linear differential equations, we now switch to the investigation of the wave equation, since thereby some of the following calculations simplify. The WKBJ-method for systems of equations is described in Lax (1957).

Let $B \subset \mathbb{R}^3$ be a bounded domain with boundary $\partial B \in C_\infty$. Let $\Omega = \mathbb{R}^3 \setminus \bar{B}$ denote the exterior of B. From the system of equations for the pressure p and the velocity v given in the first section it follows that in the case of a homogeneous medium the pressure p satisfies the wave equation

$$(\partial_t^2 - \Delta)p(t,x) = (\partial_t^2 - \sum_{i=1}^{3} \partial_{x_i}^2)p(t,x_1,x_2,x_3) = 0.$$

We choose Dirichlet boundary conditions and look for a solution $p : \mathbb{R}_0^+ \times \Omega \to \mathbb{R}$ of the wave equation satisfying the boundary and initial conditions

$$p(t,x) = 0, \qquad \text{for } x \in \partial\Omega$$

$$p(0,x) = p^0(x), \quad \partial_t p(0,x) = p^1(x), \qquad \text{for } x \in \Omega.$$

The methods of the first section show that this problem has a unique solution under certain weak assumptions for p^0 and p^1. A representation formula for this solution is obtained with the "Green's function" $R(t,x,y)$ corresponding to this initial-boundary value problem of the wave equation.

Before stating this formula, we discuss R more precisely. R is the distribution solution of

$$(\partial_t^2 - \Delta_x)R(t,x,y) = \delta(x-y)\delta(t) \tag{6.1}$$

$$R|x\in\partial B = 0 \tag{6.2}$$

$$R(t,x,y) = 0, \quad t < 0, \tag{6.3}$$

with $x, y \in \Omega$ and $t \in \mathbb{R}$. Because of the boundary condition (6.2) we must define the distribution R on the particular space of all test functions $\Psi \in C_\infty(\mathbb{R} \times \overline{\Omega})$ vanishing outside a bounded set and satis-fying the boundary conditions

$$\Delta_x^m \Psi(t,x) = 0$$

for all $m = 0,1,2,...$ and all $x \in \partial\Omega$. For, to define distribution solutions of the wave equation we need a test function space which is mapped into itself by the Laplacian. Let $\delta(x-y)\delta(t)$ be the distribution which assigns to each such test function Ψ the value $\Psi(0,y)$. R is called a distribution solution of (6.1), (6.2) if for each such test function

$$\langle R(t,x,y),(\partial_t^2 - \Delta)\Psi(t,x)\rangle = \Psi(0,y)$$

holds. Existence and uniqueness of R have been proved in Alber (1981). In the free space case we obtain as solution of (6.1), (6.3) the fundamental solution

$$r(t,x,y) = \frac{\delta(t - |x-y|)}{4\pi\ |x-y|}$$

of the wave equation. The distribution $\delta(t-|x-y|)/|x-y|$ is defined by

$$\delta(t-|x-y|)/|x-y| := \partial_t(H(t-|x-y|)/|x-y|),$$

with the Heaviside function H. Details on the initial value problem for the wave equation in the free space can be found in Leis (1986) p. 100f.

Now we can state the representation formula for the solution p of the initial-boundary value problem of the wave equation. If $p^1 \in H_1(\Omega)$ and $p^0 \in H_2(\Omega)$, then

$$p(t,y) = \partial_t\langle R(t,x,y),p^0(x)\rangle + \langle R(t,x,y),p^1(x)\rangle.$$

This formula shows that investigation of R yields information about the properties of solutions to more general problems. Therefore we shall discuss the "Green's function" R in this section and determine its singularities.

In the sequel let $y \in \Omega$ be fixed. We want to represent R in the form

$$(6.4) \qquad R(t,x,y) = r(t,x,y) + \sum_{m=0}^{\infty} H_{m-1}(t - \phi(x,y))\, z_m(x,y).$$

The series in this equation is the WKBJ-ansatz for the part of the solution contributed by the wave reflected at B. The functions ϕ, H_m and z_m are defined as follows: Let

$$H_{-1}(s) = \delta(s) \qquad s \in \mathbb{R},$$

$$(6.5) \qquad H_m(s) = \begin{cases} \frac{1}{m!} s^m & s \geq 0 \\ 0 & s < 0 \end{cases}, \quad m \geq 0.$$

The H_m satisfy $H_m' = H_{m-1}$. The functions ϕ and z_m in (6.4) must be determined such that the differential equation (6.1) and the boundary condition (6.2) are satisfied. To determine these functions, insert the series on the right hand side of (6.4) into the wave equation, compute the derivatives under the summation symbol formally, replace the appearing derivatives H_m' and H_m'' by H_{m-1} and H_{m-2}, and arrange the new series according to the functions H_m. The wave equation is satisfied if the new series vanishes. Therefore set each term of the new series equal to zero. This yields the following system of partial differential equations for the functions ϕ and z_m:

$$(6.6) \qquad |\nabla_x \phi|^2 = 1$$

$$(6.7) \qquad 2\nabla_x \phi \cdot \nabla_x z_m + (\Delta_x \phi) z_m = -\Delta_x z_{m-1}, \qquad m \geq 0,$$

with $z_{-1} = 0$. The boundary condition (6.2) is met by the function R defined in (6.4) if

$$(6.8) \qquad \phi(x,y) = |x-y|, \quad x \in \partial B$$

$$(6.9) \qquad z_m(x,y) = \begin{cases} -\frac{1}{4\pi |x-y|} & m = 0 \\ 0 & m > 0 \end{cases}, \quad x \in \partial B.$$

Equation (6.6) is called eikonal equation. The initial values for the solution of this equation are given on ∂B by (6.8). A solution ϕ can be found by the ordinary theory for partial differential equations of first order (cf. e.g. Courant & Hilbert II, p. 74). It is not unique. The function $\phi(x,y) = |x-y|$ always is a solution. Let $x \in \partial B$ and assume that the ray from y through x is not tangential to ∂B at x. Then in some neighborhood of x in $\bar{\Omega}$ there exists an arbitrarily often differentiable solution ϕ of (6.6), (6.8) different from $|x-y|$. The characteristics of (6.6) belonging to this solution coincide with the light rays issued from y and reflected at ∂B according to the laws of geometrical optics. For simplicity we assume that the following condition holds:

Condition S. *Each ray starting at y and reflected at ∂B according to the laws of geometrical optics has at most one point in common with ∂B, and the reflected light rays do not intersect.*

The second assumption means that the reflected field does not contain caustics. Under this condition we are able to construct a solution ϕ of (6.6), (6.8) existing in all of Ω, whose characteristics in the illuminated part of Ω are given by the reflected light rays, and which coincides with $|x-y|$ in the shadow. By the shadow we mean the region which is not covered by light rays starting at y and reflected at ∂B. This solution satisfies

$$\phi(x,y) \to +\infty, \quad |x| \to \infty. \tag{6.10}$$

In order to determine z_m insert the solution of (6.6), (6.8), and (6.10) into (6.7). This yields a recursively solvable system of partial differential equations for the functions z_m. In each step only one linear partial differential equation of the first order must be solved. A simple consideration shows that the characteristics of (6.7) coincide with the characteristics of (6.6). Therefore also the functions z_m are determined in all of Ω.

However the series in (6.4) thus constructed does not always converge. Therefore we study the function

$$R_\ell(t,x,y) = r(t,x,y) + \sum_{m=0}^{\ell} H_{m-1}(t-\phi(x,y))z_m(x,y), \tag{6.11}$$

where ϕ and z_m are determined as above. If R is the exact solution of (6.1) - (6.3), then it follows from (6.5) - (6.9) that

$$\begin{gathered}(\partial_t^2 - \Delta_x)(R - R_\ell)(t,x,y) = H_{\ell-1}(t-\phi(x,y))\Delta_x z_\ell(x,y)\\ (R - R_\ell)|x\in\partial B = 0\\ R - R_\ell = 0, \quad t < 0.\end{gathered} \tag{6.12}$$

Thus $R - R_\ell$ is the solution of an initial-boundary value problem for the wave equation. From regularity theorems for the wave equation it follows that the regularity of the solution $R - R_\ell$ is determined by the right hand side of (6.12). $H_{\ell-1}(t-\phi(x,y))\Delta z_\ell(x,y)$ is a $(\ell-2)$-times continuously differentiable function at each point $x \in \Omega$ with the property that the straight line segment between y and x is not tangential to ∂B. If, in addition, one uses the fact that because of the finite propagation speed those parts of the wave front $\phi(x,y) = t$ which do not belong to the tangential rays, stay unaffected by disturbances emanating from tangential rays, then it follows from regularity theorems for initial-boundary value problems for the wave equation that $(R-R_\ell)(t,x,y)$ belongs to the Sobolev space $H^{loc}_{\ell-1}$ in the neighborhood of each point (t,x) on the wave front $\phi(x,y) = t$ which does not lie on a tangential ray. From Embedding Theorems it then follows that $(R-R_\ell)(t,x,y)$ is $(\ell-3)$-times continuously differentiable in the neighborhood of each such point. If, in addition, one takes into account that the term $H_\ell(t-\phi)z_{\ell+1}$ in (6.4) determines the regularity of $R - R_\ell$, it finally follows that

$$(R - R_\ell)(t,x,y) \in C_{\ell-1} \tag{6.13}$$

in the neighborhood of each point (t,x) on the wave front $\phi(x,y) = t$, which does not lie on a tangential ray. This result can further be improved by using a result from microlocal analysis on C_∞-singularities. By C_∞-singularities of the solution of the initial-boundary value problem we mean those points at which the solution is not arbitrarily often differentiable. These C_∞-singularities are propagating according to the laws of geometrical optics. This has been proved for singularities on tangential rays in Melrose & Sjöstrand (1978, 1982). From this fact it follows that (6.13) holds for all $t \in \mathbb{R}$ and all $x \in \bar{\Omega}$, for which the straight line segment between y and x is not tangential to B. The function R_ℓ defined in (6.11) thus coincides with the exact solution R of (6.1) - (6.3) up to a $(\ell-1)$-times continuously differentiable function at all of these (t,x).

We use this result to determine the asymptotics of solutions of the Helmholtz equation $(\Delta + k^2)u(x,k) = 0$ as $k \to \pm\infty$. The Helmholtz equation describes time harmonic wave propagation. Since k is the frequency parameter we are able to calculate time harmonic wave propagation at high frequencies from the knowledge of the asymptotic behavior of solutions of the Helmholtz equation. In order to determine this asymptotic behavior, note that for $k \in \mathbb{R}$ the function defined by

$$G(x,y,k) = \int_{-\infty}^{\infty} e^{ikt} R(t,x,y)\, dt \tag{6.14}$$

is the solution of

$$-(\Delta_x + k^2)G(x,y,k) = \delta(x-y), \quad x,y \in \Omega$$
$$G(x,y,k) = 0, \quad x \in \partial B$$
$$(x_0\nabla - ik)G(\cdot,y,k) \in L^2(\{x \in \Omega \mid |x-y| > \epsilon > 0\}).$$

The Fourier transform in (6.14) is meant in the sense of the Fourier transformation of tempered distributions. To see that G defined in this way satisfies the outgoing radiation condition, replace k by $k+i\sigma$ in (6.14) and pass to the limit $\sigma \downarrow 0$. As shown in Section 3 with the Limiting Absorption Method it then follows that G meets the outgoing radiation condition. A proof of this statement as well as details regarding the definition of the Fourier transform can be found in Alber (1982b). G is the Green's function of the Dirichlet problem for the Helmholtz equation, and (6.14) shows that we obtain G by Fourier transformation of R with respect to the time t. Hence it follows that R is decomposed into the time periodic solutions

$$(2\pi)^{-1} e^{-ikt} G(x,y,k)$$

of the wave equation, and that $G/(2\pi)$ is the amplitude of this time periodic oscillation. Now let

$$G_\ell(x,y,k) = \int_{-\infty}^{\infty} e^{ikt} R_\ell(t,x,y)\, dt.$$

For $k \neq 0$ we then have

$$G_\ell(x,y,k) = g(x,y,k) + e^{ik\phi(x,y)} \sum_{m=0}^{\ell} (\tfrac{i}{k})^m z_m(x,y),$$

with

$$g(x,y,k) = \frac{e^{ik|x-y|}}{4\pi|x-y|},$$

the fundamental solution of the Helmholtz equation. ϕ and z_m are the solutions of (6.6) - (6.10), and it follows that

$$(G - G_\ell)(x,y,k) = \int_{-\infty}^{\infty} e^{ikt} (R - R_\ell)(t,x,y)\, dt.$$

Since $(R-R_\ell)(t,x,y)$ is $(\ell-1)$-times continuously differentiable with respect to t if x does not lie on a tangential ray, it follows from the properties of the Fourier transformation that

$$(G - G_\ell)(x,y,k) = O(k^{1-\ell}), \quad k \to \pm\infty,$$

provided that

$$(6.15) \qquad (t \to \partial_t^m R(t,x,y) : \{|t| \geq T\} \to \mathbb{R}) \in L^1(\{|t| > T\})$$

for all m with $0 \leq m \leq \ell-1$, and for a sufficiently large constant T. To see this, note that $R_\ell(t,x,y) = 0$ for $t < \phi(x,y)$, and that $R_\ell(t,x,y)$ is a polynomial of degree $\ell-1$ in t for $t > \phi(x,y)$. This follows from the definition of $R_l(t,x,y)$ in (6.5) and (6.11). If (6.15) is fulfilled for all $m \geq 0$, then

$$(6.16) \qquad G(x,y,k) \sim g(x,y,k) + e^{ik\phi(x,y)} \sum_{m=0}^{\infty} (\frac{i}{k})^m z_m(x,y)$$

is an asymptotic expansion of Green's function of the Dirichlet problem for the Helmholtz equation in the exterior domain Ω unless x lies on a tangential ray. However, (6.15) is not satisfied for all obstacles. The obstacle B is called nontrapping if for every ball $K \subset \mathbb{R}^3$ containing B there is a number $T > 0$ such that for each light ray reflected at ∂B according to the laws of geometrical optics the length of the section in K is bounded by T. Morawetz, Ralston & Strauss (1977) proved that if the obstacle is nontrapping, then to each ball $K \subset \mathbb{R}^3$ and to each non-negative integer m there are constants T, c, $C > 0$ with

$$|\partial_t^m R(t,x,y)| \leq C e^{-ct}$$

for all $t \geq T$ and all $x \in K \setminus B$. Therefore (6.15) is fulfilled for nontrapping obstacles, and consequently (6.16) is an asymptotic expansion of Green's function. To derive this result we needed in addition to the fact that the obstacle is nontrapping also that condition S is satisfied. Both conditions are fulfilled for convex obstacles, but of course, there exist more general obstacles for which these conditions are satisfied.

7. The asymptotic behavior near tangential rays

Up to now we excluded points on tangential rays in our investigations. In fact, the asymptotic expansion (6.16) does not hold on tangential rays. Morawetz and Ludwig (1968) and later on Melrose (1975) and Taylor (1976) gave asymptotic expansions in neighborhoods of the tangential rays, which hold if the tangential ray is tangent to the obstacle exactly of second order. These asymptotic expansions are based on an ansatz stated by Ludwig (1966, 1967), which approximates the solution in a neighborhood of the tangential rays. An asymptotic expansion on tangential rays which are tangent to the obstacle of order higher than two is still unknown. In Alber (1982b, 1983) though the first term of an asymptotic expansion on tangential rays which may be tangent to the obstacle of arbitray order was calculated and was shown to coincide with the first term of the WKBJ-expansion (6.11) and (6.16). We now state the essential results from Alber (1983).

Let $B \subset \mathbb{R}^2$ be a bounded, strictly convex obstacle with $\partial B \in C_\infty$. We do not require the curvature of ∂B to be different from zero. Let $P \in \partial B$. We choose the coordinate system with origin at P such that the x_1-axis points into the direction of the normal vector to ∂B at the point P directed from $\Omega = \mathbb{R}^2 \setminus \bar{B}$ to B. Let $u(x,k)$ be the solution of

$$(\Delta + k^2)u(x,k) = 0 \quad x \in \Omega,\ k \in \mathbb{R}$$
$$(u + u_0)|\partial B = 0$$
$$(x_0 \nabla - ik)u(\cdot,k) \in L^2(\Omega),$$

with

$$u_0(x,k) = e^{-ikx_2}.$$

$u_0 + u$ is a distorted plane wave with incoming direction $(0,-1)$. These functions were introduced in Section 5. u_0 represents the amplitude term in the plane wave $e^{-ikt}u_0(x,k)$, and for simplicity we call u_0 itself a plane wave. The directional vector $(0,-1)$ is tangential to ∂B at the point P and at a second point $\tilde{P} \in \partial B$. The solution u will be approximated by the function

$$u_\infty(x,k) = e^{ik\phi(x)} z_0(x), \tag{7.1}$$

where ϕ and z_0 are the solutions of

$$|\nabla\phi|^2 = 1, \quad \phi(x) = -x_2 \quad \text{for } x \in \partial B, \quad \phi(x) \to +\infty \quad \text{for } |x| \to \infty \tag{7.2}$$

$$2\nabla\phi\cdot\nabla z_0 + (\Delta\phi)z_0 = 0, \quad z_0(x) = -1 \quad \text{for} \quad x \in \partial B. \tag{7.3}$$

Theorem 7.1: *There is a function* $k \to c(k) > 0$ *with* $c(k) \to 0$ *for* $|k| \to \infty$ *such that*

$$\|u(\cdot,k) - u_\infty(\cdot,k)\|_{\Omega_R} \le c(k)\, R^2 \log R \tag{7.4}$$

holds for all $R > 1$.

Here again $\Omega_R = \{x \in \Omega \mid |x| < R\}$ and

$$\|v\|^2_{\Omega_R} = \int_{\Omega_R} |v(x)|^2\, dx.$$

Under certain assumptions the function $c(k)$ can be determined more precisely. To this end let s be the arclength of ∂B measured from P in direction of positive values of x_2, and let $y(s) = (y_1(s),y_2(s))$ be the parametrization of the boundary. For sufficiently small $\delta > 0$ the mappings $s \to y_1(s) \; : \; (0,\delta) \to \mathbb{R}^+$ and $s \to y_1(s) \; : \; (-\delta,0) \to \mathbb{R}^+$ are invertible, since B is strictly convex. Let s_+ and s_- be the inverses of these mappings. In the same way we define the mappings $\tilde{s}_+$ and $\tilde{s}_-$ belonging to $\tilde{P}$.

Theorem 7.2: *Assume that either*

(i) $$y_1(0) = \ldots = y_1^{(n-1)}(0) = 0, \quad y_1^{(n)}(0) \neq 0$$

and

$$\alpha = 2n/(3n+1),$$

or

(ii) $$y_1^{(n)}(0) = 0 \quad \textit{for all } n \geq 0, \quad \alpha \in (0,2/3),$$

and

$$C^{-1} \leq |y_1^{(m+1)}(s)/y_1^{(m)}(s)| \leq C\,|s|^{-\ell}$$

for all $0 \leq m \leq 3$, *for suitable positive constants* ℓ, $C > 0$, *and for all* s *with* $|s|$ *sufficiently small.*

Let one of the two conditions corresponding to (i), (ii) *be fulfilled for* $\tilde{P}$ *also. Let* $\tilde{\alpha}$ *be the constant corresponding to* α. *Then there is a constant* $C > 0$ *such that* (7.4) *is fulfilled with*

$$c(k) = C[(s_+-s_-)(|k|^{-\alpha}) + (\tilde{s}_+-\tilde{s}_-)(|k|^{-\tilde{\alpha}})]^{1/2}.$$

Note that in case (i) the curvature of the boundary vanishes at the point P of order n-2. The definition of s_+ and s_- yields in this case

$$(s_+ - s_-)(|k|^{-\alpha}) \leq \text{const } |k|^{-2/(3n+1)}.$$

An example for case (ii) is given by

$$y_1(s) = \exp(-s^{-2}).$$

It then follows that $s_\pm(t) = \pm(\ln \frac{1}{t})^{-1/2}$. If the boundary in the neighborhood of $\tilde{P}$ has a similar shape, then we obtain from Theorems 7.1 and 7.2 that

$$\|u - u_\infty\|_{\Omega_R} \leq \text{const } (\alpha \ln k)^{-1/4}, \quad k \to \infty.$$

Now let v(t,x) be the distribution solution of

$$(\partial_t^2 - \Delta)v(t,x) = 0 \quad \text{in } \mathbb{R} \times \Omega$$

$$v|\mathbb{R} \times \partial B = 0$$

$$v(t,x) = \delta(t+x_2), \quad t < -\rho,$$

where $\rho > 0$ is a constant with $B \subset \{x \in \mathbb{R}^3 \mid |x| < \rho\}$. By Fourier transformation we obtain from Theorems 7.1 and 7.2 the most singular term of v(t,x), namely

Theorem 7.3: *Let* B *be strictly convex and assume that the curvature of* ∂B *differs from zero at all points with the possible exception of the point* P, *at which point the curvature vanishes exactly of order* (n-2), $2 \leq n < \infty$. *Then*

(7.5) $$v(t,x) = \delta(t+x_2) + \delta(t-\phi(x))z_0(x) + w(t,x),$$

with $w \in L^2(\mathbb{R},H_{-\frac{1}{2}+\eta}(\Omega_R))$ *for all* $R > 0$ *and all* $\eta < 1/(3n+1)$.

The functions ϕ and z_0 are the solutions of (7.2) and (7.3). v, $\delta(t+x_2)$, and $\delta(t-\phi)z_0$ are the Fourier transforms with respect to k of the functions u, u_0, and of the function u_∞ defined in (7.1). From (7.2), (7.3) it results that in the shadow of the obstacle $\phi(x) = -x_2$ and $z_0(x) = -1$, and so the distribution

(7.6) $$\delta(t+x_2) + \delta(t-\phi(x))z_0(x)$$

vanishes in the shadow. Thus the support of this distribution is equal to the wave front reflected at ∂B. This distribution belongs to $L^2(\mathbb{R},H_{-\frac{1}{2}+\eta})$ for all $\eta < 0$, but for no $\eta \geq 0$. Therefore the distribution in (7.6) is more strongly singular than the distribution w in (7.5) and represents the most singular term in an asymptotic expansion of v.

8. High frequency asymptotics of the scattering amplitude

The methods and results of the last section can be transferred to the $\mathbb{R}^3$. Frequently however it is better to use an approximating solution in integral representation instead of the function defined in (7.1), since in this case one can allow caustics within the reflected field. We now discuss the results for $\mathbb{R}^3$ proved in Alber & Ramm (1986), where an approximating solution based on Kirchhoff's approximation is used, which is also well suited to determine the asymptotic behavior of the scattering amplitude at high frequencies.

The asymptotic behavior of the scattering amplitude $f(s,s_0,k)$ for $s \neq s_0$ has been stated by Majda (1976). In forward scattering direction, that is for s near to s_0, the asymptotic behavior of the scattering amplitude is more complicated, since it is influenced by the rays tangential to the obstacle. The scattering amplitude possesses a "diffraction peak" at this point. Under the assumption that all tangential rays are tangential to the boundary of order not greater than two, Majda & Taylor (1977) and Melrose (1980) calculated the first term and the whole asymptotic expansion of $f(s_0,s_0,k)$ for $k \to \infty$. Under the same condition Melrose & Taylor (1985) gave an asymptotic expansion for $f(s,s_0,k)$, which holds uniformly for all s in a neighborhood of the forward scattering direction s_0.

The results from Alber & Ramm (1986) to be discussed now hold without this assumption regarding the tangential rays. It will be shown that the scattering amplitude for large values of k can be decomposed into a sum of two terms, where the first term is the three-dimensional Fourier transform of the characteristic function of the obstacle, and where the second term is the two-dimensional Fourier transform of the characteristic function of the shadow projection of the obstacle. The peak in forward scattering direction is generated by the second term. Another result is that the limit of the scattering cross section of the obstacle is equal to twice the area of the shadow projection of the obstacle.

Let $B \subset \mathbb{R}^3$ be a bounded region with boundary $\partial B \in C_\infty$. Let $u(x,k)$ be the solution of

$$(\Delta + k^2)u(x,k) = 0 \quad \text{in } \Omega = \mathbb{R}^3 \backslash \bar{B}, \ k > 0 \tag{8.1}$$

$$(u + u_0)|\partial B = 0 \tag{8.2}$$

$$(x_0 \nabla - ik)u(\cdot,k) \in L^2(\Omega), \tag{8.3}$$

where

$$u_0(x,k) = \exp(ik\, s_o \cdot x) \tag{8.4}$$

is an incoming plane wave, which propagates in direction of the unit vector $s_0 \in \mathbb{R}^3$. We choose the Cartesian coordinate system (x_1,x_2,x_3) with origin in B in such a way that the x_3-axis points into the direction of s_0. It then follows that

$$u_0(x,k) = e^{ikx_3}, \quad s_0 = (0,0,1).$$

For $x = (x_1,x_2,x_3) \in \mathbb{R}^3$ let $x' = (x_1,x_2)$ be the projection of x onto the x_1,x_2-plane. B is required to meet the following two conditions:

Condition S1. *Each ray with incoming direction s_0, reflected at ∂B according to the laws of geometrical optics, has at most one point in common with ∂B.*

Condition S2. *There is a constant $\beta > 0$ with*

$$-(x \cdot n(x)) \geq \beta$$

for all $x \in \partial B$, where $n(x)$ is the unit normal vector to ∂B at x directed from Ω to B.

Condition S1 corresponds to the Condition S from Section 6. Condition S2 implies that the obstacle B is nontrapping, and in fact, instead of requiring S2 it would suffice to require that B is nontrapping, under the expense of greater technical difficulties in the proofs, however. Let

$$\mathcal{S}' = \{x' \in \mathbb{R}^2 \mid \text{there is a } x_3 \in \mathbb{R} \text{ with } (x',x_3) \in B\}$$

be the shadow projection of B onto the x_1,x_2-plane, and let

$$\partial B_+ = \{(x',x_3) \in \mathbb{R}^3 \mid x' \in \mathcal{S}',\ x_3 = \inf_{(x',t)\in B} t\}$$

$$\partial B_- = \partial B \setminus \overline{\partial B_+}.$$

$\overline{\partial B_+}$ is the illuminated part of ∂B, ∂B_- is the part of ∂B in the shadow. In order to construct an approximation for the solution u of (8.1) - (8.4), we use the representation formula

$$(8.5)\qquad u(x,k) = \frac{1}{4\pi}\int_{\partial B}\Big(\frac{e^{ik|x-y|}}{|x-y|}\frac{\partial}{\partial n}u(y,k) - \frac{\partial}{\partial n_y}\frac{e^{ik|x-y|}}{|x-y|}u(y,k)\Big)dS_y,$$

which results from Green's second formula. We have $(u + u_0)|\partial B = 0$. With Kirchhoff we assume that for large k the normal derivative of the solution at the boundary behaves as being reflected by a plane. This assumption leads to

$$\frac{\partial}{\partial n}u(x,k) \sim \begin{cases} \frac{\partial}{\partial n}u_0(x,k), & x \in \partial B_+ \\ -\frac{\partial}{\partial n}u_0(x,k), & x \in \partial B_-. \end{cases}$$

We insert these values for $\partial_n u$ into (8.5) and use

$$\int_{\partial B}\frac{\partial}{\partial n}\frac{e^{ik|x-y|}}{|x-y|}u_0(y,k)\,dS_y = \int_{\partial B}\frac{e^{ik|x-y|}}{|x-y|}\frac{\partial}{\partial n}u_0(y,k)dS_y,$$

which follows from Green's second identity, since u_0 and $e^{ik|x-y|}/|x-y|$ are both solutions of the Helmholtz equation in B. It follows

$$u(x,k) \sim \frac{1}{2\pi}\int_{\partial B_+}\frac{e^{ik|x-y|}}{|x-y|}\partial_n u_0(y,k)\,dS_y.$$

We modify the integral on the right hand side of this formula a little bit yet. To this end we choose a family $\{\chi_\alpha\}_{\alpha>0}$ of functions $\chi_\alpha \in C_\infty^\circ(\mathcal{S}')$ with

$$0 \le \chi_\alpha(y) \le 1,$$

$$\chi_\alpha(y) = 1 \quad\text{for}\quad y \in \mathcal{S}' \quad\text{with}\quad \operatorname{dist}(y,\partial\mathcal{S}') \ge \alpha^{-1},$$

and with the property that $\chi_\alpha(x)$ and $\partial_i\chi_\alpha(x)$ continuously depend on α. Then we approximate u by the function

$$(8.6)\qquad u_\alpha(x,k) := \frac{1}{2\pi}\int_{\partial B_+}\frac{e^{ik|x-y|}}{|x-y|}\chi_\alpha(y')\,\partial_n u_0(y,k)\,dS_y.$$

The function $\chi_\alpha(y')$ vanishes for all $y \in \partial B$ in a neighborhood of the set $\overline{\partial B_+} \cap \overline{\partial B_-}$ consisting of points of contact of tangential rays. This neighborhood is becoming smaller for increasing α. The proof of the following theorem is based on the same idea as the proofs of Theorems 7.1 - 7.3.

Theorem 8.1: *There are functions* $k \to \alpha(k)$ *with* $\alpha(k) \to \infty$ *and* $k \to c(k) > 0$ *with* $c(k) \to 0$ *as* $k \to \infty$ *such that*

$$\|u(\cdot,k) - u_{\alpha(k)}(\cdot,k)\|_{m,\Omega_R} \leq R^{1/2} (1 + |k|^m)c(k)$$

for $m = 0,1$ *and* $R > 1$. *Here* $\|\cdot\|_{m,\Omega_R}$ *is the usual Sobolev norm.*

With this result the scattering amplitude $f(s,s_0,k)$ introduced in Definition 5.1 can be approximated at high frequencies. For $s \in \mathbb{R}^3$ with $|s| = 1$ and $k > 0$ let

$$f_\alpha(s,s_0,k) = \lim_{r\to\infty} r\, e^{-ikr}\, u_\alpha(rs,k).$$

Using $\lim\limits_{r\to\infty} [|rs-y|-r] = -s\cdot y$, it follows from the representation of u_α in (8.6) and from (8.4) that

$$\begin{aligned} f_\alpha(s,s_0,k) &= \frac{1}{2\pi} \int_{\partial B_+} e^{-iks\cdot y}\, \chi_\alpha(y')\, \partial_n u_0(y,k)\, ds_y \\ &= \frac{ik}{2\pi} \int_{\partial B_+} e^{ik(s_0-s)\cdot y}\, s_0\cdot n(y)\, \chi_\alpha(y')\, ds_y. \end{aligned} \tag{8.7}$$

From Theorem 8.1 we obtain

Theorem 8.2: *Let* $k \to \alpha(k)$ *be the function defined in Theorem 8.1. Then*

$$\lim_{k\to\infty} \int_{|s|=1} |f(s,s_0,k) - f_{\alpha(k)}(s,s_0,k)|^2\, ds_s = 0.$$

As a corollary of this theorem we obtain

Theorem 8.3: *We have*

(i) $$\lim_{k\to\infty} \int_{|s|=1} |f(s,s_0,k)|^2\, ds_s = 2 \text{ meas } \mathcal{S}'$$

(ii) $$\operatorname{Im} f(s_0,s_0,k) = \frac{k}{2\pi} \text{ meas } \mathcal{S}' + o(k), \quad k \to \infty.$$

The first of these formulas says that in the limit the scattering cross section of the obstacle is equal to twice the area of the projection $\mathcal{S}'$. It follows from Theorem 8.2 if one applies a well known integral representation of the Bessel function $r^{-1/2}J_{1/2}(kr)$. The second formula results from (i) and from the so-called "optical theorem"

$$2 \operatorname{Im} f(s,s,k) = \frac{k}{2\pi} \int_{|\omega|=1} |f(\omega,s,k)|^2 \, d\omega$$

for the scattering amplitude. The optical theorem is a consequence of the unitarity of the scattering operator, cf. (5.9).

The next theorem throws light on the structure of the scattering amplitude at high frequencies and is basic for our investigations of the inverse scattering problem in Section 9. Let

$$B(s_0) = \{x = (x_1,x_2,x_3) \in \mathbb{R}^3 \mid x' \in \mathcal{S}',\ x_3 > \inf_{(x',t)\in B} t\},$$

and let $\chi_{B(s_0)}$ be the characteristic function of this set. Choose some constant $R > 0$ and some function $\chi_R \in C_\infty(\mathbb{R})$ with $\overline{B} \subset \{|x| < R\}$, with $0 \leq \chi_R(t) \leq 1$, and with

$$\chi_R(t) = \begin{cases} 1, & t < R \\ 0, & t > R + 1. \end{cases}$$

Application of the Divergence Theorem to (8.7) yields

$$\begin{aligned} f_\alpha(s,s_0,k) &= \frac{ik}{2\pi} \int_{\partial B_+} e^{ik(s_0-s)\cdot y} \chi_\alpha(y')\, \chi_R(y_3)\, s_0\cdot n(y)\, dS_y \\ &= \frac{1}{2\pi} k^2 (1-s\cdot s_0) \int_{B(s_0)} e^{ik(s_0-s)\cdot y} \chi_\alpha(y')\, \chi_R(y_3)\, dy \\ &\quad - \frac{ik}{2\pi} \int_{B(s_0)} e^{ik(s_0-s)\cdot y} \chi_\alpha(y')\, \chi_R'(y_3)\, dy. \end{aligned}$$

The first term on the right hand side of this equation is the Fourier transform of the function $\chi_{B(s_0)}(y)\chi_\alpha(y')\chi_R(y_3)$ up to the factor $k^2(1-s\cdot s_0)$. The following theorem shows that the second term can be replaced by the two-dimensional Fourier transform of the characteristic function $\chi_{\mathcal{S}'}$ of the shadow projection of B.

Theorem 8.4: *Let $\chi_{[1/2,1]}$ be the characteristic function of the interval [1/2,1], let* s' *be the projection of* s *onto the plane* $\{x\cdot s_0 = 0\}$, *let*

$$\hat{\chi}_{\varphi'}(\xi) = \frac{1}{2\pi} \int_{\mathbb{R}^2} e^{-i\xi\cdot y} \chi_{\varphi'}(y)\, dy, \quad \xi \in \mathbb{R}^2,$$

and finally let

$$\hat{\chi}_{\alpha,R}(\xi) = \frac{1}{2\pi} \int_{\mathbb{R}^3} e^{-i\xi\cdot y} \chi_{B(s_0)}(y)\chi_\alpha(y')\chi_R(y_3)\, dy, \quad \xi \in \mathbb{R}^3.$$

Then

$$\lim_{k\to\infty} \int_{|s|=1} |f(s,s_0,k) - k^2(1-s\cdot s_0)\, \hat{\chi}_{\alpha(k),R}(k(s-s_0))$$
$$- ik\, \hat{\chi}_{\varphi'}(ks')\, \chi_{[1/2,1]}(s\cdot s_0)|^2\, dS_s = 0.$$

The term $k^2(1-s\cdot s_0)\, \hat{\chi}_{\alpha(k),R}(k(s-s_0))$ vanishes at the point $s = s_0$, and is bounded on the set $\{s \in \mathbb{R}^3 \mid |s| = 1, |s-s_0| < 1/k\}$ with a bound independent of k. On the other hand, for large k the term $ik\, \hat{\chi}_{\varphi'}(ks')\, \chi_{[1/2,1]}(s\cdot s_0)$ represents a "peak" in this neighborhood of s_0. In forward scattering direction and for large k one therefore mainly observes the Fourier transform of the characteristic function of the shadow projection of the obstacle. We shall return to this structure of the scattering amplitude in the last section.

Finally we observe that the leading singularity of the scattering kernel introduced by Lax & Phillips (1967) is connected with the behavior of the scattering amplitude at high frequencies. This leading singularity has been investigated in Alber (1980) following papers by A. Majda (1977) and Lax & Phillips (1977).

9. Inverse problems

In this section we discuss how to compute the shape of the obstacle from the knowledge of the scattering amplitude at high frequencies, and consequently, as Theorem 5.1 shows, from the knowledge of the scattering operator. There are several possibilities how the information contained in the behavior of the scattering amplitude at high frequencies can be used to compute the obstacle. Here we discuss the procedure stated in Alber & Ramm (1986), and mention other possibilities in a short remark at the end of this section. With the notation introduced in Section 8 let

$$f_\infty(s,s_0,k) = \frac{ik}{2\pi} \int_{\partial B_+} e^{ik(s_0-s)\cdot y}\, s_0\cdot n(y)\, dS_y \tag{9.1}$$

for all $s \in S^2 = \{\omega \in \mathbb{R}^3 \mid |\omega|=1\}$. A corollary of Theorem 8.2 is

Corollary 9.1: *For each compact subset* K *of* $S^2\backslash\{s_0\}$

$$\lim_{k\to\infty} \int_K |f(s,s_0,k) - f_\infty(s,s_0,k)|^2 \, dS_s = 0.$$

In the sequel we assume that Condition S1 from Section 8 is also fulfilled for light rays with incoming direction $-s_0$. As in the proof of Theorem 8.2 and Corollary 9.1 we conclude that $f(s,-s_0,k)$ can be approximated by the function

$$(9.2) \qquad f_\infty(s,-s_0,k) = -\frac{ik}{2\pi} \int_{\partial B_-} e^{-ik(s_0+s)\cdot y} \, s_0\cdot n(y) \, dS_y$$

for $s \in S^2\backslash\{-s_0\}$. From (9.1), (9.2), and from Gauß' theorem it follows that

$$f_\infty(s,s_0,k) + \overline{f_\infty(-s,-s_0,k)} = \frac{ik}{2\pi} \int_{\partial B} e^{ik(s_0-s)\cdot y} \, s_0\cdot n(y) \, dS_y$$

$$= -\frac{ik}{2\pi} s_0\cdot ik(s_0-s) \int_B e^{ik(s_0-s)\cdot x} \, dx = k^2(1-s_0\cdot s) \, \hat{\chi}_B(k(s-s_0))$$

with

$$\hat{\chi}_B(\xi) = \frac{1}{2\pi} \int_{\mathbb{R}^3} e^{-i\xi\cdot x} \, \chi_B(x) \, dx.$$

$\hat{\chi}_B$ is the Fourier transform of the characteristic function χ_B of B. Let K be a compact subset of $S^2\backslash\{s_0\}$. If s varies in K, then -s varies in the set -K, which is a compact subset of $S^2\backslash\{-s_0\}$. By separate application of Corollary 9.1 to $f(s,s_0,k)$ and to $f(-s,-s_0,k)$ we therefore obtain the following result:

Lemma 9.2: *Let the Condition S1 be fulfilled for the incoming directions* s_0 *and* $-s_0$. *Then for each compact subset* K *of* $S^2\backslash\{s_0\}$

$$\lim_{k\to\infty} \int_K |f(s,s_0,k) + \overline{f(-s,-s_0,k)} - k^2(1-s_0\cdot s) \, \hat{\chi}_B(k(s-s_0))|^2 \, dS_s = 0.$$

This means that the Fourier transform of χ_B is approximated by the scattering amplitude up to the factor $k^2(1-s\cdot s_0)$ on that part of the sphere with radius k and center $-ks_0$, which lies outside the ball $\{\xi \in \mathbb{R}^3 \mid |\xi| < k\delta\}$. δ is an arbitrary positive constant. If Condition S1 is satis-fied for each incoming direction $s_0 \in S^2$, then we can let s_0 run through the unit sphere and determine an approximate value for $\hat{\chi}_B(\xi)$

for all ξ from the "annulus" $\{\xi \in \mathbb{R}^3 \mid k\delta < |\xi| < 2k\}$. We define $\tilde{\chi}_B(\xi)$ for every ξ from this annulus by choosing $s, s_0 \in S^2$ with $\xi = k(s-s_0)$ and set

$$\tilde{\chi}_B(\xi) = (k^2(1-s_0\cdot s))^{-1}(f(s,s_0,k) + \overline{f(-s,-s_0,k)}).$$

Lemma 9.2 then yields

$$\int_{k\delta<|\xi|<2k} |\hat{\chi}_B(\xi) - \tilde{\chi}_B(\xi)|^2 \, d\xi = k^{-1}\, o(1), \quad k \to \infty.$$

Therefore we obtain an approximation for the characteristic function of B by inverse Fourier transformation of $\tilde{\chi}_B$. Note that the behavior of $\hat{\chi}_B(\xi)$ for large values of $|\xi|$ determines the shape of the boundary of B. To see this, observe that if ψ is a function with compact support, then the inverse Fourier transform of $\hat{\chi}_B + \psi$ differs from χ_B only by an analytic function, since the inverse Fourier transform of ψ is analytic. In the above procedure it therefore suffices to know an approximation for $\hat{\chi}_B(\xi)$ at large values of $|\xi|$.

We finally note that the remainder term in the last formula can be estimated in a better way. The reason is that in the proof of this formula it suffices to approximate the scattering amplitude for all s with $|s-s_0| > \delta$. Since these directions s all lie outside a neighborhood of the forward scattering direction s_0, it is possible to give an asymptotic expansion for $f(s,s_0,k)$ with the methods of Section 6. Such expansions have been given by Majda (1976). This expansion leads to a better remainder estimate in the formula stated last.

Another possibility to determine B is to use the formula

$$\lim_{k\to\infty} |f(s,s_0,k)| = \frac{1}{2\kappa(P)^{1/2}},$$

which holds for all $s, s_0 \in S^2$ with $s \neq s_0$. Here $\kappa(P)$ is the Gaussian curvature of ∂B at the point of reflection P of the ray with incoming direction s_0 and direction s after reflection. One finds this formula for instance in Ramm (1986), p. 98, or in Majda (1976). This formula can also be found by Fourier transformation of the leading singularity of the scattering kernel computed in Lax & Phillips (1977), Majda (1977), or Alber (1980), since one obtains the scattering amplitude by Fourier transformation of the scattering kernel. The normal vector to ∂B at the point P has the direction of the vector $s-s_0$. Therefore to each direction of the normal vector the Gaussian curvature of ∂B is known. Minkowski (1903) showed that for each convex obstacle the boundary can be computed from this knowledge. Concerning this matter also confer Pogorelov (1978). However, a disadvantage of this method is that a highly nonlinear Monge-Ampère equation must be solved. Another method uses the phase function ϕ (solution of the Equations (6.6), (6.8)) to compute the convex hull of the obstacle, see Majda (1977), or Ramm (1986), pp. 94.

10. Bibliography

Agmon, S.: Lectures on elliptic boundary value problems. Van Nostrand; Princeton (1965).

Alber, H.D.: Reflection of singularities of solutions to the wave equation and the leading singularity of the scattering kernel. Proc. Roy. Soc. Edinburgh **86A**, 235-242 (1980).

Alber, H.D.: Justification of geometrical optics for non-convex obstacles. J. Math. Anal. Appl. **80**, 372-386 (1981).

Alber, H.D.: Geometrische Optik, geometrische Theorie der Beugung und ihre geschichtliche Entwicklung. SFB 72-Vorlesungsreihe Nr. 10, Universität Bonn (1982).

Alber, H.D.: Zur Hochfrequenzasymptotik der Lösungen der Schwingungsgleichung - Verhalten auf Tangentialstrahlen. Habilitationsschrift Universität Bonn (1982), SFB 72-Preprint Nr. 505.

Alber, H.D.: Highfrequency asymptotics of the field scattered by an obstacle near tangential rays. J. Reine Angew. Math. **345**, 39-62 (1983).

Alber, H.D.: Numerical computation of waves at high frequencies by an iterated WKB-method. Math. Meth. in the Appl. Sci. **6**, 520-526 (1984).

Alber, H.D. and A.G. Ramm: Scattering aplitude and algorithm for solving the inverse scattering problem for a class of non-convex obstacles. J. Math. Anal. Appl. **117**, 570-597 (1986).

Babich, V.M.: An estimate for the Green's function of the Helmholtz equation in the shadow region. (Russian, English summary). Vestnik Leningrad Univ. 20, no. 7, 5-15 (1965).

Belopolskii, A.L. and M.Sh. Birman: The existence of wave operators in the theory of scattering with a pair of spaces. Math. USSR Izv. **2**, 1117-1130 (1968).

Born, M.: Optik. Springer-Verlag; Berlin et al. (1933, 1981).

Colton, D.: The inverse scattering problem for time-harmonic acoustic waves. SIAM Review **26**, 323-350 (1984).

Courant, R. and D. Hilbert: Methoden der mathematischen Physik. Springer-Verlag; Berlin et al. (1968).

Freudenthal, H.: Über ein Beugungsproblem aus der elektromagnetischen Lichttheorie. Compositio Math. **6**, 221-227 (1938).

Eidus, D.M.: The principle of limit absorption. Math. Sb., **57** (99), 13-44 (1962) and AMS Transl. **47** (2), 157-191 (1965).

Hörmander, L.: The analysis of linear partial differential operators. Springer-Verlag; Berlin et al. Vol. I-IV (1983-1985).

Ikebe, T.: Eigenfunction expansions associated with the Schrödinger operators and their applications to scattering theory. Arch. Rat. Mech. Anal. **5**, 1-34 (1960).

Jäger, W.: Zur Theorie der Schwingungsgleichung mit variablen Koeffizienten in Außengebieten. Math. Z. **102**, 62-88 (1967).

Jensen, A. and T. Kato: Asymptotic behavior of the scattering phase for exterior domains. Comm. in Part. Diff. Equ. **3**, 1165-1195 (1978).

Kato, T.: Perturbation theory for linear operators. Springer-Verlag; Berlin et al. (1976).

Keller, J.B. and J. Papadakis (Editors): Wave propagation and underwater acoustics. Lecture Notes in Physics **70**, Springer-Verlag; Berlin et al. (1977).

Kline, M.: Electromagnetic theory and geometrical optics. Interscience; New York et al. (1965).

Kupradze, V.D.: Über das Ausstrahlungsprinzip von A. Sommerfeld. Dokl. Akad. Nauk SSSR, **1**, 55-58 (1934).

Lax, P.D.: Asymptotic solutions of oscillatory initial value problems. Duke Math. J. **24**, 627-646 (1957).

Lax, P.D. and R.S. Phillips: Scattering theory. Academic Press; New York (1967).

Lax, P.D. and R.S. Phillips: The scattering of sound waves by an obstacle. Comm. Pure Appl. Math. **30**, 195-233 (1977).

Leis, R.: Über das Neumannsche Randwertproblem für die Helmholtzsche Schwingungsgleichung. Arch. Rat. Mech. Anal. **2**, 101-113 (1958).

Leis, R.: Über die Randwertaufgaben des Außenraumes zur Helmholtzschen Schwingungsgleichung. Arch. Rat. Mech. Anal. **9**, 21-44 (1962).

Leis, R.: Zur Eindeutigkeit der Randwertaufgaben der Helmholtzschen Schwingungsgleichung. Math. Z. **85**, 141-153 (1964).

Leis, R.: Zur Dirichletschen Randwertaufgabe des Außenraumes zur Schwingungsgleichung. Math. Z. **90**, 205-211 (1965).

Leis, R.: Initial boundary value problems in mathematical physics. B.G. Teubner; Stuttgart, and John Wiley & Sons; Chichester et al.(1986).

Levy, B.R. and J.B. Keller: Diffraction by a smooth object. Comm. Pure Appl. Math. **12**, 159-209 (1959).

Ludwig, D.: Conical refraction in crystal optics and hydrodynamics. Comm. Pure Appl. Math. **14**, 113-124 (1961).

Ludwig, D.: Uniform asymptotic expansions at a caustic. Comm. Pure Appl. Math. **19**, 215-250 (1966).

Ludwig, D.: Uniform asymptotic expansion of the field scattered by a convex object at high frequencies. Comm. Pure Appl. Math. **20**, 103-138 (1967).

Majda, A.: High frequency asymptotics for the scattering matrix and the inverse problem of acoustical scattering. Comm. Pure Appl. Math. **29**, 261-291 (1976).

Majda, A.: A representation formula for the scattering operator and the inverse problem for arbitrary bodies. Comm. Pure Appl. Math. **30**, 165-194 (1977).

Majda, A. and M.E. Taylor: The asymptotic behavior of the diffractive peak in classical scattering. Comm. Pure Appl. Math. **30**, 639-669 (1977).

Melrose, R.B.: Microlocal parametrices for diffractive boundary value problems. Duke Math. J. **42**, 605-635 (1975).

Melrose, R.B.: Forward scattering by a convex obstacle. Comm. Pure Appl. Math. **33**, 461-499 (1980).

Melrose, R.B. and J. Sjöstrand: Singularities of boundary value problems I. Comm. Pure Appl. Math. **31**, 593-617 (1978).

Melrose, R.B. and J. Sjöstrand: Singularities of boundary value problems II. Comm. Pure Appl. Math. **35**, 129-168 (1982).

Melrose, R.B. and M.E. Taylor: Near peak scattering and the corrected Kirchhoff approximation for a convex obstacle. Adv. in Math. **55**, 242-315 (1985).

Minkowski, H.: Volumen und Oberfläche. Math. Ann. **57**, 447-495 (1903).

Morawetz, C.S. and D. Ludwig: An inequality for the reduced wave operator and the justification of geometrical optics. Comm. Pure Appl. Math. **21**, 187-203 (1968).

Morawetz, C.S., W. Ralston and W. Strauss: Decay of solutions of the wave equation outside nontrapping obstacles. Comm. Pure Appl. Math. **30**, 447-508 (1977).

Müller, C.: Zur Methode der Strahlungskapazität von H. Weyl. Math. **56**, 80-83 (1952).

Müller, C.: On the behaviour of solutions of the differential equation $\Delta u = F(x,u)$ in the neighbourhood of a point. Comm. Pure Appl. Math. **7**, 505-515 (1954).
Müller, C.: Grundprobleme der mathematischen Theorie electromagnetischer Schwingungen. Springer-Verlag; Berlin et al.(1957).
Pearson, D.B.: A generalization of the Birman trace theorem. J. Funct. Anal. **28**, 182-186 (1978).
Petkov, V.: High frequency asymptotics of the scattering amplitude for non-convex bodies. Comm. Part. Diff. Equ. 5, 293-329 (1980).
Picard, R. and S. Seidler: A remark on two Hilbert space scattering theory. Math. Ann. **269**, 411-415 (1984).
Pogorelov, A.V.: The Minkowski multi-dimensional problem. John Wiley & Sons; New York (1978).
Popov, G.: Some estimates of Green's functions in the shadow. Osaka J. Math. **24**, 1-12 (1987).
Ramm, A.G.: Scattering by obstacles. D. Reidel Publishing Company; Dordrecht et al. (1986).
Rellich, F.: Über das asymptotische Verhalten der Lösungen von $\Delta u+\lambda u=0$ in unendlichen Gebieten. Jber. Dt. Math.-Verein. **53**, 57-65 (1943).
Sjöstrand, J.: Analytic singularities of solutions of boundary value problems. In: H. Garnir (Editor): Singularities in boundary value problems. Proceedings of the Nato Advanced Study Institute at Maratea, Italy, 1980. D. Reidel; Dordrecht (1981).
Sleeman, B.D.: The inverse problem of acoustic scattering. IMA Appl. Math. **29**, 113-142 (1982).
Taylor, M.E.: Grazing rays and reflection of singularities of solutions to wave equations. Comm. Pure Appl. Math. **29**, 1-38 (1976).
Ursell, F.: On the short-wave asymptotic theory of the wave equation $(\nabla^2+k^2)\varphi=0$. Proc. Camb. Phil. Soc. **53**, 115-133 (1957).
Vekua, I.N.: Metaharmonic functions. Trudy Tbilisskogo matematicheskogo instituta **12**, 105-174 (1943).
Weber, C.: A local compactness theorem for Maxwell's equations. Math. Meth. Appl. Sci. **2**, 12-25 (1980).
Weck, N.: Maxwell's boundary value problem on Riemannian manifolds with nonsmooth boundaries. J. Math. Anal. Appl. **46**, 410-437 (1974).
Werner, P.: Zur mathematischen Theorie akustischer Wellenfelder. Arch. Rat. Mech. Anal. **6**, 231-260 (1960).
Werner, P.: Randwertprobleme der mathematischen Akustik. Arch. Rat. Mech. Anal. **10**, 29-66 (1962).
Werner, P.: Beugungsprobleme der mathematischen Akustik. Arch. Rat. Mech. Anal. **12**, 155-184 (1963).
Weyl, H.: Kapazität von Strahlungsfeldern. Math. Z. **55**, 187-198 (1952).
Wilcox, C.H.: Scattering theory for the d'Alembert equation in exterior domains. Lecture Notes in Mathematics 442, Springer-Verlag; Berlin et al. (1975).
Yosida, K.: Functional analysis. Springer-Verlag; Berlin et al. (1974).

On shape Optimization of a Turbine Blade

Hans Wilhelm Alt

An optimization problem for turbine blades with given velocity distribution on the blade surface is formulated as a minimum problem with free boundary. The existence of a solution and some regularity properties are proved.

1. Formulation of the problem

A turbine consists of a number of blades, periodically distributed around an axis of symmetry. In a model situation they are fixed on an inner cylinder and rotating inside on outer cylinder. We assume that radial velocities can be neglected and that the fluid is ideal and incompressible and the flow is irrotational. We then can transform a stationary flow on a cylinder of given radius by cylindrical coordinates into the two dimensional plane. This leads to the following boundary value problem. The cross section through the blades is given by disjoint compact sets $E^k \subset \mathbb{R}^2, k \in \mathbb{Z}$, with

(1.1) $$E^k = \{(y,z) \in \mathbb{R}^2; (y, z-k) \in E^0\} \quad .$$

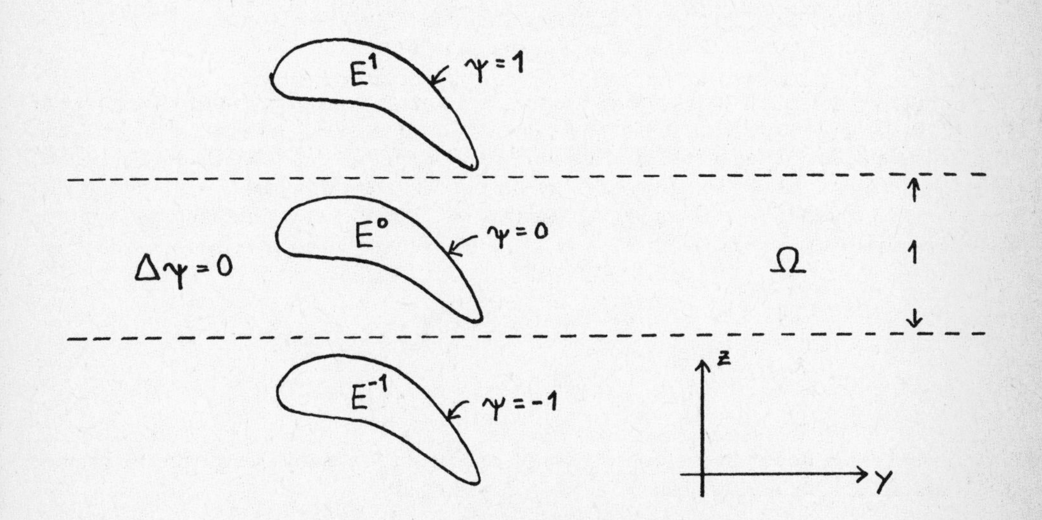

Here y is the coordinate in direction of the channel. We also write $x = (y, z)$ throughout the paper. The union of all cross sections we denote by

$$E = \bigcup_{k=\mathbb{Z}} E^k. \tag{1.2}$$

The velocity field v satisfies

$$\nabla \cdot v = 0 \quad \text{in} \quad \mathbb{R}^2 \backslash E,$$
$$v(y, z+1) = v(y, z),$$
$$v \cdot \nu = 0 \quad \text{on} \quad \partial E.$$

We assume that the flow is uniform at infinity, that is,

$$v(-\infty, z) = (s_0, s_-) \quad \text{and} \quad v(+\infty, z) = (s_0, -s_+),$$

where the two y-components must be the same. We normalize v so that $s_0 = 1$. Then the stream function ψ with $\partial_2 \psi = v_1, \partial_1 \psi = -v_2$ exists globally as a periodic function, that is,

$$\psi(y, z+1) - \psi(y, z) = 1 \quad \text{for} \quad (y, z) \in \mathbb{R}^2, \tag{1.3}$$

with prescribed gradient at infinity

$$\nabla \psi(\pm\infty, z) = (\pm s_\pm, 1).$$

Moreover, since the flow is irrotational,

$$\Delta \psi = 0 \quad \text{in} \quad \mathbb{R}^2 \backslash E. \tag{1.4}$$

Also, since we make the physical assumption that E^k consists of only one connected component we see that up to a global constant

$$\psi = k \quad \text{on} \quad E^k. \tag{1.5}$$

All considerations in this paper remain the same in the more realistic case, where the axially symmetric cross section of the channel depends on y. Then first the axially symmetric flow without blades is constructed and the transformation works on the level sets of the corresponding axially symmetric stream function (see [6], [8].) The only difference is, that the Laplacian now is replaced by $\nabla \cdot (a \nabla \psi)$ with a known function a. In the compressible case a also depends on $|\nabla \psi|$, so that the methods in [4], [5] have to be adapted to the arguments below.

The optimization problem is to choose E (that is, E^0) so that the turbine works efficiently (in applications this must be performed for severel cylindrical cross sections). As a part of this problem the following question arises: Given a pressure distribution on ∂E^0. Can one choose E^0 so that the stream function matches the given pressure profile? Since for this part of the problem the pressure is not related to any other physical quantity, we think of it as a given function in space. This means that with $\lambda := \mathbb{R}^2 \to \mathbb{R}$ given as a positive periodic function we want to find E so that

$$|\nabla \psi|^2 = \lambda^2 \quad \text{on} \quad \partial E. \tag{1.6}$$

Thus locally the free boundary problem is given by equations (1.4) - (1.6). This type of problems can be treated as a minimum problem. The general functional is

$$J(\psi) = \int (|\nabla \psi|^2 + g(\psi))$$

with a lower semicontinuous function g. In the simplest case g has just one jump, for example,

$$g(\psi) = \chi_{\{\psi > O\}} \quad .$$

For this case the general one-phase theory (that is, for solutions $\psi \geq 0$) has been developed in [2]. The two-phase case (that is, regions where $\psi > 0$ and $\psi < 0$ are seperated by the free boundary) has been studied in [3], where as an important tool a monotonicity formula has been introduced, which we shall use later in this paper. Since $g(\psi) = 0$ in $\{\psi \leq 0\}$, by the maximum principle the free boundary $\{\psi = 0\}$ will never be thick, it always shrinks to an one dimensional object.

Our situation is different. The set $\{\psi = 0\}$ should have an interior and ψ should be harmonic across the stream line $\{\psi = 0\}$ in the flow. This means that the functional should prefer the state $\psi = 0$ only, that is, the correct function g here is

$$g(\psi) = \chi_{\{\psi \neq 0\}} \quad ,$$

or, changing the functional by a constant,

$$g(\psi) = -\chi_{\{\psi = 0\}} \quad .$$

Therefore locally we are looking for stationary points of the functional

$$J(\psi, E) := \int (|\nabla \psi|^2 - \lambda^2 \chi_E)$$

under conditions (1.2) , (1.3), (1.5). Formally stationary points of J satisfy (1.4) and (1.6). However, a local functional does not feel the conditions at infinity . To achieve this we replace Dirichlet's integral by

$$\int |\nabla \psi - e|^2,$$

where e is a given periodic vector field with

$$e(\pm\infty, z) = (\pm s_\pm, 1). \tag{1.7}$$

Since this change in the functional must not destroy Euler's equation for ψ , we have to choose e so that $\nabla \cdot e = 0$. Such a divergence free vector field, of course, does not exist in the entire space (if $-s_- \neq s_+$). On the other hand, we are not allowed to define e only outside E, since E is not known in advance, but e must be specified a-priori. Therefore we choose a-priori disjoint compact sets E_*^k with

$$E_*^k = \{(y, z) \in I\!R^2; (y, z - k) \in E_*^0\}$$

and let

$$E_* := \bigcup_{k \in \mathbb{Z}} E_*^k.$$

We assume that

$$E_*^0 \quad \text{is connected and has nonempty interior} \quad , \tag{1.8}$$

although it also would be enough to assume, that E_*^0 has positive capacity.

1.1 Definition of e. If (1.8) holds there is a smooth periodic divergence free vector field e defined in $\overline{\mathbb{R}^2 \backslash E_*}$ with condition (1.7) at infinity.

Proof: Choose a rectangle

$$[y_1, y_2] \times [z_1, z_2] \subset E_*^0 \quad .$$

In the rectangle $Q := [y_1, y_2] \times [z_2, z_1 + 1]$ define

$$e_0(y, z) := (\eta(y), -\eta'(y)(z - z_2)),$$

where $\eta \in C^\infty(\mathbb{R})$ with $\eta = 1$ in a neighborhood of y_1 and $\eta = 0$ in a neighborhood of y_2. Continue e_0 periodically by

$$e_0(y, z + k) := e_0(y, z) \quad \text{for} \quad (y, z) \in Q \quad \text{and} \quad k \in \mathbb{Z}.$$

and let

$$e_0(y, z) := \begin{cases} (1, 0) & \text{for} \quad y \leq y_1, \\ (0, 0) & \text{for} \quad y \geq y_2. \end{cases}$$

Then for given speeds s_- and s_+ define

$$e(y, z) := \begin{cases} (s_+, 1) - (s_- + s_+) e_0(y, z) & \text{if} \quad (y, z) \notin E_*, \\ 0 & \text{if} \quad (y, z) \in E_*. \end{cases}$$

∎

Given a vector field e as in 1.1 we consider the global functional

$$J(\psi, E) := \int_\Omega (|\nabla \psi - e|^2 - \lambda^2 \chi_E)$$

under condition (1.2) and (1.4), where now we have to impose the additional condition $E_* \subset E$. Here Ω is any infinite strip Ω of height 1 , e.g.,

$$\Omega = \{(y, z) \in \mathbb{R}^2; 0 < z < 1\} \quad .$$

The restriction $E_* \subset E$ implies, that eventually we are unable to solve the full free boundary problem. This will occur whenever ∂E touches ∂E_* . But starting with a physically reasonable shape for E and reasonable values for λ on ∂E , then choosing E_* strictly inside E , one can hope to get a full solution to the free boundary problem. Mathematically, for example, if we extend λ so that λ increases rapidly when E shrinks, and decreases rapidly when E grows, a minimizer will exist with a perturbed blade (see also [1]).

A more striking fact is, that minimizers of this global function will not look like what we expect physically. If ∂E would be smooth, and the flow line $\psi = 0$ meets E^0 , the flow has stagnation points on ∂E^0 , with speed $|\nabla \psi| = 0$. But the free boundary condition says $|\nabla \psi| = \lambda > 0$. Therefore E^0 will form two edges, one on the left end, and one on its right end, and there the speed has the given values of λ. No doubt, this shape is optimal for the given velocities at infinity. But if we now fix the blade and pertube the flow at infinity, the optimality is destroyed totally. The stagnation points move along ∂E^0 and immediately the speed gets infinite at the edges of E^0.

We conclude that actually we are looking for a blade E which is optimal with respect to a whole range of velocities at infinity. Mathematically this means, that we have to introduce a probability measure on the space of all velocities $s = (s_-, s_+) \in I\!R^2$. We assume

$$\text{(1.9)} \qquad \begin{array}{l} \mu \quad \text{is a Radon measure on} \quad I\!R^2 \quad \text{with} \\ \text{bounded support and} \quad \mu(I\!R^2) = 1. \end{array}$$

For each $s = (s_-, s_+) \in I\!R^2$ let ψ_s the corresponding stream function and

$$\text{(1.10)} \qquad \begin{array}{l} e_s \quad \text{a vector field as in 1.1 with} \\ e_s(-\infty, z) = (-s_-, 1) \quad \text{and} \quad e_s(+\infty, z) = (s_+, 1) \quad . \end{array}$$

The function $s \mapsto \psi_s$ we denote by ψ , similar e . We call ψ admissible, if

$$\text{(1.11)} \qquad \begin{array}{l} \psi \in L^2(\mu; H^{1,2}_{loc}(I\!R^2)), \\ \psi_s(y, z+1) - \psi_s(y, z) = 1 \quad \text{for all} \quad s \in I\!R^2, (y, z) \in I\!R^2. \end{array}$$

The function space means that ψ, as a function of s, is square integrable with respect to the measure μ with values in $H^{1,2}(D)$ for any given $D \subset\subset I\!R^2$.

A set E is called admissible, if

$$\text{(1.12)} \qquad \begin{array}{l} E \subset I\!R^2 \quad \text{is measurable with} \quad E_* \subset E, \\ \text{(1.1) and (1.2) holds with disjoint measurable sets} \quad E^k, \\ \displaystyle\int_{\Omega \cap E} \lambda^2(x) dx < \infty, \end{array}$$

where we assume, that λ is a continuous positive function with $\lambda(y, z+1) = \lambda(y, z)$. Finally, a pair (ψ, E) is called admissible, if ψ and E are admissible, and if

$$\text{(1.13)} \qquad \text{for} \quad \mu - \text{almost all s we have} \quad \psi_s = k \quad \text{almost everywhere in} \quad E^k.$$

For all admissible pairs (ψ, E) the following functional is well defined (its value might be $+\infty$)

$$\text{(1.14)} \qquad J(\psi, E) := \int_{I\!R^2} \Big(\int_{\Omega} |\nabla \psi_s(x) - e_s(x)|^2 dx \Big) \ d\mu(s) - \int_{\Omega \cap E} \lambda^2(x) dx \quad .$$

The minimum problem to solve then is

1.2 Minimum problem. Find an admissible pair (ψ_0, E_0) with

$$J(\psi_0, E_0) \leq J(\psi, E) \quad \text{for all admissible} \quad (\psi, E).$$

In the next section we shall prove that, under a certain condition on λ, the variational problem admits an absolute minimum. Beside this we prove some basic regularity properties. The numerical treatment of this minimum problem will be contained in a forthcoming paper.

2. Existence and some regularity properties

We assume that the set E_* and the domain Ω is chosen as in the previous section with smooth boundary ∂E_*, and we assume that the data μ and e have the properties (1.9) and (1.10). The functional J is defined as in (1.14).

2.1 Assumption on λ. In order to ensure that the functional is finite at infinity we assume

$$e_s(y,z) = e_s^{\pm} := (\pm s_{\pm}, 1) \quad \text{for} \quad \pm y \geq R_0.$$

Here R_0 is so large that E_* is contained in $[-\frac{R_0}{2}, \frac{R_0}{2}] \times \mathbb{R}$. We also assume that λ is positive, continuous, and periodic with

$$\limsup_{y \to \pm\infty} \lambda(y,z) < ||e^{\pm}||_{L^2(\mu)} \quad \text{uniformly in} \quad z.$$

Let us then choose R_0 large enough, so that for some $\delta_0 > 0$

$$\lambda(y,z) \leq ||e^{\pm}||_{L^2(\mu)} - \delta_0 \quad \text{for} \quad \pm y \geq R_0 \quad .$$

For formal convenience we split $\lambda(x)$ into $\lambda_s(x) > 0$ with

$$\int_{\mathbb{R}^2} \lambda_s^2(x) d\mu(s) = \lambda^2(x)$$

such that

$$\lambda_s(y,z)^2 \leq |e_s^{\pm}|^2 - \delta_0 \quad \text{for} \quad s \in \mathbb{R}^2 \quad \text{and} \quad \pm y \geq R_0 \quad .$$

Finally, let

$$\lambda_0 := \sup_{x \in \mathbb{R}^2} \lambda(x) \quad .$$

We then see that the minimum problem is well posed.

2.2 Lemma.

1) $J(u,E) < +\infty$ for some admissible (u,E) .

2) J is bounded from below. To be precise, for any admissible (u,E) we have

$$\mathcal{L}^2(E \cap \Omega \cap \{|y| > R_0\}) \leq \frac{1}{\delta_0}(J(u,E) + 2R_0\lambda_0^2) \quad .$$

Proof: For 1) we choose $E = E_*$ and a smooth function v in $[-R_0, R_0] \times \mathbb{R}$, such that

$$\begin{aligned} &(y,z) \mapsto v(y,z) - z \quad \text{is periodic in} \quad z, \\ &v = 0 \quad \text{on} \quad E_*^0, \\ &v(\pm R_0, z) = z \quad \text{for} \quad z \in \mathbb{R}. \end{aligned}$$

Then for $s = (s_-, s_+) \in \mathbb{R}^2$ define

$$u_s(y,z) = \begin{cases} v(y,z), & \text{if} \quad |y| \leq R_0 \\ -s_-(y + R_0) + z, & \text{if} \quad y \leq -R_0, \\ +s_+(y - R_0) + z, & \text{if} \quad y \geq R_0. \end{cases}$$

To get 2), we rewrite

$$J(u,E) = \int_{\mathbb{R}^2} \int_{\Omega} (|\nabla u_s - e_s|^2 - \lambda_s^2 \chi_E) d\mathcal{L}^2 d\mu(s) \quad .$$

For almost all $x = (y,z)$ with $\pm y \geq R_0$ we have

$$|\nabla u_s(x) - e_s^{\pm}|^2 - \lambda_s(x)^2 \chi_E(x) = \begin{cases} |e_s^{\pm}|^2 - \lambda_s(x)^2, & \text{if } x \in E, \\ |\nabla u_s(x) - e_s^{\pm}|^2, & \text{if } x \notin E. \end{cases}$$

Hence

$$J(u,E) \geq \delta_0 \mathcal{L}^2(E \cap \Omega \cap \{|y| \geq R_0\}) - \int_{\Omega \cap \{|y| \leq R_0\}} \lambda^2 \quad .$$

∎

First we prove the existence of an absolute minimizer.

2.3 Theorem. Under assumption 2.1 there exists an absolute minimum of J.

Proof: By 2.2 there is a minimal sequence of admissible pairs (u_m, E_m). By 2.2.2 the sets $\Omega \cap E_m$ have bounded measure, and therefore

$$\int_{\mathbb{R}^2} (\int_{\Omega} |\nabla u_m - e|^2) d\mu \leq J(u_m, E_m) + \lambda_0^2 \mathcal{L}^2(\Omega \cap E_m) \leq C \quad .$$

Since $u_{m,s} = 0$ on E_*^0 for almost all $s \in \mathbb{R}^2$ and using Poincare's inequality we get that u_m are bounded in $\mathcal{L}^2(\mu; H^{1,2}_{loc}(\mathbb{R}^2))$. Therefore for a subsequence

$$\begin{aligned} u_m \to u \quad & \text{weakly in} \quad L^2(\mu; H^{1,2}_{loc}(\mathbb{R}^2)), \\ \nabla u_m - e \to \nabla u - e \quad & \text{weakly in} \quad L^2(\mu; L^2(\Omega)). \end{aligned}$$

Clearly u is admissible in the sense of (1.11) . Also by the periodicity of E_m we have

$$\mathcal{L}^2(E_m^0) = \mathcal{L}^2(E_m \cap \Omega),$$

therefore we find a function $\gamma^0 \in L^\infty(\mathbb{R}^2)$ with $0 \leq \gamma^0 \leq 1$ such that

$$\chi_{E_m^0} \to \gamma^0 \quad \text{weakly star in} \quad L^\infty_{loc}(\mathbb{R}^2).$$

Then $\chi_{E_m^k}$ converges weakly star to γ^k , where $\gamma^k(y,z) := \gamma^0(y, z-k)$.

We want the sets $E^k := \{\gamma^k > 0\}$ to be disjoint. For this pick any nontrivial measurable set $S \subset \mathbb{R}^2$ and consider

$$v_m(x) := \fint_S u_{m,s}(x) d\mu(s), \quad \text{where} \fint \quad \text{denotes the mean value} \quad ,$$

and similarly v. Then $v_m \to v$ weakly in $H^{1,2}_{loc}(\mathbb{R}^2)$, hence for a subsequence $v_m \to v$ strongly in $L^2_{loc}(\mathbb{R}^2)$ and $v_m(x) \to v(x)$ for almost all x. Since

$$(v_m - k)\chi_{E^k_m} = 0 \quad \text{almost everywhere} \quad ,$$

we obtain that

$$(v - k)\gamma^k = 0 \quad \text{almost everywhere} \quad .$$

Therefore, if $\zeta \in C_0^\infty(\mathbb{R}^2)$,

$$\int_S \Big(\int_\Omega \zeta(u_s - k)\gamma^k\Big) d\mu(s) = 0$$

for all S. We conclude that for μ-almost all $s \in \mathbb{R}^2$

$$u_s = k \quad \text{almost everywhere in} \quad E^k.$$

This implies that $E^{k_1} \cap E^{k_2}$ are null sets for $k_1 \neq k_2$, hence

$$\gamma := \sum_{k \in \mathbb{Z}} \gamma^k$$

is bounded, $0 \leq \gamma \leq 1$, and the set $E := \{\gamma > 0\}$ is, up to a null set, the disjoint union of the sets E^k . This proves that (u, E) is admissible.

Next we have to show that

$$\int_D \lambda^2 \chi_{E_m} \to \int_D \lambda^2 \gamma \quad \text{for all} \quad D \subset\subset \mathbb{R}^2.$$

For $M \in \mathbb{N}$ let

$$D_{m,M} := \bigcup_{|k| \geq M} D \cap E^k_m.$$

Then for μ-almost all s

$$|u_{m,s}| \geq M \quad \text{almost everywhere in} \quad D_{m,M}.$$

Since we know that

$$\int_{\mathbb{R}^2} \int_D |u_{m,s}(x)|^2 dx d\mu(s) \leq C(D),$$

we see that

$$\mathcal{L}^2(D_{m,M}) \leq \frac{C(D)}{M^2}.$$

Therefore

$$\begin{aligned} \sum_{|k|<M} \int_D \lambda^2 \chi_{E^k_m} &\geq \int_D \lambda^2 \chi_{E_m} - \lambda_0^2 \mathcal{L}^2(D_{m,M}) \\ &\geq \int_D \lambda^2 \chi_{E_m} - C(D)\frac{\lambda_0^2}{M^2}, \end{aligned}$$

and as $m \to \infty$ the left hand side converges to

$$\sum_{|k|<M} \int_D \lambda^2 \gamma^k \leq \int_D \lambda^2 \gamma .$$

This proves

$$\int_D \lambda^2 \gamma \geq \limsup_{m \to \infty} \int_D \lambda^2 \chi_{E_m} .$$

The estimate from above is due to semicontinuity, that is

$$\begin{aligned} \int_D \lambda^2 \sum_{|k|\leq M} \gamma^k &= \lim_{m\to\infty} \sum_{|k|<M} \int_D \lambda^2 \chi_{E^k_m} \\ &\leq \liminf_{m \to\infty} \int_D \lambda^2 \chi_{E_m} . \end{aligned}$$

Finally, if we choose $D = \Omega \cap \{|y| < R\}$ with $R \geq R_0$ the weak convergence of ∇u implies

$$\begin{aligned} &\int_{\mathbb{R}^2} \int_D (|\nabla u_s - e_s|^2 - \lambda_s^2 \gamma) d\mathcal{L}^2 d\mu(s) \\ &\leq \liminf_{m \to\infty} \int_{\mathbb{R}^2} \int_D (|\nabla u_{m,s} - e_s|^2 - \lambda_s^2 \chi_{E_m}) d\mathcal{L}^2 d\mu(s) \\ &\leq \liminf_{m \to\infty} J(u_m, E_m) \quad . \end{aligned}$$

Therefore since $\gamma \leq \chi_E$ we have found an absolute minimum. ∎

In the following we always assume 2.1. As first regularity statement we prove that minimizers are Hölder-continuous in the mean with respect to the measure μ.

2.4 Lemma. If (u, E) is an absolute minimum for J, then for almost all $x_1, x_2 \in \mathbb{R}^2$ with $|x_1 - x_2| \leq \frac{1}{2}$

$$||u(x_1) - u(x_2)||_{L^2(\mu)} \leq C \cdot |x_1 - x_2| \cdot log \frac{1}{|x_1 - x_2|} \quad .$$

Proof: First choose a ball $\overline{B_R(x_0)} \subset \mathbb{R}^2 \backslash E_*$ with $R \leq \frac{1}{2}$ and $R \geq c > 0$, where finally c will depend on ∂E^0_*. For $0 < r < R$ and $s \in \mathbb{R}^2$ let $v_{r,s}$ the harmonic function in $B_r := B_r(x_0)$ with $v_{r,s} = u_s$ on ∂B_r, and continue $v_{r,s}$ outside B_r by u_s. For almost all r the function $v_{r,s}$ is well defined for μ - almost all s and (v_r, F) is admissible, if $F := E \backslash B_r$. Then $J(u, E) \leq J(v_r, F)$, that is,

$$\int_{\mathbb{R}^2} (\int_{B_r} (|\nabla u - e|^2 - |\nabla v_r - e|^2)) d\mu \leq \int_{\mathbb{R}^2} (\int_{B_r} \lambda^2 \chi_E) d\mu \leq C \cdot r^2 \quad .$$

The left hand side equals

$$= \int_{\mathbb{R}^2} (\int_{B_r} (|\nabla u|^2 - |\nabla v_r|^2)) d\mu = \int_{\mathbb{R}^2} (\int_{B_r} |\nabla (u - v_r)|^2) d\mu \quad .$$

From this the result follows using usual techniques: If we define

$$\varphi(r) := \frac{1}{r}\Big(\int_{\mathbb{R}^2}\int_{B_r} |\nabla u|^2 d\mathcal{L}^2 d\mu\Big)^{1/2} \quad ,$$

$$\psi(r) := \frac{1}{r}\Big(\int_{\mathbb{R}^2}\int_{B_r} |\nabla (u - v_r)|^2 d\mathcal{L}^2 d\mu\Big)^{1/2} \quad ,$$

then for $0 < r < \rho$

$$\varphi(r) \leq \frac{\rho}{r}\psi(\rho) + \frac{1}{r}\Big(\int_{\mathbb{R}^2}\int_{B_r} |\nabla v_\rho|^2 d\mathcal{L}^2 d\mu\Big)^{1/2} \quad .$$

By an estimate on Dirichlet's integral for harmonic functions this is

$$\leq \frac{\rho}{r}\psi(\rho) + \frac{1}{\rho}\Big(\int_{\mathbb{R}^2}\int_{B_\rho} |\nabla v_\rho|^2 d\mathcal{L}^2 d\mu\Big)^{1/2}$$

$$\leq (1 + \frac{\rho}{r})\psi(\rho) + \varphi(\rho) \quad .$$

Since ψ is bounded we obtain

$$\varphi(r) \leq C \cdot (1 + \frac{\rho}{r}) + \varphi(\rho) \quad ,$$

in particular

$$\varphi(\frac{\rho}{2}) \leq C + \varphi(\rho) \quad ,$$

and then by recursion

$$\varphi(r) \leq C \cdot log\frac{R}{r} + \varphi(R) \quad \text{for} \quad r \leq \frac{R}{2}$$

with

$$\begin{aligned} \varphi(R) &\leq C + \frac{1}{R}\cdot\Big(\int_{\mathbb{R}^2}\int_{B_R} |\nabla u - e|^2 d\mathcal{L}^2 d\mu\Big)^{1/2} \\ &\leq C + \frac{1}{R}\cdot(J(u,E) + \lambda_0^2 \mathcal{L}^2(E \cap \Omega))^{1/2} \leq C. \end{aligned} \tag{2.1}$$

Then we proceed as in the proof of the well-known Morrey lemma: For almost all x_0, if $\overline{B_{2r}(x_1)} \subset \mathbb{R}^2 \backslash E_*$ and $x_0 \in \partial B_r(x_1)$, and $x_t := (1-t)x_0 + tx_1$,

$$\int_{\mathbb{R}^2} \Big|\frac{1}{r^2}\int_{B_r(x_1)} (u - u(x_0))\Big|^2 d\mu$$

$$= \int_{\mathbb{R}^2} \Big|\frac{1}{r^2}\int_{B_r(x_1)} \Big|\int_0^1 \nabla u((1-t)x_0 + ty)\cdot(y - x_0)dt\Big|dy\Big|^2 d\mu$$

$$\leq \int_{\mathbb{R}^2} \Big|\int_0^1 \frac{1}{rt^2}\Big(\int_{B_{tr}(x_t)} |\nabla u|\Big)dt\Big|^2 d\mu \leq \int_0^1 \frac{1}{t^2}\int_{\mathbb{R}^2}\Big(\int_{B_{tr}(x_t)} |\nabla u|^2\Big)d\mu dt$$

$$\leq C\cdot r^2 \int_0^1 log^2(\frac{1}{tr})dt \leq Cr^2(1 + log^2\frac{1}{r}) \quad .$$

Since the same holds for $\tilde{x}_o := 2x_1 - x_0$ we get the desired estimate for $u(\tilde{x}_0) - u(x_0)$.

To derive the estimate everywhere, we choose $x_0 \in \partial E^0_*$ and r small enough. We now compare u in $B_r(x_0)$ with the harmonic functions $v_{r,s}$ in $B_r(x_0)\backslash E^0_*$, that is $v_{r,s} = 0$ in $B_r(x_0) \cap E^0_*$. Due to the smoothness of ∂E^0_* we now get the perturbed estimate

$$\varphi(\frac{r}{2}) \leq C + (1 + Cr)\varphi(r) \quad ,$$

and from there again the logarithmic estimate.

∎

Next we normalize the measurable set E, where in the following (u, E) always is a minimizer for J.

2.5 Lemma. We can assume that E is a closed set. Locally the components E^k don't touch.

Proof: By 2.4 for $k \in \mathbb{Z}$ there are uniform continuous functions $w^k : \mathbb{R}^2 \to \mathbb{R}$ with

$$w^k(x) = ||u(x) - k||_{L^2(\mu)} \quad \text{for almost all} \quad x \in \mathbb{R}^2 \quad ,$$

and for $k_1, k_2 \in \mathbb{Z}$ we have $w^{k_1} = |k_2 - k_1|$ almost everywhere in E^{k_2}. From this we get, that

$$\tilde{E}^k := \{x \in \mathbb{R}^2; w^k(x) = 0\}$$

is closed, and

$$\operatorname{dist}(\tilde{E}^{k_1}, \tilde{E}^{k_2}) \geq c > 0 \quad \text{for} \quad k_1 \neq k_2 \quad .$$

Also $\chi_{E^k} \leq \chi_{\tilde{E}^k}$ almost everywhere, that is, $J(u, \tilde{E}) \leq J(u, E)$. We conclude that $\chi_{\tilde{E}} = \chi_E$ almost everywhere. ∎

In the following E will always be the closed set of 2.5. The next fundamental step is to show that E^0 is compact. This will be proved under the assumption that neither from left nor from right all prescribed velocities are the same, that is,

2.6 Assumption on μ . We assume that $\mu(L) < 1$ for all axisparallel lines L of $\mathbb{R}^2$.

2.7 Lemma. Under assumption 2.6 the set E^0 is compact.

Proof: Let us assume that $E \cap \bar{\Omega}$ is unbounded to the left (for $y \to -\infty$). Then there are points $x_0 = (y_0, z_0)$ and $x_1 = (y_1, z_1)$ in $E \cap \bar{\Omega}$ with $y_1 < y_0 \to -\infty$ and $R := y_0 - y_1 \to +\infty$. By translating Ω in z-direction we can assume that $x_0 \in E^0$. For simplicity let Ω the old domain and $z_0 = \frac{1}{2}$. Also let $x_1 \in E^k$, where k depends on the point x_1. Then by the continuity lemma 2.4 and the definition of E^k we see that for the trace of u on the boundary of $D_R := [y_1, y_0] \times [0, 1]$

$$\begin{aligned} &\int_{\mathbb{R}^2} |u_s(y_0, z)|^2 d\mu(s) \leq C, \\ &\int_{\mathbb{R}^2} |u_s(y_1, z) - k|^2 d\mu(s) \leq C \quad . \end{aligned} \tag{2.2}$$

We scale the situation by

$$v_R(y, z) := \frac{u(y_0 + Ry, z)}{R} \quad \text{and} \quad e_R(y, z) := e(y_0 + Ry, z)$$

for $(y,z) \in D := [-1,0] \times [0,1]$. Then (see(2.1))

$$C \geq \int_{\mathbb{R}^2} \int_{\Omega \cap \{y_1 \leq y \leq y_0\}} |\nabla u - e|^2 d\mathcal{L}^2 d\mu$$

$$= \int_{\mathbb{R}^2} \int_D (R|\partial_y v_R - e_R \cdot e_y|^2 + R^3 |\partial_z v_R - \frac{e_R \cdot e_z}{R^2}|^2) d\mathcal{L}^2 d\mu \quad .$$

We derive, that for $R \to \infty$

$$\text{(2.3)} \qquad \partial_y v_R \to e^- \cdot e_y \quad \text{and} \quad \partial_z v_R \to 0 \quad \text{in} \quad L^2(\mu; L^2(D)) \quad ,$$

and from (2.2) we get for almost all $0 \leq z \leq 1$

$$\text{(2.4)} \qquad \int_{\mathbb{R}^2} |v_R(0,z)|^2 d\mu \leq \frac{C}{R^2} \to 0 \quad ,$$

$$\text{(2.5)} \qquad \int_{\mathbb{R}^2} |v_R(-1,z) - \frac{k}{R}|^2 d\mu \leq \frac{C}{R^2} \to 0 \quad .$$

By (2.3) and (2.4)

$$v_R \to v \quad \text{in} \quad L^2(\mu; H^{1,2}(D)) \quad ,$$

where

$$v_s(y,z) = -s_- y \quad .$$

Then (2.5) implies, that for some $\kappa \in \mathbb{R}$

$$\frac{k}{R} \to \kappa \quad \text{for} \quad R \to \infty \quad ,$$

that is, $s_- = v_s(-1,z) = \kappa$ for $0 \leq z \leq 1$ and μ - almost all s, a contradiction to assumption 2.6. ∎

The first variation of J with respect to the function u implies immediately

2.8 Lemma. For μ - almost all s

$$\Delta u_s = 0 \quad \text{in} \quad \mathbb{R}^2 \backslash E \quad .$$

The difficulty is, that at this stage we don't know the continuity at ∂E for a single function u_s. One possibility for proving it would be by comparison, but this requires assumptions on μ. Without it we are forced to treat the notion Δu^+ carefully, in particular the product $u^+ \cdot \Delta u^+$.

2.9 Remark. Consider any ball $B_R := B_R(x_0) \subset \mathbb{R}^2$ which intersects E only at E^0 . Then for μ -almost all s and almost all $0 < r < R$ we can work with ∇u_s an ∂B_r. Therefore $\Delta u_s^+ \in H^{1,2}(B_r)^*$ is well defined by

$$< \zeta, \Delta u_s^+ >_{B_r} := \int_{\partial B_r} \zeta \partial_\nu u_s^+ - \int_{B_r} \nabla \zeta \cdot \nabla u_s^+ \quad .$$

Similar Δu_s^- , where $u_s^+ := \max(u_s, 0)$ and $u_s^- := \max(-u_s, 0)$. Then for almost all s and r

1)

$$< \zeta, \Delta u_s^+ >_{B_r} \geq 0 \quad \text{for} \quad \zeta \geq 0 \quad .$$

2)

$$< u_s^+, \Delta u_s^+ >_{B_r} = 0 \quad .$$

Both statements also hold for u_s^- .

Proof: For $\delta > 0$ define

$$\psi_\delta(z) := \begin{cases} 0 & \text{for} \quad z \leq 0, \\ \min(1, \frac{z}{\delta}) & \text{for} \quad z \geq 0. \end{cases}$$

Let $\zeta \in C_0^\infty(B_R)$, and for $\epsilon \in I\!R, S \subset I\!R^2$ compare u with v defined by

$$v_s := \begin{cases} u_s + \epsilon\zeta\psi_\delta(u_s) & \text{for} \quad s \in S, \\ u_s & \text{for} \quad s \notin S. \end{cases}$$

Then (v, E) is admissible and therefore first variation leads to

$$\begin{aligned} 0 &= \int_S \Big(\int_{B_R} \nabla u_s \cdot \nabla(\zeta\psi_\delta(u_s))\Big) d\mu \\ &= \int_S \Big(\int_{B_R} \zeta\psi_\delta'(u_s)|\nabla u_s|^2 + \int_{B_R} \psi_\delta(u_s)\nabla u_s \cdot \nabla\zeta \Big) d\mu. \end{aligned}$$

If $\zeta \geq 0$ we derive for $\delta \to 0$

$$0 \geq \int_S \Big(\int_{B_R} \nabla u_s^+ \cdot \nabla\zeta\Big) d\mu.$$

The connection to the above definition of Δu_s^+ we obtain if we replace ζ by $\zeta\eta$ and let η converge to χ_{Br} in a suitable manner. This proves 1). Replacing $\psi_\delta(u_s)$ by u_s^+ we get

$$0 = \int_S \Big(\int_{B_R} \zeta|\nabla u_s^+|^2 + \int_{B_R} u_s^+ \nabla u_s^+ \cdot \nabla\zeta \Big) d\mu \quad .$$

This shows 2), if again $\zeta \to \chi_{B_r}$. ∎

2.10 Corollary. 2.9.1 means that $\Delta u_s^\pm$ are positive measures in

$$I\!R^2 \backslash E_* \backslash \bigcup_{k \neq 0} E^k \quad .$$

In the following we denote these measures by $\mu_s^\pm$. Then, if B_R as in 2.9 and $0 < r < R$, we denote by G_x Green's function for the Laplacian in B_r with pole at x , and by $P_x := -\partial_\nu G_x$ Poisson's Kernel on ∂B_r. With these notations

1) for almost all s and every $x \in B_r \backslash E^0$

$$u_s^+(x) = \int_{\partial B_r} P_x u_s^+ - \int_{B_r} G_x d\mu_s^+ \quad .$$

2) for all $S \subset I\!R^2$ and $x \in B_r \cap E^0$

$$\int_S \Big(\int_{\partial B_r} P_x u_s^+ \Big) d\mu(s) = \int_S \Big(\int_{B_r} G_x d\mu_s^+ \Big) d\mu(s) \quad .$$

Proof: Let G_x^ϵ a suitable smooth approximation of G_x near its singularity. Then by 2.9

$$\int_{B_r} G_x^\epsilon d\mu_s^+ = -\int_{B_r} \nabla G_x^\epsilon \cdot \nabla u_s^+ = \int_{\partial B_r} P_x u_s^+ + \int_{B_r} u_s^+ \Delta G_x^\epsilon \quad .$$

Since (by 2.8) u_s is smooth in $B_r \backslash E^0$ we obtain 1) by letting $\epsilon \to 0$. For 2) we integrate over S and estimate using the definition of E^0 in 2.5

$$\int_{I\!R^2} \Big(\int_{B_r} u_s^+ |\Delta G_x^\epsilon| \Big) d\mu \le \sup_{|x'-x| \le \epsilon} ||u^+(x')||_{L^2(\mu)} \to 0 \quad \text{as} \quad \epsilon \to 0 \quad .$$

∎

Using 2.9.2. we can perform the proof of the monotoncity formula in [3] and obtain

2.11 Theorem. Let B_R as in 2.9 and for $0 < r < R$

$$\varphi_s^\pm(r) := \frac{1}{r^2} \int_{B_r} |\nabla u_s^\pm|^2 \quad .$$

Then for almost all s the product $\varphi_s(r) := \varphi_s^+(r) \cdot \varphi_s^-(r)$ is continuous and nondecreasing for $0 < r < R$.

For completeness here is the proof: First we have to check for which r both u_s^+ and u_s^- contribute on ∂B_r . Define

$$\rho_s^\pm := \sup\{r < R; \varphi_s^\pm(r) = 0\} \quad ,$$

and let $\rho_s^+ < r < R$. Then if the trace of u_s^+ on ∂B_r would vanish, we obtain from 2.9.2 , that $\nabla u_s^+ = 0$ in B_r ; a contradiction to the definition of ρ_s^+ . Similar for ρ_s^-. Hence $\varphi_s(r) = 0$ for $0 \le r \le \rho_s$, if

$$\rho_s := \max(\rho_s^+, \rho_s^-) \quad ,$$

and it suffices to prove the assertion for $\rho_s < r < R$. Moreover, for almost all r between ρ_s and R both, u_s^+ and u_s^- have nontrivial continuous traces on ∂B_r. In this situation the following "two-phase" consideration proves the theorem. Using polar coordinates r, θ and 2.9 we obtain for almost all s and r

$$\frac{d}{dr} log \varphi_s^+(s) = -\frac{2}{r} + \frac{\int_{\partial B_r} |\nabla u_s^+|^2}{\int_{\partial B_r} u_s^+ \partial_r u_s^+} \quad .$$

By scaling we can assume $r = 1$. Here scaling means, that we consider the transformed function $x \to \frac{1}{r} \cdot u_s^+(x_0 + rx)$ and call it u_s^+ again. Then the right side is

$$\ge -2 + 2 \cdot \frac{(\int_{\partial B_1} |\partial_\theta u_s^+|^2)^{1/2}}{(\int_{\partial B_1} |u_s^+|^2)^{1/2}} \ge -2 + 2 \cdot \sqrt{\lambda^+} \quad ,$$

where u_s^+ is continuous on ∂B_1, and $\lambda^+ > 0$ is the first eigenvalue of $\partial_{\theta\theta}$ on $\partial B_1 \cap \{u_s^+ > 0\}$. The comparison lemma for eigenvalues yields $\lambda^+ \geq (l^+)^{-2}$, where

$$l^+ := \frac{1}{\pi}\mathcal{H}^1(\partial B_1 \cap \{u_s^+ > 0\}) \quad .$$

We conclude

$$\frac{d}{dr} log\varphi_s(1) \geq 2(\sqrt{\lambda^+} + \sqrt{\lambda^-} - 2) \geq 2(\frac{1}{l^+} + \frac{1}{l^-} - 2) \geq 0,$$

since $l^+ + l^- \leq 2$. ∎

Now we are ready to prove Lipschitz-continuity in the mean.

2.12 Theorem. Let $B_{R_0}(x_0)$ as in 2.9 and $0 < R < R_0$ with $B_R(x_0) \subset \mathbb{R}^2 \backslash E_*$. Then, if $x_0 \in E^0$, for almost all $0 < r < \frac{R}{2}$

$$\int_{\mathbb{R}^2} (\fint_{\partial B_r} |u| d\mathcal{H}^1)^2 d\mu \leq C(R_0) \cdot r^2 \quad ,$$

where $\mathcal{H}^1$ denotes the one-dimensional Hausdorff measure, and $\fint$ the mean value.

Proof: Let $\varphi^\pm, \varphi$ as in 2.11. Then for $0 < r < R_0$

$$\int_{\mathbb{R}^2} \varphi_s(r)^{1/2} d\mu(s) \leq \int_{\mathbb{R}^2} \varphi_s(R_0)^{1/2} d\mu(s) \leq \frac{1}{R_0^2} \int_{\mathbb{R}^2} \int_{B_{R_0}} |\nabla u|^2 d\mathcal{L}^2 d\mu \leq C(1 + \frac{1}{R_0^2}) \quad .$$

The last inequality follows as in (2.1). Now we fix $r < R$ and (for convenience) scale to radius 1, keep all notations, and write φ_s instead of $\varphi_s(r)$. Then the scaled ball B_1 has center $0 \in E^0$.

For given $M > 1$ we split $\mathbb{R}^2$ into the sets

$$S_\pm := \{s \in \mathbb{R}^2; \varphi_s^\pm > M\sqrt{\varphi_s}\} \quad ,$$

and $S_0 := \mathbb{R}^2 \backslash S_+ \backslash S_-$ (one could split into two sets only). Then in B_1 we compare u with v, where

$$v_s := \begin{cases} -u_s^- + w_s^+ & \text{for} \quad s \in S_+, \\ u_s^+ - w_s^- & \text{for} \quad s \in S_-, \\ u_s & \text{for} \quad s \in S_0. \end{cases}$$

Here $w_s^\pm$ is the harmonic function in B_1 with boundary values $u_s^\pm$, therefore $v_s = u_s$ on ∂B_1 for all s. Since $J(u, E) \leq J(v, E \backslash B_1)$ (for the scaled functional), we obtain as in 2.4

$$\begin{aligned} C &\geq \int_{\mathbb{R}^2} \int_{B_1} (|\nabla u|^2 - |\nabla v|^2) d\mathcal{L}^2 d\mu \\ &= \int_{\mathbb{R}^2} \int_{B_1} \nabla(u - v) \cdot \nabla(u + v) d\mathcal{L}^2 d\mu \\ &= \int_{S_+} \int_{B_1} \nabla(u_s^+ - w_s^+) \cdot \nabla(u_s^+ - 2u_s^- + w_s^+) d\mathcal{L}^2 d\mu(s) \\ &+ \int_{S_-} \int_{B_1} \nabla(-u_s^- + w_s^-) \cdot \nabla(2u_s^+ - u_s^- - w_s^-) d\mathcal{L}^2 d\mu(s) \quad , \end{aligned}$$

and since $w_s^\pm$ is harmonic,

$$= \int\limits_{S_+}\int\limits_{B_1} (|\nabla(u_s^+ - w_s^+)|^2 - 2\nabla(u_s^+ - w_s^+)\cdot\nabla u_s^-) d\mathcal{L}^2 d\mu(s)$$

$$+ \int\limits_{S_-}\int\limits_{B_1} (|\nabla(u_s^- - w_s^-)|^2 - 2\nabla(u_s^- - w_s^-)\nabla u_s^+) d\mathcal{L}^2 d\mu(s)$$

$$\geq c\int\limits_{S_+}\int\limits_{B_1} |\nabla(u_s^+ - w_s^+)|^2 d\mathcal{L}^2 d\mu(s) + c\int\limits_{S_-}\int\limits_{B_1} |\nabla(u_s^- - w_s^-)|^2 d\mathcal{L}^2 d\mu(s)$$

$$- C\int\limits_{S_+}\int\limits_{B_1} |\nabla u_s^-|^2 d\mathcal{L}^2 d\mu(s) - C\int\limits_{S_-}\int\limits_{B_1} |\nabla u_s^+|^2 d\mathcal{L}^2 d\mu(s) \quad .$$

Since $\varphi_s^+ \cdot \varphi_s^- = \varphi_s$ we derive using the definition of $S_\pm$

$$\int\limits_{S_\pm}\int\limits_{B_1} |\nabla(u^\pm - w^\pm)|^2 d\mathcal{L}^2 d\mu \leq C(1 + \int\limits_{S_+} \varphi_s^- d\mu(s) + \int\limits_{S_-} \varphi_s^+ d\mu(s))$$

$$\leq C(1 + \frac{2}{M}\int\limits_{\mathbb{R}^2} \sqrt{\varphi_s} d\mu(s)) \leq C$$

with C depending on R_0 . Also for $s \in \mathbb{R}^2 \setminus S_+$

$$||\nabla(u_s^+ - w_s^+)||_{L^2(B_1)} \leq ||\nabla u_s^+||_{L^2(B_1)} + ||\nabla w_s^+||_{L^2(B_1)} \leq 2||\nabla u_s^+||_{L^2(B_1)} \quad ,$$

hence

$$\int\limits_{\mathbb{R}^2\setminus S_+}\int\limits_{B_1} |\nabla(u^+ - w^+)|^2 d\mathcal{L}^2 d\mu \leq 4 \int\limits_{\mathbb{R}^2\setminus S_+} \varphi_s^+ d\mu(s) \leq 4M \int\limits_{\mathbb{R}^2} \sqrt{\varphi_s} d\mu(s) \leq C \quad .$$

Similarly for u^-. We thus obtain that

$$\int\limits_{\mathbb{R}^2}\int\limits_{B_1} |\nabla(u^\pm - w^\pm)|^2 d\mathcal{L}^2 d\mu \leq C \quad .$$

Next we estimate the left hand side from below. Using 2.9 we derive for almost all s

$$\int\limits_{B_1} |\nabla(u_s^+ - w_s^+)|^2 = \int\limits_{B_1} \nabla(u_s^+ - w_s^+)\cdot\nabla u_s^+$$

$$= < w_s^+ - u_s^+, \Delta u_s^+ >_{B_1} = < w_s^+, \Delta u_s^+ >_{B_1}$$

$$\geq \inf_{B_{1/2}} w_s^+ \cdot \int\limits_{B_{1/2}} d\mu_s^+ \quad .$$

The last inequality holds, since $w_s^+ \geq 0$ and μ_s^+ (see 2.10) is a nonnegative measure. Finally we estimate w_s^+ in $B_{1/2}$ from below. Using the notation in 2.10 we get for $x \in B_{1/2}$

$$w_s(x) = \int\limits_{\partial B_1} P_x u^+ \geq c \int\limits_{\partial B_1} u^+ \quad .$$

By 2.11.2 for almost all s ($0 \in E^0$ is fixed) the last integral is

$$= \int_{B_1} G_0 d\mu_s^+ \geq c \int_{B_{1/2}} d\mu_s^+ \quad .$$

Putting things together we have proved that

$$\int_{\mathbb{R}^2} \mu_s^{\pm}(B_{1/2})^2 d\mu(s) \leq C \quad ,$$

or after rescaling

$$\int_{\mathbb{R}^2} \mu_s^{\pm}(B_r)^2 d\mu(s) \leq Cr^2 \quad \text{for} \quad 0 < r < \frac{R}{2} \quad . \tag{2.6}$$

This estimate is equivalent to the Lipschitz continuity of the solution u in mean. Here we transform the estimate to the statement of the theorem. We scale again. By 2.11

$$\int_{\mathbb{R}^2} (\fint_{\partial B_1} u^+)^2 d\mu = \int_{\mathbb{R}^2} (\int_{B_1} G_0 d\mu^+)^2 d\mu \tag{2.7}$$

Writing $G_0(x) = \psi(|x|)$, formally the right side is

$$= \int_{\mathbb{R}^2} (\int_0^1 \psi(r) \frac{d}{dr} (\int_{B_r} \Delta u_s^+) dr)^2 d\mu(s)$$

$$= \int_{\mathbb{R}^2} (\int_0^1 \psi'(r) (\int_{B_r} \Delta u_s^+) dr)^2 d\mu(s)$$

$$\leq \frac{1}{4\pi^2} \int_0^1 \frac{1}{r^2} \int_{\mathbb{R}^2} (\int_{B_r} \Delta u_s^+)^2 d\mu(s) dr \quad .$$

By (2.6) this bounded. To make the argument precise, we divide $[0,1]$ into intervals $[r_{i-1}, r_i]$ with $r_i = ih, Nh = 1$, and define

$$\psi^h(r) := \psi(r_i) \quad \text{for} \quad r_{i-1} < r < r_i, \quad i = 1, \ldots, N,$$

and $G_0^h(x) := \psi^h(|x|)$. Then (with $B_0 := \emptyset$)

$$\int_{\mathbb{R}^2} (\int_{B_1} G_0^h d\mu_s^+)^2 d\mu(s) = \int_{\mathbb{R}^2} \Big(\sum_{i=1}^{N} \psi(r_i)(\mu_s^+(B_{r_i}) - \mu_s^+(B_{r_{i-1}})\Big)^2 d\mu(s) \tag{2.8}$$

$$= \int_{\mathbb{R}^2} \Big(\sum_{i=1}^{N-1} (\psi(r_i) - \psi(r_{i+1}))\mu_s^+(B_{r_i})\Big)^2 d\mu(s)$$

$$\leq \sum_{i=1}^{N-1} \frac{1}{h} (\psi(r_i) - \psi(r_{i+1}))^2 \int_{\mathbb{R}^2} \mu_s^+(B_{r_i})^2 d\mu(s)$$

$$\leq C \sum_{i=1}^{N-1} \frac{1}{h} (\psi(r_i) - \psi(r_{i+1}))^2 r_i^2 \leq C.$$

Since for $h \to 0$ the left side of (2.8) converges (monotonically) to the right side of (2.7), and since the same holds for u^-, the theorem follows by rescaling. ∎

As a consequence we obtain

2.13 Theorem. u is globally Lipschitz continuous in mean with respect to the measure μ, that is,

$$\int_{\mathbb{R}^2} |\nabla u_s(x)|^2 d\mu(s) \le C \quad \text{for} \quad x \in \mathbb{R}^2 \quad ,$$

where C depends on $dist(E^0, E^1)$.

Proof: On E^0 almost all functions u_s vanish, therefore $\nabla u_s = 0$ almost everywhere on E^0. Similar on E^k. Now let $x_0 \in \mathbb{R}^2 \backslash E$ and $R_0 > 0$ fixed. We have two cases. If $B_{R_0} := B_{R_0}(x_0) \subset \mathbb{R}^2 \backslash E$ then u_s is harmonic in B_{R_0} for almost all s (see 2.8), therefore by elliptic estimates

$$|\nabla u_s(x_0)|^2 \le \frac{C}{R_0^2} \int_{B_{R_0}} |\nabla u_s|^2 \quad ,$$

and the assertion follows by an estimate like (2.1). For the second case we choose R_0 with

$$R_0 < \kappa_0 dist(E^0, \bigcup_{k \ne 0} E^k) \quad ,$$

where κ_0 is a small positive constant. Then for some $k \in \mathbb{Z}$

$$B_{R_0} \cap E \subset E^k \quad .$$

After a translation in z - direction we can assume that $k = 0$. Let

$$r_0 := dist(x_0, E^0) \le R_0 \quad .$$

Then u_s is harmonic in B_{r_0}, hence by Poisson's formula for $\frac{r_0}{2} < r < r_0$

$$|\nabla u_s(x_0)| \le \frac{C}{r} \fint_{\partial B_r} |u_s| \quad .$$

Integration with respect to μ gives

$$\int_{\mathbb{R}^2} |\nabla u_s(x_0)|^2 d\mu(s) \le \frac{C}{r_0^2} \operatorname*{ess\,sup}_{x \in \partial B_r} \int_{\mathbb{R}^2} |u_s(x)|^2 d\mu(s) \quad . \tag{2.9}$$

We distinguish between two cases. The first case is that the distance of x_0 to E^0_* is much bigger than the distance to E^0, say,

$$R := dist(x_0, E^0_*) \ge 9r_0 \quad .$$

We pick $x_1 \in \partial B_{r_0}(x_0) \cap E^0$, hence $B_{r_0}(x_0) \subset B_{2r_0}(x_1)$. Then for almost all $x \in B_{2r_0}(x_1) \cap E^0$ the last integral in (2.9) vanishes. For $x \in B_{2r_0}(x_1) \backslash E^0$ we have for almost all $r \in [3r_0, 4r_0]$, if we apply 2.10.1 to u_s^+ and u_s^-,

$$|u_s(x)| \le \int_{\partial B_r(x_1)} P_x |u_s| \le C \fint_{\partial B_r(x_1)} |u_s| \quad .$$

Since $B_{2r}(x_1) \subset B_{9r_0}(x_0)$ does not touch E_*, we obtain using 2.12

$$\int_{\mathbb{R}^2} |u_s(x)|^2 d\mu(s) \leq C \int_{\mathbb{R}^2} (\fint_{\partial B_r(x_1)} |u_s|^2) d\mu(s)$$
$$\leq C(R_0) \cdot r^2 \leq C(R_0) r_0^2 \quad ,$$

which proves the theorem in that case.

If R is comparable with r_0 , that is, $R \leq 9r_0$ we shall show that for almost all s

$$(2.10) \qquad |u_s(x)| \leq M_s \cdot dist(x, E_*^0) \quad \text{for} \quad x \in B_{10\cdot R_0}(E_*^0) \quad ,$$

where

$$\int_{\mathbb{R}^2} M_s^2 d\mu(s) \leq C \quad .$$

Having this we get for $x \in B_{r_0}(x_0) \subset B_{10\cdot r_0}(E_*^0)$

$$\int_{\mathbb{R}^2} |u_s(x)|^2 d\mu(s) \leq C(R + r_0)^2 \leq C r_0^2 \quad .$$

To prove (2.10), let $R_1 := 20 \cdot R_0$ and κ_0 small enough. Then by 2.10.1 for $x_1 \in B_{R_1}(E_*^0) \backslash E^0$ and $\frac{R_1}{2} \leq r \leq R_1$

$$|u_s(x_1)| \leq \fint_{\partial B_r(x_1)} |u_s| \quad .$$

Integrating over r we see that

$$|u_s(x_1)| \leq \frac{C}{R_1} (\int_{B_{2R_1}(E_*^0)} |u_s|^2)^{1/2} =: M_s$$

with

$$\int_{\mathbb{R}^2} M_s^2 d\mu(s) \leq C \cdot \int_{\mathbb{R}^2} (\int_{B_{2R_1}(E_*^0)} |\nabla u_s|^2) d\mu(s) \leq C(R_1)$$

as in (2.1). Then let $w_s^\pm$ the harmonic function in $D := B_{R_1}(E_*^0) \backslash E_*^0$ with boundary values $w_s^\pm = 0$ on ∂E_*^0 and $w_s^\pm = \pm M_s$ on $\partial B_{R_1}(E_*^0)$. We compare u with v defined by

$$v_s := \begin{cases} u_s & \text{outside} \quad B_{R_1}(E_*^0), \\ \min(u_s, w_s^+) & \text{in} \quad B_{R_1}(E_*^0), \end{cases}$$

Since for almost all s

$$v_s = u_s = 0 \quad \text{almost everywhere in} \quad B_{R_1}(E_*^0) \cap E \quad ,$$

we derive from $J(u, E) \leq J(v, E)$ that $u_s \leq w_s^+$ almost everywhere in D for almost all s . Similar $u_s \geq w_s^-$. Since

$$|w_s^\pm(x)| \leq C \cdot M_s \cdot dist(x, E_*^0) \quad \text{for} \quad x \in D$$

we have proved (2.10). ∎

Theorem 2.13 says, that in mean the functions $|u_s|$ grow away from the blade E^0 at most linearly. In the next lemma we shall show, again in the mean, that approaching E this growth does not degenerate. This non-degenerate linear growth is exactly what we expect from the formal condition

$$\int_{\mathbb{R}^2} |\nabla u_s(x)|^2 d\mu(s) = \lambda(x)^2 > 0$$

for free boundary points $x \in \partial E$. The following proof is adapted from [2; Lemma 3.4].

2.14 Lemma. Suppose $B_R(x_0) \cap E \subset E^k$ and $0 < r < R$. Then the condition

$$\int_{\mathbb{R}^2} \Big(\fint_{\partial B_r(x_0)} |u_s(x) - k| d\mathcal{H}^1(x)\Big)^2 d\mu(s) \le c \cdot r^2 \cdot \inf_{B_{r/2}(x_0)} \lambda^2$$

implies that $B_{r/2}(x_0) \subset E^k$. Here c is a universal small positive constant.

Remark: The ball $B_R(x_0)$ might intersect E^k_*. The values of u_s on $\partial B_r(x_0)$ are understood as trace from $B_r(x_0)$.

Proof: We first scale to $r = 1, x_0 = 0$, and assume that $k = 0$. Let

$$\lambda_1 := \inf_{B_\kappa} \lambda$$

with $\kappa < 1$, and for definiteness $\kappa \ge \frac{1}{4}$. Define

$$\epsilon_s := \frac{1}{\lambda_1} \fint_{\partial B_1} |u_s| \quad .$$

We have to prove that $B_\kappa \subset E$, if

$$\int_{\mathbb{R}^2} \epsilon_s^2 d\mu(s) \le \epsilon_*$$

with ϵ_* small enough. Define $u^+ := \max(u, 0)$ and $u^- := \min(u, 0)$, that is, $u = u^+ + u^-$. We let $w_s^\pm$ the harmonic function in $B_1 \backslash \overline{B}_\kappa$ with boundary values

$$w_s^\pm = \begin{cases} 0 & \text{in } \ B_\kappa, \\ u_s^\pm & \text{on } \ \partial B_1. \end{cases}$$

As comparison function for u we define

$$v_s := \begin{cases} u_s & \text{outside } \ B_1, \\ \min(w_s^+, \max(w_s^-, u_s)) & \text{in } \ B_1. \end{cases}$$

Since $v_s = 0$ on B_κ we have $J(u, E) \le J(v, E \cup B_\kappa)$ that is,

$$\int_{\mathbb{R}^2} \Big(\int_{B_1} |\nabla u_s|^2 \Big) d\mu(s) + \int_{B_\kappa} \lambda^2 (1 - \chi_E) \le \int_{\mathbb{R}^2} \Big(\int_{B_1} |\nabla v_s|^2 \Big) d\mu(s) \quad , \tag{2.11}$$

or

$$\begin{aligned} &\int_{\mathbb{R}^2} \int_{B_\kappa} (|\nabla u|^2 + \lambda^2 (1 - \chi_E)) d\mathcal{L}^2 d\mu \\ &\le \int_{\mathbb{R}^2} \int_{B_1 \backslash B_\kappa} (|\nabla v|^2 - |\nabla u|^2) d\mathcal{L}^2 d\mu \\ &= \int_{\mathbb{R}^2} \Big(\int_{B_1 \backslash B_\kappa} \nabla (v - u) \cdot \nabla ((u - v) + 2v) \Big) d\mu \\ &\le 2 \int_{\mathbb{R}^2} \Big(\int_{B_1 \backslash B_\kappa} \nabla \max(w^- - u, 0) \nabla w^- \Big) d\mu \\ &+ 2 \int_{\mathbb{R}^2} \Big(\int_{B_1 \backslash B_\kappa} \nabla \min(w^+ - u, 0) \nabla w^+ \Big) d\mu \quad . \end{aligned} \tag{2.12}$$

Since $w_s^- \leq u_s \leq w_s^+$ on ∂B_1 , this equals

$$= 2 \int_{\mathbf{R}^2} \Big(\int_{\partial B_\kappa} |u^-| \partial_{-\nu} w^- + \int_{\partial B_\kappa} |u^+| \partial_\nu w^+ \Big) d\mu \quad .$$

In the following all constants depend on κ. By elliptic theory we get the estimate

$$\leq C \int_{\mathbf{R}^2} \Big(\int_{\partial B_\kappa} |u^-| \cdot \int_{\partial B_1} |u^-| + \int_{\partial B_\kappa} |u^+| \cdot \int_{\partial B_1} |u^+| \Big) d\mu$$

$$\leq C \int_{\mathbf{R}^2} \epsilon \lambda_1 \int_{\partial B_\kappa} |u| d\mathcal{H}^1 d\mu \quad ,$$

where ϵ is the function of s defined above. By Sobolev embedding this is

$$\leq C \int_{\mathbf{R}^2} \epsilon \lambda_1 \int_{B_\kappa} (|u| + |\nabla u|) d\mathcal{L}^2 d\mu \quad .$$

Since

$$|\nabla u_s| \leq \frac{\delta}{2} |\nabla u_s|^2 + \frac{1}{2\delta} \chi_{\{u_s \neq 0\}} \quad \text{for} \quad \delta > 0 \quad ,$$

we proceed in estimating it by

$$\leq C \int_{\mathbf{R}^2} \int_{B_\kappa} (\epsilon \lambda_1 |u| + \frac{1}{\epsilon_*} (\epsilon \lambda_1)^2 \chi_{\{u \neq 0\}} + \epsilon_* |\nabla u|^2) d\mathcal{L}^2 d\mu \quad .$$

By 2.10.1 for almost all s

$$\operatorname*{ess\,sup}_{\bar{B}_\kappa} |u_s| \leq C(\kappa) \int_{\partial B_1} |u_s| \leq C \epsilon_s \lambda_1 \quad .$$

Using this we finally estimate our integral by

$$\leq \int_{\mathbf{R}^2} \int_{B_\kappa} \big((1 + \frac{1}{\epsilon_*}) (\epsilon \lambda_1)^2 \chi_{\{u \neq 0\}} + \epsilon_* |\nabla u|^2 \big) d\mathcal{L}^2 d\mu \quad .$$

Since $B_\kappa \cap \{u_s \neq 0\} \subset B_\kappa \backslash E$ for almost all s , this is estimated by

$$\leq C \cdot \epsilon_* \Big((\epsilon_* + 1) \int_{B_\kappa} \lambda_1^2 (1 - \chi_E) d\mathcal{L}^2 + \int_{\mathbf{R}^2} \int_{B_\kappa} |\nabla u|^2 d\mathcal{L}^2 d\mu \Big) \quad .$$

which we compare with the left side of (2.12). Therefore if ϵ_* is small enough, the set E must contain B_κ. ∎

3. The positive density property

In the section 2 we have proved, that the free boundary condition is satisfied qualitatively, that is, that the stream functions in mean grow linearly near the blade. As a consequence we got the global Lipschitz continuity in mean, that is, the mean speed of the flow is bounded. So far, all proofs work also in the case of one flow, that is , if μ is a Dirac measure. (There is one exception, the compactness of E was proved under condition 2.6.)

The last fundamental local property of the blade to be proved is the non-degeneracy of E in measure. This means that E has positive density at each point $x_0 \in E$. This Lebesgue density is defined by

$$\theta_*(E, x_0) := \liminf_{r \searrow 0} \theta_r(E, x_0) \quad ,$$

where

$$\theta_r(E, x_0) := \frac{\mathcal{L}^2(E \cap B_r(x_0))}{\mathcal{L}^2(B_r(x_0))} \quad .$$

We shall prove that $\theta_*(E, x_0) > 0$ under a condititon on μ , which allows us to compare certain stream functions by the maximum principle. It is clear, that the positive density property does not hold for Dirac measures μ , since then, as mentioned in section 1, the blade E forms singularities, at which its density is zero.

3.1 Remark. As an immediate consequence of non-degeneracy (see 2.14) and Lipschitz continuity (see 2.13) it follows, that the upper Lebesgue density

$$\theta^*(E, x_0) := \limsup_{r \searrow 0} \theta_r(E, x_0) \leq 1 - \kappa < 1 \quad \text{for} \quad x_0 \in \partial E \quad ,$$

where κ does not depend on x_0. In fact, if $x_0 \in \partial E^0$, by 2.14 for small r

$$\fint_{\partial B_r(x_0)} ||u(x)||^2_{L^2(\mu)} d\mathcal{H}^1(x) \geq cr^2 \quad .$$

Therefore for some point $x_1 \in \partial B_r(x_0)$

$$||u(x_1)||_{L^2(\mu)} \geq cr \quad ,$$

and then by 2.13 for $x \in B_{\kappa r}(x_1)$

$$||u(x)||_{L^2(\mu)} \geq (c - C \cdot \kappa) r \geq cr > 0$$

if κ is chosen small enough. ∎

3.2 Assumption on μ. For every set $S \subset \mathbb{R}^2$ with $\mu(S) > 0$ there are two rectangles $Q_1 :=]-\infty, \sigma_1]$ and $Q_2 := [\sigma_2, \infty[$ in $\mathbb{R}^2$, $\sigma_i = (\sigma_{i-}, \sigma_{i+})$, with $\sigma_{1-} < \sigma_{2-}$ and $\sigma_{1+} < \sigma_{2+}$, such that $\mu(Q_i) > 0$ for $i = 1, 2$, and such that $\mu(S \cap Q_1) > 0$ or $\mu(S \cap Q_2) > 0$.

Remark: This condition implies 2.6, and it is satisfied, if $d\mu = f \cdot \chi_D \cdot d\mathcal{L}^2$ with an open set D and a positive function f.

First we show the following comparison statement.

3.3 Lemma. Let Q_1, Q_2 as in 3.2. Then $u_{s_1} \leq u_{s_2}$ for almost all $s_1 \in Q_1$ and $s_2 \in Q_2$.

Proof: The idea is, that functions u_s with $s \in Q_2$ increase faster to the left and right than those with $s \in Q_1$, therefore far enough the latter should stay below. Then the lemma follows by the maximum principle.

To be precise, choose any set $S_i \subset Q_i$ with $\mu(S_i) > 0$, and consider the continuous functions

$$v_i := \fint_{S_i} u_s d\mu(s) \quad .$$

Since E is compact (see 2.7) we can choose $R > R_0$ with

$$E \subset [-R+1, R-1] \times I\!R \quad .$$

Furthermore let R so large that for a given small positive constant $\delta > 0$

$$\int_{I\!R^2} \Big(\int_{[R-1,\infty[\times[0,1]} |\nabla u - e^+|^2 \Big) d\mu \leq \delta \quad .$$

For $s \in I\!R^2$ we consider a linear Function l_s with $\nabla l_s = e_s^+$, given by

$$l_s(y,z) := u_s(R, \frac{1}{2}) + s_+(y-R) + (z - \frac{1}{2}) \quad ,$$

and we define

$$\varphi_i = \fint_{S_i} l_s d\mu(s) \quad .$$

Then for $r \geq R$

$$\int_{[r-1,r+2]\times[-1,2]} |\nabla(v_i - \varphi_i)|^2 \leq \frac{3\delta}{\mu(S_i)} := \delta_i^2 \quad ,$$

hence by elliptic estimates

$$|\nabla(v_i - \varphi_i)| \leq C\delta_i \quad \text{in} \quad [r, r+1] \times [0,1] \quad ,$$

where the constant C does not depend on r. We obtain for $0 \leq z \leq 1$ and $r \geq R+1$, using the fact that $v_i = \varphi_i$ at $(R, \frac{1}{2})$,

$$|(v_i - \varphi_i)(r,z)| \leq C\delta_i(r-R) \quad ,$$

hence

$$\begin{aligned} v_2(r,z) &\geq \varphi_2(r,z) - C\delta_2(r-R) \\ &\geq (\sigma_{2+} - C\delta_2)\cdot(r-R) + (z - \frac{1}{2}) + v_2(R, \frac{1}{2}) \quad . \end{aligned}$$

With $M := |v_2(R, \frac{1}{2}) - v_1(R, \frac{1}{2})|$ and assumption 3.2 this is

$$\begin{aligned} &\geq (\sigma_{2+} - \sigma_{1+} - C\delta_2)(r-R) + \varphi_1(r,z) - M \\ &\geq (\sigma_{2+} - \sigma_{1+} - C(\delta_1 + \delta_2))(r-R) - M + v_1(r,z) \quad . \end{aligned}$$

Therefore, if we choose δ small enough, depending on S_1 and S_2, we find that

$$v_2(r,z) > v_1(r,z) \quad \text{for} \quad 0 \le z \le 1 \quad ,$$

if r is sufficiently large. By symmetry we obtain the same for large negative r. Then, if $v_2 \ge v_1$ does not hold everywhere, by continuity we find a point x_0 at which $v_1 - v_2$ attains its positive maximum. Since $v_1 - v_2 = 0$ on E the point x_0 lies in the flow region where v_i are harmonic. This contradicts the maximum principle. Consequently $v_2 \ge v_1$ in $\mathbb{R}^2$. Since S_1 and S_2 are arbitrary sets, the assertion follows. ∎

We shall also need the following stronger version.

3.4 Lemma. Let Q_1, Q_2 as in 3.2 , and $S_i \subset Q_i$ with $\mu(S_i) > 0$, and

$$v_i := \fint_{S_i} u_s d\mu(s) \quad .$$

Then for almost all $s \in S_1$ there is a constant $\kappa_s < 1$ with

$$u_s \le \kappa_s v_2 \quad \text{in} \quad \{0 < v_2 < \frac{1}{2}\} \quad ,$$
$$\kappa_s u_s \le v_2 \quad \text{in} \quad \{-\frac{1}{2} < v_2 < 0\} \quad .$$

A similar statement holds for v_1 and $s \in S_2$.

Proof: v_2 is continuous and $u_s \le v_2$ for almost all $s \in S_1$ by 3.3. We modify the proof of 3.3. Since $\nabla u - e$ is in $L^2(\mu; L^2(\Omega))$ we know that

$$\int_\Omega |\nabla u_s - e_s|^2 < \infty$$

for almost all $s \in S_1$. Then replacing v_1 by u_s in the proof of 3.3 we see that for R large enough (depending on s) and $|y| \ge R$

$$v_2(y,z) - u_s(y,z) \ge c(|y| - R) - C(R) \quad . \tag{3.1}$$

Therefore $u_s(y,z) \le 0$ if $0 \le v_2(y,z) \le \frac{1}{2}$ with $|y|$ large enough. Since $\{0 < v_2 \le \frac{1}{2}\}$ lies outside E both, u_s and v , are harmonic in this set with $u_s \le v$ by 3.3. Consequently, if $v_2(x) = \frac{1}{2}$, by the strong maximum principle in a neighborhood of x either $u_s \le \kappa v$ for some $\kappa < 1$ or $u_s = v_2$. If $u_s = v_2$ then by continuation this is true in the connected component of $\mathbb{R}^2 \backslash E$ containing x , which by the following lemma is unbounded, a contradiction to (3.1). Thus κ exists locally on the boundary of $D := \{0 < v_2 < \frac{1}{2}\}$ and far away in $\overline{D}$. Then the assertion follows from the maximum principle. ∎

The following Lemma holds without assumption 3.2.

3.5 Lemma: $\mathbb{R}^2 \backslash E$ is connected.

Proof: Assume that $\mathbb{R}^2 \backslash E$ has a bounded connected component D'. Let D'_∞ the unbounded component of $\mathbb{R}^2 \backslash D'$ and $D_0 := \mathbb{R}^2 \backslash \overline{D}'_\infty$. Then D_0 is simply connected and ∂D_0 is connected with $\partial D_0 \subset \partial D' \subset E$. Hence ∂D_0 is contained in one component of E . We can assume $\partial D_0 \subset E_0$. Then the shifted domains

$$D_k := \{(y,z) \in \mathbb{R}^2; (y, z-k) \in D_0\}$$

do not touch D_0 . We compare (u, E) with (v, F) , where

$$F := E \cup D \quad , \quad D := \bigcup_{k \in \mathbb{Z}} D_k \quad ,$$

$$v_s(x) := \begin{cases} u_s(x) & , \text{ if } x \in \mathbb{R}^2 \backslash D \quad , \\ k & , \text{ if } x \in \overline{D}_k, k \in \mathbb{Z} \quad . \end{cases}$$

Then, if (v, F) is admissible, we conclude that $J(u, E) > J(v, F)$, a contradiction. Therefore $\mathbb{R}^2 \backslash E$ consists of unbounded component, and in fact, do to the compactness of E^0 , there is only one such component.

To prove that (v, F) is admissible we first show that $u \cdot \chi_{D_0}$ belongs to $L^2(\mu; H^{1,2}(\mathbb{R}^2))$. Let ζ a bounded test function in this space, and η the cut-off function

$$\eta(x) := \begin{cases} min(1, \frac{1}{\delta} dist(x, \partial D_0)) & , \text{for } \quad x \in D_0 \quad , \\ 0 & , \text{elsewhere.} \end{cases}$$

Then

$$\begin{aligned} 0 &= \int_{\mathbb{R}^2} \int_{\mathbb{R}^2} (u \nabla(\eta\zeta) + \eta\zeta\nabla u) d\mathcal{L}^2 d\mu \\ &= \int_{\mathbb{R}^2} \int_{D_0} \eta(u\nabla\zeta + \zeta\nabla u) d\mathcal{L}^2 d\mu + \int_{\mathbb{R}^2} \int_{D_0} u\zeta\nabla\eta d\mathcal{L}^2 d\mu \quad . \end{aligned}$$

As $\delta \to 0$ the first term converges to the desired integral. Since u is Lipschitz continuous in mean, the second term is estimated by

$$\begin{aligned} &\leq C\frac{1}{\delta} \int_{D_0 \cap B_\delta(\partial D_0)} \int_{\mathbb{R}^2} |u_s(x)| d\mu(s) d\mathcal{L}^2(x) \\ &\leq C \cdot \mathcal{L}^2(D_0 \cap B_\delta(\partial D_0)) \to 0 \quad . \end{aligned}$$

Next we have to show that $v_s = k$ on E_*^k . If these sets ly outside D_0 there is nothing to prove. Assume that $E_*^{k_0} \cap D_0$ is nonempty. If $k_0 = 0$ then $v_s = 0$ on $E_*^0 \cap D_0$ by definition. If $k_0 \neq 0$ then, since $\partial D_0 \subset E^0$ and since $E_*^{k_0}$ is connected by assumption (1.8) we conclude that $E_*^{k_0} \subset D_0$. Now, if also $E_*^{k_1} \cap D_0$ is nonempty for some $k_1 \neq k_0$, then for every $k \in \mathbb{Z}$ the set D_k intersects both $E_*^{k_0+k}$ and $E_*^{k_1+k}$. We infer for $k = k_0 - k_1 \neq 0$ that

$$\emptyset \neq E_*^{k_0} \cap D_k \subset D_0 \cap D_k = \emptyset \quad ,$$

a contradiction. Therefore the only component of E_* touching D_0 is $E_*^{k_0}$, and it is contained in D_0. Then we see that $(v + k_0, F)$ is admissible, and again $J(u, E) > J(v + k_0, F)$. ∎

As a corollary of Lemma 3.5 we obtain

3.6 Lemma. If $x_0 \in E$ then $\mathcal{L}^2(E \cap B_r(x_0))$ for every $r > 0$.

Proof: Assume that $B_{r_0}(x_0) \subset E^0 \backslash E_*^0$ with center in E^0 and measure zero. Then the functions u_s are harmonic in $B_{r_0}(x_0)$. Choose $v_i, i = 1, 2$, as in Lemma 3.4. The proof of Lemma 3.3 says that $v_1 \leq v_2$ and $v_1(y, z) < v_2(y, z)$ for large $|y|$. Now $v_i(x_0) = 0$ and since both, v_1 and v_2 , are harmonic in $B_{r_0}(x_0)$, the maximum principle gives $v_1 = v_2$ in this ball. Further, by our assumption, x_0 is a boundary point of $\mathbb{R}^2 \backslash E$. Then unique continuation and Lemma 3.5 implies that $v_1 = v_2$ in $\mathbb{R}^2 \backslash E$, in particular far to the left, a contradiction. ∎

Now we are in a position to prove

3.7 Theorem. If $x_0 \in E$ then $\theta_*(E, x_0) > 0$.

Proof: If not, then some point $x_* \in E^0 \backslash E^0_*$ has zero lower density, that is, for a sequence $\rho \searrow 0$

$$\epsilon_\rho := \rho^{-2} \mathcal{L}^2 (E \cap B_{4\rho}(x_*)) \to 0 \quad .$$

We assume that $B_R(x_*) \cap E \subset E^0 \backslash E^0_*$ and $4\rho \leq R$. By 2.13 the scaled functions

$$u_{\rho s}(x) := \frac{1}{\rho} u_s(x_* + \rho x)$$

are uniformly Lipschitz in mean, and the measure of the scaled blade

$$E_\rho := \{x \in \mathbb{R}^2; x_* + \rho x \in E\}$$

within $B_4 = B_4(0)$ tends to zero. For a subsequence

$$u_\rho \to u_0 \quad \text{weakly in} \quad L^2(\mu; H^{1,2}_{loc}(\mathbb{R}^2)) \quad .$$

We claim that the functions u_{0s} are harmonic in B_4 . To prove this choose $v_s \in \mathring{H}^{1,2}(B_4)$ so that $u_{0s} + v_s$ is harmonic in B_4 . If

$$\delta := \int_{\mathbb{R}^2} (\int_{B_4} |\nabla u_0|^2 - \int_{B_4} |\nabla (u_0 + v)|^2) d\mu > 0 \quad ,$$

then for the scaled functional J_ρ

$$\begin{aligned} & J_\rho(u_\rho, E_\rho) - J_\rho(u_\rho + v, E_\rho \backslash B_4) \\ = & - \int_{\mathbb{R}^2} (\int_{B_4} (2\nabla u_\rho + \nabla v) \cdot \nabla v) d\mu - \int_{E_\rho \cap B_4} \lambda^2 \\ \to & \delta > 0 \quad \text{as} \quad \rho \to 0 \quad , \end{aligned}$$

for small ρ a contradiction to the minimizing property of u.

The idea of the proof now is the following. We split $\mathbb{R}^2$ into sets Q_1 and Q_2 as in 3.2 and let v_i the mean of u over Q_i. Then $v_1 \leq v_2$ in $\mathbb{R}^2$, and , say, $v_1 \leq (1-\delta) v_2$ in $\{0 < v_2 < \frac{1}{2}\}$ for some $\delta > 0$. Then these estimates carry over to the limits v_{i0} of the blow-up sequence $v_{i\rho}$, which as above are harmonic functions. Then, if one of the blow-up's is nontrivial, say, $v_{20}(x_0) > 0$ at some point x_0 , we would have derived a contradiction to the strong maximum principle. In principle non-trivial blow-up's should be a consequence of the non-degeneracy of one of the sequences $v_{1\rho}$ or $v_{2\rho}$. However, the difficulty which arises is, that although non-degeneracy holds in the sense of 2.14, the functions $v_{i\rho s}$ might oscillate in space when s varies, and therefore the blow-up limits v_{i0} might vanish everywhere.

To overcome this, we replace the blow-up sequence u_ρ by

$$\tilde{u}_{\rho s} := \sigma_{\rho s} u_{\rho s} \quad ,$$

with measurable functions σ_ρ having values 0 or 1 only. Then as above, for a subsequence, $\tilde{u}_\rho$ converges weakly to $\tilde{u}_0$. Again $\tilde{u}_{0s}$ are harmonic functions. Indeed, if v_s is defined as above according to $\tilde{u}_{0s}$, then

$$\int_{\mathbb{R}^2}\int_{B_4}(|\nabla u_\rho|^2 - |\nabla(u_\rho + \sigma_\rho v)|^2)d\mathcal{L}^2 d\mu$$

$$= - \int_{\mathbb{R}^2}\int_{B_4}(2\nabla\tilde{u}_\rho + \sigma_\rho^2 \nabla v)\nabla v d\mathcal{L}^2 d\mu$$

$$\geq - \int_{\mathbb{R}^2}\int_{B_4}(2\nabla\tilde{u}_\rho + \nabla v)\nabla v d\mathcal{L}^2 d\mu \quad ,$$

and we derive a contradiction as above.

Now let us assume that

$$w_0(x_0) > 0 \quad \text{for some} \quad x_0 \in B_4 \quad , \tag{3.2}$$

where w_0 is the Lipschitz function

$$w_0 := \int_{\mathbb{R}^2} |\tilde{u}_{0s}| d\mu(s) \quad .$$

We shall prove, that this assumption leads to a contradiction. First we remark, that the similarly defined functions w_ρ are uniformly Lipschitz continuous, therefore they have a uniform limit $\tilde{w}_0$. Then the weak convergence of $\tilde{u}_\rho$ to $\tilde{u}_0$ implies that $w_0 \leq \tilde{w}_0$. We conclude, that there is a neighborhood $B_{r_0}(x_0)$ in which $w_\rho \geq c > 0$ for small ρ. In particular, all sets E_ρ ly outside this ball. Consequently all functions we consider are harmonic in this ball, and the values $\tilde{u}_{\rho s}(x_0)$ and $\tilde{u}_{0s}(x_0)$ are well defined, e.g., as mean value over $B_{r_0}(x_0)$. This implies that for every μ-measurable set $A \subset \mathbb{R}^2$

$$\int_A \tilde{u}_{\rho s}(x_0) d\mu(s) \to \int_A \tilde{u}_{0s}(x_0) d\mu(s) \quad .$$

Also our assumption (3.2) means that $\tilde{u}_{0s}(x_0) \neq 0$ for s in a set of positive measure. For definiteness we assume that

$$S_0 := \{s \in \mathbb{R}^2 \quad ; \quad \tilde{u}_{0s}(x_0) > 0\} \quad \text{has positive measure} \quad . \tag{3.3}$$

First we remark, that keeping (3.3) we can assume that

$$\sigma_{\rho s} = 0 \quad , \quad \text{if} \quad u_{\rho s}(x_0) \leq 0 \quad . \tag{3.4}$$

Indeed, if we define $u'_{\rho s} := \sigma'_{\rho s} u_{\rho s}$ with

$$\sigma'_{\rho s} := \begin{cases} \sigma_{\rho s} & ,\text{if} \quad u_{\rho s}(x_0) > 0 \quad , \\ 0 & ,\text{if} \quad u_{\rho s}(x_0) \leq 0 \quad , \end{cases}$$

then $u'_{\rho s}(x_0) \geq \tilde{u}_{\rho s}(x_0)$. Denote by u'_0 a weak limit of u'_ρ . Then for every set $A \subset S_0$

$$\int_A u'_0(x_0) d\mu = \lim_{\rho \to 0} \int_A u'_\rho(x_0) d\mu \geq \int_A \tilde{u}_0(x_0) > 0 \quad ,$$

that is, $u'_{0s}(x_0) > 0$ for almost all $s \in S_0$, or $S_0 \subset S'_0$.

Thus we can work with (3.3) and (3.4). According to (3.3) and assumption 3.2 we choose rectangles Q_1 and Q_2 such that, say,

$$\mu(S) > 0 \quad \text{with} \quad S := S_0 \cap Q_2 \quad , \tag{3.5}$$

and define

$$v := \fint_{Q_1} u_s d\mu(s) \quad .$$

We denote by v_ρ the corresponding blow-up sequence with uniform limit v_0 (choosing again a subsequence) . Since for every $w \in \mathring{H}^{1,2}(B_4)$

$$\fint_S \int_{B_4} (|\nabla u_{\rho s}|^2 - |\nabla(u_{\rho s} + w)|^2) d\mathcal{L}^2 d\mu(s) = \int_{B_4} (|\nabla v_\rho|^2 - |\nabla(v_\rho + w)|^2) d\mathcal{L}^2$$

wee see as above, that v_0 is harmonic in B_4. Now, since $v \le u_s$ for almost all $s \in S$ by Lemma 3.3 , we have

$$v_\rho \le u_{\rho s} \quad ,$$

and therefore

$$\fint_S \sigma_{\rho s} d\mu(s) v_\rho \le \fint_S \tilde{u}_{\rho s} d\mu(s) \quad .$$

We obtain

$$\theta v_0 \le \fint_S \tilde{u}_0 d\mu \quad , \tag{3.6}$$

where

$$\theta := \liminf_{\rho \to 0} \fint_S \sigma_s \quad .$$

Since both sides in (3.6) are harmonic functions vanishing at the origin, they conicide, in particular

$$\theta v_0(x_0) = \fint_S \tilde{u}_0(x_0) d\mu > 0 \quad , \tag{3.7}$$

that is, $\theta > 0$.

Using Lemma 3.4 instead of 3.3 we find $0 < \delta_s < 1$ with

$$(1 + \delta_s) v \le u_s \quad \text{in} \quad \{0 \le v \le \frac{1}{2}\} \quad ,$$

and therefore as above

$$\fint_S \sigma_{\rho s}(1 + \delta_s) d\mu(s) \cdot v_\rho \le \fint_S \tilde{u}_{\rho s} d\mu(s) \quad \text{in} \quad \{0 \le v_\rho \le \frac{1}{2\rho}\} \quad . \tag{3.8}$$

Since v_ρ converges uniformly to v_0 and $v_0(x_0) > 0$ by (3.7), we know that $v_\rho > 0$ in a fixed neighborhood of x_0 , say $B_{r_0}(x_0)$, provided ρ is small enough. We conclude that (3.8) is satisfied in this ball. Letting $\rho \to 0$ we now obtain

$$\theta' v_0 \le \fint_S \tilde{u}_{0s} d\mu(s) \quad ,$$

where

$$\theta' := \limsup_{\rho\to 0} \fint_S \sigma_{\rho s}(1+\delta_s)d\mu(s)$$

and using (3.7) we derive $\theta' \le \theta$, which is impossible. To see this we note that $\mu(S\backslash A_\epsilon) \to 0$ if

$$A_\epsilon := \{s \in S \quad ; \quad \delta_s \ge \epsilon\} \quad .$$

Then for $\rho \to 0$

$$\begin{aligned}\int_S \sigma_{\rho s}\delta_s d\mu(s) &\ge \epsilon\Big(\int_S \sigma_{\rho s} d\mu(s) - \mu(S\backslash A_\epsilon)\Big)\\ &\to \epsilon(\theta - \mu(S\backslash A_\epsilon)) \le \frac{\epsilon}{2}\theta \quad ,\end{aligned}$$

if ϵ is small enough, and therefore $\theta' \ge (1+\frac{\epsilon}{2})\theta$.

Now, in the analytic part of the proof, we have to verify (3.2), that is, we have to find appropriate functions σ_ρ. First let us consider one phase functions, that is,

$$T_\rho^\pm := \{s \in \mathbb{R}^2 \quad ; \quad \pm u_s \ge 0 \quad \text{in} \quad B_{\rho/2}(x_*)\} \quad .$$

For $s \in T_\rho^+$ we compare u_s with its harmonic continuation w_s in $B_{\rho/2}(x_*)$. As in the proof of theorem 2.12 we get

$$\begin{aligned}C\cdot\epsilon_\rho &\ge \int_{B_{1/2}} \lambda^2\chi_{E_\rho} \ge \int_{T_\rho^+}\int_{B_{1/2}} (|\nabla u_{\rho s}|^2 - |\nabla w_{\rho s}|^2)d\mu(s)\\ &= \int_{T_\rho^+}\int_{B_{1,2}} |\nabla(u_{\rho s} - w_{\rho s}|^2 d\mu(s) \ge C\cdot \int_{T_\rho^+} \mu_{\rho s}^+(B_{1/3})^2 d\mu(s)\end{aligned} \tag{3.9}$$

where $\mu_{\rho s}^+ = \Delta u_{\rho s}$ in $B_{1/2}$ in distributional sense. The last integral controls Dirichlet's integral as follows. For $0 < r_0 < \frac{1}{3}$ we use the representation in 2.10.2 for $B_{1/3}$ and B_{r_0} and obtain for almost all $s \in T_\rho^+$

$$\begin{aligned}\fint_{\partial B_{1/3}} u_{\rho s} - \fint_{\partial B_{r_0}} u_{\rho s} &= \int_{B_{1/3}} log\frac{1}{3|x|}d\mu_{\rho s}^+(x) - \int_{B_{r_0}} \log\frac{r_0}{|x|}d\mu_{\rho s}^+(x)\\ &\le log\frac{1}{3r_0}\cdot \mu_{\rho s}^+(B_{1/3}) \quad ,\end{aligned}$$

and integrating over s

$$\begin{aligned}\int_{T_\rho^+}\Big(\fint_{\partial B_{1/3}} u_{\rho s}\Big)^2 d\mu(s) &\le 2log^2\frac{1}{3r_0}\cdot\int_{T_\rho^+} \mu_{\rho s}^+(B_{1/3})^2 d\mu(s)\\ &+2\int_{\mathbb{R}^2}\Big(\fint_{\partial B_{r_0}} |u_{\rho s}|\Big)^2 d\mu(s) \quad .\end{aligned}$$

Using the Lipschitz continuity in 2.12 and (3.9) this is estimated by

$$\begin{aligned}&\le C(\epsilon_\rho\cdot log^2\frac{1}{r_0} + r_0^2)\\ &\le C\cdot\epsilon_\rho\cdot log^2\frac{1}{\epsilon_\rho} =: \tilde{\epsilon}_\rho \quad ,\end{aligned}$$

if we choose $r_0 = \sqrt{\epsilon_\rho}$. Then we estimate (using 2.9.2)

$$
\begin{aligned}
\int_{B_{1/5}} |\nabla u_{\rho s}|^2 &\leq C \int_{1/5}^{1/4} \frac{1}{r}\Big(\int_{B_r} |\nabla u_{\rho s}|^2 \Big) dr \\
&= C \int_{1/5}^{1/4} \frac{1}{r}\Big(\int_{\partial B_r} \partial_r \Big(\frac{u_{\rho s}^2}{2}\Big)\Big) dr \qquad (3.10)\\
&\leq C \int_{\partial B_{1/4}} u_{\rho s}^2 \leq C (\sup_{B_{1/4}} u_{\rho s})^2 \quad .
\end{aligned}
$$

Using that $u_{\rho s}$ is subharmonic (see 2.10.1) and combining it with the above estimates we derive that

$$
\int_{T_\rho^+} \Big(\int_{B_{1/5}} |\nabla u_{\rho s}^+|^2 \Big) d\mu(s) \leq C \cdot \tilde{\epsilon}_\rho \quad . \tag{3.11}
$$

A corresponding estimate hods for T_ρ^-. We wish to have the same estimate for nearly one-phase functions, that is, for two-phase functions with a dominating phase. By this we mean, that for given small $\kappa > 0$ we work on the set

$$
S_\rho^+ := \{ s \in \mathbb{R}^2 \backslash T_\rho^+ \backslash T_\rho^- \quad ; \quad \|\nabla u_{\rho s}^-\|_{L^2(B_4)} \leq \kappa \|\nabla u_{\rho s}^+\|_{L^2(B_1)} \} \quad ,
$$

and on the similar defined set S_ρ^-. Again following the proof of Theorem 2.12, we compare u in $B_{4\rho}(x_*)$ with v defined by

$$
v_s := \begin{cases} -u_s^- + w_s & \text{for} \quad s \in S_\rho^+ \quad , \\ u_s & \text{for} \quad s \notin S_\rho^+ \quad , \end{cases}
$$

where w_s is the harmonic continuation of u_s^+ in this ball. We obtain

$$
C \cdot \epsilon_\rho \geq c \int_{S_\rho^+} \mu_{\rho s}^+(B_3)^2 d\mu(s) - C \int_{S_\rho^+} \int_{B_4} |\nabla u_{\rho s}^-|^2 d\mu(s) \quad . \tag{3.12}
$$

Again we control the first integral on the right by Dirichlet's integral for u_ρ^+. For $s \in S_\rho^+$ the negative phase reaches $B_{1/2}$. Therefore (as in the proof of Theorem 2.11) for $1 < r < 2$ there is a point $x_r \in \partial B_r \backslash E$ with $u_{\rho s}(x_r) < 0$, that is, $u_{\rho s}^+ = 0$ in neighborhood. Then

$$
\fint_{\partial B_3} u_{\rho s}^+ \leq C \int_{\partial B_3} P_{x_r} u_{\rho s}^+ = C \int_{B_3} G_{x_r} d\mu_{\rho s}^+ \quad .
$$

Taking the mean over r we get

$$
\fint_{\partial B_3} u_{\rho s}^+ \leq C \int_{B_3} \Big(\int_1^2 G_{x_r} dr \Big) d\mu_{\rho s}^+ \leq C \mu_{\rho s}^+(B_3) \quad .
$$

Combining this with an estimate like (3.10) for $u^+_{\rho s}$ and with (3.12), we see that if κ is chosen small enough

$$\int_{S^+_\rho} \Big(\int_{B_1} |\nabla u^+_{\rho s}|^2 \Big) d\mu(s) \leq C \cdot \epsilon_\rho \leq C\tilde{\epsilon}_\rho \quad . \tag{3.13}$$

The same estimate holds for S^-_ρ .

This, together with (3.11), means that if the measure of the conicidence set is small, non-degeneracy is caused, as intuively expected, by balanced two phase flows only. Therefore, essentially we are left with

$$S_\rho := \mathbb{R}^2 \backslash T^+_\rho \backslash T^-_\rho \backslash S^+_\rho \backslash S^-_\rho \quad .$$

With the notation in 2.11 we have for $s \in S_\rho$

$$\varphi^\pm(\rho) \leq C\sqrt{\varphi(\rho)} \leq C\sqrt{\varphi(R)} \quad . \tag{3.14}$$

The last inequality is Theorem 2.11. We should remark, that this is the point, where we need that x_* is fixed, that is, does not depend on ρ. In other words, if only one-phase flows are involved, our estimate on the density is uniform with respect to x_*.

Now we make use of the non-degeneracy. First for any set $S \subset \mathbb{R}^2$, almost all $x \in \partial B_{1/5}$, and $r_0 > 0$ small

$$|u_{\rho s}(x)| \leq |x| \int_{r_0}^{1} |\nabla u_{\rho s}(rx)| dr + |u_{\rho s}(r_0 x)|$$

for almost all $s \in S$, hence

$$⨍_{\partial B_{1/5}} |u_{\rho s}| \leq \frac{1}{r_0} \int_{B_{1/5} \backslash B_{r_0}} |\nabla u_{\rho s}| + ⨍_{\partial B_{r_0}} |u_{\rho s}| \quad .$$

Thus using the Lipschitz continuity

$$\int_S \Big(⨍_{\partial B_{1/5}} |u_{\rho s}| \Big)^2 d\mu \leq \frac{C}{r_0^2} \int_S \Big(\int_{B_{1/5}} |\nabla u_{\rho s}|^2 \Big) d\mu + C r_0^2 \quad . \tag{3.15}$$

Setting $S = \mathbb{R}^2 \backslash S_\rho$ and using (3.11), (3.13), and the definition on $S^\pm_\rho$ we obtain

$$\int_{\mathbb{R}^2 \backslash S_\rho} \Big(⨍_{\partial B_{1/5}} |u_{\rho s}| \Big)^2 d\mu \leq C \cdot \frac{\tilde{\epsilon}_\rho}{r_0^2} + C r_0^2 \leq C \tilde{\epsilon}_\rho^{1/2} \quad ,$$

if $r_0 = \tilde{\epsilon}_\rho^{1/4}$. Together with the non-degeneracy in 2.14 we obtain for small ρ

$$\int_{S_\rho} \Big(⨍_{\partial B_{1/5}} |u_\rho| \Big)^2 d\mu \geq c > 0 \quad . \tag{3.16}$$

Then setting $S = S_\rho$ in (3.15) , and choosing r_0 small enough we obtain

$$\int_{S_\rho} (\int_{B_{1/5}} |\nabla u_\rho|^2) d\mu \geq c > 0 \quad . \tag{3.17}$$

We modify S_ρ a little. Since $\varphi(R)^{1/2} \in L^1(\mu)$ we can choose m large so that with

$$N := \{s \in I\!R^2 \quad ; \quad \varphi_s(R) > m^2\}$$

the integral

$$\delta_m := \int_N \varphi_s(R)^{1/2} d\mu(s)$$

is small. Then by (3.14)

$$\int_{S_\rho \cap N} (\int_{B_1} |\nabla u_\rho|^2) d\mu = \int_{S_\rho \cap N} (\varphi_s^+(\rho) + \varphi_s^-(\rho)) d\mu(s) \leq C\delta_m \quad .$$

Therefore, if m is large enough , (3.17) also holds for $\tilde{S}_\rho := S_\rho \backslash N$, and with (3.15) we obtain the same for (3.16). By definition of $\tilde{S}_\rho$ we have for $s \in \tilde{S}_\rho$

$$\int_{B_1} |\nabla u_{\rho s}|^2 \leq C \cdot \varphi_s(R)^{1/2} \leq C \cdot m \leq C \quad ,$$

since m is fixed. Now $u_{\rho s}$ has two phases down to $B_{1/2}$, that is, for $1/2 < r < 1$ both phases vanish at certain intervals on ∂B_r. Therefore we have a Poincaré inequality for $u_{\rho s}^\pm$ in B_1. Using that $u_{\rho s}^\pm$ are subharmonic we obtain

$$\begin{aligned} \fint_{\partial B_{1/5}} u_{\rho s}^\pm &\leq \sup_{B_{1/5}} u_{\rho s}^\pm \leq C ||u_{\rho s}^\pm||_{L^2(B_1 \backslash B_{1/2})} \\ &\leq C ||\nabla u_{\rho s}^\pm||_{L^2(B_1)} \leq C \quad . \end{aligned}$$

Thus the estimate (3.16) for $\tilde{S}_\rho$ also holds in $L^1(\mu)$ instead of $L^2(\mu)$, that is ,

$$\int_{\tilde{S}_\rho} (\fint_{\partial B_{1/5}} |u_\rho|) d\mu \geq c > 0 \quad .$$

Then, for each ρ, this is true either for u_ρ^+ or for u_ρ^-. For definiteness let us take a subsequence so that it holds for u_ρ^+ , that is,

$$\fint_{\partial B_{1/5}} v_\rho \geq c > 0, \quad \text{if} \quad v_\rho := \int_{\tilde{S}_\rho} u_\rho^+ d\mu \quad .$$

This estimate allows us to construct σ_ρ . We choose points $x_\rho \in \partial B_{1/5}$ with $v_\rho(x_\rho) \geq c$, and define

$$\sigma_{\rho s} := \begin{cases} 1 \quad , & \text{if} \quad s \in \tilde{S}_\rho \quad \text{and} \quad u_\rho(x_\rho) > 0 \quad , \\ 0 \quad , & \text{if not} \quad . \end{cases}$$

Further let $\tilde{u}_\rho$ as in the first part of the proof,

$$w_\rho := \int_{\mathbb{R}^2} \tilde{u}_\rho d\mu, \quad \text{and} \quad w := \int_{\mathbb{R}^2} \tilde{u}_0 \quad .$$

Then $w_\rho(x_\rho) = v_\rho(x_\rho) \geq c$, and by the Lipschitz continuity

$$w_\rho(x) \geq c - C\delta \geq c > 0, \quad \text{if} \quad |x - x_\rho| \leq \delta \quad ,$$

and δ small enough. For a subsequence x_ρ converges to $x_0 \in \partial B_{1/5}$, so that for small ρ

$$w_\rho \geq c \quad \text{in} \quad B_{\delta/2}(x_0) \quad .$$

Since $\tilde{u}_\rho$ converges weakly to $\tilde{u}_0$, the same is true for w . Thus (3.2) is satisfied. However the first part of the proof shows that this is not possible. ∎

An assumption like 3.2 is necessary for the result in 3.7. For example, if μ is the sum of two Dirac measures, say, at $s_1, s_2 \in \mathbb{R}^2$, it is possible to arrange the data and these two speeds at infinity, so that a solution (u, E) with a singular point of E exists.

3.8 Remark. As a consequence of the previous theorem almost all functions u_s are continuous in $\mathbb{R}^2$. This is well known. In fact, if $x_0 \in E^0$ and for small $\rho > 0$

$$\psi(\rho) := \fint_{\partial B_\rho} u_s^+ d\mathcal{H}^1$$

then from 2.10.2

$$\begin{aligned} c\psi(\rho) &\leq \fint_{B_{\rho/2} \cap E^0} \Big(\int_{\partial B_\rho} P_x u_s^+ \Big) dx \\ &\leq \frac{C \cdot \rho^2}{\mathcal{L}^2(B_{\rho/2} \cap E^0)} \mu_s^+(B_\rho) \leq C(x_0) \mu_s^+(B_\rho). \end{aligned}$$

Then again from 2.10 for almost all $\frac{\rho}{4} \leq r \leq \frac{\rho}{3}$

$$\psi(r) \leq \psi(\rho) - c\mu_s^+(B_{\rho/2}) \leq \psi(\rho) - c(x_0)\psi(\frac{\rho}{2}) \quad .$$

Since u_s^+ is subharmonic, ψ is nondecreasing, and we obtain that

$$\psi(\frac{\rho}{4}) \leq \frac{1}{1 + c(x_0)} \psi(\rho) \quad .$$

4. Remark on existence

In section 3 we have proved, at least in a weak sense, that the solutions we obtain lead to blades without singularities, therefore the goal of the formulation of the problem in section 1 is reached. But so far we don't know that there are solutions of the full free boundary problem, that is, with E_* in the interior of E. We even did not proof that there is a free boundary at all.

We now will give a sufficient condition, which prevents the free boundary from touching E_*. Let $\rho_0 > 0$ and η_0 a cut-off function with $\eta_0 = 0$ in $B_{\rho_0}(E_*)$ and $\eta_0 = 1$ outside $B_{2\rho_0}(E_*)$, and let

$$\lambda = \eta_0 \lambda_0 + (1 - \eta_0)\lambda_1 \quad ,$$

where λ_0 is fixed, and eventually λ_1 very large. Consider a minimizer (u, E). First we control the functional. For this choose ρ_0 small enough so that $B_{3\rho_0}(E_*^0) \cap E_* \subset E_*^0$, and let v a function as in 2.2.1 with $v = 0$ on $F := \overline{B_{2\delta_0}(E_*^0)}$. Then (v, F) is adimissible and as in 2.2

$$\begin{aligned} &\int_{\mathbb{R}^2} \Big(\int_{\Omega\cap\{|y|\le R_0\}} |\nabla u - e|^2 \Big) d\mu + \delta_0 \int_{\Omega\cap\{|y|\ge R_0\}} \lambda^2 \chi_E \\ &\le J(u, E) + \int_{\Omega\cap\{|y|\le R_0\}} \lambda^2 \chi_E \\ &\le \int_{\mathbb{R}^2} \Big(\int_{\Omega} |\nabla v - e|^2 \Big) d\mu + \int_{\Omega\cap\{|y|\le R_0\}} \lambda^2 (\chi_E - \chi_F) \\ &\le C + \int_{\Omega\setminus F\cap\{|y|\le R_0\}} \lambda_0^2 \chi_E \le C \quad , \end{aligned}$$

where C is independent of λ_1. Using Ponicare's inequality with $u = 0$ on E_*^0 we obtain that u is estimated in $L^2(\mu; H^{1,2}(\Omega \cap \{|y| \le R_0\}))$ uniformly in λ_1. Then, if $x_0 \in \partial E_*^0$,

$$\int_{\mathbb{R}^2} \Big(\fint_{\partial B_{\rho_0}(x_0)} u_s^{\pm} \Big)^2 d\mu(s) \le C \int_{\mathbb{R}^2} \Big(\fint_{B_{2\rho_0}(x_0)} |u_s^{\pm}|^2 \Big) d\mu(s) \le C(\rho_0) \quad .$$

Therefore, if λ_1 is large enough, Lemma 2.14 implies $B_{\rho_0/2}(E_*^0) \subset E^0$, that is, the free boundary stays away from E_*.

References

[1] N.Aguilera, H.W.Alt, L.A.Caffarelli: An Optimization problem with volume constraint. SIAM J. Control Opt. 24, 191 - 198 (1986)

[2] H.W.Alt, L.A.Caffarelli: Existence and regularity for a minimum problem with free boundary. J. Reine Angew.Math. 105, 105 - 144 (1981)

[3] H.W.Alt, L.A.Caffarelli, A.Friedman: Variational problems with two phases and their free boundaries. Trans. Amer. Math. Soc. 282, 431 - 461 (1984)

[4] H.W.Alt, L.A.Caffarelli, A.Friedman: A free boundary problem for quasilinear elliptic equations. Ann. Scuola Norm. Sup. Pisa 11(4), 1 - 44 (1984)

[5] H.W.Alt, L.A.Caffarelli, A.Friedman: Compressible flows of jets and cavities. J. Diff. Equ. 56, 82 - 141 (1985)

[6] M.Feistauer: On irrotational flows through cascades of profiles in a layer of variable thickness. Aplikace Matematiky 29, 423 - 458 (1984)

[7] M. Feistauer: Finite element solution of n-viscous flows in cascades of blades. Z. Angew. Math. u. Mech. 65, T191 - T194 (1985)

[8] M.B. Wilson, R.Mani, A.J.Acosta: A note on the influence of axial velocity ratio on cascade performance. Proceedings of "Theoretical prediction of two - and three - dimensional flows in turbomachinery ", Pennsylvania State Univ. (1974)

FREE BOUNDARY PROBLEMS FOR THE NAVIER-STOKES EQUATIONS

Josef Bemelmans

Abstract: A free boundary problem for the Navier-Stokes equations describes the flow of a viscous, incompressible fluid in a domain that is unknown or partially unknown. In this paper several results for flows in drops or in vessels are presented. The free boundary is governed by self-attraction or surface tension, and dynamic contact angles may occur.

AMS-Classification: 76 D 05 , 35 R 35

§ 1. The Equations of Motion

To determine the shape of a fluid body is a classical problem in mathematical physics. If the liquid rotates about a fixed axis and is moreover subject to self-attraction the problem was already investigated by I.Newton as a model for the figure of the earth. Since it was treated for the first time in the Philosophiae Naturalis Principia Mathematica 300 years ago it has stimulated research in various branches of mathematical analysis as for example potential theory, bifurcation theory for nonlinear integral equations, and more recently it was taken up again in connection with variational methods for free boundary problems, see e.g. Friedman [F2] Chap.4.
According to Newton's law the force of self-attraction equals

$$DU(x) = D \int_{\Omega} \frac{\rho g}{|x-y|}\, dy ,$$

where ρ = const in the density, $\Omega \subset \mathbb{R}^3$ the domain occupied by the fluid, and g the gravitational constant. If the body rotates about the x^3-axis the centrifugal force is

$$DR(x) = D \frac{\omega^2}{2} r^2(x) ,$$

where ω denotes the angular velocity and $r(x) = \left[(x^1)^2 + (x^2)^2\right]^{1/2}$ the distance of a point x from the axis of rotation. With no other forces present the boundary Σ of Ω must be an equipotential surface:

$$(1.1)\qquad \int_{\Omega} \frac{\rho g}{|x-y|}\, dy + \frac{\omega^2}{2} r^2(x) = \text{const} \quad \forall x \in \Sigma .$$

As the total mass is prescribed, too, we have the side condition

$$(1.2)\qquad \text{meas}\, \Omega = V_o .$$

Relation (1.1) can easily be modified to cover other physical situations, too, like compressible fluids, figures with prescribed angular momentum or variable angular velocity. In this sense (1.1) can be regarded as the basis for all investigations on equilibrium figures if treated as problems in hydrostatics.

A related problem concerns rotating drops that are held together by surface tension rather than self-attraction. The boundary is now determined by

$$(1.3)\qquad 2\kappa H(x) + \rho\omega^2 r^2(x) = \text{const} \quad \forall x \in \Sigma ,$$

where $H(x)$ denotes the mean curvature of Ω at x, and κ is a material constant. If Σ is assumed to be of a specific topological type, (1.3) can be transformed into a differential equation for a scalar function whose graph is Σ. Solutions of the topological type of the sphere were investigated by E.Hölder [H]; toroidal figures are treated by R.Gulliver [G].

In §§3,4 we present some of the author's work on free boundary problems for the Navier-Stokes equations that can be regarded as dynamical versions of (1.1) or (1.3) because now we allow relative motions inside the fluid body. For a viscous and incompressible fluid a stationary flow inside the unknown domain Ω is governed by the following equations

$$(1.4)\qquad \begin{aligned} -\nu\Delta v + Dp + (v\cdot D)v &= f \qquad \text{in } \Omega \\ \text{div}\, v &= 0 \end{aligned}$$

$$(1.5)\qquad v\cdot n = 0 , \quad t_k\cdot T(v,p)\cdot n = 0 \qquad \text{on } \Sigma$$

together with one of the following conditions on the free boundary

$$(1.6)\qquad n\cdot T(v,p)\cdot n = 2\kappa H \qquad \text{on } \Sigma ,$$

or

$$(1.7)\qquad n\cdot T(v,p)\cdot n = 0 \qquad \text{on } \Sigma ,$$

depending whether surface tension is present or not. Here v and p denote the velocity and the pressure at x, ν is the kinematic viscosity, and $T(v,p)$ the stress tensor

$$(1.8)\qquad T_{ij}(v,p) = -p\delta_{ij} + \nu\{D_i v^j + D_j v^i\} .$$

The exterior normal to Σ is denoted by n, and t_1, t_2 span the tangent plane. The exterior force density f that generates the motion will be specified later.

It is easily checked that for $v = 0$ (with respect to a suitable reference frame) the free boundary problem (1.4) - (1.6) reduces to

(1.1) if we only set $f = DU$, and similarly for (1.4), (1.5) and (1.7). The physical assumption then is that the hydrostatic pressure is replaced by the normal stress if we pass from a static to a dynamic problem.

In the analogues to (1.1) and (1.3) the fluid occupies a bounded domain Ω , and its boundary Σ is a closed surface. The methods can be extended to treat also a layer of fluid where the capillary surface is a graph over all of $\mathbb{R}^2$. In these problems there is no contact between the free boundary and a rigid wall, which means that contact angle phenomena are excluded. For such a situation, namely the steady flow in a capillary tube that is partly filled with liquid the free boundary value problem was first solved by D.H.Sattinger [S]; he assumed the contact angle under which the free surface meets the wall of the cylindrical tube to be $\pi/2$. More general angles were studied by V.A.Solonnikov [SO]. In §5 we present some of the recent results obtained by D.Kröner [K] who studies the following two-dimensional problem:

$$(1.9)\qquad \begin{cases} -\nu\Delta v + Dp + (v\cdot D)v = 0 \\ \operatorname{div} v = 0 \end{cases} \qquad \text{in } G$$

$$(1.10)\qquad v\cdot n = 0 \qquad \text{on } \partial G$$

$$(1.11)\qquad \begin{cases} t\cdot T(v,p)\cdot n = 0 \\ n\cdot T(v,p)\cdot n = -\kappa H \end{cases} \qquad \text{on } \Sigma$$

$$(1.12)\qquad \nu D_1 v^2 + \gamma_o v^2 = 0 \qquad \text{on } \Gamma_o$$

$$(1.13)\qquad \nu D_2 v^1 + \gamma v^1 = -\gamma S \qquad \text{on } \Gamma$$

$$(1.14)\qquad g(1) = 0 \ , \ g(0) = 0 \ .$$

The domain G which is occupied by the fluid is given as

$$(1.15)\qquad G = \{(x^1,x^2) \in \mathbb{R}^2 : 0 < x^2 < 1 \ ; \ g(x^2) < x^1 < x^1_o\} \ ,$$

and its boundary consists of

$$(1.16)\qquad \begin{cases} \Gamma_o = \{(x^1,x^2) \in \mathbb{R}^2 : 0 < x^2 < 1 \ , \ x^1 = x^1_o\} \\ \Gamma = \{(x^1,x^2) \in \mathbb{R}^2 : 0 < x^1 < x^1_o \ , \ y \in \{0,1\}\} \\ \Sigma = \{(x^1,x^2) \in \mathbb{R}^2 : 0 < x^2 < 1 \ , \ x^1 = g(x^2)\} \ . \end{cases}$$

The domain G is bounded by a capillary surface Σ that is given as the graph of a function g , and by rigid walls Γ and Γ_o . It is assumed that Γ_o moves with constant velocity S through the infinite cylinder $\{(x^1,x^2) \in \mathbb{R}^2 : -\infty < x^1 < \infty \ , \ 0 < x^2 < 1\}$ and pushes the fluid into the negative x^1-direction; the boundary-value problem (1.9) - (1.14) then describes this flow in a coordinate system that moves with fluid. Of particular interest is the contact angle φ at (0,0) and (0,1) , especially how it depends on S and on the boundary conditions on Γ . In (1.9) γ is a friction coefficient, and therefore (1.13) is derived under the assumption that there is a force on Γ which is proportional to the tangential velocity. If one imposes Dirichlet data on Γ , i.e. $v^1 = -S$, then only for $\varphi = 0$ and $\varphi = \pi$ physically reasonable solutions $v \in H_2^1(G)$ may exist, cf. V.V.Puchnachev - V.A.Solonnikov [PS] .

§ 2. Approximation schemes

Free boundary problems to the Navier-Stokes equations have been solved so far only under the assumption of small data. This contrasts the situation in fixed domains where according to Leray's existence theorem at least one solution exists to arbitrary data; the proof is based on an a priori bound for Dirichlet's integral $\int_\Omega |Dv|^2 dx \le C(\nu,\Omega,f,v^*)$ which holds for any solution to (1.4) that satisfies $v = v^*$ on $\partial\Omega$. If instead of Dirichlet data v^* a condition of Neumann type is imposed, Dirichlet's integral may not be finite any longer; a counter-example has been given by T.A.McCready [M] who showed the existence of solutions (v_n,p_n) to the Navier-Stokes equations $-\Delta v_n + D_{p_n} + \lambda_n(v_n\cdot D)v_n = 0$, $\operatorname{div} v_n = 0$ in $\Omega = \{(x^1,x^2) \in \mathbb{R}^2 : r^2 < (x^1)^2 + (x^2)^2 < \mathbb{R}^2\}$ such that $\int_\Omega |Dv_n|^2 dx \longrightarrow \infty$ as $\lambda_n \longrightarrow \infty$.

Although this example depends strongly on the fact that the underlying domain is an annulus and therefore does not apply to the situations considered in this paper it suggests that a global estimate for $\int_\Omega |Dv|^2 dx$ does not hold in general. Also physically it seems quite plausible because Dirichlet's integral measures the deformation energy, and by imposing $v = v^*$ on $\partial\Omega$ one assumes that the rigid boundary can resist arbitrary large stresses. For a drop as considered before, however, large stresses might result in large deformations of the shape and eventually the drop might break, a phenomenon that was already investigated by J.Plateau [P]. Clearly this implies that certain norms become unbounded.

Therefore we investigate solutions that are perturbations of a known static configuration. As examples we may take a spherical drop held together by surface tension or self-attraction but without any interior relative motion. We then construct a sequence of successive approximations $\{(v_n,p_n,\Sigma_n)\}$, starting with the static figure $v_o \equiv 0$, $p_o \equiv \text{const}$, $\Sigma_o = S = \{x \in \mathbb{R}^3 : |x| < 1\}$. In the first step we solve

(1.4) - (1.5) in Ω_o , the domain that is bounded by Σ_o ; this solution (v_1,p_1) is then inserted into $n\cdot T(v,p)\cdot n$ in (1.6) or (1.7), and from this equation we determine Σ_1 . Then we solve (1.4) - (1.5) in Ω_1 and obtain in this way the sequence $\{(v_n,p_n,\Sigma_n)\}$.
As a solution is sought in a neighborhood of (v_o,p_o,Σ_o) we can restrict Σ to be a graph over Σ_o . For $\Sigma_o = S$ the free boundary will then be of the form

$$\Sigma = \{(\xi,\rho) \in \mathbb{R}^3 : \xi \in S \; , \; \rho = 1 + \zeta(\xi) \; , \; \zeta : S \longrightarrow \mathbb{R}\} \; . \tag{2.1}$$

$\overline{\Omega} = \{(\xi,\rho) : \xi \in S \; , \; 0 \le \rho \le \zeta(\xi)\}$ can then be mapped onto $\overline{B} = \{x \in \mathbb{R}^3 : |x| \le 1\}$ by the transformation

$$y = \sigma(x) = \sigma(\xi,\rho) = \left[\xi, \frac{\rho}{1+\zeta(\xi)} \right] . \tag{2.2}$$

On $\overline{B}$ we introduce the new independent variables

$$\begin{cases} u^i(y) = \left[\det \dfrac{\partial\sigma^i}{\partial x^j}\right]^{-1} \dfrac{\partial\sigma^i}{\partial x^j}(x) \; v^j(x) \\ \\ q(y) \;\; = p(x) \end{cases} \tag{2.3}$$

where $x \in \Omega$ and $y \in B$ are related by $y = \sigma(x)$. For two boundaries Σ_n and Σ_{n-1} that are graphs of functions ζ_n and ζ_{n-1} their difference $\Sigma_n - \Sigma_{n-1}$ is defined to be $\{(\xi,\rho) : \xi \in S, \; \rho = 1 + \zeta_n(\xi) - \zeta_{n-1}(\xi)\}$; and furthermore we choose $u_n(y) - u_{n-1}(y)$ as difference between $v_n\left[\sigma_n^{-1}(y)\right]$ and $v_{n-1}\left[\sigma_{n-1}^{-1}(y)\right]$.
A straightforward calculation gives for the transformed Navier-Stokes equations

$$Lu^i + \bar{a}_{ij} D_j q + N_i(u,Du) = \tilde{a}_{ij} f^j \qquad \text{in } B \tag{2.4}$$

$$D_j u^j = 0 \qquad \text{in } B \tag{2.5}$$

with

$$Lu^i = -\nu D_k(a_{kl} D_l u^i) + b_{ikl} D_k u^l + c_{ij} u^j \tag{2.6}$$

$$N_i(u,Du) = a^{-1} u^j D_j u^i + \tilde{b}_{ikl} u^k u^l \; , \tag{2.7}$$

where D^j means now partial differentiation with respect to the new variables y^j . The coefficients depend on σ and its derivatives, namely

$$
(2.8)\quad \begin{cases}
a_{ij} = \dfrac{\partial\sigma^i}{\partial x^h}\dfrac{\partial\sigma^j}{\partial x^h}\ ,\ a = \left[\det \dfrac{\partial\sigma^i}{\partial x^j}\right]^{-1}\ ,\ \bar{a}_{ij} = a a_{ij} \\
b_{ikl} = \delta_{li}\dfrac{\partial}{\partial y^n} a_{nk} - \delta_{li}\Delta_x\sigma^k - 2\dfrac{\partial\sigma^i}{\partial x^r}\dfrac{\partial\sigma^k}{\partial x^s} a\dfrac{\partial}{\partial x^s}\left[a^{-1}\dfrac{\partial(\sigma^{-1})^r}{\partial y^l}\right] \\
c_{ij} = -\dfrac{\partial\sigma^i}{\partial x^n} a\Delta_x\left[a^{-1}\dfrac{\partial(\sigma^{-1})^n}{\partial y^j}\right] \\
\tilde{a}_{ij} = \dfrac{\partial\sigma^i}{\partial x^j} a\ ,\ \tilde{b}_{ikl} = \dfrac{\partial\sigma^i}{\partial x^n}\dfrac{\partial(\sigma^{-1})^m}{\partial y^k}\dfrac{\partial}{\partial x^m}\left[a^{-1}\dfrac{\partial(\sigma^{-1})^n}{\partial y^l}\right]
\end{cases}
$$

To indicate that L and its coefficients depend on σ (and therefore on ζ) we sometimes write $L(\zeta)$, $a_{ij}(\zeta)$ etc.

It is understood that the coefficients a_{ij}, a, and α_{ijk} etc. (see below) depend on σ and its first derivatives, b_{ikl}, β_{ij} etc. on σ, $D\sigma$, $D^2\sigma$, and finally c_{ij} on σ and its derivatives up to order three. As new boundary conditions we obtain

$$(2.9)\qquad \alpha_i u^i = 0\ ,\ \alpha_{ijk}D_i u^j + \beta_{jk}u^j = 0 \qquad \text{on } S, k = 1,2\ .$$

In principle, this reasoning applies also to (1.9) - (1.14), but due to the corners of the domain some modifications have to be made. If G is given and $\hat{G}$ is a domain of the same type, i.e. $\hat{G} = \{(x^1,x^2) \in \mathbb{R}^2 : 0 < x^2 < 1\ ,\ \hat{g}(x^2) < x^1 < x^1_o\}$, again with $\hat{g}(0) = \hat{g}(1) = 0$, then one can use the transformation $\sigma : \hat{G} \longrightarrow G$, defined by

$$(2.10)\qquad (y^1,y^2) = \sigma(x^1,x^2) = (x^1 + \chi(x^1)[g(x^2) - \hat{g}(x^2)], x^2)$$

$\forall (x^1,x^2) \in \hat{G}$, with a cut-off function $\chi \in C(-\infty, x^1_o)$ that satisfies $\chi(x^1) \equiv 1$ on $(-\infty, \frac{1}{4}x^1_o)$ and $\chi(x^1) \equiv 0$ for $x^1 > \frac{3}{4}x^1_o$. (It is assumed without loss of generality that $\inf g$, $\inf \hat{g} < \frac{1}{4}x^1_o$.) As the equations of motion in (1.9) - (1.14) are two-dimensional, one can reduce the problem to a fourth-order equation for the stream function ψ; therefore we can define $\varphi = \psi\circ\sigma$ as transformation of the dependent variable. In the corner $(0,0)$ for instance one uses a local transformation such that the free boundary becomes a straight line segment. Then ψ can be controlled up to the boundary in suitably weighted Hölder spaces, which are defined as

$$C^k_s(G,M) := \{u : \Omega \longrightarrow \mathbb{R} : \|u\|_{C^k_s(G,M)} := \sum_{|\beta|\le k}\sup_{x\in\Omega} |\rho(x)|^{-s+|\beta|}\cdot|D^\beta u(x)| < \infty\}$$

$$C^{k+\mu}_s(G,M) := \{u \in C^k_s(G,M) : \|u\|_{C^{k+\mu}_s(G,M)} = \|u\|_{C^{k+\mu}_s(G,M)}$$

$$+ \sum_{|\beta|=k}\sup_{x\in G} |\rho(x)|^{-s+k+\mu} \sup_{|x-x'|\le\frac{\rho(x)}{2}} \frac{|D^\beta u(x) - D^\beta u(x')|}{|x-x'|^\mu} < \infty\}.$$

As usual we have $k \in \mathbb{N}$, $\mu \in (0,1)$, β is a multi-index, and $\rho(x) = \mathrm{dist}(x,M)$, $s \in \mathbb{R}$.

We now state in what spaces the successive approximations will converge to a solution. If the free boundary is governed by surface tension we can solve (1.4) - (1.6) in $C^{2+\mu} \times C^{1+\mu} \times C^{3+\mu}$ because in this case we have the estimates

$$(2.11)\qquad \begin{cases} \|u_{n+1}-u_n\|_{C^{2+\mu}} + \|q_{n+1}-q_n\|_{C^{1+\mu}} \le C\|\zeta_n-\zeta_{n+1}\|_{C^{3+\mu}} \\ \|\zeta_n-\zeta_{n-1}\|_{C^{3+\mu}} \le C^*\{\|u_n-u_{n-1}\|_{C^{2+\mu}} + \|q_n-q_{n-1}\|_{C^{1+\mu}}\} . \end{cases}$$

For small data there holds $C \cdot C^* < 1$, and therefore we have convergence for $\{(u_n,q_n,\zeta_n)\}$. A solution (u,q) to (2.4) - (2.9) can be estimated as in (2.11) because on the right hand side of the Schauder estimates the $C^{0+\mu}$-norm of the coefficients of L occurs, and this clearly contains third derivatives of ζ; similarly the $C^{2+\mu}$-norm of the coefficients in the Dirichlet boundary condition (2.9) enters into it and this again leads to $\|\zeta\|_{C^{3+\mu}}$. On the other hand, the equation for the free boundary (1.6) is of the form

$$(2.12)\qquad \frac{1}{\sqrt{g}}\left\{ D_i \frac{g^{ij}D_j\zeta}{\sqrt{1+|\mathfrak{D}\zeta|^2}} - \frac{\partial}{\partial\zeta}\sqrt{g}\sqrt{1+|\mathfrak{D}\zeta|^2} \right\} = n\cdot T\cdot n ,$$

$\forall \xi \in S$, where $g_{ij}(\xi,\rho)$ is the metric on $\partial B_\rho(0)$, $g = \det g_{ij}$, and g^{ij} is the matrix of the adjoints; $|\mathfrak{D}\zeta|^2 := g^{ij}D_i\zeta D_j\zeta$. Equation (2.12) is a non-uniformly elliptic equation of second order, and $u \in C^{2+\mu}$, $q \in C^{1+\mu}$ implies $T \in C^{1+\mu}$, and consequently $\|\zeta\|_{C^{3+\mu}}$ can be estimated by $\|u\|_{C^{2+\mu}} + \|q\|_{C^{1+\mu}}$ as stated in (2.11). In this way the successive approximations all lie in the same function space, hence for small data $\{(v_n,p_n,\zeta_n)\}$ forms a Cauchy sequence.

In the problem (1.9) - (1.14) we proceed basically in the same way; it becomes necessary, however, to give additional estimates of the behaviour in the corners. The stream function ψ satisfies

$$(2.13)\qquad \nu\Delta^2\psi = D_2\psi D_1\Delta\psi - D_1\psi D_2\Delta\psi \qquad \text{in } G$$

$$(2.14)\qquad \psi = 0 \qquad \text{on } \partial G$$

$$(2.15)\qquad \nu D_2^2\psi + \gamma D_2\psi = -\gamma S \qquad \text{on } \Gamma$$

$$(2.16)\qquad \nu D_1^2\psi + \gamma_o D_1\psi = 0 \qquad \text{on } \Gamma_o$$

$$(2.17)\qquad D_2^2\psi - HD_2\psi = 0 \qquad \text{on } \Sigma$$

$$(2.18) \qquad \frac{1}{\sqrt{1+|g'|^2}}(-\kappa H + \beta g') = D_2\psi \frac{\partial}{\partial n} D_1\psi - D_1\psi \frac{\partial}{\partial n} D_2\psi$$

$$+ \nu \frac{\partial}{\partial n} \Delta\psi + 2\nu \frac{\partial^2}{\partial t^2} \frac{\partial}{\partial n}\psi \quad \text{on } \Sigma$$

$$(2.19) \qquad g(1) = g(0) = 0$$

The estimate which is analogues to (2.11) is

$$(2.20) \qquad \|\psi_{n+1} - \psi_n\|_{\bar{C}^{4+\mu}_{1+\delta}(G,M)} \leq C\|g_{n+1} - g_n\|_{\bar{C}^{4+\mu}_{1+\delta}((0,1),M)} ,$$

where M consist of the corners of ∂G. If $\omega(x^2) := g(x^2) - g_o(x^2)$ denotes the deviation of Σ from the static configuration Σ_o (that is the graph of a function g_o) then (2.18) implies for ω

$$(2.21) \qquad \begin{aligned} -(\omega' F'(g_o'))'' + \beta\omega' &= Q_o' + Q && \text{in } (0,1) \\ \omega(x^2) &= 0 && \text{for } x^2 = 0,1 \end{aligned}$$

where $F(t) = \frac{t}{\sqrt{1+t^2}}$ and Q_o depends on ω and g_o and their derivatives up to second order; Q is a nonlinear function in ψ and its derivatives up to third order, too. The solution to (2.21) with E as its right-hand side satisfies

$$(2.22) \qquad \|\omega\|_{\bar{C}^{4+\mu}_{1+\delta}} \leq C\|E\|_{C^{1+\mu}_{-2+\delta}} ,$$

and apart from the weights which we will discuss later the estimates (2.20) and (2.22) are to be expected from Schauder's theory for elliptic equations and the fact that (2.22) is a third order equation for ω. If we insert $Q_o + Q(\psi,\ldots,D^3\psi)$ into (2.22) then the estimates show that also for the free boundary problem (2.13) - (2.19) the successive approximations converge to a solution.

In case there is no surface tension force the scheme from before will not yield approximations that lie all in the same space. If (1.7) can be solved at all for given $v \in C^{2+\mu}$, $p \in C^{1+\mu}$, its solution ζ will not be more regular than $T(v,p)$, i.e. $\zeta \in C^{1+\mu}$, and consequently we encounter the "loss of derivatives", a phenomenon to which hard implicit function theorems are especially suited. But (1.7) will generally not admit any solution, as it describes Σ only as a level set. We can turn (1.7) into an integral equation, however, for which existence can be shown, if we assume f to be of the form $f = f_o + h$ with $f_o(x) = DU(x)$. As the force of self-attraction f_o can be absorbed into the pressure, (1.7) becomes

$$(2.23) \qquad \int \frac{g}{|x-y|}\, dy = -p(x) + \nu\left[D_i v^j + D_j v^i\right](x) .$$

Now the unknown ζ appears in the domain of integration Ω. According to Lichtenstein [L] the integral can be written in the form

$$c_o \zeta(\xi) + \oint_S \frac{\zeta(\eta)}{d(\xi,\eta)}\, d\sigma(\eta) + N(\zeta)(\xi) \tag{2.24}$$

where c_o = const and $d(\xi,\eta)$ denotes the Euclidean distance between two points $\xi,\eta \in S$.
$N(\zeta)$ is a nonlinear operator which we will discuss in §3. If f_o is the dominating force, i.e. $\|h\| << \|f_o\|$, then $n\cdot T\cdot n$ will be small, too, and the solvability of (1.8) follows from the fact that $c_o\zeta + \oint \frac{\zeta}{d}\, d\sigma$ is invertible. In this way the introduction of f_o as dominating force leads to an equation for the free boundary that can be handled. But also for physical reasons f must be regarded as necessary. Self-attraction tends to hold the drop togethter and therefore balances other forces that possibly act in the opposite way.

§ 3. Equilibrium figures with self-attraction

In [B4] we proved the following result.

Theorem 1: Let $f_o(x) = DU(x)$ be the force of self attraction. For $f = f_o + h$, $h \in C^{\lambda+\mu}$ with $\lambda > 6$ and (in cylindrical coordinates r,θ,x^3)

$$\begin{aligned} h^\theta &= h^\theta(r,x^3) = -h^\theta(r,-x^3)\ , \quad r^2 = (x^1)^2 + (x^2)^2 \\ h^3 &= h^r = 0\ , \end{aligned} \tag{3.1}$$

$\|h\|_{C^{\lambda+\mu}}$ small enough, there exists a unique solution $v \in C^{5+\mu}(\bar\Omega)$, $p \in C^{4+\mu}(\bar\Omega)$, and $\Sigma \in C^{6+\mu}$ to the free boundary problem (1.4), (1.5), (1.7); v and p are small in the sense that

$$\|v\|_{C^{5+\mu}} + \|p-U\|_{C^{4+\mu}} \leq C\|h\|_{C^{\lambda+\mu}}\ , \tag{3.2}$$

and Σ lies in a $C^{6+\mu}$ neighborhood of the unit sphere S ; the $C^{6+\mu}$-norm of the distance of Σ from S can again be estimated by $C\|h\|_{C^{\lambda+\mu}}$.

The proof is based on a suitable version of the hard implicit function theorem; we regard (1.4), (1.5), (1.7) as a nonlinear mapping $F:\mathcal{Y}_o \times \mathcal{Z}_o \longrightarrow \mathcal{X}_o$ which is defined by associating to $z = (u,q,\zeta) \in \mathcal{Z}_o := C^{2+\mu}(\bar B;\mathbb{R}^3) \times C^{3+\mu}(S;\mathbb{R})$ the right-hand side of these equations which then is an element of $\mathcal{X}_o := C^{0+\mu}(\bar B;\mathbb{R}^3) \times C^{0+\mu}(\bar B;\mathbb{R}) \times C^{0+\mu}(S;\mathbb{R})$. Here we have identified functions (v,p) and (u,q) that are related by the transformation σ as in (2.3). In this way F is defined on a set which admits an affine structure, and consequently one can compute the linearisation $DF(f,z)$ of F with respect to the second argument, cf. [B4] (40) - (42).

$$DF^i(f,z)\tilde{z} = L(\zeta)\tilde{u}^i + \bar{a}_{ij}(\zeta)D_j\tilde{q} + l_{ij}(u,\zeta)\tilde{u}^j$$

$$(3.3) \qquad + l_j(u,\zeta)D_j\tilde{u}^i + \sum_{|\gamma|\leq 3} l_\gamma(u,q.\zeta)D^\gamma\tilde{\sigma}$$

$$+ \sum_{|\gamma|\leq 1} m_\gamma(f,\zeta)D^\gamma\tilde{\sigma}, \qquad i = 1,2,3$$

$$(3.4) \qquad DF^4(f,z)\tilde{z} = D_j\tilde{u}^j$$

$$DF^5(f,z)\tilde{z} = M\tilde{\zeta} + m_o(\zeta)\tilde{\zeta} + \sum_{|\gamma|\leq 2} r_\gamma(u,q,\zeta)D^\gamma\tilde{\zeta}$$

$$(3.5) \qquad + m_{ij}(\zeta)D_i\tilde{u}^j + m(\zeta)\tilde{q}.$$

The boundary conditions for $(\tilde{u},\tilde{q})$ are of the form (2.9). We note that $DF^i(f,z)$, $i = 1,\dots,4$ is not just the Stokes linearization of (2.4), (2.5) - which would consist only of $D_jF_i(f,z)$, $i,j = 1,\dots,4$ - but contains the derivative of F^i with respect to $z^5 \equiv \zeta$, too; this results then in the third order operators in $\tilde{\sigma}$, which is the transformation belonging to $\tilde{\zeta}$. Similarly, in $DF^5(f,z)$ also operators in $\tilde{u}$ and $\tilde{q}$ occur. So in contrast to the approximation scheme from before the equations (3.3) - (3.5) no longer split into a boundary value problem for the velocity and the pressure and in a separate equation for the free boundary. On the other hand, it is not known how to invert $DF(f,z)$, and therefore we will use a variant of Moser's implicit function theorem which is due to E.Zehnder [Z], and which allows to work with (2.4), (2.5), (2.9), (2.24) instead of (3.3) - (3.5). It requires only the existence of an approximate inverse $H(f,z)$ to $DF(f,z)$ in the sense that

$$DF(f,z_n)\circ H(f,z_n) \longrightarrow \mathbb{1},$$

as z_n tends to the solution z of the nonlinear equation. More precisely, the hypotheses for this implicit function theorem are as follows. Let $\{\mathcal{Z}_t\}_{t\geq 0}$ be a one parameter family of Banach spaces with norms $|\cdot|_t$ such that for all t,t' with $0 \leq t' \leq t < \infty$ there holds

$$\mathcal{Z}_o \supset \mathcal{Z}_{t'} \supset \mathcal{Z}_t \supset \mathcal{Z}_\infty \equiv \bigcap_{t>0}$$

and

$$|z|_{t'} \leq |z|_t \qquad \forall z \in \mathcal{Z}_{t'}, \ t' \leq t.$$

The same properties are assumed to hold for $\{\mathcal{Y}_t\}_{t\geq 0}$ and $\{\mathcal{X}_t\}_{t\geq 0}$. $z_o = (0,U_o(x),o)$, where $U_o(x)$ is the gravity potential of $\Omega_o = B(0)$, satisfies $F(f_o,z_o) = 0$. We then postulate

(H.1) F is continuous in (f,z) and two times differentiable in z ; in $\mathcal{B}_o = \{(f,z): |f-f_o|_o + |z-z_o|_o < 1\}$ these derivatives are bounded.

(H.2) F is Lipschitz continuous in the first argument.

(H.3) F is of order s , where s is related to the loss of derivatives in (H.4); this means that if (f,z) becomes more regular its image $F(f,z)$ is more regular, too: $F(\mathcal{B}_o \cap (\mathcal{Y}_t \times \mathcal{Z}_t)) \subset \mathcal{X}_t$ $\forall\, t \in [1,s]$.

Hypotheses (H.1) - (H.3) can easily be verified because they are consequences of the regularity of the coefficients in (2.4), (2.5), (2.9), (2.24) .

(H.4) For every $(f,z) \in \mathcal{B}_\gamma$ there exists a linear map $H(f,z): \mathcal{X}_\gamma \longrightarrow \mathcal{Z}_o$ such that $|H(f,z)(\varphi)|_o \le M_o|\varphi_\gamma| \ \forall\ \varphi \in \mathcal{X}_\gamma$; $H(f,z)$ is furthermore continuous from $\mathcal{X}_t$ into $\mathcal{Z}_{t-\upsilon}$. $H(f,z)$ is an approximate inverse in the sense that

(3.6) $$|[D_2F(f,z) \circ H(f,z) - \mathbb{1}](\varphi)|_o \le M_o|F(f,z)|_\gamma|\varphi|_\gamma$$

for all $\varphi \in \mathcal{X}_\gamma$.

<u>Theorem 2:</u> (E.Zehnder [Z]) Let F satisfy (H.1) - (H.4). Then there exists an open neighborhood $\mathcal{D}_\lambda = \{f \in \mathcal{Y}_\lambda: |f-f_o|_\lambda < C\}$ and a mapping $\psi: \mathcal{D}_\lambda \longrightarrow \mathcal{Z}_\rho$ such that for all $f \in \mathcal{D}_\lambda$

(3.7) $$F(f,z) = 0 \qquad \text{with} \quad z = \psi(f)$$

and

(3.8) $$|z-z_o|_\rho \le C^{-1}|f-f_o|_\lambda .$$

The numbers λ and ρ can be chosen to be $\rho = 3$, $\lambda > 6$.

To verify (H.1) - (H.4) we choose first the underlying function spaces to be

$$\mathcal{X}_t = C^{t+\mu}(\bar{B};\mathbb{R}^3) \times C^{t+\mu}(\bar{B};\mathbb{R}^3) \times C^{t+\mu}(S;\mathbb{R})$$
$$\mathcal{Y}_t = C^{t+\mu}(\mathbb{R}^3;\mathbb{R}^3)$$
$$\mathcal{Z}_t = C^{t+2+\mu}(\bar{B};\mathbb{R}^3) \times C^{t+1+\mu}(\bar{B};\mathbb{R}^3) \times C^{t+3+\mu}(S;\mathbb{R}) .$$

To define the approximate inverse $H(f,z)$ we first consider the operator $D^*F(f,z)$ which consists of the linearized equations (2.4), (2.5), (2.9) in its first four components and of the linearization of (2.24) in $\tilde{\zeta}$. It is of the form (3.3) - (3.5), but with $l_\gamma D^\gamma\tilde{\sigma}$, $m_\gamma D^\gamma\tilde{\sigma}$, $r_\gamma D^\gamma\tilde{\zeta}$, $m_{ij}D_i\tilde{u}^j$ and $m\tilde{q}$ left out. $D^*F(f,z)$ is invertible, and we call its inverse $H(f,z)$. To prove the estimate (3.6) we showed in [B4] that the terms $l_\gamma D^\gamma\tilde{\sigma}$ etc. tend to zero if $\{z_n\}$ approaches its limit; the main idea in doing so consists in choosing suitable representations: when z_{n+1} is constructed we choose Ω_{n-1} as reference domain, such that ζ_{n+1} measures the distance between Σ_{n+1} and Σ_{n-1} along the normals to Σ_{n-1}.

That $D^*F(f,z)$ is invertible or equivalently that the approximations (u_n,q_n) and ζ_n can be constructed as claimed in §2 follows from Lemma 3 and Lemma 4.

Lemma 3: Let $u \in C^{2+\mu}(\bar\Omega)$, $\zeta \in C^{3+\mu}(\partial\Omega)$ be given, when Ω is a domain with boundary of class $C^{3+\mu}$. Then the boundary value problem

$$(3.9)\quad \begin{cases} L(\zeta)\tilde u^i + \bar a_{ij}(\zeta)D_j\tilde q + l_{ij}(u,\zeta)\tilde u^j + l_j(u,\zeta)D_j\tilde u^i = \varphi^i \\ D_j\tilde u^j = 0 \quad \text{in } \Omega, \\ i=1,2,3 \end{cases}$$

$$(3.10)\quad \alpha_i(\zeta)\tilde u^i = 0\,,\ \alpha_{ijk}(\zeta)D_i\tilde u^j + \beta_{kj}(\zeta)u^j = 0 \quad \text{on } \partial\Omega\,,\ k = 1,2$$

with operators as defined in (2.6), (2.8), admits to $\varphi \in C^{0+\mu}$ a classical solution $\tilde u \in C^{2+\mu}(\bar\Omega)$, $\tilde q \in C^{1+\mu}(\bar\Omega)$ as long as $\|u\|_{C^{2+\mu}}$ and $\|\zeta\|_{C^{3+\mu}}$ are small enough. The solution can be estimated by

$$(3.11)\quad \|\tilde u\|_{C^{k+2+\mu}} + \|\tilde q\|_{C^{k+1+\mu}} \le C\Big[\nu, k, \|\partial\Omega\|_{C^{k+3+\mu}}, \|\zeta\|_{C^{k+3+\mu}}, \|u\|_{C^{k+2+\mu}}\Big]\, \|\varphi\|_{C^{k+\mu}}$$

for all $k \ge 0$.

The lemma states essentially that the Stokes equations are solvable if mixed boundary conditions as in (1.5) are prescribed rather than Dirichlet data; for a proof see [SS], [B1].

Lemma 4: Let Σ be a closed surface in a $C^{2+\mu}$-neighborhood of S. Then

$$(3.12)\quad \psi_\Sigma(\xi)\tilde\zeta(\xi) + \oint_\Sigma \frac{\tilde\zeta(\eta)}{d(\xi,\zeta)}\, d\sigma(\eta) = 0$$

for at most 6 eigensolutions $\tilde\zeta_1,\dots,\tilde\zeta_6$. Here ψ_Σ is the normal derivative of the Newtonian potential of the body Ω that is bounded by Σ.

$\tilde\zeta_1,\tilde\zeta_2,\tilde\zeta_3$ are the infinitesimal translations in the directions of the coordinate axes, and $\tilde\zeta_4,\tilde\zeta_5,\tilde\zeta_6$ are the rotations about these axes. If Σ is rotationally symmetric with respect to the x^3-direction, then $\tilde\zeta_{3+i}$ is not an eigensolution.

The proof is classically known for equilibrium figures, and requires therefore only some perturbation arguments.

<u>Remark:</u> (i) Lemma 4 is not restricted to surfaces near S. Actually the proof only uses that the sphere is an equilibrium figure to a value ω for which no bifurcation occurs. So if Σ lies near an equilibrium figure that is locally unique, Lemma 4 holds, too.
(ii) The restriction (3.1) on h in Theorem 1 guarantees that there holds $\int_\Omega h dx = 0$. Physically this means that the resultant of the forces vanishes which is obviously a necessary condition for the existence of stationary configurations. We will investigate this question again in § 4.

§ 4. Closed surfaces of prescribed mean curvature and free boundaries governed by surface tension

The solution to the free boundary problem (1.4) - (1.6) where the free boundary is now determined by surface tension can be obtained by the approximation scheme that we outlined in §2. Therefore it remains to prove existence for solutions (v_n, p_n) to (1.4), (1.5) in a fixed domain Ω_{n-1} and Σ_n to (1.6) or equivalently ζ_n to (2.12) where T is evaluated at (v_n, p_n). To show existence, uniqueness and regularity of solutions to the Navier-Stokes equations with mixed boundary conditions one can proceed in a way that is very close to the case of Dirichlet data.
To indicate the main difficulty in the problem of closed surfaces with prescribed mean curvature we start with the following formula for integration by parts on the surface:

$$(4.1) \qquad -2 \oint_\Sigma H g n d\sigma = \oint_\Sigma \delta g d\sigma \qquad \forall\ g \in C_c^1(U(\Sigma)) .$$

Here $U(\Sigma)$ is a three-dimensional neighborhood of Σ, and $\delta g = Dg - (Dg \cdot n)n$ denotes the tangential part of the gradient of g. If Σ is a closed surface we can choose g to be one on Σ, and hence

$$(4.2) \qquad \oint_\Sigma H n d\sigma = 0 .$$

It turns out that (4.2) poses a restriction to the data H, for which a surface with this H is its mean curvature exists.[1)] For $H = -1 + \epsilon x^3$, which is only a perturbation to the mean curvature of the unit sphere, equation (4.2) obviously cannot hold.

If H, however, can be interpreted in physical terms as in (1.6), condition (4.2) is always satisfied, cf. [B2]. As in the remark to Lemma 4 we require the force f in (1.4) that generates a motion inside of Ω to be balanced: $\int_\Omega f(x)dx = 0$. Therefore

$$0 = \int_\Omega -\nu\Delta v + Dpdx + \int_\Omega (v\cdot D)vdx$$
$$= \oint_\Sigma T(v\cdot p)\cdot n \, d\sigma \, .$$

As the tangential part of $T\cdot n$ vanishes pointwise on Σ, cf. (1.5), this means $0 = \oint_\Sigma (n\cdot T\cdot n)nd\sigma \equiv 2\kappa \oint_\Sigma Hnd\sigma$. The restriction (4.2) to the purely geometric problem of constructing a closed surface of prescribed mean curvature is eventually quite natural if interpreted in physical terms. As (4.2) involves the data H and the solution Σ it still remains to find conditions on H alone such that (2.12) is solvable.

As there is a volume constraint (1.2) to be satisfied by the solution to (2.12) we will apply variational methods. (2.12) is the Euler-Lagrange equation to

$$I(\zeta) = \oint_S \sqrt{1+|\mathscr{D}\zeta|^2}\,\sqrt{g}\, d\xi + \oint_S \mathcal{H}(\xi,\zeta)\,\sqrt{g^*}\, d\xi \tag{4.3}$$

where

$$\mathcal{H}(\xi,\zeta) = \int_0^{\zeta(\xi)} - 2h(\xi,t)t^2dt \; ; \tag{4.4}$$

for h we have to insert the prescribed mean curvature $n\cdot t(v,p)\cdot n$ which after it is calculated for a specific approximation $(v_k,p_k)|\Sigma_k$ we may extend to be constant along rays.

The area integral $A(\zeta) = \oint_S \sqrt{1+|\mathscr{D}\zeta|^2}\,\sqrt{g}\, d\xi$ in (4.3) grows linearly in $|D\zeta|$, but not uniformly with respect to ζ:

$$c_o\zeta^2 + c_1\zeta^2|D\zeta| \le \sqrt{g}\,\sqrt{1+|\mathscr{D}\zeta|^2} \le c_o\zeta^2 + c_1\zeta^2 + c_1\zeta^2|D\zeta| \, .$$

Therfore the space of functions of bounded variation does not seem to

---

1) The rôle of (4.2) and an example for H that is even constant on rays such that there is no graph over S whose mean curvature is H was communicated to me by Henry C. Wente.

be appropriate as in the case of the Euclidean area functional $\int_\Omega \sqrt{1+|Du|^2}\,dx$. Hence we introduce another function space which is a variant of $BV(\Omega)$ by exploiting the following (formal) relation

$$\sqrt{g(\xi,\zeta)}\,\sqrt{1+g^{ij}(\xi,\zeta)D_i\zeta D_j\zeta} = \sqrt{\zeta^4+\zeta^2 g^{*ij}D_i\zeta D_j\zeta}\,\sqrt{g^*}$$
$$= \sqrt{(\zeta^2)^2+\tfrac{1}{4}g^{*ij}D_i(\zeta^2)D_j(\zeta^2)}\ .$$

If we now regard ζ^2 as the new dependent variable we can extend $A(\zeta)$ in terms of ζ^2 onto the function space

$BV_R(S) = \{\zeta\in L_2(S): \oint_S |\mathscr{D}^*(\zeta^2)|\,\sqrt{g^*} < \infty\}$ where

$$\oint_S |\mathscr{D}^*(\zeta^2)|\,\sqrt{g^*} = \sup\left\{\oint_S \zeta^2 D_i(\sqrt{g^*}\, g^{*ij}\varphi^j)\,d\xi : \varphi^1,\varphi^2\in C^1(S)\ ,\ \sqrt{g^{*ij}\varphi^i\varphi^j}\le 1\right\}. \tag{4.5}$$

On $BV_R(S)$ we now define the area integral to be

$$\oint_S \sqrt{\zeta^4+\tfrac{1}{4}g^{*ij}D_i\zeta^2 D_j\zeta^2}\,\sqrt{g^*} = \sup\left\{\oint_S \zeta^2\varphi^o_*\sqrt{g^*} + \tfrac{1}{2}\zeta^2 D_i(g^{*ij}\varphi^j)\,d\xi : \varphi^o,\varphi^1,\varphi^2\in C^1(S)\ ,\ (\varphi^o)^2+g^{*ij}\varphi^i\varphi^j\le 1\right\}. \tag{4.6}$$

The approximations to the free boundary can now be obtained by the following variational problem: minimize $I(\zeta)$ in the class of functions

$$C = BV_R(S)\cap\left\{\zeta : \tfrac{1}{3}\oint_S \zeta\,\sqrt{g^*}\,d\xi = V_o\right\}\cap\left\{\zeta : \oint_S (\zeta-1)\zeta_i\,\sqrt{g^*}\,d\xi = 0\right\}$$

where ζ_i , $i = 1,2,3$ are the eigenfunctions to the Laplace-Beltrami operator Δ^* on S to the eigenvalue 2.

<u>Remarks:</u> (i) Because we use $BV_R(S)$ instead of BV , the volume constraint $\frac{1}{3}\oint \zeta^3\,\sqrt{g^*}\,d\xi = V_o$ becomes a compact side condition; for according to the Sobolev embedding theorem $BV_R(S)$ is continuously embedded in $L_4(S)$ and hence compactly in $L_p(S)$, $p < 4$.

(ii) The side condition $\oint (\zeta-1)\zeta_i\,\sqrt{g^*}\,d\xi = 0$ guarantees that the center of mass of the fluid body stays in the origin, for ζ_i are the infinitesimal translations in the coordinate axes. This side condition leads to Langrange multipliers but as we will restrict the exterior forces to be balanced, the corresponding Lagrange multipliers will vanish if (v_n, p_n, Σ_n) tends to the solution.

(iii) The introduction of spaces of BV-type where instead of the function u itself an expression $\varphi(u)$ has bounded variation turns out to be useful in other variational problems, too. See e.g. [BD], where the degenerate variational integral $\int_\Omega u\sqrt{1+|Du|^2}\,dx$, $u \geq 0$ a.e. in $\Omega \subset \mathbb{R}^n$, is studied.

For fluid bodies whose free boundary is governed by surface tension we obtain the following result.

<u>Theorem 5:</u> The free boundary problem (1.4) - (1.5) admits a unique solution $v \in C^{2+\mu}(\bar{\Omega})$, $p \in C^{1+\mu}(\bar{\Omega})$, $\Sigma \in C^{3+\mu}$, if the force density f is of class $C^{o+\mu}$ and satisfies (3.1).

One can easily extend this result to the case of two immiscible fluids of the same density, where the drop Ω is immersed in a second fluid that fills a fixed container.

Higher regularity $v \in C^{k+2+\mu}$, $p \in C^{k+1+\mu}$, $\Sigma \in C^{k+3+\mu}$ can be shown easily if the forces are more regular, too, like $f \in C^{k+\mu}$. Furthermore, in [BF] analyticity is proved.

<u>Theorem 6:</u> Let (v,p,Σ) be a solution to (1.4) - (1.6) and f an analytic force density. Then v,p and Σ are real analytic, provided $\|v\|_{C^{1+\mu}}$ is small.

Standard techniques for proving analyticity in free-boundary problems do not seem to apply and therefore the proof had to be based on Friedman's method to show analyticity for solutions of elliptic and parabolic equations, cf. [F1]. All derivatives are estimated successively, and this required the smallness of v .

In [B3] we investigated the problem of a drop Ω that falls down under its own weight in an unbounded reservoir of a second viscous fluid of smaller density. In this case the flow can be stationary only in a reference frame that is attached to Ω . Its speed γ relative to a fixed Galilean frame is a further unknown of the problem. The condition that in the moving frame the net weight of the drop Ω is balanced by the viscous forces determines γ uniquely.

If (v,p) , (u,q) denote the velocity and the pressure in Ω and its complement $\mathcal{E}$, resp., we obtain

<u>Theorem 7.</u> If the difference of the densities of the two fluids is small then exists a unique solution (v,p,u,q,Σ) to the problem of a falling drop. The regularity of the velocities v,u , the pressures

p,q and the free boundary Σ is the same as in Theorem 5. The solution is axially symmetric with respect to the direction of the (uniform) gravitational field.
The proof uses results of H.F.Weinberger [W] on the steady fall of a rigid body in a Navier-Stokes fluid; there a weak formulation is given by which also γ can be determined.

§ 5. A free boundary problem with a dynamic contact angle.

In the free boundary value problem (1.9) - (1.14) the core of the investigation by D.Kröner [K] lies in the estimates near the singular points of the boundary. As in the theorems of §§ 3,4 existence for the Navier-Stokes equations with boundary conditions rather than Dirichlet data in smooth domains poses no particular difficulty, also after the perturbation of the operators by the transformations onto a fixed domain.
In this context the first goal is to establish precise asymptotic estimates for the function g that represents the free boundary under the following hypotheses

(5.1) $$v \in H_2^1(\Omega) \text{ and } v \text{ is smooth in } \bar{\Omega} \setminus M ,$$

where M denotes the set of singular boundary points,

(5.2) $$g \in C_1^{4+\upsilon}([0,1])$$

and

(5.3) $$\|g(y) - g'(0)y\|_{C_1^{4+\upsilon}([0,a])} \longrightarrow 0 ,$$

as a tends to zero.
The assumption on v means that the energy of the flow is finite. In addition to the weighted Hölder spaces defined in (2.10) we need to work also in Sobolev spaces with weights. They are defined as

(5.4) $$W_\mu^{k,p}(\Omega;M) = \Big\{u:\ \|u\|_{W_\mu^{k,p}(\Omega;M)} := \sum_{|\beta|\le k} \int_\Omega \rho^{p(\mu-k+p)} |D^\beta u|^p dx < \infty\Big\}$$

where (ρ,φ) are polar coordinates with the singular point as its center. The main result on the regularity of ψ is contained in

<u>Theorem 8:</u> Let $\{\psi,g\}$ be a solution of (2.13) - (2.19) and let (5.1) - (5.3) be satisfied. Then there holds

(5.5) $$\psi \in C_{1+\delta}^{4+\upsilon}(\Omega)$$

(5.6) $$k \in C^4_{3-\mu_o}([o,1])$$

where $k(y) = g'(0)y - g(y)$ and

(5.7) $$\mu_o > \begin{cases} 3 - \Pi/\varphi_o & \text{if } \frac{\Pi}{2} < \varphi_o < \Pi \\ 1 & \text{if } 0 < \varphi_o < \Pi . \end{cases}$$

φ_o denotes the contact angle, i.e. $\varphi_o = \frac{\Pi}{2} - \arctan g'(0)$.

As a consequence one gets the following asymptotic expansion.

<u>Theorem 9:</u> Under the assumptions of Theorem 8 the stream function ψ is of the form $\psi = \psi_{as} + \psi_o$ with

(5.8) $$\psi_o \in W^4_\sigma(\Omega)$$

$$\text{with } \sigma > \begin{cases} \frac{3}{2} - \frac{\Pi}{2\varphi_o} & \text{if } 0 < \varphi_o \leq \frac{\Pi}{2} \\ \frac{7}{2} - \frac{3\Pi}{2\varphi_o} & \text{if } \frac{\Pi}{2} < \varphi_o < \Pi , \end{cases}$$

and

(5.9) $$\psi_{as}(r,\varphi) = \begin{cases} u_3(r,\varphi) & \text{if } 0 < \varphi_o \leq \frac{2\Pi}{5} \\ u_3(r,\varphi) + r^{\frac{\Pi}{\varphi_o}} \sum_{s=0}^{s_o} \log r\hat{P}_s(r \log^q r,\varphi) & \text{if } \frac{2\Pi}{5} < \varphi_o \leq \frac{3\Pi}{5} \\ r^{\frac{\Pi}{\varphi_o}} \sum_{s=0}^{s_o} \log rP_s(r \log^q r,\varphi) & \text{if } \frac{3\Pi}{5} < \varphi_o < \Pi \end{cases}$$

with $r \in (0,1)$, $\varphi \in [0,\varphi_o]$ and an integer q . P_s and $\hat{P}_s$ are polynomials in $r \log^q r$ with smooth coefficients and $u_3(r,\varphi) = \frac{\gamma S}{4\upsilon} r^2 a(r,\varphi)$ with

$$a(r,\varphi) = \begin{cases} \frac{\varphi-\varphi_o}{\varphi_o} - \frac{\sin 2(\varphi-\varphi_o)}{\sin 2\varphi_o} , & \text{if } \varphi_o \neq \frac{\Pi}{2} \\ \frac{2}{\Pi} \{(\varphi-\varphi_o)(\cos 2(\varphi-\varphi_o)+1)+\sin 2(\varphi-\varphi_o)\log r , & \text{if } \varphi_o = \frac{\Pi}{2} \end{cases}$$

To show the result of Theorem 8 one improves on the regularity of ψ in several steps. The first estimate is $v \in C^o_{-1+\delta}$ for all $\delta \in (0,1)$. Because of the assumptions on g one can perform the transformation

$$w(z) = v(rz) , q(z) = rp(rz) , \gamma(y) = \frac{1}{\gamma} g(ry)$$

and then apply L_p-stimates in a fixed annular domain. Due to the regularity of g the transformed equations have smooth coefficients which finally gives $\psi \in C^{4+\upsilon}_{1-\delta}(\Omega)$. To improve on the regularity of ψ and to obtain (5.5) one applies a method used by V.A.Kondratev which allows to estimate solutions of linear elliptic equations in corners of

the domain. As g and ψ are related by the boundary condition (2.18) this results in an improved estimate for g in the singular point as stated in (5.6).
The existence of a unique solution $\{\psi,g\}$ to (2.13) - (2.19) which has the properties stated in theorems 8 and 9 is shown to exist in a neighborhood of a hydrostatic capillary problem with a contact angle φ_s provided the data are small which means that the fluid is pushed through the tube such that $|\gamma S|$ in (2.15) is sufficiently small.

Theorem 10: Let g_o be the parametrization of a static configuration with contact angle $\frac{\Pi}{2}$-arctan $g_o'(0) = \varphi_s \in (0,\Pi)$. Let $\alpha = \Pi - \varphi_s > 0$, $0 < \delta < \min\{\frac{\alpha}{2\Pi-\alpha}, 1\}$, and $\rho,\beta,\nu,\gamma,\gamma_o \in \mathbb{R}^+$. Then there exists an $\epsilon_o > 0$ such that for all values of γS with $|\gamma S| \leq \epsilon_o$ the free boundary problem (2.13) - (2.19) is uniquely solvable. The solution $\{\psi,g\}$ satisfies $\psi \in \overline{C}^{4+\beta}_{1+\delta}(\Omega;M)$, $g \in \overline{C}^{4+\beta}_{1+\delta}([0,1],M)$ for all $\beta \in (0,1)$.
The proof is based on solving the equations of motion in fixed domains and the equation for the free boundary to given data, as outlined in §2. One starts with weak solutions to the linearization of (2.13) - (2.17) ; then estimates up to the boundary are given in weighted Hölder classes, as required in (2.20) such that eventually a fixed point argument gives the solution to the free boundary problem.

References

[B1] Bemelmans, J.: Gleichgewichtsfiguren zäher Flüssigkeiten mit Oberflächenspannungen, Analysis 1 (1981) 241-282

[B2] -: A note on the interpretation of closed H-surfaces in physical terms, manuscripta math. 36 (1981) 347- 354

[B3] -: Liquid drops in a viscous fluid under the influence of gravity and surface tension, manuscripta math. 36 (1981) 105-123

[B4] -: On a free boundary problem for the stationary Navier-Stokes equations, Ann. Inst. Henri Poincaré, Analyse non linéaire 4 (1987) 517-547 .

[BD] Bemelmans, J. - Dierkes, U.: On a singular variational integral with linear growth, I: existence and regularity of minimizers, Arch. Rat. Mech. Analysis 100 (1987) 83-103

[BF] Bemelmans, J. - Friedman, A.: Analyticity for the Navier-Stokes Equations Governed by Surface Tension on the Free Boundary, J. Differential Equations 55 (1984) 135-150

[F1] Friedman, A.: On the regularity of solutions of nonlinear elliptic and parabolic systems of partial differential equations, J. Math. Mech. 7 (1958) 43-60

[F2] -: Variational Principles and Free Boundary Problems, New York, 1982

[G] Gulliver, R.: Tori of prescribed mean curvature and the rotating drop, Astérisque 118 (1984) 167-179

[H] Hölder, E.: Gleichgewichtsfiguren rotierender Flüssigkeiten mit Oberflächenspannung, Math. Z. 25 (1926) 188-208

[K] Kröner, D.: Asymptotische Entwicklungen für Strömungen von Flüssigkeiten mit freiem Rand und dynamischem Kontaktwinkel, Habilitationsschrift, Bonn, 1986

[L] Lichtenstein, L.: Zur Theorie der Gleichgewichtsfiguren rotierender Flüssigkeiten, Math. Z. 39 (1935) 639-648

[M] McCready, T.A.: The interior Neumann problem for stationary solutions of the Navier-Stokes equations, Diss. Stanford Univ., 1968

[P] Plateau, J.J.: Recherches expérimentales et théoriques sur les figures d'equilibre d'une masse liquide sans pesanteur, Mem. Acad. Roy. Belgique, tom. 16,23,30,31

[PS] Pukhnachev, V.V. - Solonnikov, V.A.: On the problem of dynamic contact angle, Prikl.Mat.Mekh. 46 (1983) 771-779

[S] Sattinger, D.H.: On the free surface of a viscous fluid motion, Proc. R. Soc. Lond. A. 349 (1976) 183-204

[SO] Solonnikov, V.A.: Solvability of a problem on the plane motion of a heavy viscous incompressible capillary liquid partially filling a container, Math. USSR Izvestija 14 (1980) 193-221

[SS] Solonnikov, V.A. - Ščadilov, V.E.: On a boundary value problem for a stationary system of Navier-Stokes equations, Proc. Steklov Inst. Math. 125 (1973) 186-199

[Z] Zehnder, E.: Generalized Implicit Function Theorems with Applications to Some Small Divisor Problems, I, Comm. Pure Appl. Math. 28 (1975) 91-140.

A GEOMETRIC MAXIMUM PRINCIPLE, PLATEAU'S PROBLEM FOR SURFACES OF PRESCRIBED MEAN CURVATURE, AND THE TWO DIMENSIONAL ANALOGUE OF THE CATENARY

Ulrich Dierkes

In the first part of our paper we shall derive an inclusion theorem for surfaces f of prescribed mean curvature H in a three-dimensional Riemannian manifold M. The decisive quantities which are involved in our result are the absolute values of both, the prescribed mean curva-ture H and the mean curvature $\mathcal{H}$ of the boundary S of some inclu-ding set $\bar{J}$, the area of the surface f and the distance from the boundary of f to S. To be more precise, if $f\colon \Omega \to J \cup S \subset M$ is some conformally parametrized surface which is of prescribed mean curvature H in the interior J, then there exists some constant $c = c(\Lambda,\tau,\kappa,R)$ depending only on $\Lambda := \max\{|H|_0, |\mathcal{H}|_0\}$, the injectivity radius τ, an upper bound for the sectional curvature κ and the distance $R := \mathrm{dist}_M(f(\partial\Omega),S)$, such that $f(\bar{\Omega}) \subset J$ provided the area of f is smaller than c.

Thus the main emphasis of the theorem, which also distinguishes this result from the $\mathcal{H} - \Lambda$ maximum principle by Hildebrandt [H1], and Gulliver-Spruck [GS], is the fact that the inward mean curvature $\mathcal{H}$ of the boundary S need not be greater than the absolute value of the prescribed mean curvature H. In particular we allow an obstacle S which has negative (oriented) mean curvature $\mathcal{H}$. Exterior domains are therefore typical examples which fit in our framework.

The analytic tool for the proof of our inclusion theorem is an estimate by Grüter [G] who used a method from Michael-Simon [MS] to prove a pulled back version of the standard monotonicity formula from geometric measure theory.

We want to explicitly remark, that our result is applicable to solutions f of the following two-dimensional elliptic variational problem, minimize

$$\int_B \{g_{ij}(f_u^i f_u^j + f_v^i f_v^j) + Q(f)(f_u \wedge f_v)\}\, du\, dv$$

in the class

$$\mathscr{C}(\Gamma,J) := \{f \in H_2^1(B,\mathbb{R}^3):\ f(\bar{B}) \subset \bar{J},$$

$$f_{|\partial B}: \partial B \to \Gamma \text{ is continuous and (weakly) monotonic}\},$$

$$\text{where } B = \{(u,v):\ u^2 + v^2 < 1\} \quad .$$

Here $\bar{J} \subset \mathbb{R}^3$ is some closed (quasi-) regular set and $\Gamma \subset J$ denotes some closed, rectifiable Jordan arc, cf. [HK, H2, H3]. Combining the existence-regularity results of Hildebrandt and Kaul [HK] and Hildebrandt [H2] with the inclusion argument of our paper, we immediately obtain a new existence result for minimal surfaces in sets $J \subset \mathbb{R}^3$, the boundary $S = \partial J$ of which is not $\mathscr{H}$-convex (i.e. $\mathscr{H} \geq 0$ is not satisfied). If $\mathscr{C}(\Gamma,J)$ is not empty, $S = \partial J$ is sufficiently regular and $A_{\Gamma,J}$ = infimum of area in $\mathscr{C}(\Gamma,J)$, then the condition is that

$$|\mathscr{H}|_{0,S} < \left\{-\frac{1}{4R^2} + \frac{\pi}{2A_{\Gamma,J}}\right\}^{\frac{1}{2}} - \frac{1}{2R} \quad .$$

Two examples illustrate this result.

In a second paragraph we apply our inclusion argument to the following Plateau problem: Given some closed Jordan arc $\Gamma \subset J$, $J \subset \mathbb{R}^3$ some domain, and some function $H: \bar{J} \to \mathbb{R}$, then find a surface $x \in C^2(B,\mathbb{R}^3) \cap C^0(\bar{B},\mathbb{R}^3)$ which has the following properties,

(1) $x(\bar{B}) \subset \bar{J}$,

(2) $\Delta x = 2H(x)\ (x_u \wedge x_v)$, $\Delta := \frac{\partial^2}{\partial u^2} + \frac{\partial^2}{\partial v^2}$, $x_u := \frac{\partial}{\partial u}x$ etc.,

(3) $x_u^2 - x_v^2 = x_u x_v = 0$,

(4) $x_{|\partial B}: \partial B \to \Gamma$ is homeomorphic.

If $|H(\xi)|$ is smaller than the (inward) mean curvature $\mathscr{H}$ of $S = \partial J$ in ξ, different sufficient conditions on H, Γ and J are available which guarantee the solvability of the problem (1) - (4), cf. the papers by Gulliver-Spruck [GS1, 2], Heinz [Hz], Hildebrandt [H4, 5, 6], Hildebrandt-Kaul [HK], Steffen [S1, 2, 3], Wente [W] and Werner [Wr].

However note that neither of the above results holds if the condition $|H(\xi)| < \mathscr{H}(\xi)$ for all $\xi \in S$ is not satisfied. On the other hand there are interesting examples available where even the condition $\mathscr{H}(\xi) > 0$ is violated, e.g., if J equals the inside of a suitable torus of revolution.

It is precisely our aim here to find sufficient conditions for the solvability of (1) - (4) also in cases where $|H(\xi)| < \mathscr{H}(\xi)$ does not hold for all $\xi \in S$. In fact we are able to modify the geometric measure theoretical approach by Steffen [S2,3]. The resulting condition covers the non "$\mathcal{H}$-convex" case, see theorems 3 and 4 for details.

In a final section we apply the argument from part one to the parametric variational problem, $E(x) = \int_B x_3|\nabla x|^2 du\, dv \to$ minimum in $\mathcal{C}(\Gamma,J)$, where $J \subset \{\xi \in \mathbb{R}^3 : \xi_3 \geq \epsilon\}$ for some $\epsilon > 0$. The integral $E(x)$ describes the x_3-coordinate of the center of gravity for the surface x, whence extremals for E appear as two dimensional analogues of a heavy chain which is suspended from two points. Further applications in architecture give a special interest to the problem, cf. [BHT].

Different existence results for extremals of $E(x)$ were proved by Böhme, Hildebrandt and Tausch [BHT], cf. also Dierkes [D1]. Our propositions 1 and 2 improve the corresponding results thms 12, 13 in [BHT].

Using ideas from minimal surface theory, we are able to show that every parametric solution is in fact of non-parametric type, provided of course the boundary Γ has this property and, in addition, projects onto a convex curve. In particular we have thus settled some existence results for the singular elliptic equation

$$\frac{\partial}{\partial x}\left(\frac{uu_x}{\sqrt{1+|\nabla u|^2}}\right) + \frac{\partial}{\partial y}\left(\frac{uu_y}{\sqrt{1+|\nabla u|^2}}\right) = \sqrt{1+|\nabla u|^2}\,, \tag{5}$$

$u = u(x,y) > 0$, which is equivalent to an equation of mean curvature type

$$\frac{\partial}{\partial x}\left(\frac{u_x}{\sqrt{1+|\nabla u|^2}}\right) + \frac{\partial}{\partial y}\left(\frac{u_y}{\sqrt{1+|\nabla u|^2}}\right) = \frac{1}{u\sqrt{1+|\nabla u|^2}}\,, \quad u > 0\,. \tag{6}$$

The last equation shows that the mean curvature of $u = u(x,y)$ is given by

$$H(u,\nabla u) = \frac{1}{2u\sqrt{1+|\nabla u|^2}}\,, \quad u > 0\,.$$

Note that there is no a priori bound on the mean curvature $H(u,\nabla u)$ in (6), nor is equation (5) amenable to standard nonlinear elliptic theory, since it lacks the regularity assumption for the coefficients in an essential way.

It is a pleasure for me to express my gratitude for the generous support of the Sonderforschungsbereich 72 at the University of Bonn. Especially I want to thank Professor Stefan Hildebrandt for introducing me into the subject which is treated in this article.

§ 1 A geometric maximum principle

We shall adopt here the definition of H-surfaces in Riemannian manifolds given by Hildebrandt and Kaul [HK], but, in short, repeat the basic concept.

Let M be a complete, connected and orientable Riemannian manifold of differentiability class three and $\Omega \subset \mathbb{R}^2$ be an open, connected and bounded set with Lipschitz-boundary $\partial\Omega$ and with standard euclidian metric, put $w = u + iv$, and $u = u_1$, $v = u_2$.

The Levi-Civita connection on M will be denoted by D, furthermore $d: M \times M \to \mathbb{R}$ stands for the distance function on M and $\|v\|$ denotes the norm of $v \in T_pM$.

In the following let M be three-dimensional and $\varphi: U \to \mathbb{R}^3$ denote some chart of an open set $U \subset M$. Then, with respect to these coordinates, g_{ik} and Γ^k_{ij} denote the coefficients of the metric and the Christoffel symbols, respectively. Put $g := \det g_{ik}$, $g^{ik} := (g_{ik})^{-1}$, and let $J \subset M$ be some open set with boundary S of class C^2. The mean curvature of S at p with respect to the interior normal will be denoted by $\mathcal{H}(p)$. Consider a mapping $f \in H^2_{2,loc}(\Omega,M) \cap H^1_2(\Omega,M)$ and let $H = H(f)$ be a function of class $L_\infty(\Omega,\mathbb{R})$. Then f is called an H-surface if it satisfies the equation

$$(*)\quad \begin{aligned} &D_{U_\alpha} f_*(U_\alpha) = 2H(f(w))f_*(U_1) \times f_*(U_2) \\ \text{and} \quad &\|f_*(U_1)\| = \|f_*(U_2)\| \quad , \quad \langle f_*(U_1), f_*(U_2)\rangle = 0 \end{aligned}$$

a.e. in Ω.

Here U_1, U_2 denote the basis fields with respect to u_1, u_2 and $f_*: T\Omega \to TM$ is the induced mapping of the tangent bundles. Moreover "$\times$" denotes the cross product and $\langle\ ,\ \rangle$ is the scalar product on T_pM.

If $x = x(u,v) = (x^1,x^2,x^3)$ is the representation of f with respect to some coordinate neighbourhood, then (*) implies[1)]

---

1) Here and in the sequel we agree to sum over repeated Latin indices $i, j, k \ldots$ from 1 to 3 and over α, β from 1 to 2.

$$\Delta x^l + \Gamma^l_{ij}\, x^i_{u^\alpha}\, x^j_{u^\alpha} = 2H(x(w))(x_u \times x_v)^l \quad ,$$

$$\text{a.e. on } \Omega_1 \subset \Omega \quad \text{for } l = 1, 2, 3$$

and $g_{ij}(x)\, x^i_u\, x^j_u = g_{ij}(x)\, x^i_v\, x^j_v$, $g_{ij}(x)\, x^i_u\, x^j_v = 0$ a.e. on Ω_1, where $f(\Omega_1)$ is supposed to be contained in that coordinate neighbourhood. Note that H-surfaces are also weak H-surfaces in the sense of [G], (3.5) Def.

Moreover we use the abbreviation $D(f) =: \int_\Omega g_{ij}(x) D_\alpha x^i D_\alpha x^j \, dudv$ and $\Gamma := f(\partial\Omega)$ to denote the Dirichlet integral and the boundary of f respectively. Finally, put

$$R := \operatorname{dist}(\Gamma, S) = \inf_{\substack{\xi\in\Gamma \\ \eta\in S}} d(\xi,\eta) \quad , \quad A(f) = \text{area of } f$$

$$\Lambda := \max\{|H|_{0,\Omega}, |\mathscr{H}|_{0,S}\} \quad ,$$

let τ be the injectivity radius on $f(\Omega)$ and κ denote an upper bound for the sectional curvature on $f(\Omega)$.

Theorem 1: *Let* Ω, M, σ, J, S *be defined as above. Assume that* $f \neq$ const. *is some surface of class* $H^1_2(\Omega,M) \cap C^0(\bar\Omega,M) \cap H^2_{2,loc}(\Omega,M)$ *with the following properties*

(i) $f(\Omega) \subset J \cup S$

(ii) $D_{U_\alpha} f_*(U_\alpha) = 2H(f) f_*(U_1) \times f_*(U_2)$

a.e. on $\Omega' := \Omega - \Omega^*$, *where* $\Omega^* = f^{-1}(S)$

(iii) $\|f_*(U_1)\| = \|f_*(U_2)\|$, $\langle f_*(U_1), f_*(U_2)\rangle = 0$

a.e. on Ω.

Then $f(\bar\Omega) \subset J$ *provided that either of the cases* I) *or* II) *holds.*

I) $\kappa \leq 0$ *and*

(7) $$A(f) < \frac{\pi\rho^2}{1 + 2\Lambda\rho + \frac{1}{2}(2\Lambda\rho)^2} \quad , \text{ where } \rho := \min\{R,\tau\} \;;$$

II) $\kappa > 0$ *and*

(8) $$A(f) < \frac{\pi\kappa^{-1}}{\dfrac{1}{\sin^2(\rho\sqrt{\kappa})} + \dfrac{2\Lambda\rho}{\sin^2(\rho\sqrt{\kappa})} + \dfrac{(2\Lambda)^2}{\kappa}}$$

with $\rho := \min\{R, \tau, \frac{\pi}{2\sqrt{\kappa}}\}$.

Furthermore f *is of class* $C^{k,\alpha}(\Omega,M)$ *if* M *belongs to* $C^{k+1,\alpha}$ *and* H *is of class* $C^{k-2,\alpha}(M,\mathbb{R})$, $k \geq 2$.

Remarks:

1. Since f is supposed to be continuous on $\bar{\Omega}$ we have $\tau > 0$ and $\kappa < \infty$.
2. If, in addition to the other hypotheses M is simply connected then case I) holds with $\rho = R$ provided that $\kappa \leq 0$. In fact, this is a consequence of a theorem of Hadamard and Cartan, cf. [GKM, 7.2 Satz].
3. In view of iii) we find that $\frac{1}{2} D(f)$ = area of f.
4. Theorem 1 may be applied to certain variational problems, cp. also theorem 2.

The following corollaries are simple consequences of the theorem.

Corollary 1: *Suppose that* M *is a simply connected, complete and orientable Riemannian manifold of class* C^3 *with non-positive sectional curvature and let* $f \in C^0(\bar{\Omega},M) \cap H^2_{2,loc}(\Omega,M) \cap H^1_2(\Omega,M)$ *satisfy* i) - iii) *of theorem* 1 *with* $H \equiv 0$. *Then* $f(\bar{\Omega}) \subset J$ *is a minimal surface in* M *provided that, in addition,* $A(f) < \dfrac{\pi R^2}{1+2\Lambda R+\frac{1}{2}(2\Lambda R)^2}$ *where* $\Lambda = |\mathscr{H}|_{0,S}$.

Remark: Note that $M = \mathbb{R}^3$ is possible in corollary 1.

Corollary 2: *Let the assumption of theorem* 1 *hold with* $\kappa \leq 0$ *and assume*

$$\Lambda < \sqrt{\frac{\pi}{2A(f)}} \quad . \tag{9}$$

Then $f(\Omega) \subset J$ *provided that*

$$\sqrt{D}\,\frac{\Lambda\sqrt{D}+\sqrt{2\pi-\Lambda^2 D}}{2\pi-2\Lambda^2 D} < \rho \quad , \quad D := 2A(f) \quad . \quad \square$$

Theorem 1 leads to the following existence result for minimal surfaces in domains $J \subset \mathbb{R}^3$.

Theorem 2: *Let* $\Gamma \subset J$ *be a closed Jordan curve with* $\mathscr{C}(\Gamma,J) \neq \emptyset$ *and suppose* $S = \partial J$ *is of class* C^3 *has bounded principal curvatures and a global parallel surface in* J. *If* $\Lambda := |\mathscr{H}|_{0,S}$ *satisfies*

$$\Lambda < \left\{ -\frac{1}{4R^2} + \frac{\pi}{2A_{\Gamma,J}} \right\}^{\frac{1}{2}} - \frac{1}{2R}$$

then there exists a minimal surface x *in* J*, i.e.* (1)-(4) *holds with* $H \equiv 0$.

Note that theorem 2 appears as special case of theorem 4 with $a = 0$.

<u>Example 1:</u> Let J be the torus of revolution which is generated by revolving the disk $(\xi_1-a)^2 + \xi_2^2 < r^2$ about the ξ_2-axis and assume Γ permits $\mathscr{C}(\Gamma,J) \neq \emptyset$. For $r < a < 2r$ the torus $S = \partial J$ has regions of negative inward mean curvature and thus the "$\mathscr{H}$-Λ maximum principle" by Hildebrandt [H1] and Gulliver-Spruck [GS1] cannot be applied to solutions of the variational problem

$$D(x) = \int_B |\nabla x|^2 \, du \, dv \to \text{minimum in } \mathscr{C}(\Gamma,J) \quad .$$

On the other hand the maximum absolute value of the mean curvature of S is given by

$$\Lambda_0 = \frac{1}{2} \max \left\{ \frac{a+2r}{r(a+r)} , \frac{|a-2r|}{r(a-r)} \right\} \quad .$$

Theorem 2 gives the existence of a minimal surface spanned by Γ in J provided $A_{\Gamma,J}$ and R satisfy

$$\Lambda_0 < \left\{ -\frac{1}{4R^2} + \frac{\pi}{2A_{\Gamma,J}} \right\}^{\frac{1}{2}} - \frac{1}{2R} \quad .$$

To obtain a numerical example one may assume further that Γ is contained in the torus of revolution that is generated by the disk $(\xi_1-a)^2 + \xi_2^2 \leq (\frac{4}{5}r)^2$ and that $r = 2$, $a = 3$. Then $R = \frac{2}{5}$ leads to the sufficient condition $A_{\Gamma,J} \leq 0.41$.

<u>Example 2:</u> Let $J = \{\xi \in \mathbb{R}^3 : |\xi| > 1\}$ be the exterior of the unit ball. Then $\mathscr{H} = -1$, $\Lambda = 1$ and for $R \geq 1$ theorem 2 gives the existence of a minimal surface spanned by Γ in J if $A_{\Gamma,J} < \frac{\pi}{5}$. Note that the critical value for $A_{\Gamma,J}$ in this configuration is 3π, since the disk spanned by the circle $\{\xi_3 = 1\} \cap \{|\xi| = 2\}$ has this area and touches $|\xi| = 1$ in $(0,0,1)$.

We now turn to the proof of theorem 1.

Let $\chi(w)$ denote the characteristic function of $\Omega^* = f^{-1}(S)$ and put

$$\Lambda^*(w) := \chi(w)\, \mathscr{H}(f(w)) + (1 - \chi(w))\, H(f(w)) \quad .$$

Following an observation of Hildebrandt [H1], we claim that

$$(10) \qquad D_{U_\alpha} f_*(U_\alpha) = 2\Lambda^* f_*(U_1) \times f_*(U_2) \quad \text{a.e. on } \Omega \quad .$$

In fact, (10) is obvious on $\Omega - \Omega^*$, while it is a consequence of the conformality relations on Ω^*. We refer to [H1], [D1] for explicit calculations.
Introduce local coordinates $\varphi: U \to \mathbb{R}^3$, choose $\Omega_1 \subset \Omega$ such that $f(\Omega_1) \subset U \subset M$, and let $x(w) = \varphi \circ f(w)$. Then (10) yields

$$(11) \qquad \Delta x^l + \Gamma^l_{ij}\left\{x^i_u x^j_u + x^i_v x^j_v\right\} = 2\Lambda^*(w)\sqrt{g}\, g^{ln}(x_u \wedge x_v)^n$$

$$\text{a.e. on } \Omega_1 \text{ and for } l = 1,2,3 \quad .$$

By virtue of $|\Lambda^*|_{0,\Omega} \leq \Lambda < \infty$ and arguments from L_p-theory one immediately infers

$$f \in H^2_{p,loc}(\Omega,M) \cap C^{1,\alpha}(\Omega,M) \quad ,$$

for all $p < \infty$ and $\alpha \in (0,1)$.
In view of $f \in C^1(\Omega,M)$ and (11) we see that

$$(12) \qquad |\Delta x| \leq \text{const.}|\nabla x|$$

holds a.e. in $\tilde{\Omega}$ for every $\tilde{\Omega} \subset\subset \Omega_1$.
Hence a technique of Hartman and Wintner is applicable, cf. [HW]. In particular one obtains the asymptotic expansion

$$(13) \qquad 2x_w(w) := x_u - ix_v = (a-ib)(w-w_o)^\upsilon + o(|w-w_o|^\upsilon)$$

$$\text{for } w \text{ close to } w_o \in \Omega \quad .$$

Here the vectors $a,b \in \mathbb{R}^3$ fulfill the conformality conditions $\|a\| = \|b\|$, $\langle a,b\rangle = 0$ and $\upsilon = \upsilon(w_o)$ stands for a non negative integer. It is now clear that the density estimate

$$(14) \qquad \limsup_{\rho\to 0} \frac{1}{\rho^2} \int_{K_\rho(w_o)} g_{ij}(x)\, D_\alpha x^i\, D_\alpha x^j \, du\, dv \geq 2\pi(\upsilon+1) \quad ,$$

$$\text{where } K_\rho(w_o) = \{w \in \Omega,\ d(f(w),f(w_o)) < \rho\} \quad ,$$

holds for every $w_o \in \Omega$, and for some $\upsilon \geq 0$.

We are thus in a position to carry over a result of Grüter, compare (3.10) theorem in [G]. (Note that the left hand side in [G] (3.11) has to be replaced by $\frac{2\pi}{\kappa}$.)

Lemma 1: *Let* f *be as above, then the following assertions hold:*

a) *If* $\kappa \geq 0$ *and if for some* $w_o \in \Omega$, $\inf_{w\in\partial\Omega} d(f(w),f(w_o)) \geq r$, *where* $0 < r \leq \tau$, *then*

$$(15) \qquad (\upsilon+1)\, 2\pi r^2 \leq D(f)\{1 + 2\Lambda r + \tfrac{1}{2}(2\Lambda r)^2\} \quad .$$

b) *If* $\kappa > 0$ *and if for some* $w_o \in \Omega$, $\inf_{w \in \partial\Omega} d(f(w), f(w_o)) \geq r$, *where* $0 < r \leq \min\left\{\tau, \frac{\pi}{2\sqrt{\kappa}}\right\}$, *then*

$$(16) \qquad \frac{2\pi(\nu+1)}{\kappa} \leq D(f)\left\{\frac{1}{\sin^2(r\sqrt{\kappa})} + \frac{2\Lambda r}{\sin^2(r\sqrt{\kappa})} + \frac{(2\Lambda)^2}{\kappa}\right\} .$$

Observe that the proof of (3.10) theorem in [G] applies to our situation even if w_o is a branch point, i.e. $\nabla x(w_o) = 0$. In fact in this case w_o may not belong to the class of "good" points, compare the definition of the set A in [G]. However, in view of what was said before, especially relation (14), it is clear that, in our case, branch points are even "better" points, since $\nu \geq 1$ then. This, in turn leads to the estimates (15) and (16), as follows now from a repetition of Grüter's argument.

Proceeding with the proof of our theorem, we now assume on the contrary to the assertion that there exists some $w_o \in \Omega^*$.
Since $R = \mathrm{dist}(\Gamma, S)$ we obtain

$$\inf_{w \in \partial\Omega} d(f(w), f(w_o)) \geq R \geq \rho \quad .$$

Putting $r := \rho$ and $\nu = 0$ in the previous lemma one immediately derives the desired contradiction. We have thus proved that $f(\bar{\Omega}) \subset J$. The remaining assertions will follow from potential theory.
Theorem 1 is completely proved.

To prove theorem 2, we note that there exists a solution $x \in C^0(\bar{B}, \mathbb{R}^3) \cap H^2_{2,loc}(B) \cap H^1_2(B)$ to the variational problem

$$\int_B |\nabla x|^2 \, du \, dv \longrightarrow \text{minimum in } \mathscr{C}(\Gamma, J) \quad .$$

Moreover x satisfies (i)-(iii) of theorem 1 with $M = \mathbb{R}^3$ and $H = 0$. But then corollary 1 implies the assertion.

§ 2 Plateau's problem for surfaces of prescribed mean curvature in given regions

We adopt the notations of the preceding paragraph with the agreement that now M equals the euclidean space $\mathbb{R}^3$. Concerning the Plateau problem (1) - (4) the following existence result holds.

Theorem 3: *Let* $\Gamma \subset J$ *be a closed rectifiable Jordan curve with* $\mathcal{C}(\Gamma,J) \neq \emptyset$ *and suppose* $S = \partial J$ *is of class* C^3*, has bounded principal curvatures and a global parallel surface in* J*. If* $H: \bar{J} \to \mathbb{R}$ *is locally Hölder continuous on* $\bar{J}$ *and* $\Lambda = \max\{|H|_{0,J}, |\mathcal{H}|_{0,S}\}$ *satisfies*

$$(17) \qquad \Lambda \leq \left\{-\frac{1}{4R^2} + \frac{\pi}{2c_0 A_{\Gamma,J}}\right\}^{\frac{1}{2}} - \frac{1}{2R} ,$$

where c_0 *is determined by*

$$(18) \qquad 1 + \frac{1}{\sqrt{18c_0}} (1+c_0)^{\frac{3}{2}} \leq c_0 , \text{ and } \infty > R \geq \left\{\frac{c_0}{\pi} A_{\Gamma,J}\right\}^{\frac{1}{2}} ,$$

then there exists a surface $x \in \mathcal{C}(\Gamma,J)$ *solving the Plateau problem (1) - (4) corresponding to* H *and* Γ *in* J.
(The constant $c_0 = 1.83$ is admissible in (17).)

Remark: In view of (18), the condition (17) is much more restrictive then the one given by Steffen [S2,3], $|H|_{0,J} < \left\{\frac{2\pi}{3A_{\Gamma,J}}\right\}^{\frac{1}{2}}$. However, note that we are in particular dealing with the non $\mathcal{H}$-convex case.

In certain cases it is possible to relax the assumption on Λ, however one then has to impose strong additional conditions on H.

Theorem 4: *Let* $\Gamma \subset J$ *be rectifiable with* $\mathcal{C}(\Gamma,J) \neq \emptyset$ *and* ∂J *of class* C^3 *with bounded principal curvatures. Suppose* ∂J *possesses a global parallel surface in* J *at distance* $\epsilon > 0$*. If* $H \in C^1(A,\mathbb{R})$ *for some open set* $A \supset \bar{J}$ *then the Plateau problem corresponding to* H *and* Γ *in* J *is solvable, provided there exists some number* $a \in (0,1)$ *such that one of the following conditions holds,*

$$(19) \qquad \Lambda < \left\{-\frac{1}{4R^2} + \frac{\pi(1-\sqrt[3]{a})}{2(1+\sqrt[3]{a})A_{\Gamma,J}}\right\}^{\frac{1}{2}} - \frac{1}{2R} \text{ and}$$

$$\int_J |H(\xi)|^3 d\xi \leq a\,\frac{9\pi}{2} ,$$

$$(20) \qquad \Lambda < \left\{-\frac{1}{4R^2} + \frac{\pi(1-\sqrt{a})}{2(1+\sqrt{a})A_{\Gamma,J}}\right\}^{\frac{1}{2}} - \frac{1}{2R} , \text{ and}$$

$$\operatorname*{ess\,sup}_{\xi^3\in\mathbb{R}} \int_{J_{\xi^3}} |H(\xi^1,\xi^2,\xi^3)|^2 \, d\xi^1 \, d\xi^2 \le a\cdot\pi$$

(here $J_{\xi^3} := \{(\xi^1,\xi^2) \in \mathbb{R}^2 : (\xi^1,\xi^2,\xi^3) \in J\}$),

(21) $$\Lambda < \left\{-\frac{1}{4R^2} + \frac{\pi(1-\sqrt[3]{a})}{2(1+\sqrt[3]{a})A_{\Gamma,J}}\right\}^{\frac{1}{2}} - \frac{1}{2R} \text{ and for all } t > 0,$$

$$\operatorname{meas}\{\xi \in J : |H(\xi)| \ge t\} \le a\,\frac{4\pi}{3}\,t^{-3} ,$$

(22) $$\Lambda < \left\{-\frac{1}{4R^2} + \frac{\pi(1-\sqrt{a})}{2(1+\sqrt{a})A_{\Gamma,J}}\right\}^{\frac{1}{2}} - \frac{1}{2R} \text{ and for every } \xi^3 \in \mathbb{R},\ t > 0,$$

$$\operatorname{meas}\{(\xi^1,\xi^2) : (\xi^1,\xi^2,\xi^3) \in J,\ |H(\xi)| \ge t\} \le a\,\frac{\pi}{4}\,t^{-2} ,$$

Remark: 1. Note that conditions (19), (21) or (20), (22) are stronger than (17) in case that $\frac{1-\sqrt[3]{a}}{1+\sqrt[3]{a}} \le \frac{1}{c_0}$ or $\frac{1-\sqrt{a}}{1+\sqrt{a}} \le \frac{1}{c_0}$ respectively. Thus we only obtain new information if a is smaller than $\left\{\frac{c_0-1}{c_0+1}\right\}^3$ or $\left\{\frac{c_0-1}{c_0+1}\right\}^2$ in the respective cases.

2. Clearly R must not be smaller than $\left\{\frac{(1+\sqrt[3]{a})A_{\Gamma,J}}{\pi(1-\sqrt[3]{a})}\right\}^{1/2}$ or $\left\{\frac{(1+\sqrt{a})A_{\Gamma,J}}{\pi(1-\sqrt{a})}\right\}^{1/2}$ in the cases (19), (21) or (20) and (22) respectively. □

Proofs of theorem 3 and 4.
Put $\tilde{K} := \mathbb{R}^3$, $C := 2c_o A_{\Gamma,J}$ and extend $H = 0$ outside J. In view of

(17) $|H|_{0,\mathbb{R}^3} < \left\{\frac{\pi}{2c_o A_{\Gamma,J}}\right\}^{1/2}$ and taking proposition 2.1 of [S3] into account we infer that H and $\tilde{K}$ satisfy an isoperimetric condition of

type $c_o\, A_{\Gamma,J}$, $|H|_{0,\mathbb{R}^3}\left\{\frac{c_o A_{\Gamma,J}}{36\pi}\right\}^{1/2} \leq \frac{1}{2}$.

Using theorem 1.1 in [S3] we find a solution x of the variational problem

$$E_H(\tilde{x}) = \int_B |\nabla\tilde{x}|^2\, du\, dv + V_H(\tilde{x},y) \rightsquigarrow \min$$

on $\mathcal{C}(\Gamma,J,C) := \{x \in \mathcal{C}(\Gamma,J) : D(x) \leq C\}$, $y \in \mathcal{C}(\Gamma,J,C)$ denoting some fixed surface. Here $V_H(\tilde{x},y)$ stands for the H-volume functional which was introduced and investigated by Steffen [S2,3]. Every solution x is locally Hölder continuous on B, continuous on $\overline{B}$ and satisfies a Morrey condition

$$D_{B_r(w_o)}(x) \leq D_{B_{r'}(w_o)}(x) \cdot \left(\frac{r}{r'}\right)^{2\alpha} \tag{23}$$

for all $w_o \in B$ and $0 < r \leq r' \leq \min\{r_o, 1-|w_o|\}$ for some $r_o > 0$ and $\alpha \in (0,1)$, cf. Prop. 5.1 in [S2].

Furthermore the parameter invariance of H-volume may be exploited to show that x in fact satisfies the conformality relations (3) a.e. in B.

We claim that

$$D(x) < C = 2c_o\, A_{\Gamma,J} \; . \tag{24}$$

Indeed, let $h_{\Gamma,J} \in \mathcal{C}(\Gamma,J,C)$ denote some surface with $D(h_{\Gamma,J}) = 2A_{\Gamma,J}$. Comparing $E_H(x)$ with $E_H(h_{\Gamma,J})$ one finds $D(x) + 4V_H(x,y) \leq D(h_{\Gamma,J}) + 4V_H(h_{\Gamma,J},y)$.

Now (see prop. 3.2 in [S2])

$$V_H(h_{\Gamma,J},y) - V_H(x,y) = V_H(h_{\Gamma,J},x)$$

which implies

$$D(x) \leq D(h_{\Gamma,J}) + 4|V_H(h_{\Gamma,J},x)| \; .$$

The isoperimetric inequality theorem 2.10 [S2] yields

$$|V_H(h_{\Gamma,J},x)| = \left|\int_{\mathbb{R}^3} (i_{h_{\Gamma,J},o} - i_{x,o})\, H(\xi)\, d\xi\right|$$

$$\leq |H|_{0,J}(36\pi)^{-\frac{1}{2}} \{\text{area}(h_{\Gamma,J}) + \text{area}(x)\}^{\frac{3}{2}}$$

$$\leq \frac{1}{12\sqrt{2\pi}}\, |H|_{0,J}\{D(h_{\Gamma,J}) + D(x)\}^{\frac{3}{2}} \; .$$

Therefore

$$D(x) \leq D(h_{\Gamma,J}) + \frac{1}{3\sqrt{2\pi}}\, |H|_{0,J}(1+c_o)^{\frac{3}{2}}\, D(h_{\Gamma,J})^{\frac{3}{2}} \; .$$

The last expression is smaller than $c_o D(h_{\Gamma,J})$ if and only if

$$1 + \frac{1}{3\sqrt{2\pi}} |H|_{0,J}(1+c_o)^{\frac{3}{2}} D(h_{\Gamma,J})^{\frac{1}{2}} < c_o \quad .$$

In view of $\infty > R$ this will be true if

$$1 + \frac{1}{3\sqrt{2\pi}} \left\{\frac{\pi}{2c_o A_{\Gamma,J}} D(h_{\Gamma,J})\right\}^{\frac{1}{2}} (1+c_o)^{\frac{3}{2}} \leq c_o \quad ,$$

that is if (18) holds, and (24) follows.

The following lemma is an immediate consequence of 5.3 theorem in [S2].

Lemma 2: *Let* $x \in \mathscr{C}(\Gamma,J,C)$ *or* $\mathscr{C}(\Gamma,J)$ *be a solution to the problem* $E_H \rightsquigarrow \min$ on $\mathscr{C}(\Gamma,J,C)$ or $\mathscr{C}(\Gamma,J)$ *resp. and suppose that* H *is bounded and continuous on* J. *If* $B' \subset B$ *is open then the variational inequality*

(25) $$\int_B \{\nabla x \nabla\phi(0,w) + 2H(x)\,(x_u \wedge x_v)\,\phi(0,w)\}\,du\,dv \geq 0$$

holds for every $\phi(0,w) \in \overset{\circ}{H}{}^1_2(B') \cap L_\infty(B')$ *with the following properties,*

(26) $x_\epsilon(u,v) := x(u,v) + \epsilon\phi(\epsilon,u,v) \in \mathscr{C}(\Gamma,J,C)$ *or* $\mathscr{C}(\Gamma,J)$ *respectively for* $\epsilon \in (0,\epsilon_o)$, *where* $\epsilon_o > 0$ *is suitably small,*

(27) $\int_{B'} |\nabla\phi(\epsilon,u,v)|^2\,du\,dv \leq$ *constant independent of* ϵ

(28) $H^1_2(B') \cap L_\infty(B') \ni \phi(\epsilon,w) \longrightarrow \phi(0,w) \in \overset{\circ}{H}{}^1_2(B') \cap L_\infty(B')$ *f.a.e.* $w \in B'$.

(29) $|\phi(\epsilon,w)|_{L_\infty(B')} \leq$ *constant independent of* ϵ.

Moreover the variational equality

$$\int_{B'} \{\nabla x \nabla\varphi + 2H(x)(x_u \wedge x_v)\varphi\}\,du\,dv = 0$$

holds for every $\varphi \in \overset{\circ}{H}{}^1_2(B') \cap L_\infty(B')$ *such that* $x + \epsilon\varphi \in \mathscr{C}(\Gamma,J)$ *or* $\mathscr{C}(\Gamma,J,C)$ *for all* ϵ *with* $|\epsilon| < \epsilon_o(\varphi)$.

We now have to show that x is indeed of class $H^2_{2,loc}(B,\mathbb{R}^3)$, where we have to assume for the moment, and in addition to the hypotheses stated in the theorem that $H \in C^1(\tilde{J},\mathbb{R})$ for some open set $\tilde{J} \supset J \cup S$. Suppose first that $w_o \in B$ is some point with $x(w_o) \in S = \partial J$. Then there exists some neighborhood $U \subset \mathbb{R}^3$ containing $x(w_o)$ and a class C^3-diffeomorphism $g: \mathbb{R}^3 \to \mathbb{R}^3$ with the properties

$$g(U \cap J) = B_1^+ := \{x \in \mathbb{R}^3 : |x| < 1,\ x^3 \geq 0\} \quad ,$$

$$g(U \cap \partial J) = B_1^o := \{x \in \mathbb{R}^3 : |x| < 1,\ x^3 = 0\} \quad \text{and}$$

$$g(x(w_o)) = 0 \quad .$$

Let $h = g^{-1}$ denote the inverse of g and put $x(w) = h \circ y(w)$. The continuity of x implies that there exists some number $\rho_o > 0$ with $y(w) = g \circ x(w) \in B^+_{\frac{1}{2}}$ for all $w \in B_{\rho_o}(w_o) \subset B$.

Define $A^\alpha_\ell(y,\nabla y)$ and $B_i(y,\nabla y)$ by

$$A^\alpha_\ell(y,\nabla y) := h^j_{y^i}(y)\, h^j_{y^\ell}\, y^i_{u^\alpha} \quad \text{and}$$

$$B_i(y,\nabla y) := h^j_{y^\ell}(y)\, h^j_{y^i y^k}(y)\, y^\ell_{u^\alpha} y^k_{u^\alpha} + 2\det(h_y(y))H(h(y))(y_u \wedge y_v)^i \quad .$$

Then we can state

__Lemma 3:__ *Under the hypotheses described above the variational inequality*

$$\int_{B_{\rho_o}(w_o)} \left\{ A^\alpha_\ell(y,\nabla y)\, \varphi^\ell_{u^\alpha} + B_i(y,\nabla y)\varphi^i \right\} du\, dv \leq 0 \tag{30}$$

holds for every $\varphi \in C^o_c(B_{\rho_o}(w_o),\mathbb{R}^3) \cap H^1_2(B_{\rho_o}(w_o))$ *with* $y^3(u,v) - \epsilon\varphi^3(u,v) \geq 0$, *for all* $(u,v) \in B_{\rho_o}(w_o)$ *and* $0 < \epsilon < \epsilon_o$, *where* $\epsilon_o = \epsilon_o(\varphi)$ *is suitably small.*

Proof of Lemma 3. Put $x_\epsilon(u,v) := h(y - \epsilon\varphi)$. Using the properties of φ and (24) we infer that $x_\epsilon \in \mathscr{C}(\Gamma,J,C)$, for all $\epsilon \in (0,\epsilon_o(\varphi))$. Furthermore

$$x_\epsilon(u,v) = x(u,v) + \epsilon\phi(\epsilon,u,v)$$

where $\phi(\epsilon,u,v) := \frac{1}{\epsilon}\{h(y-\epsilon\varphi) - h(y)\}$. A straightforward computation shows that $\phi(0,w) = -h_y(y)\cdot\varphi(w)$ is admissible in lemma 2. Inserting $\phi(0,w)$ into (25) one finds (30). □

One recognizes that the functions A^α_ℓ, B_i fulfill the growth conditions which are needed for the estimation of the H^2_2-norm given by Hildebrandt in [H2], cf. conditions D and D^+ in that paper. Moreover the Dirichlet growth condition (23) also holds for y, where now, of

course, other constants may appear. With the help of lemma 1 in [H2] we conclude that $y \in H_2^2(B_{\rho_o}(w_o),\mathbb{R}^3)$, whence also $x \in H_2^2(B_{\rho_o}(w_o),\mathbb{R}^3)$, provided $x(w_o) \in \partial J$ and $\rho_o > 0$ is sufficiently small. The remaining case $x(w_o) \in \operatorname{int} J$ may be handled in a similar way and finally using a covering argument, we see that $x \in H^2_{2,loc}(B,\mathbb{R}^3)$.
Hence we have

(31) $$\Delta x = 2\Lambda^*(u,v)\,(x_u \wedge x_v) \quad \text{a.e. in } B \quad ,$$

where $\Lambda^*(u,v) := H(x(u,v))(1-\chi) + \chi\mathcal{H}(x(u,v))$, χ denoting the characteristic function of $\mathcal{T} := \{w \in B\colon x(w) \in \partial J\}$. Note that (31), $\Lambda < \infty$, and $x \in H^1_{s,loc}(B,\mathbb{R}^3)$, $s < \infty$, imply that $x \in H^2_{s,loc}(B,\mathbb{R}^3)$ taking the L_p-estimates into account.
We assert that

(32) $$D(x) < \frac{2\pi R^2}{1+(2\Lambda R)+\frac{1}{2}(2\Lambda R)^2} \quad .$$

In fact, in view of (24), (32) will be an immediate consequence of (17). Note that this also implies, that $h_{\Gamma,J}$ is a minimal surface in J taking corollary 1 of §1 into account.
Now assume there existed some $w_o \in \mathcal{T}$. In that case we apply lemma 1 with $\Omega = B$ and $r = R$.
The resulting inequality, however, contradicted (32), whence $x(\overline{B}) \subset J$. Moreover $x \in C^{1,\alpha}(B,\mathbb{R}^3)$ satisfies (2) weakly. A standard argument yields that $x \in C^{2,\alpha}(B,\mathbb{R}^3)$ and (2) holds classically. The strong monotonicity of the boundary values $x|_{\partial B}$ will be shown as in [H5]. Thus x furnishes a solution to the Plateau problem determined by H and Γ in J.

Finally we have to get rid of the additional assumption $H \in C^1(\tilde{J},\mathbb{R})$, $\overline{J} \subset \tilde{J}$ open. This will be done by an approximation procedure similarly as described in [S3], [H6] and [D2]. We omit the details.

Proof of theorem 4. According to the assumptions (19), (20), (21) and (22) define functions $\mu(H)$ as follows,

(19') $$\mu(H) := \left[\frac{1}{36\pi}\int_J |H(\xi)|^3\, d\xi\right]^{\frac{1}{3}}$$

(20') $\mu(H) := \operatorname{ess\,sup}_{\xi^3 \in \mathbb{R}} \left\{ \frac{1}{4\pi} \int_{J_{\xi^3}} |H(\xi^1,\xi^2,\xi^3)|^2 \, d\xi^1 \, d\xi^2 \right\}^{\frac{1}{2}}$

(21') $\mu(H) := \frac{1}{2} \sup_{t>0} \left[\frac{3}{4\pi} t^3 \operatorname{meas}\{\xi \in J\colon |H(\xi)| \geq t\} \right]^{\frac{1}{3}}$ and

(22') $\mu(H) := \operatorname{ess\,sup}_{t>0} \sup_{\xi^3 \in \mathbb{R}} \left[\frac{t^2}{\pi} \operatorname{meas}\{(\xi^1,\xi^2) : (\xi^1,\xi^2,\xi^3) \in J,\ |H(\xi)| \geq t \right]^{\frac{1}{2}}$.

Then (19), (20), (21) and (22) imply

(19") $\mu(H) \leq \frac{\sqrt[3]{a}}{2}$

(20") $\mu(H) \leq \frac{\sqrt{a}}{2}$

(21") $\mu(H) \leq \frac{\sqrt[3]{a}}{2}$

(22") $\mu(H) \leq \frac{\sqrt{a}}{2}$

in the respective cases (19'), (20'), (21') and (22').

As before we extend $H = 0$ outside J. Using propositions 3.1, 4.3, 5.1 and (5.13) of the paper [S3] by Steffen we infer that H and $\mathbb{R}^3$ satisfy an isoperimetric condition of type ∞, $\mu(H)$. (This means that $|\int_A H(\xi)\, d\xi| \leq \mu(H)\, M(\partial A)$, whenever $A \subset \mathbb{R}^3$ is some set of finite perimeter $M(\partial A)$ in $\mathbb{R}^3$.) Since H is bounded, $\mu(H) < \frac{1}{2}$ we are in a situation to apply theorem 1.2 in [S3], taking $\tilde{K} = \mathbb{R}^3$. (The condition $\inf_{H' \in L_\infty(\mathbb{R}^3)} \mu(H-H') = 0$ will either be obvious or follows from the boundedness assumption on H.) Let x denote some solution of the variational problem $E_H \rightsquigarrow \min$ on $\mathscr{C}(\Gamma,J)$. We may then argue as in the proof of theorem 3 that x in fact belongs to the class $H^2_{S,loc}(B,\mathbb{R}^3) \cap C^0(\bar{B})$ and satisfies the conformality relations $|x_u|^2 - |x_v|^2 = x_u x_v = 0$ classically on B. Let $M(\partial f)$ denote the total variation of the first order distributional derivatives for some function $f \in L_{1,loc}(\mathbb{R}^3,\mathbb{R})$, then it follows from the proof of theorem 1.2 in [S3], cf. also [S2]

$$|V_H(h_{\Gamma,J},x)| = \left|\int_{\mathbb{R}^3} (i_{h_{\Gamma,J},0} - i_{x,0})\, H(\xi)\, d\xi\right| \leq$$

$$\mu(H)\, M(\partial(i_{h_{\Gamma,J},0} - i_{x,0})) \leq \frac{\mu(H)}{2} \left\{ D(h_{\Gamma,J}) + D(x) \right\} .$$

Moreover $V_H(h_{\Gamma,J},y) - V_H(x,y) = V_H(h_{\Gamma,J},x)$ and using the minimum property of x,

$$D(x) + 4V_H(x,y) = E_H(x) \le E_H(h_{\Gamma,J}) = D(h_{\Gamma,J}) + 4V_H(h_{\Gamma,J},y) \quad ,$$

whence

$$D(x) \le D(h_{\Gamma,J}) + 2\mu(H)\,\{D(h_{\Gamma,J}) + D(x)\} \quad .$$

Concluding we find

$$D(x) \le \frac{1+\sqrt[3]{a}}{1-\sqrt[3]{a}}\, D(h_{\Gamma,J}) \quad , \quad \text{or}$$

$$D(x) \le \frac{1+\sqrt{a}}{1-\sqrt{a}}\, D(h_{\Gamma,J}) \quad ,$$

holding in the cases (19"), (21") or (20"), (22") respectively.
Now the inequalities

$$\frac{1+\sqrt[3]{a}}{1-\sqrt[3]{a}}\, D(h_{\Gamma,J}) < \frac{2\pi R^2}{1+(2\Lambda R)+\frac{1}{2}(2\Lambda R)^2} \quad \text{and}$$

$$\frac{1+\sqrt{a}}{1-\sqrt{a}}\, D(h_{\Gamma,J}) < \frac{2\pi R^2}{1+(2\Lambda R)+\frac{1}{2}(2\Lambda R)^2}$$

are immediate consequences of

$$\Lambda < \left\{-\frac{1}{4R^2} + \frac{\pi(1-\sqrt[3]{a})}{2(1+\sqrt[3]{a})A_{\Gamma,J}}\right\}^{\frac{1}{2}} - \frac{1}{2R} \quad \text{and}$$

$$\Lambda < \left\{-\frac{1}{4R^2} + \frac{\pi(1-\sqrt{a})}{2(1+\sqrt{a})A_{\Gamma,J}}\right\}^{\frac{1}{2}} - \frac{1}{2R}$$

respectively.

We are thus in a situation to apply lemma 1 which yields that the "touching set" $\mathcal{T}$ must be empty, whence $x(\bar{B}) \subset J$.
The rest of the assertion now follows as in the proof of theorem 3.

§ 3 Non-parametric solutions for the two-dimensional analogue of the catenary

Given a conformal surface $x \in \mathscr{C}(\Gamma,J)$, $J \subset \mathbb{R}^3 \cap \{\xi_3 > 0\}$ then the potential energy under gravitational forces is given by the integral

$$E(x) = \int_B x_3\, |\nabla x|^2 \,du\,dv \quad , \quad x = (x_1,x_2,x_3) \; . \tag{33}$$

In the following we are interested in surfaces of minimal potential energy (or at least stationary energy!), which, in addition, do not touch the obstacle $S = \partial J$. In other words we are looking for solutions x of the variational problem

$$E(\cdot) \to \text{minimum on } \mathscr{C}(\Gamma, J) \tag{34}$$

such that $x(\bar{B}) \subset J$.

We show here that the supergraph

$$J_\delta := \{(\xi_1, \xi_2, \xi_3) : \xi_3 > z_\delta(\xi_1, \xi_2)\}$$

corresponding to a $C^4(\mathbb{R}^2)$-solution $z_\delta(x_1, x_2)$ of

$$\sum_{i=1}^{2} \frac{\partial}{\partial x_i} \left\{ \frac{z \cdot z_{x_i}}{\sqrt{1+|\nabla z|^2}} \right\} \geq \sqrt{1+|\nabla z|^2} \quad , \tag{35}$$

with $z_\delta \geq \delta > 0$, determines a domain of inclusion for the problem (34) with $J = J_\delta$. Note that z_δ is a subsolution to the Euler-equation of the corresponding non-parametric functional

$$G(z) = \int z \cdot \sqrt{1+|\nabla z|^2}\, dx_1\, dx_2 \quad , \quad z = z(x_1, x_2) \quad .$$

<u>**Theorem 5:**</u> *Let* z_δ, J_δ *be as above and assume* $S_\delta = \partial J_\delta$ *has bounded principal curvatures.*[2] *If* $x \in \mathscr{C}(\Gamma, J_\delta)$ *solves* (34) *with* $J = J_\delta$, *then* $x(\bar{B}) \subset J_\delta$. *In conclusion,* x *is analytic and solves the Euler equations*

$$\Delta x = \frac{1}{2} \frac{\omega'}{\omega} |\nabla x|^2 - x_u\left(\frac{\omega'}{\omega} x_u\right) - x_v\left(\frac{\omega'}{\omega} x_v\right) \tag{36}$$

with $\omega = x_3$, $\omega' = (0,0,1)$.

<u>**Corollary 1:**</u> (cf. [BHT] theorem 11). *The one sheeted hyperboloid* $z_\delta = +\sqrt{\delta^2 + (x_1)^2 + (x_2)^2}$ *is admissible in theorem 5 for every* $\delta > 0$.

pf. One only has to check that z_δ satisfies (35).

Consider now rotationally symmetric solutions $z(r)$, $r^2 = x_1^2 + x_2^2$, for the Euler equations of the non-parametric integral $G(z)$. They may be obtained as extremals for the one-dimensional integral

$$I = \int r\, u(r) \sqrt{1+[u'(r)]^2}\, dr$$

which has been investigated by Keiper [Ke]. It turns out that extremals $z_\delta(r)$ exist for all $r \geq 0$ with $z_\delta(0) = \delta > 0$. Moreover $z_\delta(r)$ oscillates along the cone $z = r$ and when r approaches infinity also

---

2) this means that J_δ is quasiregular, cf. [GS1, lemma 2.4]

$z_\delta(r) \to r$. Since $z_\delta(r)$ even furnishes a solution to the Euler equation for G we have

Corollary 2: *Theorem 5 holds for* $z_\delta(x_1,x_2) = z_\delta(r)$, *where* z_δ, $\delta > 0$, *denotes some extremal for* I. □

Now we would like to bring the idea of section 1 into play. Our propositions 1 and 2 improve the corresponding theorems 12 and 13 in the paper [BHT][3])

Proposition 1: *Let* $x = x(u,v)$ *be a solution of* (34), *where* $J = J_\epsilon = \{\xi \in \mathbb{R}^3: \xi_3 > \epsilon\}$, $\epsilon > 0$. *Assume also that*

$$\text{area}(x) < \frac{\pi R^2 \epsilon^2}{\epsilon^2 + \epsilon R + \frac{1}{2} R^2}, \tag{37}$$

$R = \text{dist}\,(\Gamma, S_\epsilon)$.

Then $x = x(u,v)$ *is contained in the open halfspace* $\{\xi_3 > \epsilon\}$. *Furthermore* x *furnishes an analytic solution of* (36).

Proposition 2: *Denote by* $h(\Gamma) := \sup_{\xi \in \Gamma} \xi_3$ *the height of* Γ *and suppose that*

$$A_{\Gamma, J_\epsilon} < \frac{\epsilon}{h(\Gamma)} \frac{\pi R^2 \epsilon^2}{\epsilon^2 + \epsilon R + \frac{1}{2} R^2}.$$

Then the conclusion of proposition 1 holds. □

Proof of theorem 5. There exists a solution $x \in C^0(\overline{B}) \cap H^2_{S,loc}(B,\mathbb{R}^3)$ for all $s \in [1,\infty)$, of problem (34) with $J = J_\delta$. Let $\mathcal{T} = \{(u,v) \in B: x(u,v) \in \partial J_\delta\}$ denote the closed coincidence set, and suppose that $\partial\mathcal{T} \subset\subset B$ does not only contain branch points. If $(u_o,v_o) \in \partial\mathcal{T}$ is regular, then there exist neighborhoods U, V of (u_o,v_o) and $(x_o,y_o) := (x_1(u_o,v_o), x_2(u_o,v_o))$ and a function $z \in C^1(V,\mathbb{R})$, such that for all $(u,v) \in U$, $x_3(u,v) = z(x_1(u,v),x_2(u,v))$. $z = z(x_1,x_2)$

---

3) The constants $\pi e^{-2}(4\epsilon)^2$ and $\frac{\epsilon}{h}\pi\,(\frac{4\epsilon}{e})^2$ which appear in [BHT] theorems 12, 13 have to be replaced by $\pi e^{-2}(2\epsilon)^2$ and $\frac{\epsilon}{h}\pi\,(\frac{2\epsilon}{e})^2$, because in lemma 7 of that paper $\mathcal{H}$ denotes 2-times the mean curvature which is actually used by these authors.

satisfies $z \geq z_\delta$ on V and $z(x_o,y_o) = z_\delta(x_o,y_o)$. Moreover $z = z(x_1,x_2)$ is a supersolution to the Euler equation of the integral

$$\int z \sqrt{1+|\nabla z|^2}\, dx_1\, dx_2 \quad .$$

Hence we have

$$\int_V \left\{ \frac{\sum_{i=1}^{2} z\, z_{x_i} \phi_{x_i}}{\sqrt{1+|\nabla z|^2}} + \sqrt{1+|\nabla z|^2}\, \phi \right\} dx_1\, dx_2 \geq 0 \quad ,$$

for all $\phi \in \mathring{H}^1_2(V,\mathbb{R})$, $\phi \geq 0$ a.e.

Now put $w := z - z_\delta \geq 0$; then w satisfies the linear inequality

$$\int_V \sum_{i,j=1}^{2} \left[\phi_{x_i} \left\{ a_{ij} w_{x_j} + b_i w \right\} + \phi b_i w_{x_i} \right] dx_1\, dx_2 \geq 0$$

for suitable regular coefficients a_{ij}, b_i and all $\phi \in \mathring{H}^1_2(V,\mathbb{R})$, $\phi \geq 0$ a.e. The weak Harnack inequality then yields $w = 0$, which contradicts the fact that $(u_o,v_o) \in \partial\mathcal{T}$. Therefore $\partial\mathcal{T}$ only contains isolated branch points, and since $\mathcal{T}$ is bounded we obtain that $\mathcal{T} = \partial\mathcal{T}$. In particular, the Euler equations (36) hold almost everywhere on B. Standard regularity results then imply that $x \in C^\omega(B,\mathbb{R}^3)$.

Suppose now that $(u_o,v_o) \in \partial\mathcal{T}$ is a branch point. W.l.o.g. we assume $(u_o,v_o) = (0,0)$ and $x(u_o,v_o) = (0,0,0)$. There exists a neighborhood U of $(0,0)$ and a function $\varphi \in C^2(U,\mathbb{R})$ such that by a suitable choice of coordinates (x_1,x_2,x_3) we obtain (cf. [Gr] Lemma 2.1, 2.2)

$$(38) \qquad \begin{aligned} & x_1 + ix_2 = w^m \quad , \quad w = u + iv = u_1 + iu_2 \quad , \quad m \geq 2 \\ & x_3 = \varphi(w) \quad \text{where} \quad |D^k\varphi| = O(|w|^{m+1-k}) \\ & \text{for } k = 0,1,2, \text{ as } |w| \to 0 \quad . \end{aligned}$$

Moreover, with respect to these coordinates,

$$z_\delta(x_1,x_2) = O(|x_1|^2 + |x_2|^2) \quad , \quad \text{as} \quad (x_1,x_2) \to (0,0) \quad ,$$

and we have the non-parametric representation

$$(39) \qquad \begin{aligned} & x_1 + ix_2 = w^m \\ & z^\delta(u,v) = z_\delta(\mathrm{Re}\, w^m, \mathrm{Im}\, w^m), \text{ with } z^\delta \in C^2(u,\mathbb{R}) \text{ and} \\ & |D^k z^\delta| = O(|w|^{2m-k}),\ k = 0,1,2, \text{ as } |w| \to 0 \quad . \end{aligned}$$

Each point $P \in U - \{(0,0)\}$ has a neighborhood V_1 which is mapped diffeomorphically onto an open set V_2 by the mapping (38.1). Then over V_2 the image of V_1 under (x_1,x_2,x_3) may be described non-parametrically as $x_3 = \psi(x_1,x_2)$ and ψ satisfies

$$\sum_{i=1}^{2} \frac{\partial}{\partial x_i} \left[\frac{\ell(x_1,x_2,\psi)\cdot\psi_{x_i}}{\sqrt{1+|\nabla\psi|^2}}\right] - \sqrt{1+|\nabla\psi|^2} = 0$$

for some linear function ℓ. In view of $\varphi(w) = \psi(\mathrm{Re}w^m, \mathrm{Im}w^m)$ we find for $w \in U - \{(0,0)\}$ the relation

$$\sum_{i=1}^{2} \frac{\partial}{\partial u_i} \left\{\frac{\ell(x_1,x_2)\varphi)\varphi_{u_i}}{\sqrt{1+\frac{|\nabla\varphi|^2}{A^2}}}\right\} - A^2 \sqrt{1+\frac{|\nabla\varphi|^2}{A^2}} = 0 \quad , \tag{40}$$

where $A^2 = m^2|w|^{2(m-1)}$, while a corresponding differential inequality holds for $z^\delta(u,v)$. By virtue of the asymptotic behavior of $|D^k\varphi|$ and $|D^k z^\delta|$ as $|w| \to 0$ it is readily seen that $\frac{|\nabla\varphi|^2}{A^2}$ and $\frac{|\nabla z^\delta|^2}{A^2}$ are of class $C^1(U)$. Of course this means that (40) holds in U and, finally, it again follows from Harnack's inequality that $\varphi(u,v) - z^\delta(u,v) = 0$, which is impossible. This proves theorem 5.

Propositions 1 and 2 are immediate consequences of theorem 1 applied to solutions of (34) with $J = J_\epsilon$. Since the argument is similar to the proof of theorem 2 we may omit the details.

So far we have investigated the parametric problem, but what is really desirable is to show that solutions constructed above are indeed graphs $z = z(x_1,x_2)$ provided that Γ is a graph over the boundary of a convex region. In fact in case of the integral $E(x)$ this conjecture is fostered by the physical interpretation of the problem.

Theorem 6: *Let* $x = x(u,v) \in C^\omega(B) \cap C^0(\bar{B})$ *denote some solution for the system* (36). *Suppose that the boundary values* $x|_{\partial B}$ *map* ∂B *topologically onto* Γ *and that* Γ *has the representation*

$$\Gamma = \{(\xi_1,\xi_2,\xi_3): \xi_3 = \varphi(\xi_1,\xi_2), (\xi_1,\xi_2) \in \partial K, \; K \subset \mathbb{R}^2 \text{ some closed, convex set}\}.$$

Then $x = x(u,v)$ *possesses the nonparametric representation* $z = z(x_1,x_2)$, *which is an analytic solution of*

$$\sum_{i=1}^{2} \frac{\partial}{\partial x_i} \left\{\frac{z z_{x_i}}{\sqrt{1+|\nabla z|^2}}\right\} = \sqrt{1+|\nabla z|^2} \quad , \quad (x_1,x_2) \in \mathring{K} .$$

<u>Proof:</u> We have to show that the mapping

$$r: \bar{B} \to K$$
$$(u,v) \mapsto (x_1(u,v), x_2(u,v))$$

defines a C^1-diffeomorphism from the open unit disc onto the interior of K, which, in addition, maps $\bar{B}$ topologically onto K. Here we can use ideas of Radó [R] and Kneser [Kn], cf. also Nitsche [N1, § 398].

First observe that $r: \partial B \to \partial K$ is a homeomorphism, since $x: \partial B \to \Gamma$ has this property and ∂K is a simple covered projection of Γ. Thus it remains to show that the Jacobian

$$J(u,v) = \left|\frac{\partial(x_1,x_2)}{\partial(u,v)}\right|$$

is different from zero for all points $(u,v) \in B$. Suppose on the contrary that for some $(u_0,v_0) \in B$, $J(u_0,v_0) = 0$. Then there exist numbers $a,b \in \mathbb{R}$, such that

(41)
$$a \frac{\partial x_1}{\partial u} + b \frac{\partial x_2}{\partial u} = 0$$
$$a \frac{\partial x_1}{\partial v} + b \frac{\partial x_2}{\partial v} = 0 \quad \text{at} \quad (u_0,v_0) \quad .$$

By virtue of (36.1), (36.2) the function

$$f(u,v) := ax_1(u,v) + bx_2(u,v) - c \quad ,$$

where $c = ax_1(u_0,v_0) + bx_2(u_0,v_0)$, is seen to fulfill

(42)
$$\Delta f + \frac{1}{x_3} \nabla x_3 \nabla f = 0 \quad .$$

Note that since $r: \partial B \to \partial K$ is topological and K is convex, there are at most two distinct points $\eta_1, \eta_2 \in \partial B$ with

(43) $f(\eta_1) = f(\eta_2) = 0$, and $f > 0$ resp. $f < 0$ on the arcs

$\Sigma_1, \Sigma_2 \subset \partial B$ determined by these points.

Since $f(u_0,v_0) = 0 = \nabla f(u_0,v_0)$ there exists $\mathbb{N} \ni \nu \geq 1$ and $c, d \in \mathbb{R}$ such that with $w = u+iv$, $w_0 = u_0+iv_0$ and for some real analytic ψ

(44)
$$f_u(w) - if_v(w) = c(w-w_0)^\nu + \psi(w), \text{ where}$$
$$D^k\psi(w) = o(|w-w_0|^{\nu-k}) \ , \ 0 \leq k \leq 2, \ w \to w_0 \ .$$

In fact (44) follows from a well-known result due to Hartman and Wintner. (44) leads to

(45)
$$f(w) = \mathrm{Re}\,\{T(w)^{\nu+1}\}$$

with some C^1-diffeomorphism $T: B_\tau(w_0) \to V$ from the open ball $B_\tau(w_0)$ with center w_0 and radius τ onto some neighborhood of zero. Moreover $T(w_0) = 0$, and T is conformal at zero, cf. [Gr, lemma 2.2, 3.3].

Hence there exist C^1-arcs $\sigma_1,\ldots,\sigma_{2(\nu+1)}$ in $B_\tau(w_0)$ emanating from w_0 into $B_\tau(w_0)$ with the property that $f(w) = 0$ for all $w \in \sigma_i$, $i = 1,\ldots 2(\nu+1)$. Moreover f is strictly positive or negative in the regions

$$Q_1,\ Q_3,\ \ldots\ ,\ Q_{2\nu+1} \quad \text{and}$$
$$Q_2,\ Q_4,\ \ldots\ ,\ Q_{2\nu+2} \quad \text{respectively}$$

which are determined by these arcs.
The open set $P = \{w \in B\colon f(w) \neq 0\}$ splits into denumerable many open connected and distinct sets. Let

$$P_1,\ P_3\ \ldots\ ,\ P_{2\nu+1} \quad \text{and} \quad P_2,\ P_4,\ \ldots\ ,\ P_{2\nu+2}$$

denote the components of P which contain the respective regions Q_j. Since $f > 0$ in P_1 e.g. there must exist some point $\xi_1 \in \partial P_1 \cap \partial B$ with $f(\xi_1) > 0$ (because otherwise $f \equiv 0$ in P_1 by virtue of the maximum principle for (42), and hence f had to vanish identically on B which is impossible).

We thus obtain the existence of points $\xi_1,\ \xi_2,\ \ldots\ ,\ \xi_{2\nu+2}$ on ∂B, which are ordered in the indicated way, such that

$$f(\xi_{2j+1}) > 0 \quad , \quad j = 0,\ldots,\nu \quad \text{and}$$
$$f(\xi_{2j+2}) < 0 \quad , \quad j = 0,\ldots,\nu \quad .$$

Since f is continuous there are at least $2\nu + 2 \geq 4$ zeroes of f on ∂B which contradicts (43). The Jacobi determinant $J(u,v)$ is therefore different from zero and by the inverse function theorem r is locally invertible. Now since the boundary mapping $r\colon \partial B \to \partial K$ is a homeomorphism, it follows that $r\colon \bar{B} \to K$ is also a homeomorphism, which is a diffeomorphism from B onto $\mathring{K}$. Thus the equations

$$x_1 = x_1(u,v) \qquad x_2 = x_2(u,v)$$

can be inverted. Moreover the inverse

$$u = u(x_1,x_2) \qquad v = v(x_1,x_2)$$

is continuous on K and analytic in the interior. Also $z = z(x_1,x_2) = x_3(u(x_1,x_2),v(x_1,x_2))$ fulfills

$$\sum_{i=1}^{2} \frac{\partial}{\partial x_i} \left\{ \frac{z z_{x_i}}{\sqrt{1+|\nabla z|^2}} \right\} = \sqrt{1+|\nabla z|^2} \quad ,$$

which is the Euler equation for the non-parametric functional

$$G(z) = \int z \sqrt{1+|\nabla z|^2}\, dx_1\, dx_2 \quad . \qquad \square$$

Remark: The singular variational problem which is associated with the integral G and its n-dimensional counterpart is the subject of an investigation by Bemelmans and Dierkes, cp. [BD].

References

[BD] Bemelmans, J., Dierkes, U.: On a singular variational integral with linear growth, I: existence and regularity of minimizers. Arch. Rat. Mech. Analysis, **100**, 83-103 (1987).

[BHT] Böhme, R., Hildebrandt, S., Tausch, E.: The two dimensional analogue of the catenary. Pacific Journal Math **88**, 247-278 (1980).

[D1] Dierkes, U.: Singuläre Variationsprobleme und Hindernisprobleme. Dissertation Bonn 1984, Bonner Math. Schriften 155.

[D2] Dierkes, U.: Plateau's problem for surfaces of prescribed mean curvature in given regions. manuscripta math. **56**, 313-331 (1986).

[D3] Dierkes, U.: A geometric maximum principle for surfaces of prescribed mean curvature in Riemannian manifolds. To appear in Zeitschrift für Analysis und ihre Anwendungen.

[GKM] Gromoll, D., Klingenberg, W., Meyer, W.: Riemannsche Geometrie im Großen. Berlin, Heidelberg, New York, Springer 1968.

[G] Grüter, M.: Regularity of weak H-surfaces. Journ. Reine Angew. Math. **329** (1981), 1-15.

[Gr] Gulliver, R.: Regularity of minimizing surfaces of prescribed mean curvature. Ann. Math. **97**,1 275-305 (1973).

[GS1] Gulliver, R., Spruck, J.: Existence theorem for parametric surfaces of prescribed mean curvature. Indiana Univ. Math. J. **22**, 445-472 (1972).

[GS2] Gulliver, R., Spruck, J.: The Plateau problem for surfaces of prescribed mean curvature in a cylinder. Inventiones Math. **13**, 169-178 (1971).

[HW] Hartmann, P., Wintner A.: On the local behavior of solutions of non-parametric partial differential equations. Am. J. Math. **75**, 149-476 (1953).

[Hz] Heinz, E.: Über die Existenz einer Flächer konstanter mittlerer Krümmung bei vorgegebener Berandung. Math. Ann. 127, 258-287 (1954).

[H1] Hildebrandt, S.: Maximum principle for minimal surfaces and for surfaces of continuous mean curvature. Math. Z. 128, 253-269.

[H2] Hildebrandt, S.: On the regularity of solutions of two-dimensional variational problems with obstructions. CPAM 25, 479-496 (1972).

[H3] Hildebrandt, S.: Interior $C^{1+\alpha}$-regularity of solutions of two dimensional variational problems with obstacles. Math. Z. 131, 233-240 (1973).

[H4] Hildebrandt, S.: Randwertprobleme für Flächen vorgeschriebener mittlerer Krümmung und Anwendungen auf die Kapillaritätstheorie. Math. Z. 112, 205-213 (1969).

[H5] Hildebrandt, S.: Einige Bemerkungen über Flächen beschränkter mittlerer Krümmung. Math. Z. 115, 169-178 (1971).

[H6] Hildebrandt, S.: Über einen neuen Existenzsatz für Flächen vorgeschriebener mittlerer Krümmung. Math. Z. 119, 267-272 (1971).

[HK] Hildebrandt, S., Kaul, H.: Two-dimensional variational problems with obstructions, and Plateau's problem for H-surfaces in a Riemannian manifold. CPAM 25 (1972), 187-223.

[K] Kaul, H: Isoperimetrische Ungleichung und Gauss-Bonnet Formel für H-Flächen in Riemannschen Mannigfaltigkeiten. Arch. Rat. Mech. Analysis 45 (1972), 194-221.

[Ke] Keiper, J.: The axially symmetric n-tectum. Preprint, Toledo University (1980).

[Kn] Kneser, H.: Lösung der Aufgabe 41. Jahresbericht der DMV 35, 123-124 (1926).

[MS] Michael, J.H., Simon, L.M.: Sobolev and mean-value inequalities on generalized submanifolds of $\mathbb{R}^n$. CPAM 26 (1973), 361-379.

[N1] Nitsche, J.C.C.: Vorlesungen über Minimalflächen. Springer Grundlehren 199, Berlin-Heidelberg-New York 1975.

[R] Radó, T.: The problem of least area and the problem of Plateau. Math. Z. 32 (1930), 763-796.

[S1] Steffen, K.: Ein verbesserter Existenzsatz für Flächen konstanter mittlerer Krümmung. manuscripta math. 6, 105-139 (1972).

[S2] Steffen, K.: Isoperimetric inequalities and the problem of Plateau. Math. Ann. 222, 97-144 (1976).

[S3] Steffen, K.: On the existence of surfaces with prescribed mean curvature and boundary. Math. Z. 146, 113-135 (1976).

[W] Wente, H.: An existence theorem for surfaces of constant mean curvature. J. math. Analysis Appl. 26, 318-344 (1969).

[Wr] Werner, H.: Das Problem von Douglas für Flächen konstanter mittlerer Krümmung. Math. Ann. **133**, 303-319 (1957).

Ulrich Dierkes
Fachbereich Mathematik
Universität des Saarlandes

D-6600 Saarbrücken

and

Dept. of Mathematics
Research School of Physical Sciences
Australian National University
GPO Box 4, Canberra, ACT 2601
Australia

FINITE ELEMENTS FOR THE BELTRAMI OPERATOR ON ARBITRARY SURFACES

Gerhard Dziuk
Institut für Angewandte Mathematik,
Wegelerstraße 6, D-5300 Bonn, FRG

Abstract: We develop a Finite Element Method for elliptic differential equations on arbitrary two–dimensional surfaces. Global para metrizations are avoided. We prove asymptotic error estimates. Numerical examples are calculated.

Keywords: Finite Elements, Beltrami Operator, Elliptic Equations on Surfaces

Classification Numbers: 65 N 30, 35 A 40

§ 1. INTRODUCTION

Our aim is to develop a Finite Element Method for elliptic differential equations on arbitrary two–dimensional surfaces - not necessarily embedded - in $\mathbb{R}^3$. We shall avoid global parametrizations, and think of surfaces just given by splines. The most important point in our method is that we write down the Laplace-Beltrami operator in terms of the tangential gradient. In order to present the idea we shall confine ourselves to the most simple equation

$$-\Delta_S u = f \text{ on } S .$$

$-\Delta_S$ is the Laplace-Beltrami operator on S . Let us assume for the moment that $\partial S = \emptyset$. We approximate the surface S by a polyhedron S_h and solve

$$-\Delta_{S_h} u_h = f \text{ on } S_h$$

weakly. We use linear elements on the surface S_h , i. e. u_h is a linear polynomial on each triangle of S_h and globally continuous. The Laplace-Beltrami operator on S_h is defined by

$$\int_{S_h} \nabla_{S_h} u \, \nabla_{S_h} \varphi = \int_{S_h} f \varphi$$

for all φ in the Sobolev space $H^1(S_h)$ where

$$\nabla_{S_h} u = \nabla u - (\nabla u \cdot n_h) n_h$$

is the tangential gradient on S_h , ∇ is the three-dimensional gradient and n_h is the normal vector to S_h .

Practically this means that $\nabla_{S_h} u_h$ is constant on each triangle of S_h if u_h is linear. If we take $\varphi_{h1},\ldots,\varphi_{hN}$ (N = number of vertices $x_{(k)}$ of S_h) to be those piecewise linear functions on S_h which are globally continous and $\varphi_{hj}(x_{(k)}) = \delta_{jk}$ then

$$u_h(x) = \sum_{j=1}^{N} u_j \varphi_{hj}(x)$$

and we have to solve the linear system

$$\sum_{j=1}^{N} u_j \int_{S_h} \nabla_{S_h} \varphi_{hj} \nabla_{S_h} \varphi_{hk} = \int_{S_h} f \varphi_{hk}$$

$(k = 1,\ldots,N)$. Thus the numerical scheme is just the same as in a plane two-dimensional problem. The only difference is that in our case the computer has to memorize three-dimensional nodes instead of two-dimensional ones. Since the triangles of S_h can be parametrized via the unit triangle in $\mathbb{R}^2$ the method is fairly easy.

We prove that the order of convergence is the same as in plane problems.

Let us mention that error-estimates on surfaces have been proved by J.C.Nedelec in [N] for the boundary element method.

In [BF] and in [S] the authors construct spherical Finite Elements in order to solve problems on $S = S^2$.

§ 2. CONTINUOUS PROBLEM

We consider a compact $C^{k,\alpha}$-hypersurface S $(k \in \mathbb{N} \cup \{0\}$, $0 \leq \alpha \leq 1)$ in $\mathbb{R}^3$. For simplicity we assume that S can be represented globally by some oriented distance function d which is defined on some open subset U of $\mathbb{R}^3$.

$$S = \{x \in U \mid d(x) = 0\}$$

d is in $C^{k,\alpha}(U)$, $\nabla d \neq 0$. Almost everywhere the normal to S in the direction of growing d is given by

(1) $$\int_S \nabla_S u \, \nabla_S \varphi = \int_S f \varphi ,$$

and

$$\|u\|_{H^2(S)} \leq c \, \|f\|_{L^2(S)} .$$

b) $\partial S = \emptyset$. For every $f \in L^2(S)$ with $\int_S f \, do = 0$ there exists a weak solution $u \in H^1(S)$ of

$$-\Delta_S u = f \quad \text{on} \quad S ,$$

i.e. (1) holds for all $\varphi \in H^1(S)$, and u is unique up to a constant , and

$$\|u\|_{H^2(S)} \leq c \left[\|f\|_{L^2(S)} + \|u\|_{L^2(S)} \right] .$$

§ 3. DISCRETE PROBLEM

We shall approximate the smooth surface S by a surface S_h which globally is of class $C^{0,1}$. For example S_h is a polyhedron consisting of triangles T_h of size proportional to h^2 with corners on S . The conclusions of 1. Theorem hold as long as H^2 is not involved. Let X_h be a finite-dimensional subspace of $H^1(S_h)$.

2. THEOREM:

a) $\partial S_h \neq \emptyset$. For every $f_h \in L^2(S_h)$ there exists a unique weak solution $u_h \in X_h \cap \overset{\circ}{H}{}^1(S_h)$ of

$$-\Delta_{S_h} u_h = f_h \quad \text{on} \quad S_h \, , \; u_h = 0 \quad \text{on} \quad \partial S_h .$$

b) $\partial S_h = \emptyset$. For every $f_h \in L^2(S_h)$ with $\int_{S_h} f_h \, do_h = 0$ there exists a weak solution $u_h \in X_h$ of

$$-\Delta_{S_h} u_h = f_h \quad \text{on} \quad S_h$$

which is unique up to a constant.

The <u>proof</u> is a simple application of the usual Hilbert space methods.

§ 4. PROJECTION

We shall first have a look at the case where $S \in C^3$ is approximated

$$\int_S \nabla_S u \, \nabla_S \varphi = \int_S f \varphi \,, \tag{1}$$

and

$$\|u\|_{H^2(S)} \leq c \, \|f\|_{L^2(S)} \,.$$

b) $\partial S = \emptyset$. For every $f \in L^2(S)$ with $\int_S f \, do = 0$ there exists a weak solution $u \in H^1(S)$ of

$$-\Delta_S u = f \quad \text{on} \quad S \,,$$

i.e. (1) holds for all $\varphi \in H^1(S)$, and u is unique up to a constant , and

$$\|u\|_{H^2(S)} \leq c \left[\|f\|_{L^2(S)} + \|u\|_{L^2(S)} \right] .$$

§ 3. DISCRETE PROBLEM

We shall approximate the smooth surface S by a surface S_h which globally is of class $C^{0,1}$. For example S_h is a polyhedron consisting of triangles T_h of size proportional to h^2 with corners on S . The conclusions of 1. Theorem hold as long as H^2 is not involved. Let X_h be a finite-dimensional subspace of $H^1(S_h)$.

2. THEOREM:

a) $\partial S_h \neq \emptyset$. For every $f_h \in L^2(S_h)$ there exists a unique weak solution $u_h \in X_h \cap \overset{\circ}{H}{}^1(S_h)$ of

$$-\Delta_{S_h} u_h = f_h \quad \text{on} \quad S_h \,, \quad u_h = 0 \quad \text{on} \quad \partial S_h \,.$$

b) $\partial S_h = \emptyset$. For every $f_h \in L^2(S_h)$ with $\int_{S_h} f_h \, do_h = 0$ there exists a weak solution $u_h \in X_h$ of

$$-\Delta_{S_h} u_h = f_h \quad \text{on} \quad S_h$$

which is unique up to a constant.

The *proof* is a simple application of the usual Hilbert space methods.

§ 4. PROJECTION

We shall first have a look at the case where $S \in C^3$ is approximated by a polyhedron $S_h \in C^{o,1}$ which is the union of triangles T_h with diameter $\leq c_1 h$ and inner radius $\geq c_2 h$ and corners on S. Refering to our considerations in 2. we define

$$T = \{a(x) \in S \mid x \in T_h\}$$

(see figure 1.)

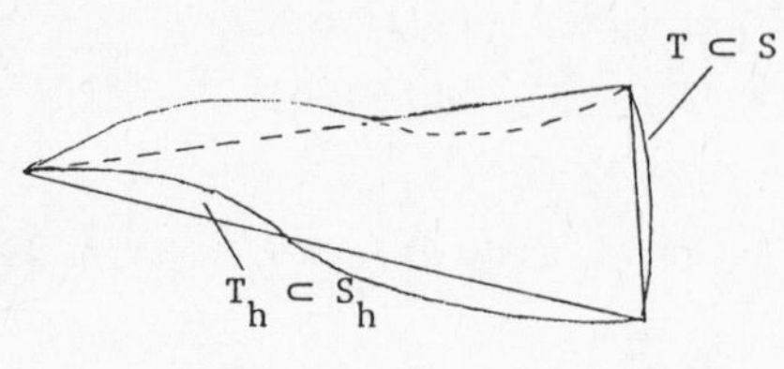

Fig. 1

In order to compare the discrete solution on S_h with the continuous solution on S we lift a function v_h defined on S_h onto S by

(2) $$v_h(x) = v(x-d(x)n(x)) \qquad (x \in T_h) .$$

<u>3. LEMMA</u>

$$\frac{1}{c} \|v_h\|_{L^2(T_h)} \leq \|v\|_{L^2(T)} \leq c \, \|v_h\|_{L^2(T_h)}$$

$$\frac{1}{c} |v_h|_{H^1(T_h)} \leq |v|_{H^1(T)} \leq c \, |v_h|_{H^1(T_h)}$$

$$|v_h|_{H^2(T_h)} \leq c \left[|v|_{H^2(T)} + h|v|_{H^1(T)} \right]$$

<u>Proof:</u> It is obvious that for $\mu_h = do \,/\, do_h$ we have

$$0 < \frac{1}{c} \leq \mu_h \leq c < \infty .$$

Thus

$$\frac{1}{c} \|v\|_{L^2(T)} \leq \|v_h\|_{L^2(T_h)} \leq c \, \|v\|_{L^2(T)} .$$

On each triangle with normal n_h

$$\nabla_{S_h} v_h = \nabla v_h - (n_h \cdot \nabla v_h) \, n_h$$

$$= P_h \, \nabla v_h$$

with $$P_{hik} = \delta_{ik} - n_{hi} \, n_{hk} \quad (i,k = 1,2,3) \text{ , and}$$

$$\nabla v_h = (P - dH) \nabla v$$

where $P_{ik} = \delta_{ik} - n_i n_k$ and $H_{ik} = d_{x_i x_k} = n_{i x_k} = n_{k x_i}$. But since $PH = HP = H$ we get

$$\nabla_{S_h} v_h = P_h(I - dH)\, P\nabla v = P_h(I - dH)\, \nabla_S v .$$

Because of $|n - n_h| \leq ch$ and $|d| \leq ch^2$, for $h \leq h_o$:

$$|\nabla_S v \cdot n_h| \leq \frac{1}{2} |\nabla_S v| ,$$

$$\frac{1}{c} |\nabla_S v| \leq |\nabla_{S_h} v_h| \leq c\, |\nabla_S v|$$

on T , i.e.

$$\frac{1}{c} |v|_{H^1(T)} \leq |v_h|_{H^1(T_h)} \leq c\, |v|_{H^1(T)}$$

A short calculation delivers

$$|D_{S_h i} D_{S_h k} v_h| \leq c \left[\sum_{|\mu|=2} |D_S^{\mu} v| + (|n_i - (n \cdot n_h)\, n_{hi}| + |d|)\, |\nabla_S v| \right]$$

$$\leq c \left[\sum_{|\mu|=2} |D^{\mu} v| + h |\nabla_S v| \right]$$

which proves the Lemma.

§ 5. ENERGY ESTIMATE

We now are ready to prove the energy estimate. Let us first of all treat the case $\partial S = \emptyset$, $\partial S_h = \emptyset$. So we have got $u \in H^2(S)$, $u_h \in X_h \subset H^1(S_h)$, $f \in L^2(S)$, $\int_S f = 0$, $f_h \in L^2(S_h)$, $\int_{S_h} f_h = 0$ with

$$\int_S \nabla_S u\, \nabla_S\, \varphi do = \int_S f\, \varphi do \qquad (\varphi \in H^1(S)) \tag{3}$$

and

$$\int_{S_h} \nabla_{S_h} u_h \nabla_{S_h} \varphi_h\, do_h = \int_{S_h} f_h\, \varphi_h\, do_h \qquad (\varphi_h \in X_h) \tag{4}$$

According to (2) we define

$$u_h(x) = U_h(x - d(x) n(x)) \qquad (x \in S_h)$$

and

$$\varphi_h(x) = \Phi_h(x - d(x) n(x)) \qquad (x \in S_h).$$

With these transformations we get from (4)

$$\int_S P_h\, (I - dH)\, \nabla_S\, U_h\, P_h\, (I - dH)\, \nabla_S\, \Phi_h \frac{1}{\mu_h}\, do = \int_S F_h\, \Phi_h\, do$$

where F_h is the transformed f_h times $1/\mu_h$. If we define $A_h = \frac{1}{\mu_h} P(I - dH)\, P_h(I - dH) P$, since P is a projection this reads

$$\int_S \nabla_S\, U_h\, \nabla_S\, \Phi_h\, do = \int_S F_h\, \Phi_h\, do + \int_S (A_h - I)\, \nabla_S\, U_h\, \nabla_S\, \Phi_h\, do$$

This together with (3) gives us

$$\int_S \nabla_S (u - U_h)\, \nabla_S \phi_h \, do = \int_S (I - A_h)\, \nabla_S U_h \, \nabla_S \phi_h \, do + \int_S (f - F_h)\, \phi_h \, do$$

for all projections $\phi_h \in H^1(S)$ of testfunctions $\varphi_h \in X_h$. So,

$$\begin{aligned}
\|\nabla_S (u - U_h)\|^2_{L^2(S)} &= \int_S \nabla_S (u - U_h)\, \nabla_S (u - \phi_h)\, do \\
&\quad + \int_S (A_h - I)\, \nabla_S U_h \, \nabla_S (U_h - \phi_h)\, do \\
&\quad - \int_S (f - F_h)\,(U_h - \phi_h)\, do \\
&\le \|\nabla_S (u - U_h)\|_{L^2(S)} \, \|\nabla_S (u - \phi_h)\|_{L^2(S)} \\
&\quad + \|(A_h - I)P\|_{L^\infty(S)} \, \|\nabla_S U_h\|_{L^2(S)} \, \|\nabla_S (U_h - \phi_h)\| \\
&\quad + \|f - F_h\|_{L^2(S)} \, \|U_h - \phi_h\|_{L^2(S)} .
\end{aligned}$$

Here it is important for later use that we have to estimate $\|(A_h - I)P\|_{L^\infty(S)}$ instead of $\|(A_h - I)\|_{L^\infty(S)}$.

Without loss of generality we assume that

$$\int_S U_h - \phi_h \, do = 0$$

and so, by Poincaré's inequality on S and elementary operations we achieve

$$\|\nabla_S (u - U_h)\|_{L^2(S)} \le c \Big[\|\nabla_S (u - \phi_h)\|_{L^2(S)} + \|(A_h - I)P\|_{L^\infty(S)} \, \|\nabla_S U_h\|_{L^2(S)} + \|f - F_h\|_{L^2(S)} \Big]$$

Now we observe that

$$|1 - \mu_h| \le ch^2 \;, \; |d| \le ch^2$$

and

$$|(A_h - I)P| \le ch^2 .$$

We shall prove the last inequality only.

$$\begin{aligned}
(A_h - I)\, P &= \Big(\frac{1}{\mu_h} P(I - dH)P_h(I - dH)P - I\Big)P \\
&= (P(I - dh)P_h(I - dH)P - I)P + O(h^2) \\
&= PP_hP - P + O(h^2)
\end{aligned}$$

$$|(A_h - I)P| \le |n \wedge n_h|^2 + ch^2 \le ch^2 .$$

In addition to that we have as an a priori bound from the discrete problem

$$\|\nabla_S U_h\|_{L^2(S)} \le c \, \|f_h\|_{L^2(S_h)} .$$

This altogether implies the inequality

$$(5)\quad \|\nabla_S(u - U_h)\|_{L^2(S)} \leq c \left[\inf \|\nabla_S(u - \phi_h)\|_{L^2(S)} + \|f_h\|_{L^2(S_h)} h^2 + \|f - F_h\|_{L^2(S)} \right].$$

Now we have to make clear which f_h we shall chose with respect to f. The most simple choice would be to take

$$f_h(x) = f(x - d(x)n(x))\mu_h(x) \qquad (x \in S_h) ,$$

but numerically it is not easy to compute μ_h. So, let us take

$$(6)\qquad f_h = \tilde{f}_h - \int_{S_h} \tilde{f}_h \, do_h$$

where $\tilde{f}_h$ is the lifted f. But then it is clear that in (5)

$$\|f_h\|_{L^2(S_h)} \leq c\|f\|_{L^2(S)}$$

and

$$\|f - F_h\|_{L^2(S)} \leq ch^2 \|f\|_{L^2(S)} .$$

Up to now we have proved:

<u>4. LEMMA:</u>

Let $\partial S = \emptyset$, $S \in C^3$. If u is a continuous solution as in 1. Theorem b and u_h is a discrete solution as in 2. Theorem b with respect to f_h defined in (6), then

$$(7)\qquad |u - U_h|_{H^1(S)} \leq c \left[\inf_{\phi_h \in Y_h} |u - \phi_h|_{H^1(S)} + h^2 \|f\|_{L^2(S)} \right]$$

where U_h is defined by

$$u_h(x) = U_h(x - d(x)n(x)) \qquad (x \in S_h)$$

and $\quad Y_h = \{\phi_h(x - d(x)n(x)) = \varphi_h(x) \quad (x \in S_h) , \varphi_h \in X_h\}$.

This means that we have estimated the consistency error which stems from the approximation of S by S_h. It remains to define an interpolation operator from $H^2(S)$ to Y_h.

<u>5. LEMMA:</u>

For $S \in C^3$, $\partial S = \emptyset$ let

$$X_h = \{\varphi_h : S_h \longrightarrow \mathbb{R} \mid \varphi_h|_{T_h} \text{ linear polynomial, } \varphi_h \in C^0(S_h)\}$$

and Y_h the transformed space as in 4. Lemma. Then for given

$u \in H^2(S)$ there exists a unique $I_h u \in Y_h$ such that

$$|u - I_h u|_{H^1(S)} \leq ch \left[|u|_{H^2(S)} + h|u|_{H^1(S)} \right].$$

Proof: According to Sobolev's theorem u is in $C^o(S)$, and so the linear interpolation $\tilde{I}_h u \in X_h$ is well defined by

$$\tilde{I}_h u(a_j) = u(a_j)$$

where a_j $(j=1,\dots,N)$ are the nodes of S_h. It is well known, [C], that for $\tilde{u} = u$ lifted onto S_h:

$$|\tilde{u} - \tilde{I}_h \tilde{u}|_{H^1(T_h)} \leq ch |\tilde{u}|_{H^2(T_h)}.$$

But with 3. Lemma this implies

$$\begin{aligned} |u - I_h u|_{H^1(T)} &\leq c |\tilde{u} - \tilde{I}_h \tilde{u}|_{H^1(T_h)} \\ &\leq ch |\tilde{u}|_{H^2(T_h)} \\ &\leq ch \left[|u|_{H^2(T)} + h |u|_{H^1(T)} \right]. \end{aligned}$$

Let us summarize the results.

6. LEMMA:

Let the situation be as in 4. Lemma. Then

$$(8) \qquad |u - U_h|_{H^1(S)} \leq ch \|f\|_{L^2(S)}$$

if X_h is as in 5. Lemma.

§ 6. L^2-ESTIMATE AND RESULT

We employ the Aubin-Nitsche-trick in order to get quadratic asymptotic convergence in the $L^2(S)$-norm. Let us confine ourselves to surfaces S without boundary.

7. LEMMA:

Let $\partial S = \emptyset$. Then

$$\|u - U_h\|_{L^2(S)/\mathbb{R}} \leq ch^2$$

Proof: We solve the problem

$$-\Delta_S v = u - U_h - m \quad \text{on} \quad S, \quad \int_S v\,do = 0,$$

where

$$m = \int_S u - U_h \, do .$$

Due to 1. Theorem there exists a unique solution $v \in H^2(S)$ and

$$\|v\|_{H^2(S)} \leq c \, \|u - U_h\|_{L^2(S)/\mathbb{R}} .$$

$$\begin{aligned}
\|u - U_h\|^2_{L^2(S)/\mathbb{R}} &= \int_S \nabla_S(u - U_h) \, \nabla_S \, v \, do \\
&= \int_S \nabla_S(u - U_h) \, \nabla_S(v - I_h v) \, do \\
&\quad + \int_S (I - A_h) \, P\nabla_S \, U_h \, \nabla_S \, I_h \, v \, do \\
&\leq \|\nabla_S(u - U_h)\|_{L^2(S)} \, \|\nabla_S(v - I_h v)\|_{L^2(S)} \\
&\quad + ch^2 \, \|f\|_{L^2(S)} \, \|u - U_h\|_{L^2(S)/\mathbb{R}} \\
&\leq ch^2 \, \|u - U_h\|_{L^2(S)/\mathbb{R}} .
\end{aligned}$$

We remark that we never seriously used that S had no boundary. This means that the energy estimate (8) remains valid for $\partial S \neq 0$ as long as $u \in H^2(S)$ although $\partial S \in C^{0,1}$ only. S is the projection of a polyhedron S_h and thus has piecewise smooth boundary. We assume in the case $\partial S \neq \emptyset$ that for every $f \in L^2(S)$ the weak solution is in $H^2(S)$ and the corresponding a priori bound holds. So we can summarize our results.

8. THEOREM

Let $\partial S = \emptyset$, $S \in C^3$. If u is a continuous solution as in 1. Theorem b and u_h is a discrete solution with respect to f_h defined in (6), then for U_h as in 4. Lemma

$$\|u - U_h\|_{L^2(S)/\mathbb{R}} + h \, |u - U_h|_{H^1(S)} \leq ch^2 .$$

If $S \in C^3$, $\partial S \neq \emptyset$ is the projection of a polyhedron such that for every $f \in L^2(S)$ the bound $\|u\|_{H^2(S)} \leq c \, \|f\|_{L^2(S)}$ holds, then

$$\|u - U_h\|_{L^2(S)} + h \, |u - U_h|_{H^1(S)} \leq ch^2 .$$

§ 7. NUMERICAL EXAMPLE

To illustrate the method and to assess the sharpness of the convergence rate given in the preceeding section, we present numerical results for a simple test problem. The surface S is taken to be

$$S = \{x \in \mathbb{R}^3 \mid (x_1 - x_3^2)^2 + x_2^2 + x_3^2 = 1\}$$

and we consider the problem $-\Delta_S u = f$ on S whose exact solution is given by $u(x) = x_1 x_2$. Let us remark that the right hand side f is not that simple since

$$f = -\nabla_S \cdot v \ , \ v = \nabla_S u = \nabla u - (\nabla u \cdot n) n$$

$$\nabla_S \cdot v = \nabla \cdot v - \sum_{j=1}^{3} (\nabla v_j \cdot n) n_j$$

where

$$n(x) = (x_1 - x_3^2, x_2, x_3(1 - 2(x_1 - x_3^2))) \ / \ (1 + 4x_3^2(1 - x_1 - x_2^2))^{1/2}$$

We start with a very crude six-node approximation of S . If h_j is the largest diameter of the jth grid we determine the experimental order of convergence by

$$\ln \frac{E(h_j)}{E(h_{j+1})} \ / \ \ln \frac{h_j}{h_{j+1}} \qquad (j=1,\ldots,5)$$

where E is the relative error in the L^2-norm

$$E(h) = \|u - U_h\|_{L^2(S)} \ / \ \|u\|_{L^2(S)} \ .$$

The results are given in table 1.

triangulation level	nodes	triangles	h	relative L^2-error	experimental order of convergence
1	6	8	2.236	0.8120	
					1.06
2	18	32	1.399	0.4930	
					1.93
3	66	128	0.8426	0.1855	
					2.10
4	258	512	0.4613	0.5227 E-1	
					2.04
5	1026	2048	0.2384	0.1356 E-1	
					1.99
6	4098	8192	0.1233	0.3664 E-2	

Table 1 Results for the test problem

In order to give an impression of the discretization we plot the approximation surface S_h and some limes of the discrete solution on S_h in Figure 2

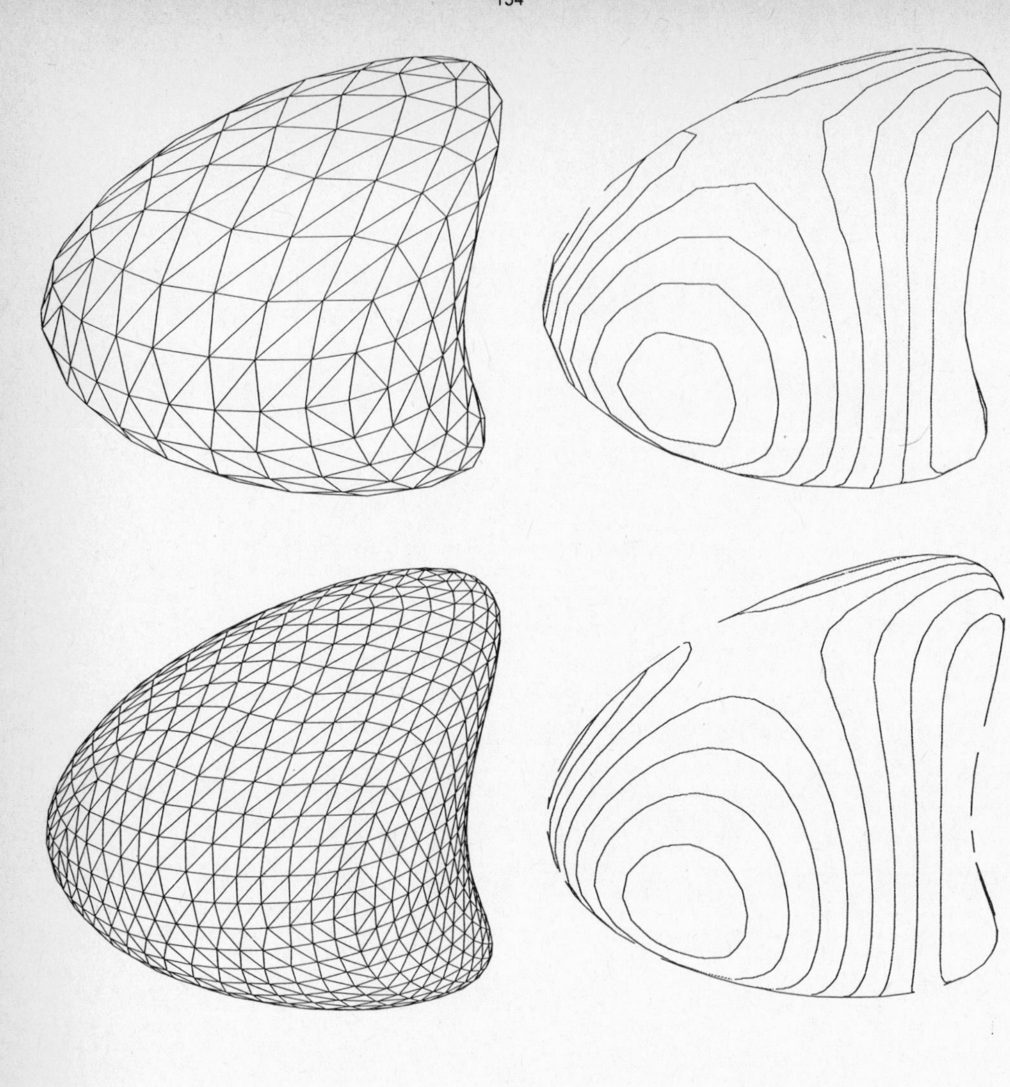

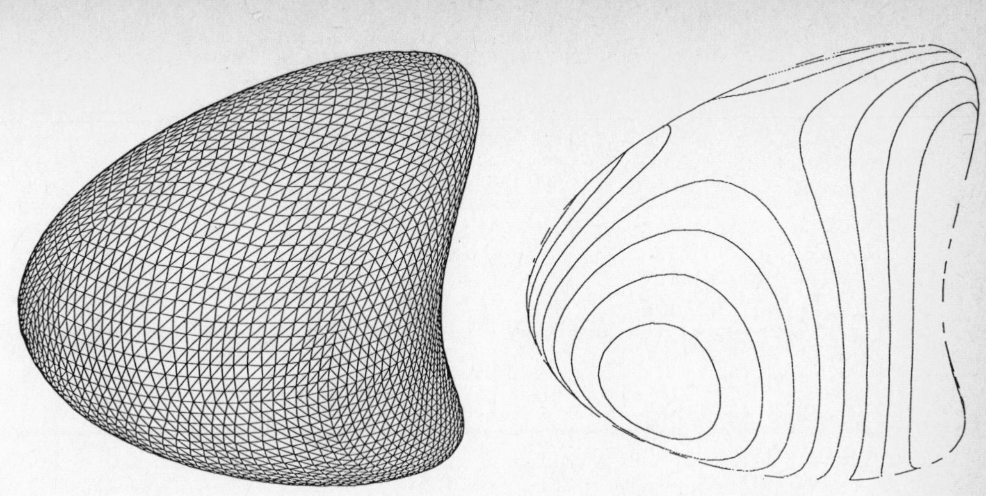

Fig. 2. S_h and level lines of u_h on S_h for triangulations 1-6 .

[A] Aubin, Th.: Nonlinear Analysis on Manifolds. Monge-Ampère Equations (1982) New York-Heidelberg-Berlin

[BF] Baumgartner, J.R.; Frederickson, P.O.: Icosahedral Discretization of the Two-Shere. SIAM J. Numer. Anal. 22 (6), (1985), 1107-1115

[C] Ciarlet, P.G.: The Finite Element Method for Elliptic Problems. (1987) Amsterdam-New York-Oxford

[N] Nedelec, J.C.: Curved Finite Element Methods for the Solution of Integral Singular Equations on Surfaces in $\mathbb{R}^3$. Computing Methods in Appl. Sciences and Engeneering (1976), 374-390, Lecture Notes in Economics and Math. Systems

[S] Steger, L.K.: Sphärische Finite Elemente und ihre Anwendung auf Eigenwertprobleme des Laplace-Beltrami-Operators. Dissertation München (1983)

[W] Wloka, J.: Partielle Differentialgleichungen. Stuttgart (1982)

COMPARISON PRINCIPLES IN CAPILLARITY

Robert Finn

Contents

The study of capillary surfaces can be traced to times of antiquity, however a formal theory capable of quantitative prediction first appears in the writings of Young [1] and of Laplace [2] in the early nineteenth century. The theory was later put onto a firm conceptual foundation by Gauss [3], who obtained both the equation and the boundary condition as a consequence of the principle of virtual work. For background details, see [4], Chapter 1. We take here as starting point that an equilibrium capillary surface $\mathcal{S}$ separating two fluid (or fluid and gas) regions in $\mathbb{R}^3$ is determined by the conditions:

a) the mean curvature H of $\mathcal{S}$ is to satisfy a relation

$$2H = \varphi(p) + \lambda \tag{1}$$

where $\varphi(p)$ is the potential of an external force field and λ a Lagrange parameter determined by an eventual volume constraint, and

b) $\mathcal{S}$ meets prescribed rigid bounding walls S locally in an angle γ which depends only on the materials (and not on the shape of $\mathcal{S}$ or of S or on the thickness of S or on volumes of the fluids). In some cases (that will not be considered here) part or all of the boundary of $\mathcal{S}$ is prescribed.

If $\mathcal{S}$ is smooth then locally it can be represented as a graph $z = u(x,y)$, and we obtain the equation

$$\text{div } Tu = \varphi(p) + \lambda \tag{2}$$

with

$$Tu = \frac{\nabla u}{\sqrt{1+|\nabla u|^2}} \quad . \tag{3}$$

Our interest will center on the case for which $\varphi(p)$ is a vertically directed gravitational potential; then

$$\text{div } Tu = \varphi(u) + \lambda \quad . \tag{4}$$

We find $\varphi'(u) > 0$ (< 0) according as the denser fluid lies locally below (above) $\mathcal{S}$. On the earth's surface

$$\text{div } Tu = \kappa u + \lambda \quad , \quad \kappa = \frac{\rho g}{\sigma} \; , \tag{5}$$

with ρ = density change across $\mathcal{S}$, σ = surface tension, g = gravitational acceleration. If $\kappa \neq 0$ it is always possible to choose coordinates so that $\lambda = 0$. For a cylindrical tube inserted vertically into an infinite reservoir we find in this case that $z = 0$ corresponds to the asymptotic level surface of the reservoir.

In the absence of gravity (or if $\rho = 0$) we obtain a surface of constant mean curvature $H = \frac{1}{2}\lambda$. The height $u(x,y)$ then satisfies, locally,

$$\text{div } Tu = \lambda = 2H \equiv \text{const} \quad . \tag{6}$$

In this latter case two families of rotation surfaces satisfying (6) can be determined explicitly in terms of elliptic integrals. These are designated as *unduloids* and as *nodoids*. Both families include the sphere and the circular cylinder as limiting configurations. Except in the special case $H \equiv 0$, these are the only explicit solutions known. (It now seems in principle feasible to construct further solutions by means of a remarkable representation formula due to Kenmotsu [5], however none of particular interest has yet been produced).

If $\kappa \neq 0$, one dimensional solutions of (5) corresponding to the immersion of a vertical plate (or of two parallel vertical plates) into a reservoir, with constant contact angle γ (or angles γ_1, γ_2) on the vertical walls, can be given by an explicit formula, again involving elliptic integrals (cf. [6] p. 77); however no other explicit solutions have been found. Even in the simplest case of a cylindrical capillary tube with circular section, no solutions in closed form are known to exist. Nevertheless, in 1806 Laplace [2], using a method that has come to be known as "matching expansions" obtained his now celebrated formula

$$u_0 \approx \frac{2\cos\gamma}{\kappa a} - \left[\frac{1}{\cos\gamma} - \frac{2}{3}\,\frac{1-\sin^3\gamma}{\cos^3\gamma}\right] a \tag{7}$$

for the height at the center of a narrow tube of radius a in an infinite reservoir. Laplace did not define what he meant by "narrow", although (7) clearly provides no information if a is large enough. Nor did he prove that (7) is asymptotic as $a \to 0$. We return to these questions in §1, where we will settle them by the method of *comparison of volumes*.

Although a satisfactory theory was not available prior to the time of Laplace, capillarity problems were actively studied and experiments were made for various geometrical configurations. In 1712 Brook Taylor [7] examined the configuration of water in a wedge between two vertical glass plates meeting in an angle $2\alpha \approx 2.5^{\circ}$. He wrote that the contact curve where the upper water surface meets the glass "seems to approach very near to the common hyperbola". His observation was substantiated with more precise measurements by Hawksbee [8], who also used other liquids, among them "spirits of wine".

Musschenbroek [9] and also Ferguson and Vogel [10] have provided "proofs" of this behavior for any contact angle in the range $0 \le \gamma < \frac{\pi}{2}$, while Princen [11] gave a "proof" for $\gamma = 0$ and stated that other contact angles could be studied similarly. The "hyperbolic" behavior is illustrated in the Encyclopaedia article of Minkowski [12] and the book by Boys [13] and has become popular for college laboratory demonstrations (cf. the beautifully made film on surface tension by Trefethen).

In §2 we introduce a form of the maximum principle (designated CP) that is in some ways characteristic for the nonlinearity in the mean curvature operator. On applying this principle (in §3) to the wedge configuration just described, we find that the behavior can be quite different from that envisaged in the above discussions and experiments. Specifically, we find that if $\alpha + \gamma \ge \frac{\pi}{2}$ then every solution is bounded in the corner, and thus the "hyperbolic" behavior does not occur. The behavior does become hyperbolic (asymptotically) whenever

$\alpha + \gamma < \frac{\pi}{2}$. Thus, there is a discontinuous dependence on the boundary data (and/or domain of definition). The discontinuity is reflected in reality and can be verified by simple experiments.

The above result would seem to suggest that for fixed γ the behavior at a corner should improve with increasing α. Not so. Tam [14] has in fact shown that if $\alpha + \gamma \geq \frac{\pi}{2}$, $\alpha < \frac{\pi}{2}$ then the normal to $\mathcal{S}$ is continuous up to the vertex V; however Korevaar [15] in a striking application of the above comparison principle, showed that if $\alpha > \frac{\pi}{2}$ the solution can be discontinuous at V. We describe his contribution in §4, where we also show, by a similar method, that even on a smooth boundary a discontinuity in contact angle can lead to a discontinuity in the contact curve.

In §4bis we apply CP to show that if $\varphi'(u) \geq 0$ then every isolated singularity of a solution of (4) is removable. This theorem can fail if $\varphi'(u) < 0$ (negative gravity), see, e.g. [52, 54, 55, 69].

Solutions of equations of the form (4) are bounded a priori, depending only on domain, under much weaker conditions on $\varphi(u)$ than are used in §3. In §5 we employ a "touching principle" to show that a bound from above holds whenever $\lim_{z\to\infty} \varphi(z) = \infty$, and a bound from below if $\lim_{z\to-\infty} \varphi(z) = -\infty$. The result holds under the same geometrical conditions as those of §3, and does not require $\varphi'(z) \geq 0$; thus it holds under conditions for which the maximum principle can fail.

In §6 we describe "comparison by monotone symmetry". We use the method to show that if a capillary tube of arbitrary section $\Omega^{(i)}$ lies interior to a circular one $\Omega^{(0)}$ of the same material, then $\Omega^{(i)}$ raises liquid at each point to a greater height than does $\Omega^{(0)}$ at that point. The same result holds for certain other choices of $\Omega^{(0)}$ (e. g. an infinite strip); however in general the situation can be quite different. In fact, it can happen that $\Omega^{(i)}$ raises less volume of fluid than does $\Omega^{(0)}$ over the section $\Omega^{(i)}$.

A reverse form of the above principle, due to Siegel [27] is also described in §6.

For functions of one variable global inequalities on derivatives can be inferred from inequalities on curvature. In §7 we indicate the use of this simple principle to obtain global estimates on rotationally symmetric pendant drops.

Forms of the maximum principle can be used to bound derivatives for capillary surfaces $u(x)$, without symmetry restriction. This was first shown by Spruck [16] who showed that interior gradient estimates together with the maximum principle can yield boundary gradient estimates, see §8. Korevaar [17, 18, 19] and, independently, Lieberman [20, 21], showed by the rather surprising device of using the solution itself to construct a barrier, that the maximum principle can also yield the interior gradient bounds, see §9. These authors later extended the method to obtain estimates in the closed domain, also in higher dimension.

The methods give interior information depending only on a bound for the solution, and then global information depending only on regularity of $\partial\Omega$. For the equation (4) with $\lim_{|z|\to\infty} (\text{sgn } z)\varphi(z) = \infty$, the solution can be bounded a priori as in §3, and if $\Sigma = \partial\Omega$ is smooth the bound extends up to Σ. This enables an existence proof for a solution regular up to Σ, when $\varphi' > \delta > 0$ and $0 < \gamma < \pi$. If $\Sigma \in C^2$ the bound on $|u|$ is independent of γ; Spruck used this information to obtain for $\gamma = 0$ or π the existence of a solution $u(x) \in C^2(\Omega) \cap C^0(\partial\Omega)$ which is Lipschitz continuous on $\partial\Omega$. These cases, which correspond to perfect wetting or non wetting, are necessarily singular in the sense that $|\nabla u| \to \infty$ as Σ is approached, which in turn entails loss of uniformity in the ellipticity of (4); in this sense Spruck's result could not have been predicted.

The most singular case of all is that in which both $\varphi'(z) \equiv 0$, so that the surface $u(x,y)$ is governed by the equation (6), and also $\gamma = 0$ or π on the boundary. In general it cannot be expected that solutions will exist (cf. [4], Chapter 6), however under some conditions the existence of a "variational solution" that assumes the boundary data in a weak sense can be shown (cf. [4]; Chapter 7). We take up such cases in §11 with the aid of CP and the "principle of nth order division". It turns out the (singular) data must be assumed strictly at every smooth boundary point, and a precise estimate can be given for the rate of approach of $\nu \cdot Tu$ to its limiting value.

The same principle was used by Chen and Huang [22] to prove strict convexity of the capillary surfaces over a convex Ω in the absence of gravity when $\gamma = 0$; by Chen [23] and by Siegel [37] to prove uniqueness of the minimal point when Ω is convex, and by Huang [24] to estimate the distance of the minimal point from $\partial\Omega$, see §12. Korevaar [25] used the classical maximum principle to prove convexity also in a gravity field, see §13, but he required stricter hypotheses on $\partial\Omega$ and on the solution near $\partial\Omega$. Convexity can fail for any $\gamma \neq 0, \frac{\pi}{2}$, see [26].

In §14 we show that a sessile drop can wet a convex domain on a plane and not itself be convex. In the final §15 we discuss some applications of the E. Hopf boundary point lemma and its generalizations. The lemma was used by Siegel [27] to obtain conditions under which an "inner" domain $\Omega^{(i)}$ will raise fluid higher than will an "outer" $\Omega^{(0)} \supset \Omega^{(i)}$; it was used by Aleksandrov [28], by Serrin [29] and then by Wente [30] to prove deep and elegant symmetry results, and most recently by Vogel [31] to obtain conditions under which a capillary surface projects simply onto its base plane.

1. <u>Comparison of Volumes</u>

In the interest of making quantitative Laplace's concept of "narrow" tube in the case $\varphi(z) = \kappa z$ considered by him, we introduce a representative length a and replace u, x, y in (5) by ua, xa, ya. With g, ρ, σ, κ as before, (5) takes the non dimensional form

$$\text{div}\, Tu = Bu + \lambda a \quad ; \tag{8}$$

further replacement of u by $u - \lambda a B^{-1}$ yields

$$\text{div}\, Tu = Bu \quad . \tag{9}$$

Here $B = \kappa a^2$, known as the Bond number (cf. [32]), serves to distinguish "small" from "large" configurations. The boundary condition is homogeneous in the variables and remains unchanged.

The equivalence of (8) with (9) shows that *the manifold of all solutions of* (8) *is encompassed with the single parameter* B; that is, any solution of (8) can be found also among the solutions of (9), and can be achieved by addition of a constant (this property was crucial for the uniqueness theorem for sessile drops, see [33]). The equivalence of (5) with (9) shows that *in any situation for which uniqueness holds, solutions* $u^{(1)}$ *and* $u^{(2)}$ *of* (5) *are geometrically similar if and only if they correspond to the same value of* B. Thus, the nondimensional equation (9) groups the solution surfaces of (5) into equivalence classes of geometrically similar surfaces that need not be distinguished from each other.

For a capillary tube with section Ω, Laplace discovered the formula for the fluid volume lifted above rest level

$$(10) \qquad \mathcal{V} = B^{-1} \oint_{\Sigma} \cos \gamma \, ds$$

giving $\mathcal{V}$ explicitly in terms of the prescribed data on $\Sigma = \partial\Omega$; if γ is constant, as we assume in this section, then $\mathcal{V} = B^{-1}|\Sigma|\cos \gamma$. Thus if a comparison surface $v(x)$ can be found with explicitly known volume greater or less than $\mathcal{V}$, it should be possible to obtain useful information by means of (10).

Let us apply this idea to a symmetric configuration in a circular tube [34, 35]. We compare the exact solution $u(x,y) \equiv u(r)$ of (9) with two spherical caps $v^{(1)}$ and $v^{(2)}$. Here $v^{(1)}$ is chosen so that $v^{(1)}(0) = u(0) = u_0$ and so that the mean curvatures are equal at the point. It can then be shown (cf. §7) that $v^{(1)}(r) < u(r)$, all $r > 0$. Thus $v^{(1)}$ raises the smaller volume. Next we choose $v^{(2)}$, also passing through u_0, such that $v^{(2)}$ meets the outer boundary $r = 1$ at the prescribed angle γ. It can then be shown that $v^{(2)}(r) > u(r)$, all $r > 0$. Formal calculation of the comparison volumes, together with (10), then yields

$$(11) \qquad \mathcal{L}(B;\gamma) < u_0 < \mathcal{L}(B;\gamma) + \frac{B}{6} \frac{\cos \gamma}{(1+\sin \gamma)^4} (1+2 \sin \gamma) + O(B^2)$$

with

$$(12) \qquad \mathcal{L}(B;\gamma) \equiv \frac{2 \cos \gamma}{B} - \frac{1}{\cos \gamma} + \frac{2}{3} \frac{1-\sin^3\gamma}{\cos^3\gamma} .$$

We recognize the right side of (12) as exactly the original Laplace estimate (7) in nondimensional form. *Thus* (11) *provides a formal and quantitative proof of the asymptotic correctness of* (7); *it shows in addition that the Laplace estimate provides in every case a strict lower bound for* u_0.

Analogous considerations lead to the estimate for the height u_1 on the contact curve $r = 1$

$$(13) \qquad \frac{2}{B} \cos \gamma < u_1^- < u_1 < \mathcal{F}(B;\gamma)$$

with

$$(14) \qquad \mathcal{F}(B;\gamma) \equiv \mathcal{L}(B;\gamma) + \frac{1-\sin \gamma}{\cos \gamma}$$

and u_1^- the solution of a certain transcendental equation.

By observing the global majorizing properties of the spherical caps used in the above proofs, Siegel [36] obtained the elegant uniform estimate

$$\mathcal{S}(B;\gamma;r) - f(\gamma)B < u(r) < \mathcal{S}(B;\gamma;r) + f(\gamma)B \tag{15}$$

valid throughout the trajectory. Here

$$\mathcal{S}(B;\gamma;r) \equiv 2\,\frac{\cos\gamma}{B} + \frac{2}{3}\,\frac{1-\sin^3\gamma}{\cos^3\gamma} - \frac{1}{\cos^2\gamma}\sqrt{1-r^2\cos^2\gamma} \tag{16}$$

$$f(\gamma) \equiv \frac{1-\sin\gamma}{\cos^2\gamma}\left[\frac{1}{3}\,\frac{1-\sin^3\gamma}{\cos^3\gamma} - \frac{1}{2}\tan\gamma\right] . \tag{17}$$

It should be remarked that the first proof of asymptotic correctness of (7) was due to Siegel [27] and was based on other comparison principles. Siegel later obtained [38] a form of (15) that holds for general Ω, whenever a zero-gravity solution (in the present case spherical cap) can be found. In general, it cannot be expected that such a solution will exist, see [4], Chapter 6. Tam [39] improved Siegel's results, and also obtained information for some situations in which there is no zero-gravity solution.

2. The CP principle

Let $u(x,y)$ be a smooth function in Ω, let Σ' be a relatively open smooth subarc of $\Sigma = \partial\Omega$. *We will say that* $u(x,y)$ *is a variational solution of* (4) *relative to data* γ *on* Σ' *if*

$$\int_\Omega [\nabla\eta\cdot Tu+(\varphi(u)+\lambda)\eta]dx = \oint_{\Sigma'} \eta\cos\gamma\, ds \tag{18}$$

for any $\eta \in H^{1,1}(\Omega)$ *whose support lies in the complement of* $\Sigma \setminus \Sigma'$. We verify easily that if u is a strict solution of (4) in Ω, smooth up to Σ', then u is a variational solution relative to γ on Σ'.

Let $u_1, u_2 \in H^{1,1}_{loc}(\Omega)$. *We will say that* u_1 *majorizes* u_2 *relative to data* γ_1,γ_2 *on* $\Sigma' \subset \Sigma$ *if for any nonnegative* $\eta \in L^\infty(\Omega) \cap H^{1,1}_{loc}(\Omega)$ *whose support lies in the complement of* $\Sigma \setminus \Sigma'$, *there is a sequence* $\Sigma'_n \subset \Omega$ *of smooth curves joining the end points of* Σ' *and tending uniformly to* Σ' *together with normal directions, such that*

$$\liminf_{n\to\infty} \int_{\Omega\setminus\Omega'_n} \{\nabla\eta\cdot(Tu_1-Tu_2)+(\varphi u_1-\varphi u_2)\eta\}dx \geq 0 \quad , \tag{19}$$

Ω'_n *being the part of* Ω *bounded between* Σ' *and* Σ'_n.

In terms of these definitions, we have

Comparison principle CP: *Suppose* Ω *is connected and bounded and that* $\Sigma = \partial\Omega$ *admits a decomposition*

$$\Sigma = \Sigma_\alpha \cup \Sigma_\beta \cup \Sigma'_\beta \cup \Sigma_0 \tag{20}$$

such that Σ'_β *consists of smooth arcs, and* Σ_0 *has one dimensional Hausdorff measure zero. Let* u_1 *majorize* u_2 *with respect to data* γ_1, γ_2 *on* Σ'_β *and suppose* $\varphi'(u) \geq 0$ *for all* u *in the interval* $[u_1(x), u_2(x)]$ *at each* $x \in \Omega$. *Suppose there is a sequence* $\Sigma_\beta^{(n)}$ *of smooth curves of bounded length joining points of* Σ, *and corresponding domains* $\Omega_\beta^{(n)}$ *bounded between* $\Sigma_\beta^{(n)}$ *and arcs of* Σ *containing no points of* Σ_β , *such that* $\Omega_\beta^{(n)} \uparrow \Omega$, *and such that for any* $\epsilon > 0$ *there holds*

$$\lim_{n\to\infty} \mu\{x \in \Sigma_\beta^{(n)} : \nu\cdot Tu_1 - \nu\cdot Tu_2 < -\epsilon\} = 0 , \tag{21}$$

μ = *Lebesgue measure in arc. Finally, suppose there exists* $A \geq 0$ *such that* $\liminf_{x\to\Sigma_\alpha} (u_1-u_2) \geq -A$ *for any approach to* Σ_α. *Then if* $\Sigma_\alpha = \emptyset$ *and* $\varphi'(z) \equiv 0$ *in the interval between* u_1 *and* u_2, *there holds* $u_1 \equiv u_2 + \text{const}$. *Otherwise* $u_1 \geq u_2 - A$, *equality holding at a single point if and only if equality holds everwhere.*

The essential content of CP is that if $\operatorname{div} Tu_1-\varphi(u_1) \leq \operatorname{div} Tu_2-\varphi(u_2)$ in Ω, if Σ consists of two parts on which, respectively $u_1 \geq u_2 - A$ and $\nu\cdot Tu_1 \geq \nu\cdot Tu_2$, then $u_1 \geq u_2 - A$ in Ω, ecxept for the exceptional case in which $u_1 \equiv u_2 + c$. The more elaborate statement we have made is however not simply idle generality. The properties of Σ_0 will play a crucial role in the applications directly below, and reflect basic differences that distinguish the behavior of the operators considered here from that of linear elliptic operators; similarly, the weak hypotheses on boundary behavior for the derivatives on Σ_β, Σ'_β are essential for the discussion in §11, and also in other contexts.

The proof of CP is obtained by adjoining the considerations of Lemma 4 in [40] to the proof of Theorem 5.1 in [4].

3. Applications I

We apply CP to a domain whose boundary includes a wedge with opening angle 2α, as indicated in Figure 1. We introduce a disk B_δ of radius δ as shown, and observe that a lower hemisphere $v(x)$ over B_δ satisfies

(22) $$\operatorname{div} Tv = \frac{2}{\delta}$$

in $B_\delta \cap \Omega$, and meets the vertical walls over Σ in the constant angle $\gamma_0 = \frac{\pi}{2} - \alpha$. Let $u(x)$ be a solution of

(23) $$\operatorname{div} Tu = \kappa u \quad , \qquad \kappa > 0$$

in Ω, with boundary angle γ satisfying $\gamma_0 \le \gamma \le \frac{\pi}{2}$. Then $\nu \cdot Tv \ge \nu \cdot Tu$ on $\Sigma \cap B_\delta$, while $\nu \cdot Tv = 1$ on $\partial B_\delta \cap \Omega$. We observe that $\nu \cdot Tu < 1$ on this set in virtue of the finiteness of $|\nabla u|$, and we choose Σ_β as the union of the two boundary sets, with the intersection points deleted. We then set $\Sigma_\beta' = \Sigma_\alpha = \emptyset$ and let Σ_0 be the union of the two points just deleted and the vertex. We adjust v by an additive constant, so that its minimum $v_0 = \frac{2}{\kappa\delta}$. We then have by (22)

$$\operatorname{div} Tv = \frac{2}{\delta} = \kappa v_0 \le \kappa v$$

and thus $\operatorname{div} Tu - \kappa v \le \operatorname{div} Tu - \kappa u$ in $\Omega \cap B_\delta$. One verifies readily that the conditions of CP for a suitable sequence $\Sigma_\beta^{(n)}$ are verified, with $A = 0$. *We conclude from* CP *that* $v \ge u$, *and since* $v \le v_0 + \delta$ *we find*

(24) $$u < \frac{2}{\kappa\delta} + \delta$$

throughout $\Omega \cap B_\delta$. *This inequality holds for any solution* u *of* (23), *for which* $\alpha + \gamma \ge \frac{\pi}{2}$ *on* Σ. *In particular, it holds in the limiting configuration, for which* $\alpha + \gamma = \frac{\pi}{2}$.

Let us now adopt polar coordinates, centered at the vertex, with $\theta = 0$ bisecting the wedge. Set $k = \frac{\sin\alpha}{\cos\gamma}$. It is shown in [41] (see also [4], Theorem 5.5) by another use of CP that *there exists* $A < \infty$ *such that for any solution* u *of* (23) *in* $\Omega \cap B_\delta$ *for which* $\alpha + \gamma < \frac{\pi}{2}$ *there holds*

(25) $$\left| u - \frac{\cos\theta - \sqrt{k^2 - \sin^2\theta}}{k\kappa r} \right| < A$$

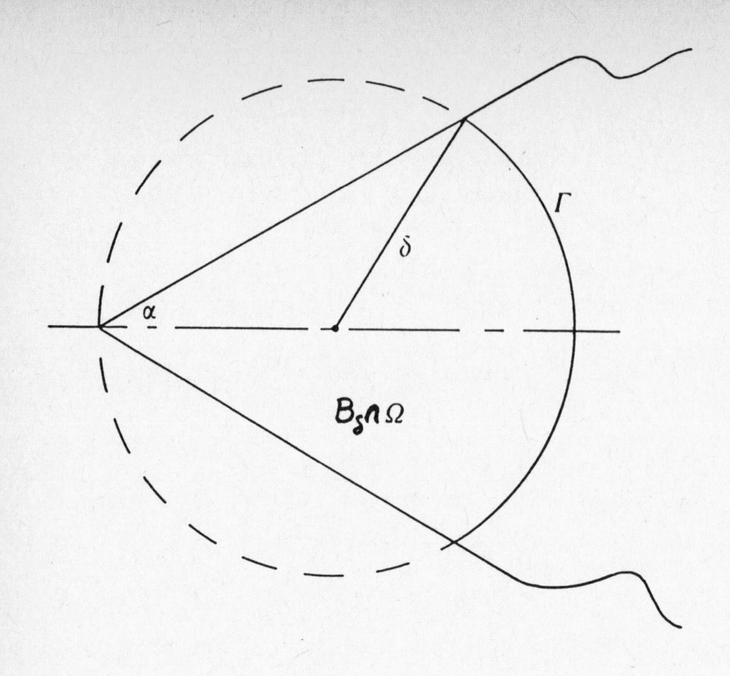

Figure 1: Wedge domain, comparison disk

throughout $\Omega \cap B_\delta$. *Thus, the solutions depend discontinuously on* γ; *if* $\alpha + \gamma \geq \frac{\pi}{2}$ *then by* (24) *all solutions lie below a fixed bound, while if* $\alpha + \gamma < \frac{\pi}{2}$ *then by* (25) *all solutions become infinite at the vertex.*

We have already remarked the result of Tam [14], that *if* $\alpha + \gamma \geq \frac{\pi}{2}$, $\alpha < \frac{\pi}{2}$, *then all solutions are continuous together with the surface normal up to the vertex.*

We remark also an independent heuristic reasoning due to Langbein and Rischbieter [42] which indicates that for the case $\alpha + \gamma < \frac{\pi}{2}$ solutions should become infinite with order $O(\frac{1}{r})$ at the vertex.

4. Further applications.

Given $\gamma \in [0, \frac{\pi}{2}]$, it is clear that by opening the wedge sufficiently in the above example, we can achieve $\alpha + \gamma \geq \frac{\pi}{2}$ and hence a uniform a priori bound on height for any solution. It is however not true that the solutions must continue to improve in smoothness with increasing α. Korevaar [15] considered the configuration of Figure 2, for which $\alpha = \frac{3\pi}{4}$ at V, V'. From general existence theorems (cf. [4], Chapter 7) it is known that for any $\gamma \in [0, \frac{\pi}{2}]$ there is a unique variational solution of (23) in Ω, with data γ on $\Sigma = \partial\Omega$. For fixed $\gamma \neq \frac{\pi}{2}$ we consider that solution in its dependence on the separation 2ϵ of the parallel segments.

From the result of the above section, we have the bound (24) throughout B_δ, uniformly as $\epsilon \to 0$. Now consider a circular arc Γ_ϵ (of radius $\frac{\epsilon}{\cos\gamma}$) making equal angles γ with L, L', as indicated in the figure. We choose Γ_ϵ as generator to construct half of the underside of a torus $\mathcal{T}$, with principal radii $R, \frac{\epsilon}{\cos\gamma}$ and lying above $\Gamma_\epsilon, \Gamma'_\epsilon$, which meets the "strip" region again in a reflection Γ'_ϵ of Γ_ϵ. This torus has mean curvature H satisfying

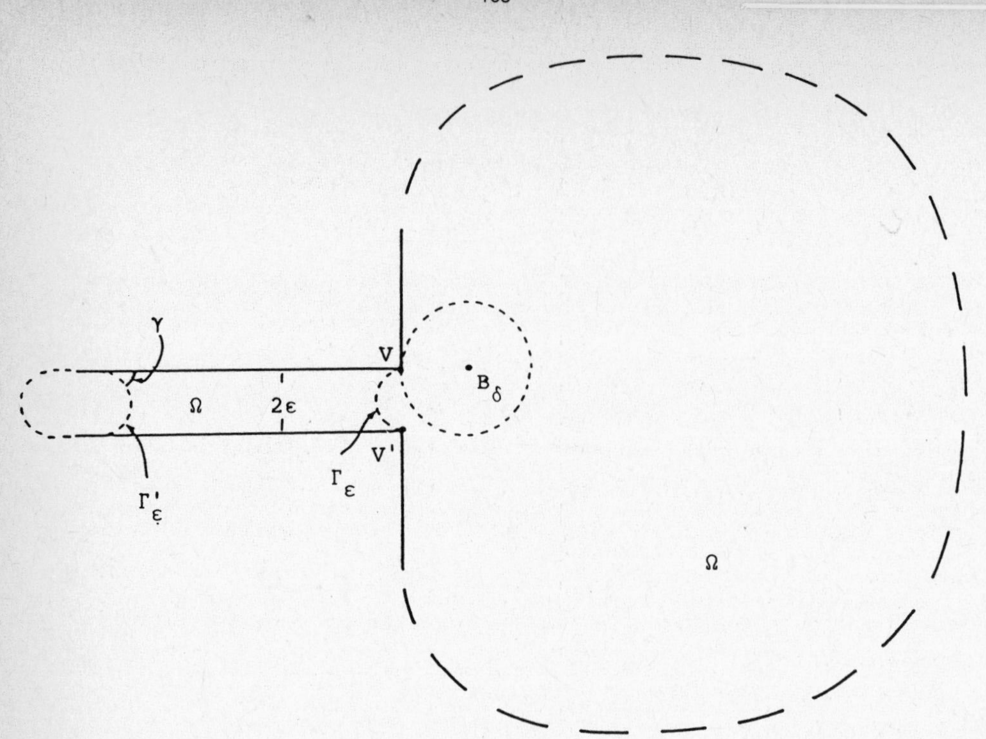

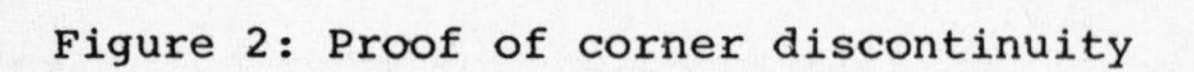
Figure 2: Proof of corner discontinuity

$$H \geq \frac{1}{2}\left[\frac{\cos\gamma}{\epsilon} - \frac{1}{R-\epsilon}\right] \quad ;$$

thus, if we describe $\mathcal{T}$ as a graph v_ϵ over the strip, we will have

$$\operatorname{div} Tv_\epsilon = 2H \geq \left[\frac{\cos\gamma}{\epsilon} - \frac{1}{R-\epsilon}\right] \quad .$$

If we move $\mathcal{T}$ vertically until its maximum height v_ϵ^M satisfies $v_\epsilon^M = \frac{1}{\kappa}\left[\frac{\cos\gamma}{\epsilon} - \frac{1}{R-\epsilon}\right]$, we will then have

$$\operatorname{div} Tv_\epsilon \geq \kappa v_\epsilon^M \geq \kappa v_\epsilon$$

throughout the region Ω_ϵ of definition of v_ϵ. Since $\mathcal{T}$ meets the bounding walls over the open segments L, L' in the constant angle γ, we have $\nu\cdot Tv_\epsilon = \cos\gamma$ on these lines. Further, on the open arcs Γ_ϵ, Γ'_ϵ we have $\nu\cdot Tv_\epsilon = -1$.

In Ω_ϵ we have $\operatorname{div} Tv_\epsilon - \kappa v_\epsilon \geq \operatorname{div} Tu - \kappa u$. We choose $\Sigma_\beta = L \cup L' \cup \Gamma_\epsilon \cup \Gamma'_\epsilon$, $\Sigma'_\beta = \Sigma_\alpha = \emptyset$, Σ_0 to be the four intersection points of Γ_ϵ, Γ'_ϵ with L, L'. As in §3, we conclude from CP that $u > v_\epsilon$ in Ω_ϵ. But $v_\epsilon \geq v_\epsilon^M - R$, and thus

$$u > \frac{1}{\kappa}\left[\frac{\cos\gamma}{\epsilon} - \frac{1}{R-\epsilon}\right] - R \tag{26}$$

throughout Ω_ϵ. *From* (24) *and* (26) *we see that for small enough* ϵ, u *must be discontinuous at* V, V'.

We may show by a variant of the method that *even for smooth boundaries, discontinuities in prescribed data can lead to discontinuities in the contact curve.* As domain Ω we now take the infinite strip $|y| < \epsilon$; we prescribe data γ_1 for $x < 0$, γ_2 for $x > 0$. Corresponding to γ_1 we introduce a surface v_ϵ as above, obtaining (26) again with $\gamma = \gamma_1$ over $\Omega_\epsilon^{(1)}$. For γ_2 we introduce half the outer (lower) part w_ϵ of a torus, chosen to meet the bounding walls in the angle γ_2. For this surface $\mathcal{T}_2$ the mean curvature satisfies

$$H \leq \frac{1}{2}\left[\frac{\cos\gamma_2}{\epsilon} + \frac{1}{R+\epsilon}\right] \quad .$$

We now move $\mathcal{T}_2$ vertically so that its minimum height $w_\epsilon^m = \frac{1}{\kappa}\left[\frac{\cos\gamma_2}{\epsilon} + \frac{1}{R+\epsilon}\right]$. Essentially a repetition of the above reasoning, with the roles of u and v reversed, yields

$$u < \frac{1}{\kappa}\left[\frac{\cos\gamma_2}{\epsilon} + \frac{1}{R+\epsilon}\right] + R$$

in the corresponding $\Omega_\epsilon^{(2)}$. *If* $\gamma_1 < \gamma_2$, *we find* $\max_{\Omega_\epsilon^{(2)}} u < \min_{\Omega_\epsilon^{(1)}} u$ *for all sufficiently small* ϵ, *and thus the contact curves must be discontinuous at* $x = 0$.

4bis. Isolated singularities

Consider an equation (4) with $\varphi' \geq 0$ (thus equations (6) are included). Let $u(x)$ be a solution of (4) in a disk D with center $x = p$ deleted. Let h be a constant vector, set $v(x) = u(x-h)$. Then $v(x)$ will be a solution in a disk D^h with center $x = p + h$ deleted. Let $h_0 > 0$ be such that if $|h| < h_0$ then $D \cap D^h$ contains both centers, and let $\mathscr{C} \subset D \cap D^h$ be a simple closed convex curve surrounding the points p and $p + h$. The difference quotient

$$\delta^h u = \frac{u(x)-v(x)}{h}$$

then remains uniformly bounded on $\mathscr{C}$ as $h \to 0$; we may write

$$|\delta^h u| < A < \infty \quad \text{on } \mathscr{C}.$$

Setting $\Sigma_0 = p \cup p + h$, $\Sigma_\alpha = \mathscr{C}$, $\Sigma_\beta = \Sigma_\beta' = \emptyset$, the comparison principle of §2 assures us that $|\delta^h u| < A$ throughout $D \cap D^h$. Letting $h \to 0$, we obtain that $|\nabla u| < A$ in D, and thus u is bounded and (4) uniformly elliptic near p. From standard theory of elliptic equations we conclude (cf. [67]):

Theorem: *If* $\varphi'(u) \geq 0$, *any isolated singularity of a solution of* (4) *is removable.*

We remark that CP (and hence also the above theorem) extends without change to any equation of the form

$$\operatorname{div} A(\nabla u) = \varphi(u)$$

in any number n of dimensions whenever $|A| < A_0 < \infty$ and $\varphi'(u) \geq 0$, and extends to still more general equations under suitable hypotheses. As particular case, we retrieve with a very much simpler proof a theorem of Bers [68] on minimal surfaces for $n = 2$.

If $\varphi'(u) < 0$ isolated singularities can appear. See [52, 54, 55] for the (physical) case $\varphi(u) \equiv -\kappa u$, and [69] for a more general case.

5. The touching principle

The inequality (24), which we derived for a corner situation, holds as well - with the same proof - in any configuration in which the lower hemisphere over B_δ meets the bounding walls over Σ in angles $\leq \gamma$ at each contact point. We proved it in §3 using CP, however the underlying behavior is not basically a maximum principle property, but arises rather as a geometrical consequence of the mean curvature requirement [41]. We may state the essential property as a *touching principle:*

If two surfaces $u(x)$, $v(x)$ *have a first order contact at* p, *and if* $v \geq u$ *in a neighborhood of* p, *then the corresponding mean curvatures satisfy* $H_v \geq H_u$ *at* p.

The proof is immediate, as each normal curvature based on the common normal to the surfaces at p must satisfy that inequality, and the mean curvature is the average of the normal curvatures in any two orthogonal directions.

We consider in general an equation

$$\operatorname{div} Tu = \varphi(u) \tag{27}$$

with the single requirement that

$$\lim_{z\to\infty} \varphi(z) = \infty \quad . \tag{28}$$

Monotonicity of φ is not assumed, and thus CP does not apply as stated.

Let $u(x)$ satisfy (27) in Ω, with $\nu\cdot Tu = \cos\gamma$ on Σ. To fix the ideas, we assume at first that Σ is smooth. We consider again a lower hemisphere v_δ over B_δ, and we suppose B_δ situated so that $\nu\cdot Tv_\delta \geq \cos\gamma$ on $\Sigma\cap B_\delta$. Then for $\delta' < \delta$ we will have $\nu\cdot Tv_{\delta'} > \cos\gamma$ on $\Sigma\cap B_{\delta'}$, or else $\Sigma\cap B_{\delta'} = \emptyset$.

We now add a constant to $v_{\delta'}$ until $v_{\delta'} > u$ in $\overline{(\Omega\cap B_{\delta'})}$, and then lower $v_{\delta'}$ until a first point of contact p of the two surfaces occurs. This point cannot lie over Σ, since $\nu\cdot Tv_{\delta'} > \nu\cdot Tu$ on Σ. For the same reason, p does not lie over $\Omega\cap\partial B_{\delta'}$. Thus, p lies over a point $x_p \in \Omega \cap B_{\delta'}$, and therefore the two tangent planes coincide at p. Since $v_{\delta'} \geq u$, we infer from the touching principle that the two mean curvatures satisfy the same inequality at p; that is

$$\frac{2}{\delta'} \geq \operatorname{div} Tu\Big]_{x_p} = \varphi(u_p) \quad .$$

Therefore

$$u_p \leq M_{\delta'} = \max \{u: \varphi(u) \leq \frac{2}{\delta'}\}$$

and since u lies below the hemisphere $v_{\delta'}$ we have $u(x) \leq M_{\delta'} + \delta'$ in $B_{\delta'}$. Letting $\delta' \uparrow \delta$ we obtain

$$u(x) \leq M_\delta + \delta \tag{29}$$

throughout $B_{\delta'}$ which establishes the a priori bound.

Similarly, a bound from below holds if $\lim_{z\to-\infty} \varphi(z) = -\infty$. The reasoning can be extended to include situations such as the one discussed in §3, for which Σ may have isolated corners.

6. Monotone symmetry (MS)

Consider a solution $u(x)$ in Ω of the capillary problem

$$\operatorname{div} Tu = \varphi(u) \quad , \quad \varphi'(u) \geq 0 \tag{30}$$

with

$$\nu \cdot Tu = \cos\gamma \quad , \quad 0 \leq \gamma < \frac{\pi}{2} \tag{31}$$

on $\Sigma = \partial\Omega$, $\gamma \equiv$ const. Suppose Ω can be enclosed in another domain $\Omega^{(0)}$, over which there is a corresponding function $u^{(0)}$ that satisfies (31) on $\Sigma^{(0)}$ with the same γ, and which has the properties

i) each component of $\Sigma^{(0)} = \partial\Omega^{(0)}$ is a level curve for $u^{(0)}$
ii) each point $p \in \Sigma$ is joined to a point $p^{(0)} \in \Sigma^{(0)}$ by a curve of steepest ascent of $u^{(0)}$, on which $|\nabla u^{(0)}|$ is monotone non decreasing
iii) in Ω there holds $\operatorname{div} Tu^{(0)} - \varphi(u^0) \geq 0$.

Then either $u^{(0)} \leq u$ *in* Ω, *equality holding at any point if and only if* $\Omega \equiv \Omega^{(0)}$ *and* $u \equiv u^{(0)}$, *or else* $u \equiv u^{(0)} + c$ *and* $\varphi'(t) \equiv 0$ *in the interval* $[u,u^{(0)}]$ *at each* $x \in \Omega$.

Proof: Corresponding to a curve of steepest ascent joining $p \in \Sigma$ to $p^{(0)} \in \Sigma^{(0)}$ we have

$$\nu \cdot Tu^{(0)}\Big]_p = \frac{\nu \cdot \nabla u^{(0)}}{\sqrt{1+|\nabla u^{(0)}|^2}}\Big]_p \le \frac{|\nabla u^{(0)}|}{\sqrt{1+|\nabla u^{(0)}|^2}}\Big]_p$$

$$\le \frac{|\nabla u^{(0)}|}{\sqrt{1+|\nabla u^{(0)}|^2}}\Big]_{p^{(0)}} = \nu \cdot Tu^{(0)}\Big]_{p^{(0)}} = \cos \gamma = \nu \cdot Tu\Big]_p$$

and thus $\nu \cdot Tu^{(0)}\big]_p \le \nu \cdot Tu\big]_p$. The result then follows from CP.

As an application, we find that if $\varphi(0) = 0$, $\varphi'(u) > 0$, and if a capillary tube with section Ω lies strictly interior to a tube with circular section $\Omega^{(0)}$, then Ω raises liquid at each point of its section to a greater height than does $\Omega^{(0)}$ at that point. In fact, taking for $u^{(0)}$ the (symmetric) solution in $\Omega^{(0)}$ with data γ on $\Sigma^{(0)}$, the curves of steepest ascent are the meridian curves relative to the axis of symmetry, and it is a formal exercise to prove the requisite monotonicity property (cf. [48] p. 13 for the case $\varphi(u) = \kappa u$).

By another use of CP and of the result of §5, it can be shown ([49], see also [4], §5.4) that *the above inclusion property does not hold in general when* $\Omega^{(0)}$ *is not circular*. However, Siegel [50] applied the E. Hopf maximum principle to show that *if* Ω *is circular and can contact any* $p^{(0)} \in \Omega^{(0)}$ *from within* $\Omega^{(0)}$, *then the inclusion property will hold*, see §15.

Emmer [50] proved that *if* Ω *is smooth*, $0 < \gamma < \frac{\pi}{2}$, $\varphi(u) = \kappa u$, *then* $\lim_{\kappa \to 0} u(x) = \infty$ *at each* $x \in \Omega$ (that is, the liquid in a capillary tube rises to infinity as gravity $\to 0$). By using MS we obtain a simpler proof under weaker conditions. In fact, the inequality (11) leads to an explicit lower bound for the height u in terms of g in the case considered by Emmer. Similar inequalities can be obtained for more general $\varphi(u)$.

Siegel ([27], Theorem 20) gave the following counterpart (MSS) of MS, in which the inequalities are reversed:

Let $u(x)$ *satisfy* (30), (31) *in* Ω, *with* $\varphi(u) = \kappa u$, $\kappa > 0$, *and let* B_R *be a disk of radius* R *lying in* Ω. *Let* $v(x)$ *be the (symmetric) solution in* B_R *of* (30), (31), *and let* $\bar{R}$ *be the radius of the largest concentric disk into which* $v(x)$ *can be extended (thus,* $\nu \cdot Tv = 1$ *on* $\partial B_{\bar{R}}$*). Suppose* $\Omega \cap B_{\bar{R}}$ *is convex and piecewise smooth. Then* $u \leq v$ *in* $\Omega \cap B_{\bar{R}}$.

The proof is not significantly more difficult than that of MS, and the result also extends to more general $\varphi(u)$. Siegel used the result to prove uniqueness of capillary surfaces in wedge domains, also to prove uniqueness of the minimal point when Ω is convex, see §12.

7. Gradient bounds from curvature bounds I.

The procedure of §§2-4 can be adapted to situations in which $\varphi'(u) < 0$. This was done in [51] to obtain estimates for the (symmetric) pendent liquid drop. All that is necessary is to satisfy the curvature inequality in the appropriate direction. In [51] the configuration of a pendent drop for which $\varphi(u) = -\kappa u$, $\kappa > 0$ is estimated. It is shown that if the drop contains a minimum point $(0,u_0)$ for which $u_0 \leq -2\sqrt{2}$, then the generating curve $u(r)$ must become vertical at some $r_1 > 0$ and at a height

$$u_1 < \frac{1}{2} u_0 (1 + \sqrt{1-8u_0^{-2}}) \quad .$$

We now compare $u(r)$ with the circular arc $v(r)$ of (constant) curvature

$$\beta^{-1} = -\frac{1}{4} u_0 (1 + \sqrt{1-8u_0^{-2}})$$

centered at $(u_0, u_0+\beta)$, so that $u(0) = v(0)$. In the interval $0 < r < r_1$ we have $H[u] > H[v]$, hence $u > v$, $u' > v'$ until the vertical is reached, and we find

$$r_1 < \beta = -\frac{1}{2} u_0 (1 - \sqrt{1-8u_0^{-2}}) \quad .$$

Next we introduce a circular arc

$$w(r) = u_0 - \frac{2}{u_0} - \sqrt{4u_0^{-2} - r^2}$$

which has constant curvature $\frac{-u_0}{2}$ equal to that of $u(r)$ at $r = 0$, where the curves are tangent. Since the curvature of $u(r)$ decreases in u, we find $u'(r) < w'(r)$ in $0 < r < 1$, so that $r_1 > -\frac{2}{u_0}$. Denoting the inclination angles of $u(r)$ and of $w(r)$ by ψ and by φ,

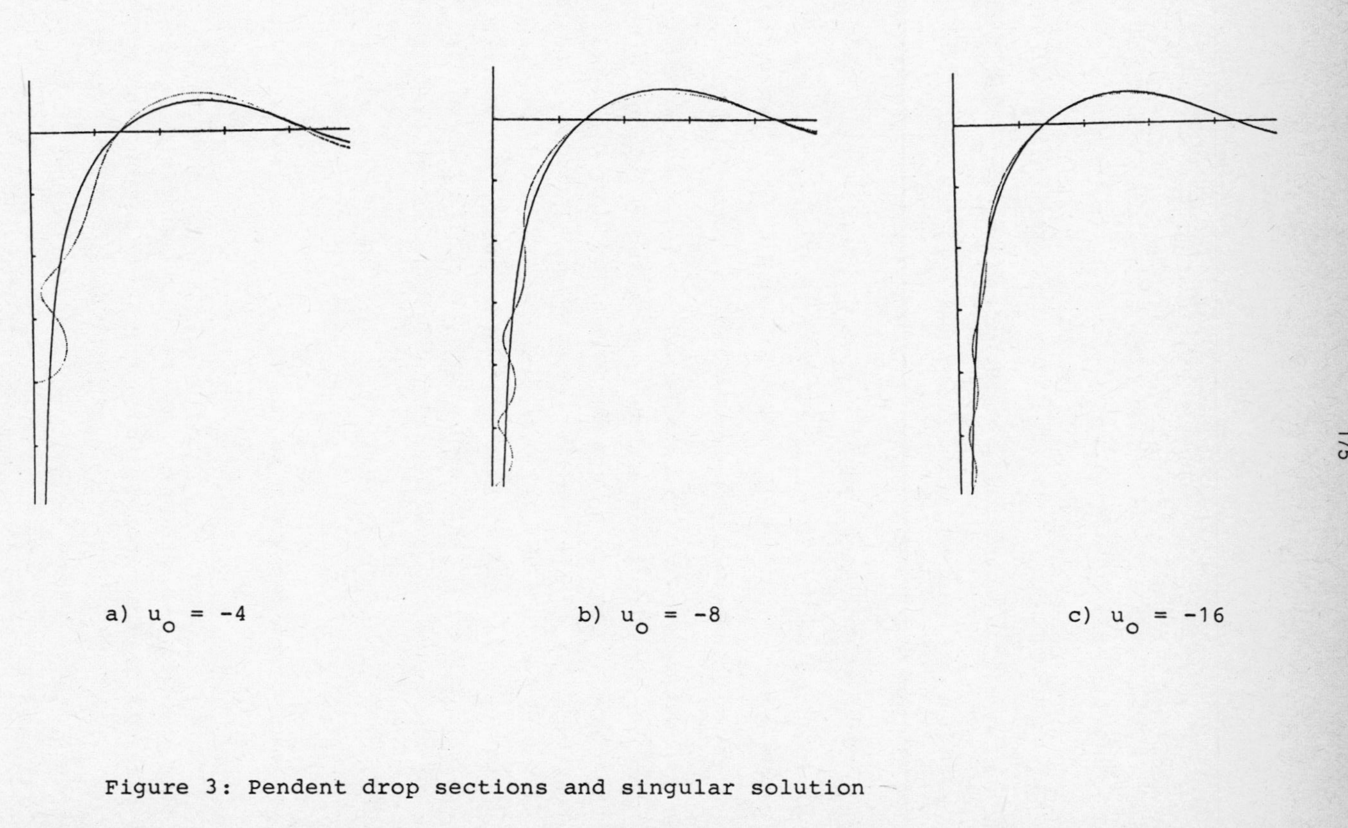

Figure 3: Pendent drop sections and singular solution

we have

$$k_m = (\sin\psi)_r = -(\cos\psi)_u < -\frac{1}{2}u_0 = -(\cos\varphi)_u$$

and thus $\psi < \varphi$ at corresponding heights. It follows that $u(r)$ can be continued until the vertical height of $w(r)$, so that $u_1 > u_0 - 2u_0^{-1}$.

The above four inequalities for r_1, u_1 localize the initial vertical point of $u(r)$, asymptotically exactly as $u_0 \to -\infty$. The further traverse is then controlled by introducing unduloids as comparison surfaces, with curvatures determined by u_1 and an eventual next vertical height u_2, and which are vertical at (r_1,u_1). In this way, the configuration of the drop section can be traced over a large portion of its traverse, the accuracy in any interval $[u_0,u_0+h]$ improving as $u_0 \to -\infty$. Three such sections, corresponding to vertex heights $u_0 = -4,-8,-16$ respectively, are shown in Figure 3a,b,c. Independently, it can be shown [52-55] that there exists a singular solution $U(r)$, as shown in the figures. The comparison procedure leads to the result [51] that *in any fixed compact region of* (r,u) *space, the drop profiles converge to within a narrow strip surrounding the singular solution curve*. Computer calculations suggest that there is in fact uniform convergence to that curve. In this connection it should be remarked that uniqueness of the singular solution has not yet been completely proved.

8. Gradient bounds from curvature bounds II.

Under some conditions, the classical maximum principle can be used to obtain a priori gradient bounds in a general situation, analogous to the familiar bounds for harmonic functions. Under appropriate smoothness hypotheses, these bounds extend to the entire closed domain of definition. This was first shown by Spruck [16] for capillary tubes with smooth section Ω in positive gravity field (heavier fluid below the interface). The underlying equation can be normalized to read

$$\operatorname{div} Tu = u \tag{32}$$

in Ω, with boundary condition

$$\nu\cdot Tu \equiv \frac{\partial u/\partial n}{\sqrt{1+|\nabla u|^2}} = \cos\gamma \tag{33}$$

on $\Sigma = \partial\Omega$. It is assumed that u is suitably smooth in $\bar{\Omega}$ and that $0 < \gamma \leq \frac{\pi}{2}$. It suffices to obtain a tangential derivative estimate $|u_s| < M$ on Σ. In conjunction with (33) one then finds a bound for $|\nabla u|$ on Σ; this estimate then extends to all of $\bar{\Omega}$ since the derivatives of any solution of (32) satisfy the maximum principle.

To obtain the bound on $|u_s|$, Spruck rewrites (32) and (33) in a neighborhood of Σ in terms of normal and tangential coordinates (r,s):

$$(32') \qquad u = \frac{1}{1-\kappa r}\left\{\frac{\partial}{\partial r}(1-\kappa r)\frac{u_r}{W} + \frac{\partial}{\partial s}(1-\kappa r)^{-1}\frac{u_s}{W}\right\} ,$$

$$W^2 = 1 + u_r^2 + \frac{1}{(1-\kappa r)^2}u_s^2 ,$$

κ being the curvature of Σ, and

$$(33') \qquad \frac{1}{W}u_r = -\cos\gamma .$$

Consider the point p in $0 \leq r \leq \epsilon$, at which $|u_s|$ achieves its maximum. Then $u_{ss}(p) = 0$. If $r(p) = 0$, we differentiate (33') to obtain $u_{rs} = -\frac{u_s u_{ss}}{\sqrt{1+u_s^2}}\cot\gamma = 0$. Thus, p must be a critical point of u_s. From (32') one obtains at p

$$(34) \quad u_s = -\left[\frac{u_r}{W} + \frac{u_r u_s^2}{W^2}\right] + \frac{1}{W^3}\left\{(1+u_s^2)u_{srr} - 2u_r u_s u_{srs} + (1+u_r^2)u_{sss}\right\} .$$

Since p is both a maximum and a critical point of u_s, its Hessian must be negative semidefinite. Hence the last term in (34), as the trace of the product of the Hessian with a positive definite matrix, is non positive. Using (33') we thus find

$$|u_s(p)| \leq 2|\kappa_s(p)| \cos\gamma .$$

If $r(p) = \epsilon$, the procedure of §5 gives an a priori bound for $|u|$ in an ϵ neighborhood of p, and a bound for $|u_s|$ then follows from general interior gradient estimates for solutions of (32), cf. [56, 57, 58].

Finally, if $0 < r(p) < \epsilon$. then p must be a critical point for u. A reasoning analogous to that of the first case again yields the desired bound.

Remarks: The reasoning is not entirely self contained within the framework of the maximum principle, as it uses at one point a priori gradient estimates that must be obtained by other methods. Global smoothness of the boundary had to be assumed, and the proof relies heavily on the hypothesis of non zero gravity. In these respects a new approach, due to Korevaar and to Lieberman, provides improved -- if not yet complete -- results, see §9.

Although the proof as we have given it works only for constant γ, Spruck was able to extend it to allow varying γ.

Spruck used his estimates to obtain an existence proof for smooth domains Ω. Specifically, *if* $0 < \gamma < \pi$ *he obtained a solution smooth in* $\bar{\Omega}$. Since the interior bounds on $|u|$ do not depend on γ (see §5) he could show by a limiting argument *the existence, for the case* $\gamma = 0$ *(or* π*), of a solution in* $C^2(\Omega) \cap C^0(\bar{\Omega})$ *that is Lipschitz continuous on* Σ. For a corresponding result in the zero gravity case, see §11.

9. Barriers by normal perturbation.

Korevaar [17-19] and, independently, Lieberman [20, 21] made the remarkable discovery that for surfaces of prescribed mean curvature, the surface itself can be used as a barrier to obtain gradient bounds. Although the method was developed by these authors to include boundary estimates and more general equations, we outline here only the simplest case [17] of interior bounds for a solution surface S_u: $u(x)$ of an equation

$$\text{div}\, Tu = h(x,u) \tag{35}$$

in Ω, under the hypothesis

$$|h| + |D_x h| < M < \infty \quad , \quad h_u \geq 0 \quad . \tag{36}$$

It will be shown that *if* $u(x) < 0$ *in* Ω, *then for any* $x_0 \in \Omega$, $|\nabla u(x_0)|$ *is bounded depending only on* M, *on* $u(x_0)$, *and on distance from* x_0 *to* $\Sigma = \partial\Omega$.

We introduce a nonnegative function $\eta(x,z)$ whose support lies in a ball centered on S_u, and consider a perturbation normal to S_u, of the form $\epsilon\eta N$, N = unit normal to S_u. This leads to a new surface $S_{\bar{u}}$: $\bar{u} = u - \epsilon \frac{1}{W} \eta$, $W = \sqrt{1+|\nabla u|^2}$, which will be smooth over Ω if ϵ is small enough.

The height difference achieves its maximum at some point $\bar{x} \in \Omega$. The two tangent planes are parallel, also $W = \bar{W}$ at $\bar{x}$, and we have the mean curvature inequality

$$H(\bar{x},\bar{u}(\bar{x})) \geq H(\bar{x},u(\bar{x})) = h(\bar{x},u(\bar{x})) \ . \tag{37}$$

Korevaar derives the estimate

$$H(\bar{x},\bar{u}(\bar{x})) = H(x,u(x)) - \epsilon\{\eta\Sigma\kappa_i^2+\Delta_T\eta-N\cdot\nabla\eta H(x,u(x))\} + O(\epsilon^2) \ . \tag{38}$$

Here κ_i are the principal curvatures on S_u, and $\Delta_T\eta$ is the Laplacian with respect to tangent plane coordinates at $(x,u(x))$.

Using (38) on the left side of (37) and Taylor's theorem on the right, we obtain

$$\begin{aligned} h(x,u(x)) &- \epsilon\{\eta\Sigma\kappa_i^2+\Delta_T\eta-N\cdot\nabla\eta h(x,u(x))\} \\ &\geq h(x,u(x)) + \epsilon\eta \ \tfrac{1}{W}D_x u\cdot\{D_x h+h_u D_x u\} + O(\epsilon^2) \ , \end{aligned}$$

and since η, κ_i^2, $h_u \geq 0$ there follows

$$\Delta_T\eta - |\nabla\eta|M - \eta M \leq O(\epsilon) \ . \tag{40}$$

The method hinges on a judicious choice for η. Set $u_0 = |u(0)|$,

$$\varphi(x,z) = \max\left\{\frac{1}{2u_0} z + (1-|x|^2),0\right\} \tag{41}$$

and choose $\eta = f \circ \varphi$, f to be determined. Then

$$\Delta_T\eta = |\nabla_T\varphi|^2 f'' + (\Delta_T\varphi)f' \tag{42}$$

while if e is the direction of steepest ascent on S we find

$$\begin{aligned} |\nabla_T\varphi| &\geq e\cdot\nabla\varphi = \left(D_x\varphi,\frac{1}{2u_0}\right) \cdot \frac{1}{W}\left[\frac{D_x u}{|D_x u|},D_x u\right] \\ &\geq \frac{1}{W}\left(-2 + \frac{1}{2u_0}|D_x u|\right) \end{aligned} \tag{43}$$

and if $|D_x u| \geq \max(12u_0,3)$ we obtain

$$|\nabla_T\varphi| \geq \frac{1}{\sqrt{10}\,u_0} \ . \tag{44}$$

From (40) and (42) we have

$$|\nabla_T\varphi|^2 f'' + (\nabla_T\varphi-M|\nabla\varphi|)f' - Mf \leq O(\epsilon) \ . \tag{45}$$

Here the coefficient of f' is bounded. For large C_1, the choice $f = e^{C_1\varphi} - 1$ leads to a contradiction when ϵ is small, unless $|Du|$ is bounded at $\bar{x}$. Thus $W(\bar{x}) \leq C_2$.

The height difference $u - \bar{u} = \epsilon\eta W + O(\epsilon^2)$ and is maximized at $\bar{x}$. Thus, throughout Ω we have

$$\eta W \leq C_2 e^{C_1} + O(\epsilon)$$

which yields the desired gradient bound.

10. Principle of nth order division.

The following lemma can be traced at least to [59]. It has since been used in varying contexts by a number of authors, cf. e. g. [22, 23, 24, 56, 60-66].

Let $S: u(x,y)$ *and* $S': v(x,y)$ *be distinct solutions of* (4) *in* Ω. *Suppose there exists* $p \in \Omega$ *at which* S *and* S' *have* n^{th} *order contact,* $n \geq 0$. *Then there are* $n + 1$ *smooth curves through* p, *dividing a (sufficiently small) disk centered at* p *into* $2(n+1)$ *regions sharing* p *as common boundary point, in which alternately* $u > v$, $u < v$. We apply this principle in two different contexts in the following two sections.

11. Boundary behavior in the most singular case.

We consider the capillary problem in zero gravity. The governing equation becomes

$$\text{div}\, Tu = 2H \equiv \text{const.} \tag{46}$$

in Ω, with again

$$\nu \cdot Tu \equiv \cos\gamma \tag{47}$$

on Σ. Integration of (46) over Ω yields

$$2H|\Omega| = \oint_\Sigma \cos\gamma\, ds \quad , \tag{48}$$

thus H is determined by the value distribution of γ over Σ. In what follows we suppose $H \neq 0$; we can then normalize by a similarity

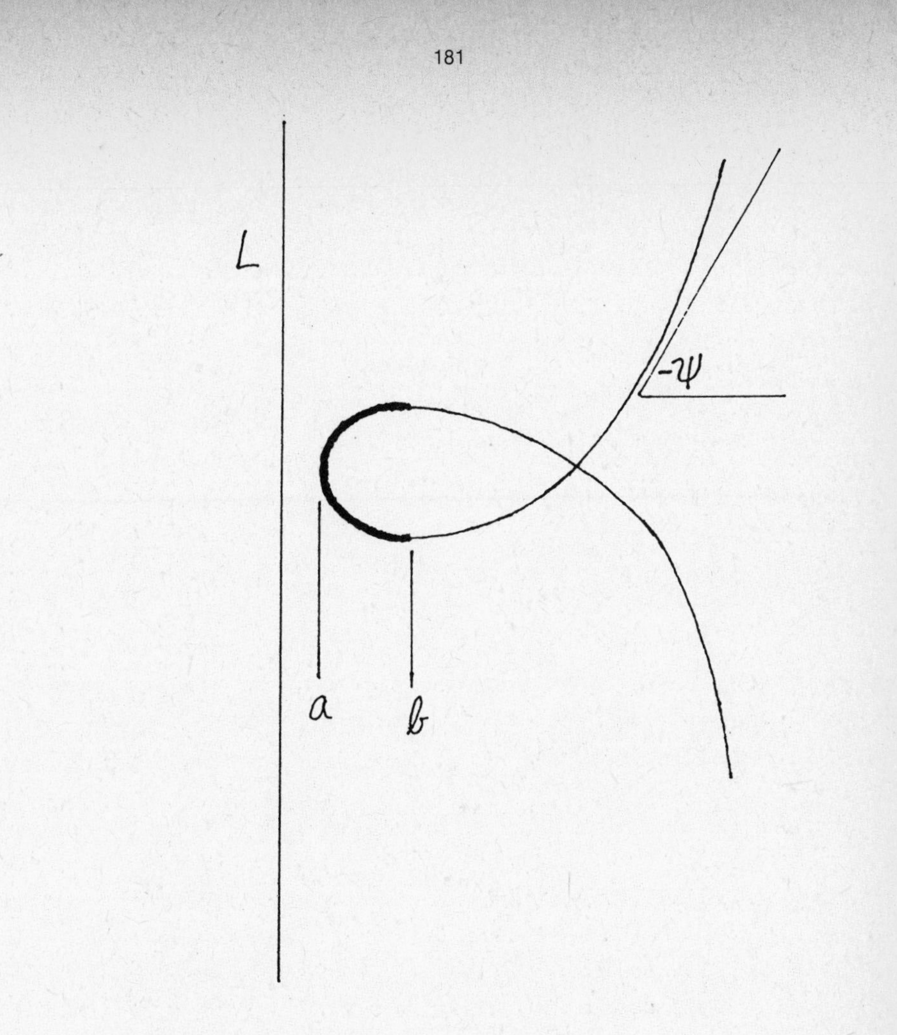

Figure 4: Generating curve for nodoid

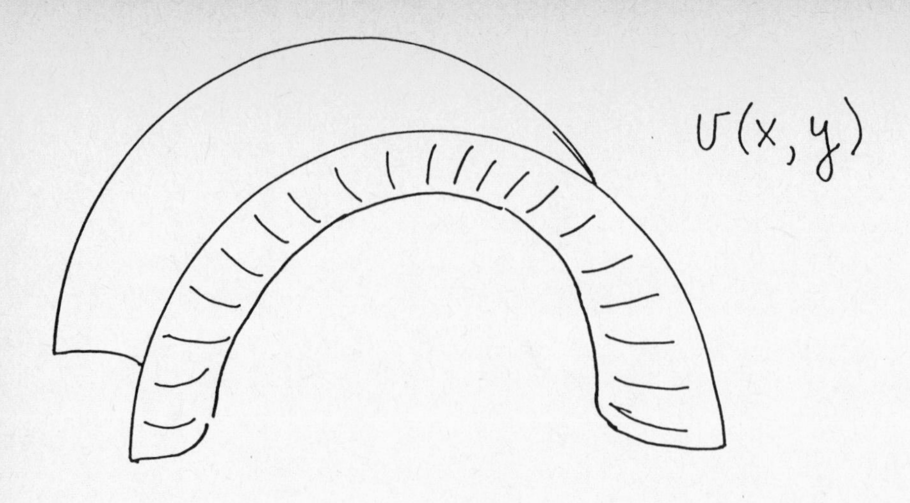

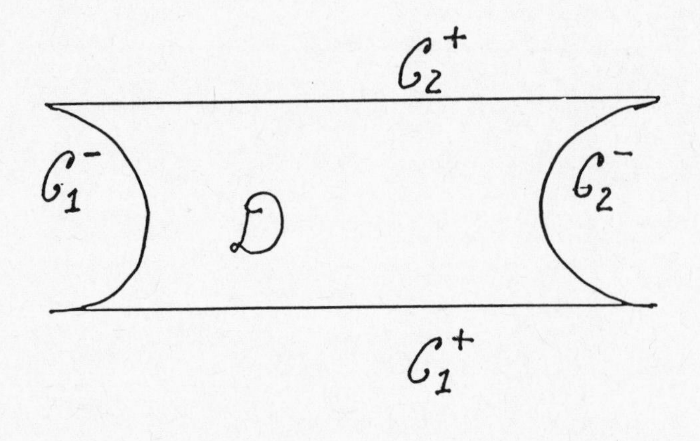

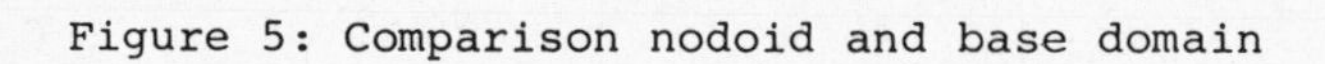
Figure 5: Comparison nodoid and base domain

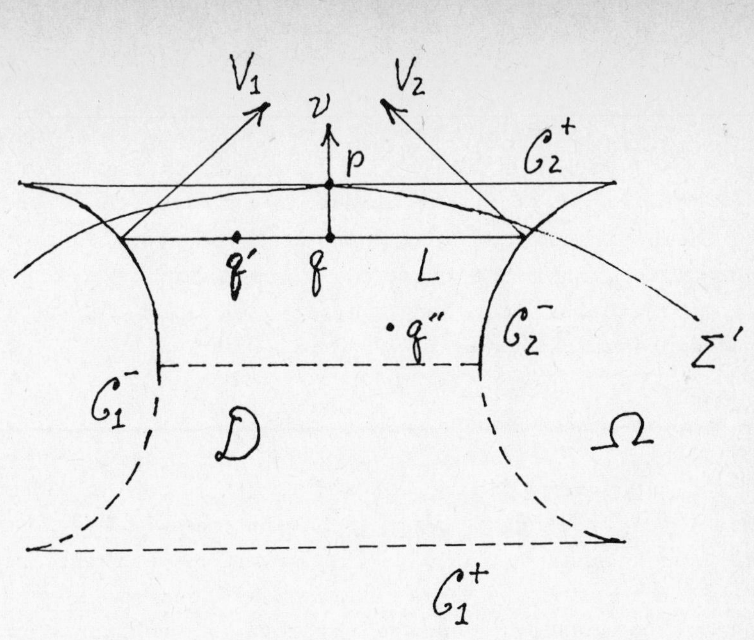

Figure 6: Proof of boundary estimate I

transformation so that $H \equiv 1$, and we suppose that done.

In general, if γ is prescribed on Σ it cannot be expected that the problem will have a solution; in zero gravity the fluid has a propensity to flow out to infinity along the walls of the bounding cylinder. Nevertheless, under certain conditions (see [4], Chapters 6 and 7) it is possible to show the existence of a unique (up to an additive constant) "variational solution" $u(x,y)$ that is smooth in Ω and achieves the prescribed data in the weak sense

$$\int_{\Omega} (\nabla\eta\cdot Tu+2\eta)\, dx\, dy = \oint_{\Sigma} \eta \cos\gamma\, ds \tag{49}$$

for every $\eta \in H^{1,1}(\Omega)$ (recall that now $H \equiv 1$). We consider such a solution corresponding to data for which $\gamma = 0$ on a subarc $\Sigma' \subset \Sigma$, and we seek to characterize the behavior near Σ'. To do so, we introduce as reference surfaces the rotationally symmetric solutions of (46), known as unduloids and as nodoids. We consider specifically nodoids, which are surfaces generated by rotating certain periodic self-intersecting curves (determined by elliptic integrals) about a line, see Figure 4. We focus attention on the darkened portion of the curve in the figure. Introducing coordinates with the z-axis directed upwards from the (x,y) plane of the paper, the portion of the rotation surface lying __above__ the plane (as shown in Figure 5) determines a solution $v(x,y)$ of (46), defined over a domain $\mathcal{D}$ as shown. The boundary conditions

$$\nu\cdot Tv = +1 \quad \text{on} \quad \mathcal{C}_1^+,\ \mathcal{C}_2^+$$
$$\nu\cdot Tv = -1 \quad \text{on} \quad \mathcal{C}_1^-,\ \mathcal{C}_2^-$$

are extremal, in the sense that $|Tf| < 1$ for any differentiable function.

Let $p \in \Sigma'$. Consider first the case in which Σ' is concave to Ω at p. We place a domain $\mathcal{D}$ as shown in Figure 6, and consider ∇u at the point q at distance d from p along the normal. We assert:

a) *the direction of* $\nabla u(q)$ *lies between the extreme directions of* ∇v *on* L

b) *if* q' *is the (unique) point on* L *at which the two gradients have the same directions, then* $|\nabla u(q)| > |\nabla v(q')|$.

For if not, there would be a point q'' as indicated, at which $\nabla u(q) = \nabla v(q'')$. We could then translate $\mathcal{D}$ so that $q'' \to q$. Adding a constant to v, we can arrange to have $v(q'') = u(q)$, thus the two surfaces would have first order contact at q, hence by the lemma of §10 there must be at least four regions sharing q as common boundary

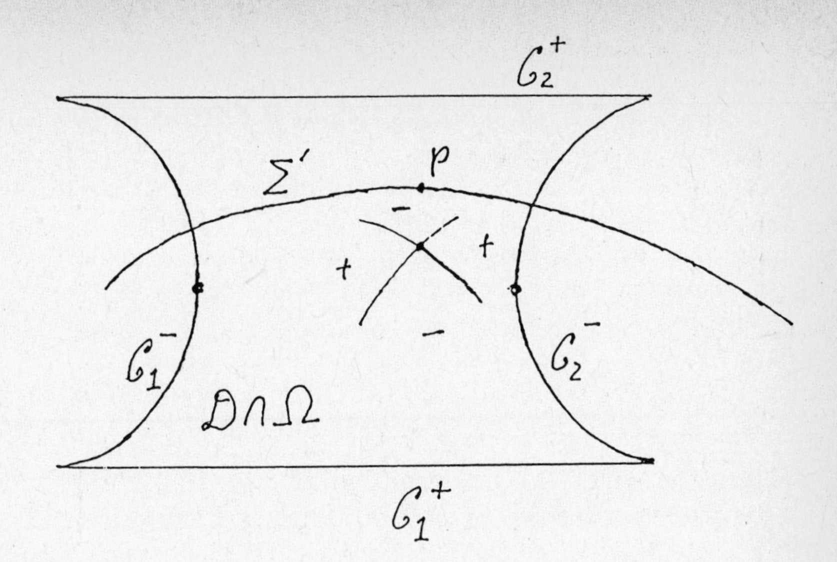

Figure 7: Proof of boundary estimate II

point, as shown in Figure 7.

Now in view of the variational condition (49), we can show that u majorizes v with respect to Σ' (see §2). On $\mathscr{C}_1^-$, $\mathscr{C}_2^-$ we have $\nu\cdot Tu > \nu\cdot Tv$ in a clear limiting sense. On $\mathscr{C}_1^+$ we have $\nu\cdot Tv > \nu\cdot Tu$. An examination of the possibilities shows that at least one of the four regions emanating from q must lead to a contradiction with the maximum principle CP.

Thus $|\nabla u(q)| > |\nabla v(q')|$. But $|\nabla v(q')| \geq |\nabla v(q)|$, which is just the slope of $\partial\mathfrak{D}$ at $L \cap \partial\mathfrak{D}$. Similarly, the inclinations of V_1, V_2 with respect to the normal at p are just the inverse of that slope. Using the explicit knowledge of v, *we obtain that for any* $\epsilon > 0$,

$$|\nabla u| > \tfrac{1}{2}(1-\epsilon)d^{-\frac{1}{2}} \tag{50}$$

$$|\theta| < 2(1+\epsilon)d^{\frac{1}{2}} \tag{51}$$

for all small enough d. Here θ = angle between ∇u and normal to Σ at p. *From* (50) *and* (51) *we obtain*

$$|\nu\cdot Tu - \cos 0| < 4(1+\epsilon)d \quad , \tag{52}$$

that is, the prescribed datum $\gamma = 0$ *is assumed strictly, at the indicated rate.*

This case in which Σ' is convex to Ω, and that in which $\gamma = \pi$ on Σ', can also be studied, with similar results, see [40].

12. <u>Convexity I.</u>

When is a capillary surface convex? Chen and Huang [22] consider a tube with convex section Ω in zero gravity, such that a solution $u(x,y)$ of (46) exists over Ω, corresponding to $\gamma = 0$ on Σ. Since u must then achieve a minimum interior to Ω, there exists $p \in \Omega$ at which the Gaussian curvature $K(p) \geq 0$. Suppose $K(q) = 0$ for some $q \in \Omega$, and let L be the line containing $(q,u(q))$ and tangent to a zero curvature direction on the solution surface S. Consider a lower half cylinder Z of radius $\frac{1}{2H}$, tangent to S at $(q,u(q))$, of which L is a generator. Since S and Z have equal mean curvatures, the two surfaces have second order contact over q. Thus at least six do-

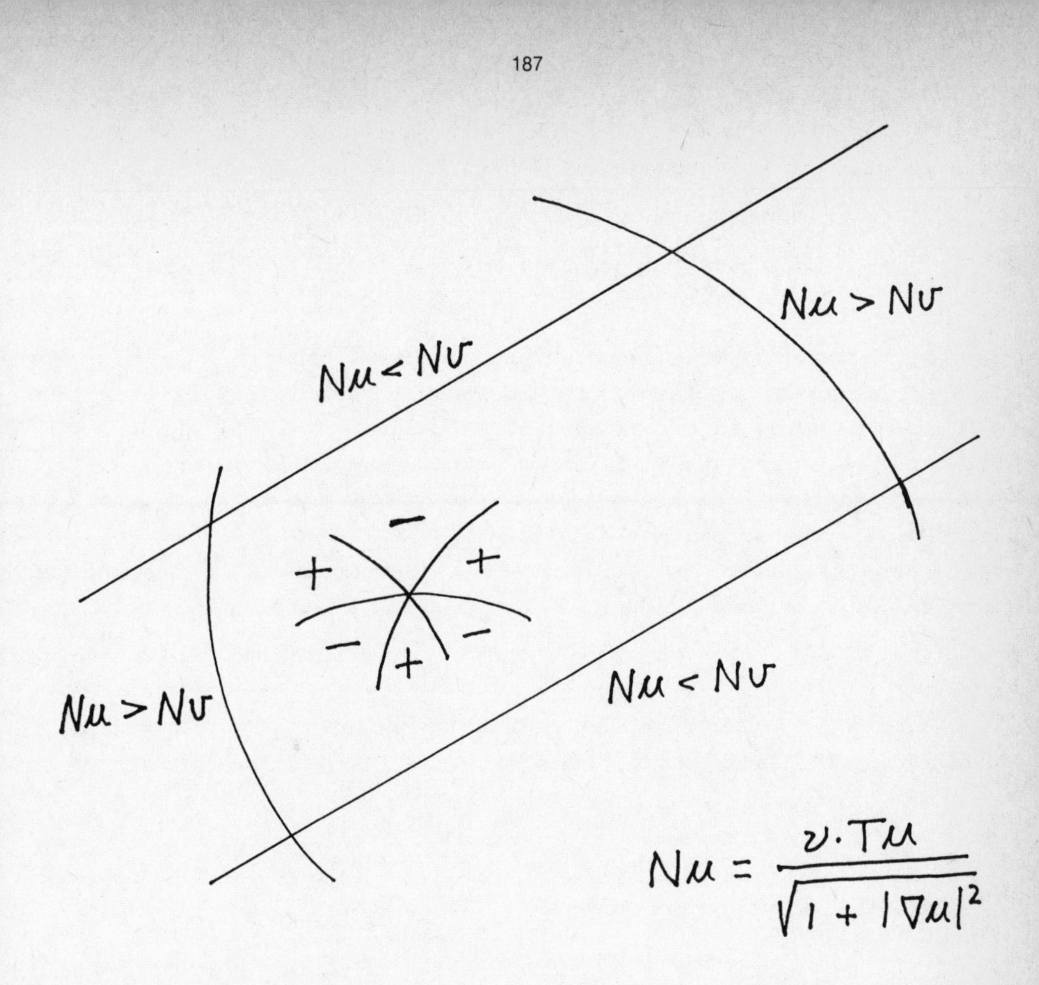

Figure 8: Proof of convexity

mains (as in §10) share q as common boundary point. The two sufaces share a common domain, bounded by (at most) two projections L_1, L_2 of the equatorial generators of Z, and by (at most) two convex subarcs Σ', Σ'' of Σ, see Figure 8. As in §11 we come into conflict with the maximum principle; *we conclude* $K > 0$ *over the entire surface.*

Remarks: 1. Chen and Huang suppose the boundary datum $\gamma = 0$ to be assumed strictly on Σ. As in §11, it would have sufficed to know that $\gamma = 0$ on Σ holds in a weak variational sense.

2. If $\gamma \neq 0$, the result fails, a counterexample is given in [26].

3. The hypothesis of convexity of Σ is clearly essential, however strict convexity need not be assumed.

4. Chen [23] used similar ideas to show uniqueness of the minimum point for any convex Ω, when $0 \leq \gamma < \frac{\pi}{2}$. The reasoning was later improved by Siegel [37], who also extended the result to non zero gravity.

5. Huang [24] again used the method to show that the minimum point has distance at least $|\Omega|/|\Sigma|$ from $\partial\Omega$.

13. Convexity II.

Korevaar [25] considered the equation

$$\text{div}\, Tu = \kappa u \quad , \quad \kappa > 0 \tag{52}$$

(positive gravity) in a strictly convex Ω, again with $\gamma = 0$ on $\Sigma = \partial\Omega$. For points $x_1, x_3 \in \bar{\Omega}$ and $\lambda \in [0,1]$ he sets $x_2 = \lambda x_3 + (1-\lambda)x_1$ and introduces the "concavity function"

$$\mathcal{X}(x_1,x_3;\lambda) = u(x_2) - \lambda u(x_3) - (1-\lambda)u(x_1) \quad .$$

There holds $\mathcal{X} \leq 0$ for all admissible arguments if and only if $u(x)$ determines a convex surface over Ω.

The first of the following two lemmas is easy to prove; the second is not much more difficult:

i) Let $u \in C(\bar{\Omega})$ and suppose that the surface $\mathcal{S}$ determined by u is not convex. Then $\mathcal{X}$ attains its positive supremum for points x_1, $x_3 \in \bar{\Omega}$, such that $x_1 \neq x_3$, $0 < \lambda < 1$.

ii) Let u satisfy (52) in Ω. Then $\mathcal{K}$ cannot have a positive local maximum for points $x_1, x_2 \in \Omega$, $0 < \lambda < 1$.

From the two lemmas, Korevaar concludes that $\mathcal{S}$ *is strictly convex.* For if not, then by i) a positive maximum is achieved by $\mathcal{K}$, while by ii) at least one of the points x_1, x_3 must lie on Σ. Suppose $x_1 \in \Sigma$ and let s denote arc length along the segment $\overline{x_1x_3}$. Since $\nu \cdot Tu = 1$ on Σ, we find $\left.\frac{\partial u}{\partial s}\right|_{x_1} = \infty$, so that u increases as x_1 is approached. Thus, $\mathcal{K}$ could be increased by moving x_1 a small distance into Ω along $\overline{x_1x_3}$, with λ, x_3 held fixed. This contradiction establishes the result.

<u>Remark:</u> Korevaar later improved his result so as to cover the zero gravity case. Also, his result extends to higher dimensions, which the result of §12 does not. Nevertheless for the case they consider, Chen and Huang require considerably weaker boundary hypotheses, both for the domain and for the solution.

14. Convexity III.

We consider a different but related question, that of a liquid drop resting on a horizontal surface. If the contact angle γ is constant, then the drop is symmetric and convex, see §15.

If γ is not constant, convexity cannot in general be expected; however the question has been raised, whether convexity of the wetted surface on the support plane implies convexity of the drop. We show now by example that the answer is negative.

We consider a situation without gravity, and focus attention on an unduloid, which is a rotation surface of constant mean curvature H generated by a curve $r(y)$ as indicated in Figure 9. We will take $H \equiv 1$. The curve is determined by the relations

$$r = \frac{1}{2}\,(\sin\psi \pm \sqrt{4c^2-\cos^2\psi}) \tag{53}$$

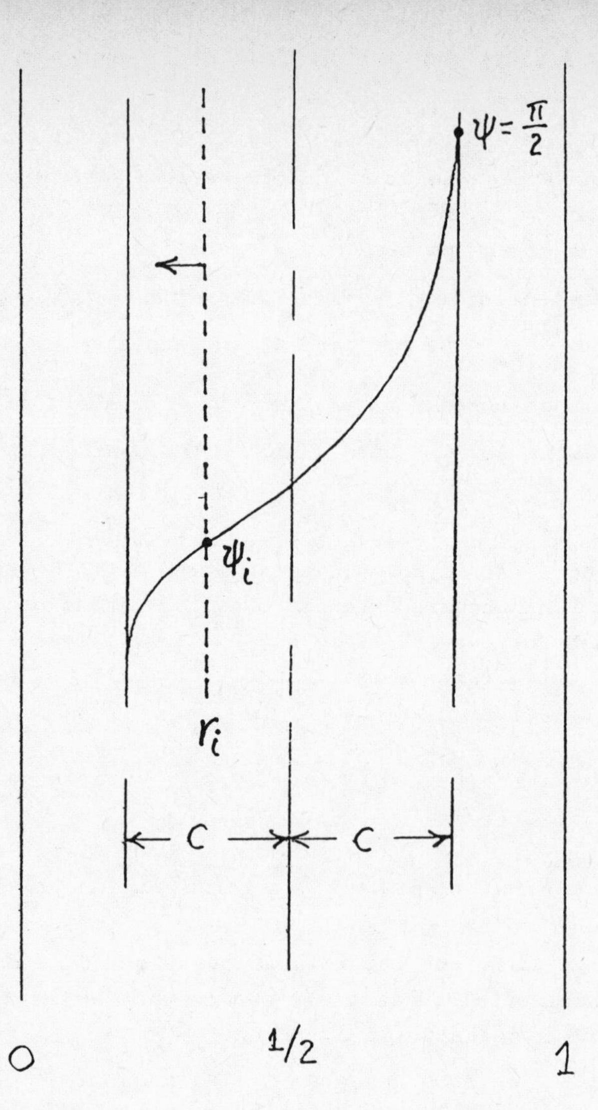

Figure 9: Proof of non-convexity

$$y = \frac{1}{2}\int_{\psi_i}^{\psi}\left[1 \pm \frac{\sin\psi}{\sqrt{4c^2-\cos^2\psi}}\right]\sin\psi\, d\psi \tag{54}$$

with $0 < c < \frac{1}{2}$, ψ = inclination angle. Here the sign is to be taken positive or negative according as $\psi > \psi_i$ or $\psi < \psi_i$, the value of ψ at the inflection. We have $\cos\psi_i = 2c$, $r_i = \frac{1}{2}\sqrt{1-4c^2}$, $y_i = 0$. We restrict attention to the range $\psi_i \lesssim \psi \lesssim \frac{\pi}{2}$ indicated in the figure.

We consider a plane Π orthogonal to the plane of the figure and passing through (r_i, y_i). If we take Π as plane of support for the drop determined by the portion of the unduloid for which $r > r_i$, then clearly the drop will be convex; however, convexity will be lost on any translation of Π in the direction of decreasing r. Introducing a coordinate function $z(y)$ to describe the boundary of the wetted surface, we compute

$$z_{yy} = \frac{2\sec^2\psi \pm \sqrt{4c^2-\cos^2\psi}}{(\pm\sqrt{4c^2-\cos^2\psi}+\sin\psi)\cos\psi}\,\frac{y^2}{r_0^2} + \frac{r^2}{r_0^2}\tan\psi\ . \tag{55}$$

Since the second term on the right in (55) is bounded from zero, we conclude that if the displacement of Π is small enough, the intersection curve with the unduloid will remain convex. Nevertheless the surface cannot be convex, in view of the inflection at ψ_i.

15. The E. Hopf boundary point lemma.

This effective weapon, which is sufficiently well known to need no repetition here, has striking applications in capillarity as in other contexts. We outline first the proof of the inclusion theorem of Siegel, that was stated in §6. We consider solutions of (30), subject to (31). We suppose $0 < \gamma < \frac{\pi}{2}$; if $\gamma = 0$ a more general result holds, see [4], Theorem 5.9.

Suppose the theorem were false. Then there would be a disk $\Omega^{(i)} \subset \Omega^{(0)}$, which could contact $\Sigma^{(0)}$ at any point from within $\Omega^{(0)}$, and solutions $u^{(i)}$, $u^{(0)}$ of (30), (31) in $\Omega^{(i)}$, $\Omega^{(0)}$, such that $w = u^{(0)} - u^{(i)} \geq 0$ at some $q \in \bar{\Omega}^{(i)}$. Since w satisfies a maximum principle, we may assume $q \in \Sigma^{(i)}$.

Since $u^{(0)}$ satisfies the maximum principle, there exists $p \in \Sigma^{(0)}$ such that $u^{(0)}(p) - u^{(i)}(q) > 0$. We now observe that $u^{(i)}$ is constant on $\Sigma^{(i)}$, and translate $\Omega^{(i)}$ so as to contact $\Sigma^{(0)}$ at p. We then have $u^{(0)}(p) - u^{(i)}(p) > 0$.

If we decrease the radius of $\Omega^{(i)}$, we can maintain contact, while $u^{(i)}(p) \uparrow \infty$. Hence a configuration can be achieved, at which $u^{(0)}(p) = u^{(i)}(p)$.

Again using the maximum principle for $u^{(0)}$, we have $u^{(0)}(q)-u^{(i)}(q) < 0$ for $q \neq p$ on $\Sigma^{(i)}$. Hence $u^{(0)}-u^{(i)} < 0$ in $\Omega^{(i)}$, while $u^{(0)} - u^{(i)} = 0$ at p. By the Hopf lemma, $\frac{\partial u^{(0)}}{\partial n}(p) > \frac{\partial u^{(i)}}{\partial n}(p)$, contradicting the boundary condition (31).

The Hopf lemma was used by Aleksandrov [28] to prove that the only embedded closed surface $\mathcal{S}$ of constant mean curvature (soap bubble) is a sphere. The underlying idea is to start with a tangent plane Π at $p \in \mathcal{S}$, then translate Π continuously into $\mathcal{S}$ until the reflection in Π of the part of $\mathcal{S}$ containing p contacts $\mathcal{S}$ at p'. The Hopf lemma then shows that $\mathcal{S}$ is identical to its reflection. Since this holds for all $p \in \mathcal{S}$, the result follows. The method was adapted by Serrin [29] to show that if a capillary surface $z = u(x,y)$ has constant boundary height and constant γ, then the domain of definition of $u(x,y)$ is a disk. Serrin's proof requires a delicate extension of the Hopf lemma that has independent interest. This proof was in turn adapted by Wente [30] to the case of liquid drops on a support plane Π (for which the projection onto Π need not be simple). Wente showed that every connected drop with constant boundary angle $\gamma \in (0,\pi]$ is generated by an interval of disks centered on a line segment orthogonal to Π.

We remark parenthetically the analytic proof in [33] that γ and the volume V uniquely determine the drop up to translation. It would be desirable to have an independent proof based on comparison principles. Such a proof was obtained by Spruck for the case in which $\mathcal{S}$ projects simply onto Π.

The following elegant application of the Hopf lemma is due to Vogel [31]. Let $\mathcal{S}$ be a surface lying over a smooth domain Ω, whose mean curvature H satisfies

$$H \equiv \varphi(z) \quad , \quad \varphi'(z) \geq 0 \tag{56}$$

and which satisfies a contact angle boundary condition on $\Sigma = \partial\Omega$. Then

$\mathcal{S}$ is a graph over Ω. For if not, $\mathcal{S}$ could be lowered to a surface $\mathcal{S}'$ lying entirely below $\mathcal{S}$, and then raised until a first point of contact occurs. At this point the Hopf lemma can be applied, leading to a contradiction.

As in the other proofs, the case in which p lies over Σ requires special and detailed attention. It should be noted here particularly that $\mathcal{S}$ is not required to be energy minimizing.

References

[1] T. Young: An essay on the cohesion of fluids. Philos. Trans. Roy. Soc. London 95 (1805), 65-87.

[2] P.S. Laplace: Traité de mécanique céleste; suppléments au Livre X, 1805 and 1806 resp. in Oeuvres Complete Vol. 4. Gauthier-Villars, Paris; see also the annotated English translation by N. Bowditch (1839); reprinted by Chelsea, New York, 1966.

[3] C.F. Gauss: Principia Generalia Theoriae Figurae Fluidorum. Comment. Soc. Regiae Scient. Gottingensis Rec. 7 (1830). Reprinted as "Grundlagen einer Theorie der Gestalt von Flüssigkeiten im Zustand des Gleichgewichtes", in Ostwald's Klassiker der exakten Wissenschaften, vol. 135. W. Engelmann, Leipzig, 1903.

[4] R. Finn: Equilibrium Capillary Surfaces. Springer-Verlag, New York, 1986. (Grundlehren der Mathem. Wiss. # 284).

[5] K. Kenmotsu: Weierstrass formula for surfaces of prescribed mean curvature. Math. Ann. 245 /1979) 89-99.

[6] G. Bakker: Kapillarität und Oberflächenspannung. In: Handbuch der Experimentalphysik, Band 6. Akademische Verlagsgesellschaft, Leipzig, 1928.

[7] B. Taylor: Concerning the ascent of water between two glass planes. Philos. Trans. Roy. Soc. London 27 (1712), 538.

[8] F. Hauksbee: Some further experiments, showing the ascent of water between two glass planes in an hyperbolic curve. Philos. Trans. Roy. Soc. London 28 (1713), 153.

[9] Petrus van Musschenbroek: Introductio ad Philosophiam Naturalem. Tom 1. S.J. et Luchtnams, Leiden, 1762, p. 376.

[10] A. Ferguson and I. Vogel: On the "hyperbola" method for the measurement of surface tensions. Phys. Soc. Proc. 38 (1926) 193-203.

[11] H.M. Princen: Capillary phenomena in assemblies of parallel cylinders. II. Capillary rise in systems with more than two cylinders. J. Colloid Interface Sci. 30 (1969), 359-371.

[12] H. Minkowski: Kapillarität. Encycl. Mathem. Wiss. VI, Teubner, Leibzig, 1903-1921, pp. 559-613.

[13] C.V. Boys: Soap Bubbles, and the Forces Which Mould Them. Society for Promoting Christian Knowledge, London 1902; revised edition, 1916. Reprinted by Doubleday & Co., Garden City, 1959.

[14] L.-F. Tam: Regularity of capillary surfaces over domains with corners. Pac. J. Math. 124 (1986), 469-482.

[15] N.J. Korevaar: On the behavior of a capillary surface at a re-entrant corner. Pacific J. Math. 88 (1980), 379-385.

[16] J. Spruck: On the existence of a capillary surface with a prescribed angle of contact. Comm. Pure Appl. Math. 28 (1975), 189-200.

[17] N.J. Korevaar: The normal variations technique for studying the shape of capillary surfaces. Astérisque 118 (1984), 189-195.

[18] N.J. Korevaar: An easy proof of the interior gradient bound for solutions to the prescribed mean curvature equation. Proc. Symp. Pure Math. 45 (1986) Part 2, 81-89.

[19] N.J. Korevaar: Maximum principle gradient estimates for the capillary problem. To appear.

[20] G.M. Lieberman: Gradient estimates for capillary-type problems via the maximum principle. To appear.

[21] G.M. Lieberman: Boundary behavior of capillary surfaces via the maximum principle. In: Proc. Conf. Var. Methods, ed. P. Concus & R. Finn, Vallombroso 1985; Springer-Verlag, 1986.

[22] Jin-Tzu Chen and W.-S. Huang: Convexity of capillary surfaces in the outer space. Invent. Math. 67 (1982), 253-259.

[23] J.-T. Chen: Uniqueness of minimal point and its location of capillary free surfaces over convex domains. Astérisque 118 (1984), 137-143.

[24] W.-H. Huang: Level curves and minimal points of capillary surfaces over convex domains. Bull. Inst. Math. Acad. Sin. 2 (1983), 390-399.

[25] N.J. Korevaar: Capillary surface convexity above convex domains. Indiana Univ. Math. J. 32 (1983), 73-81.

[26] R. Finn: Existence criteria for capillary free surfaces without gravity. Indiana Univ. Math. J. 32 (1983), 439-460.

[27] D. Siegel: Height estimates for capillary surfaces. Pacific J. Math. 88 (1980), 471-516.

[28] A.D. Aleksandrov: Uniqueness theorems for surfaces in the large. V. Vestnik Leningrad Univ. 13 (1958) 5-8. Amer. Math. Soc. Translations (Series 2) 21, 412-416.

[29] J.B. Serrin: A symmetrty problem in potential theory. Arch. Rational Mech. Anal. 43 (1971), 304-318.

[30] H.C. Wente: The symmetry of sessile and pendent drops. Pacific. J. Math. 88 (1980), 387-397.

[31] T.I. Vogel: Weak conditions for capillary surfaces. Preprint, Texas A & M University; to appear.

[32] W.N. Bond and D.A. Newton: Bubbles, drops, and Stokes' Law (Paper 2). Phil. Mag. Ser. 7, no. 5 (1928), 794-800.

[33] R. Finn: The sessile liquid drop I: Symmetric Case. Pacific J. Math. 88 (1980), 541-587.

[34] R. Finn: On the Laplace formula and the meniscus height for a capillary surface. Z. Angew. Math. Mech. 61 (1981), 165-173.

[35] R. Finn: Addenda to my paper "On the Laplace formula and the meniscus height for a capillary surface". Z. Angew. Math. Mech 61 (1981), 175-177.

[36] D. Siegel: Height estimates for the narrow capillary tube. Preprint, New Mexico Inst. of Mining and Technology.

[37] D. Siegel: Uniqueness of the minimum point for capillary surfaces over convex domains. To appear.

[38] D. Siegel: The behavior of a capillary surface for small Bond number. Proc. Conf. Var. Methods, Vallombrosa 1985, ed. P. Concus & R. Finn, Springer-Verlag 1986.

[39] L.-F. Tam: The behavior of capillary surfaces as gravity tends to zero. Comm. P.D.E. 11 (1986), 851-901.

[40] R. Finn: Moon surfaces, and boundary behavior of capillary surfaces for perfect wetting and non wetting. Preprint, Univ. Bonn 1987, to appear.

[41] P. Concus and R. Finn: On capillary free surfaces in a gravitational field. Acta Math. 132 (1974), 207-223.

[42] D. Langbein and F. Rischbieter: Form, Schwingungen und Stabilität von Flüssigkeitsgrenzflächen. Forschungsbericht BMFT, Battelle Inst. Frankfurt/M., 1986.

[43] M. Emmer: Esistenza, unicita e regolarita nelle superfici di equilibrio nei capillari. Ann. Univ. Ferrara Sez. VII 18 (1973), 79-94.

[44] N.N. Ural'tseva: Solution of the capillary problem (Russian). Vestnik Leningrad Univ. 19 (1973), 54-64.

[45] C. Gerhardt: Existence and regularity of capillary surfaces. Boll. Un. Mat. Ital. 10 (1974), 317-335.

[46] R. Finn and C. Gerhardt: The internal sphere condition and the capillary problem. Ann. Mat. Pura App. 112 (1977), 13-31.

[47] C. Gerhardt: Boundary value problems for surfaces of prescribed mean curvature. J. Math. Pures Appl. 58 (1979), 75-109.

[48] R. Finn: Global size and shape estimates for symmetric sessile drops. J. Reine Angew. Math. 335 (1982), 9-36.

[49] R. Finn: Some comparison properties and bounds for capillary surfaces. In Complex Analysis and its Applications (Russian). Moscow Math. Soc.: volume dedicated to I.N. Vekua, Scientific Press, Moscow, 1978.

[50] M. Emmer: On the behavior of the surfaces of equilibrium in the capillary tubes when gravity goes to zero. Rend. Sem. Mat. Univ. Padova 65 (1981), 143-162.

[51] P. Concus and R. Finn: The shape of a pendent liquid drop. Philos. Trans. Roy. Soc. London Ser. A 292 (1979), 307-340.

[52] P. Concus and R. Finn: A singular solution of the capillary equation. I. Existence. Invent. Math. 29 (1975), 143-148.

[53] P. Concus and R. Finn: A singular solution of the capillary equation. II: Uniqueness. Invent. Math. 29 (1975), 149-160.

[54] M.-F. Bidaut-Veron: Global existence and uniqueness results for singular solutions of the capillarity equation. Pacific. J. Math., 125 (1986) 317-334.

[55] M.-F. Bidaut-Veron: New results concerning the singular solutions of the capillarity equation. Proc. Conf. Var. Methods, Vallombrosa, 1985, ed. P. Concus & R. Finn, Springer-Verlag 1986.

[56] J.B. Serrin: The Dirichlet problem for surfaces of constant mean curvature. Proc. Lon. Math. Soc. (3) 21 (1970), 361-384.

[57] O.A. Ladyzhenskaya, N.N. Ural'tseva: Local estimates for gradients of solutions of non-uniformly elliptic and parabolic equations. Comm. Pure Appl. Math. 23 (1970), 677-703.

[58] E. Heinz: Interior gradient estimates for surfaces $z = f(x,y)$ of prescribed mean curvature. J. Diff. Geom. 5 (1971), 149-157.

[59] R. Finn: On equations of minimal surface type. Annals of Math. 60 (1954), 397-416, esp. p. 397.

[60] R. Finn and E. Giusti: Nonexistence and existence of capillary surfaces. Manuscripta Math. 28 (1979), 13-20.

[61] Fei-tsen Liang: On nonparametric surfaces of constant mean curvature. Dissertation, Stanford University, 1986; to appear.

[62] R. Finn: New estimates for equations of minimal surface type. Arch. Rat. Mech. Anal. 14 (1963), 337-375.

[63] R.D. Gulliver, II: Regularity of minimizing surfaces of prescribed mean curvature. Annals of Math. 97 (1973), 275-305.

[64] J. Spruck: Infinite boundary value problems for surfaces of constant mean curvature. Arch. Rat. Mech. Anal. 49 (1972/3), 1-31.

[65] W.H. Fleming: On the oriented Plateau problem. Rend. Palermo 11 (1962), 69-90.

[66] J.B. Serrin: A priori estimates for solutions of the minimal surface equation. Arch. Rat. Mech. Anal. 14 (1963), 376-383.

[67] R. Finn: On partial differential equations whose solutions admit no isolated singularities. Scripta Math. 26 (1961), 107-115.

[68] L. Bers: Isolated singularities of minimal surfaces. Ann. of Math. 53 (1951), 364-386.

[69] M.-F. Bidaut-Veron: A singular solution of a quasilinear partial differential equation. Technical report, Univ. Tours, 1986.

REMARKS ON DIAGONAL ELLIPTIC SYSTEMS

Jens Frehse

§1 Introduction and discussion of known results and open problems

In this paper, we study diagonal elliptic systems of the type

(1.1) $$-\Delta u_\nu + \alpha_o u_\nu = F_\nu(x,u,\nabla u) \quad , \quad \nu = 1,\ldots,N \,,$$

where

(1.2) $$u = (u_1,\ldots,u_N) \in H^1_o(\Omega;\mathbb{R}^N) \,.$$

Here Ω is a bounded domain of $\mathbb{R}^N$ and $H^{1,p}(\Omega;\mathbb{R}^N)$, $H^{1,p}_o(\Omega;\mathbb{R}^N)$ denote the usual Sobolev spaces of N-vector functions on Ω . We abbreviate $H^1_o = H^{1,2}_o$.

Equation 1 has to be understood in the weak sense; we assume a *quadratic growth* assumption for F_ν , i.e.

(1.3) $$|F_\nu(x,u,\nabla u)| \le a|\nabla u|^2 + K \quad , \quad |u| \le C \text{ , a.e. in } \Omega$$

for all $u \in H^1(\Omega;\mathbb{R}^N)$, *with constants* $a = a(C)$, $K = K(C)$.

The number α_o satisfies

(1.4) $$\alpha_o \ge 0 \,.$$

(Sometimes the strict inequality $\alpha_o > 0$ is assumed.)

Our discussion applies with a few modifications to diagonal elliptic systems

(1.5) $$Lu_\nu = F_\nu(x,u,\nabla u) \quad , \quad \nu = 1,\ldots,N$$

where

$$Lu_\nu = -\mathrm{div}\left(A(x,u)\nabla u_\nu\right) + \alpha_o u_\nu$$

satisfies the condition of uniform ellipticity. In this contribution we discuss only the case that L does not depend on ν .

Throughout the paper we assume that

(1.6) $F(x,\nu,\eta)$ *is measurable in* $x \in \Omega$ *and continuous in* $(v,\eta) \in \mathbb{R}^N \times \mathbb{R}^{nN}$.

An analogous condition is assumed for the N×N-matrices $A(x,u)$.

Furthermore the major part of our results is valid for non homogeneous boundary conditions

(1.7) $$u \in g + H^1_o(\Omega;\mathbb{R}^N) .$$

The above boundary value problem (1.1) or (1.5) with the quadratic growth assumption (1.3) occurs in many applications, for example, see Hildebrandt's survey [4] for applications in differential geometry, in particular the theory of harmonic mappings, or the work of Bensoussan and the author [1] for applications to stochastic differential games.

The Hamilton-Jacobi-Bellman equation of stochastic differential games has (in the elliptic case) the form

$$-\Delta u + \alpha_o u = H(x,\nabla u) + f$$

where H is computed in the following way. We restrict ourselves to the case $N = 2$ and do not treat the most general case which would lead to more complicated functions H and different (diagonal) principal part.

For the computation of $H = (H_1,H_2)$ we need the Lagrange functions

$$L_i(x,p_1,p_2,v_1,v_2) = l_i(x,v_1,v_2) + p_i \cdot g(x,v_1,v_2)$$

where $x \in \Omega$, $p_1,p_2 \in \mathbb{R}^n$, $v_1,v_2 \in \mathbb{R}^n$. The real valued functions l_i (as well as $f = (f_1,f_2) \in L^\infty$) come from the integrands of the payoff functionals of the original stochastic game problem, the function $g : \Omega \times \mathbb{R}^{2n} \to \mathbb{R}^n$ comes from the stochastic differential equation

$$dy = g\big(y(t),v_1(t),v_2(t)\big)\,dt + \sigma(y(t))dw(t) \quad , \quad y(0) = x .$$

The term $\sigma(y(t))dw(t)$ corresponds to the "white noise", and v_1,v_2 are the control variables. (See [1] for details.)

Under appropriate convexity conditions there exists a unique Nash point $V = (V_1,V_2)$ of (L_1,L_2) , that means

$$L_1(x,p_1,p_2,V_1,V_2) \le L_1(x,p_1,p_2,v_1,V_2) \quad \forall v_1 \in \mathbb{R}^n$$

$$L_2(x,p_1,p_2,V_1,V_2) \le L_2(x,p_1,p_2,V_1,v_2) \quad \forall v_2 \in \mathbb{R}^n .$$

Clearly, V depends on p_1,p_2 and x . Using this Nash point V one defines

$$H(x,p_1,p_2) = L(x,p_1,p_2,V) .$$

It is surprising that even "trivial" cases of stochastic differential equations like

$$dy = \big(v_1(t) + v_2(t)\big)\,dt + \sigma(y(t))dw(t)$$

and simple integrands of the payoff functionals like $l_i(v) = |v_i|^s$, e.g. $s = 2$, lead to elliptic systems which are non trivial from the point of view of classical analysis. In the above example one has to compute a Nash point (V_1,V_2) of the two functions

$$|v_1|^s + p_1(v_1 + v_2)$$
$$|v_2|^s + p_2(v_1 + v_2) \;.$$

This yields $s|v_i|^{s-2}\, v_i = -p_i$, $i = 1,2$ and hence $|v_i| = |p_i|^{1/(s-1)}\, s^{-1/(s-1)}$ and

$$v_i = -c_s\, p_i |p_i|^{-\frac{s-2}{s-1}} \;, \quad c_s = s^{-1/(s-1)} \;.$$

For H we obtain

$$H_1(x,p_1,p_2) = c_s^s |p_1|^{\frac{s}{s-1}} - c_s\, p_1 \left(p_1|p_1|^{-\frac{s-2}{s-1}} + p_2|p_2|^{-\frac{s-2}{s-1}}\right)$$
$$H_2(x,p_1,p_2) = c_s^s |p_2|^{\frac{s}{s-1}} - c_s\, p_2 \left(p_1|p_1|^{-\frac{s-2}{s-1}} + p_2|p_2|^{-\frac{s-2}{s-1}}\right) .$$

For $s = 2$ we have merely

$$H_1(x,p_1,p_2) = \tfrac{1}{4}\, p_1^2 - \tfrac{1}{2}\, p_1(p_1 + p_2)$$
$$H_2(x,p_1,p_2) = \tfrac{1}{4}\, p_2^2 - \tfrac{1}{2}\, p_2(p_1 + p_2) \;.$$

This case is already "difficult" from the point of view of the theory of elliptic systems since we have no smallness condition for the constant a in (1.3) which controls the quadratic growth behaviour. However, it is possible to overcome this difficulty due to the special structure of H in this case. In the case $1 < s < 2$ the function H has *super quadratic* growth in p which corresponds to a *super quadratic* growth of the function F in (1.1) with respect to ∇u . In the case of elliptic *systems* nobody knows how to treat such equations concerning existence and regularity of solutions although the corresponding stochastic differential game seems to be a nice, convex and coercive problem which should have a solution. Even in the quadratic case, where $l_i(x,v_1,v_2)$, $i = 1,2$, are positive definite quadratic forms in v_1, v_2, has not yet been treated in the general case.

For the problem of proving existence of weak solutions of (1.1) it is important to establish H^1-compactness criteria. By this we mean the following:

Let $F^m : \Omega \times \mathbb{R}^N \times \mathbb{R}^{nN} \to \mathbb{R}^N$ be functions which satisfy condition (1.3) *uniformly* in $m \in \mathbb{N}$, with F replaced by F^m , and let $F^m \to F$ pointwise a.e.. The Carathéodory condition (1.6) is assumed for F^m as well.

Let u^m be a weak solution of (1.1) with F replaced by F^m such that $u^m \to u$ weakly in H^1, $(m \to \infty\ ,\ m \in \mathbb{N})$ and $\|u^m\|_{L^\infty} \leq K$ uniformly. *Does there exist a subsequence* $\Lambda \subset \mathbb{N}$ *such that* $u^m \to u$ $(m \in \Lambda,\ m \to \infty)$ *strongly in* H^1 ?

If such a compactness criterion were true one could prove the existence of weak solutions provided that L^∞- and H^1-estimates are available for solutions to approximate problems.

For example, one can approximate problem (1.1) by setting

$$F^m = mF/(m + |F|)$$

and consider (1.1) with F replaced by F^m. Under additional conditions for F (examples see below) we have a uniform $L^\infty \cap H^{1,2}$- bound for the approximate solutions u^m. We may select a subsequence Λ such that $u^m \to u\ ,\ m \in \Lambda\ ,\ m \to \infty$, weakly and if we knew that $u^m \to u$ strongly in H^1 we would have $F^m(x,u^m,\nabla u^m) \to F(x,u,\nabla u)$ in L^1 and would obtain that u is a weak solution of (1.1). Unfortunately, such *a compactness theorem is false*, in general.

In §2 we present a sequence of diagonal elliptic systems of type (1.1) whose right hand sides F^m have the appropriate growth and convergence properties, but where the strong H^1-compactness fails to hold.

In this situation, it is of interest to find "structure conditions" for the right hand side F in (1.1) such that strong H^1-compactness for approximate solutions u^m of the system

$$-\Delta u_\nu^m + \alpha_o u_\nu^m = mF_\nu(x,u^m,\nabla u^m) \ / \ \big(m + |F(x,u^m,\nabla u^m)|\big)$$

(or other approximations) can be established. In [3] a survey about possible analytical techniques has been given.

The most common way to establish H^1-compactness for solutions u^m of diagonal elliptic systems of type (1.1) consists in proving C^α-a-priori estimates for u^m. In the scalar case $N = 1$ this is true due to the theory of Ladyženskaza-Ural'ceva [5]. In the case of systems $(N \geq 2)$, however, it has been shown by counterexamples that uniform C^α-estimates for the solutions of (1.1) are not true, in general. See the survey [4]. It turned out that a smallness condition for the growth constant a in (1.3) is necessary and essentially also sufficient for obtaining C^α-estimates. The condition reads

$$a \cdot \sup|u| < \lambda \tag{1.8}$$

where λ is the ellipticity constant of the principal part ($\lambda = 1$ if the principle part is Δ). If (1.8) holds with strict inequality we have $u \in C^\alpha$; on the other hand there are examples of diagonal elliptic systems with discontinuous bounded H^1-solutions such that (1.8) fails. In many situations a smallness condition like (1.8) is unsatisfactory. Let us discuss some cases where it is *not* needed.

Clearly one can avoid smallness conditions of type (1.8) if equation (1.1) is the Euler equation of a variational problem

$$\int G(x,u,\nabla u)\,dx = \min! \quad , \quad u \in g + H_o^1(\Omega;\mathbb{R}^N) .$$

For example we can take for G the function

$$G(x,u,\nabla u) = \frac{1}{2}\, a(x,u)\,|\nabla u|^2 + \frac{1}{2}\,\alpha_o u^2 - fu$$

where $|\nabla u|^2 = \Sigma_{\nu=1}^N |\nabla u_\nu|^2$, $\alpha_o \geq 0$

which leads to a diagonal elliptic system of type (1.1). Here we assume that $a(x,u)$ satisfies Carathéodory conditions and $0 < \gamma_o \leq a < \gamma_1$, $f \in L^\infty$.

Under these conditions there exists a minimum of the above variational integral and hence a solution of the system (1.1). This is a well known and standard result of the calculus of variations where very general existence theorems are known.

The discussion of this paper is only of relevance for the non variational and non minimal case.

Another situation where a smallness condition was avoided comes from the elliptic system arising in the theory of stochastic differential games. It turns out that the following "structure condition" for F is sufficient to prove C^α-a-priori estimates (see [1] and [3]).

(1.9) There exists a constant K such that for all $u \in H^{1,2}(\Omega;\mathbb{R}^2)$

$$|F_1(x,u,\nabla u)| \leq K|\nabla u|^2 + K$$

$$|F_2(x,u,\nabla u)| \leq K|\nabla u_2|^2 + K|\nabla u_1|\,|\nabla u_2| + K$$

(No "smallness" for K is assumed).

In order to obtain the C^α-a-priori estimate an L^∞-estimate is needed. Two possibilities are (alternatively)

(1.10) $F(x,v,\nabla v)\cdot v \leq \lambda|\nabla v|^2 + K$, $\alpha_o \geq 0$, $K = o(v^2) + K_o$

where λ is the ellipticity constant or the following conditions (1.11a)-(1.11c) (simultaneously)

(1.11a) $\alpha_o > 0$

(1.11b) $F \in C^\alpha$

(1.11c) $v_\nu F_\nu(x_o, v, \vec{0}) \le K$.

THEOREM 1.1: *Let* Ω *be a bounded domain of* $\mathbb{R}^n$ *with Lipschitz boundary,* $n \ge 1$ *, and* $F^m = (F_1^m, F_2^m)$ *be functions on* $\Omega \times \mathbb{R}^2 \times \mathbb{R}^{2n}$ *which satisfy the Carathéodory conditions and the growth condition* (1.9). *Then every sequence* (u^m) *of solutions* $u^m \in L^\infty \cap H_o^1(\Omega, \mathbb{R}^2)$ *with* $\|u^m\|_\infty \le K$ *uniformly has a subsequence which converges strongly in* H^1 *and* $|u^m|_{C^\alpha} \le \overline{K}_\alpha$ *uniformly as* $m \to \infty$, $0 \le \alpha \le 1$.

As a consequence we have:

THEOREM 1.2: *Let* Ω *be a bounded domain of* $\mathbb{R}^n$ *with Lipschitz boundary,* $n \ge 1$ *, and* $F = (F_1, F_2)$ *be a function on* $\Omega \times \mathbb{R}^2 \times \mathbb{R}^{2n}$ *which satisfies the Carathéodory conditions and the growth condition* (1.9) *and one of the two conditions* (1.10) *or* (1.11a-c). *Then there exists a solution* $u \in H_o^{1,2}(\Omega, \mathbb{R}^2) \cap H_{loc}^{2,p}$ *of* (1.1)

For the proof we refer to [1] and [3].

Let us note that the L^∞-estimates are based on maximum principle methods and that the derivation of H^1-estimates is *not* trivial, one has to use complicated testfunctions which use iterated exponentials of the components of u . The C^α-estimate is derived very similarly as the H^1-estimates. Morrey's lemma can be used. We believe that the case $N \ge 3$ (i.e. more than two equations) can be treated in a similar manner just by using more complicated testfunctions.

Let us remark that elliptic systems of type

$$-\Delta u_\nu + \alpha_o u_\nu = l_\nu(x, u, \nabla u) \cdot \nabla u_\nu + f_\nu(x) \quad , \quad \nu = 1, \dots, N$$

where l_ν has *linear* growth in ∇u can be considered as scalar equations in u_ν , namely

$$-\Delta u_\nu + \alpha_o u_\nu = g_\nu(x) \cdot \nabla u_\nu + f_\nu(x) \ .$$

This gives rise to the following question concerning *scalar equations*.

Let $v^m \in H_o^1(\Omega, \mathbb{R}^1)$ and $v^m \to v$ weakly in H^1 , $\|v^m\|_\infty \le K$ where v^m solves

$$-\Delta v^m + \alpha_o v^m = g^m(x)\cdot\nabla v^m + f^m(x)$$

and $g^m \to g$ weakly in L^2 and strongly in $L^{2-\epsilon}$, $\forall\epsilon > 0$, and $\|f^m\|_\infty \leq K$, $f^m \to f$ weakly in L^2. *Does* v *satisfy the limiting equation* $-\Delta v + \alpha_o v = g\cdot\nabla v + f$?

Let us remark that the following related question has to be answered in a *negative* way:

Let $u^m \in H^1_o(\Omega) \cap L^\infty(\Omega)$ be a sequence of *scalar* functions such that $u^m \to u$ weakly in $L^2(\Omega)$ and

$$-\Delta u^m + u^m f^m = g^m \tag{1.12}$$

where $\|f^m\|_{L^1} \leq K$ uniformly, $f^m \geq 0$ and $f^m \to f$ a.e., further $\|g^m\|_{L^\infty} \leq K$, $g^m \to g$ a.e.

Then, in general, u^m does *not converge strongly* in H^1. A counterexample is constructed in the following way. In [2] we have constructed a sequence of capacitary potentials φ^m of certain sets $E^m \subset \mathbb{R}^3$ which are contained in a fixed bounded set $Q \subset \mathbb{R}^3$ such that φ^m does *not* converge strongly in H^1. (An independent construction is due to Murat.)

Let $w^m *$ be the usual convolution operation with non negative mean functions $w^m \in C^\infty_o$ such that w^m converges sufficiently fast to the Dirac functional. Set

$$u^m = (2 - \varphi^m) * w^m .$$

Then $1 \leq u^m \leq 2$ and $\Delta u^m \geq 0$ in a neighbourhood of E^m. Setting $f^m = \tau\cdot(\Delta u^m)/u^m$ and $g^m = (1 - \tau)\Delta u^m$ where $0 \leq \tau \leq 1$, $\tau \in C^\infty_o$, and $\tau = 1$ in a uniform neighbourhood of E^m we have

$$-\Delta u^m + u^m f^m = g^m .$$

By the properties of the capacitary potentials φ^m we have the required conditions for f^m and g^m. The convergence $f^m \to 0$ a.e. follows also from the construction in [2]. The zero boundary condition for u^m can be achieved by considering τu^m and changing g^m.

Problem (1.12) is important for the question whether the passage to the limit can be justified for the following elliptic systems

$$-\Delta u^m_\nu = F^m_\nu(x,u^m,\nabla u^m) + g^m(x) \tag{1.13}$$

where u^m is bounded uniformly in $L^\infty \cap H^1(\Omega,\mathbb{R}^N)$, $u^m \to u$ weakly in $H^1_o(\Omega)$, $F^m_\nu \to F$ a.e. $\|g^m\|_\infty \leq K$ uniformly, $g^m \to g$ a.e., and F^m satisfies

$$(1.14) \qquad |F^m(x,v,\nabla v)| \le a|\nabla v|^2 + K$$

uniformly and the "one sided" condition, say

$$(1.15) \qquad F^m(x,v,\nabla v)\cdot v \le (\lambda - \epsilon)|\nabla v|^2$$

(λ = ellipticity constant).

Note that no smallness for a is assumed. We hope that this problem can be solved by "capacity methods" (i.e. methods which employ the notion of "quasi-uniform convergence") but up to now it is only known that strong convergence $u^m \to u$ in H^1 holds if it is *already known* that u satisfies the limiting equation, see [2]. We remark that the problem of justifying the passage to the limit and thus the solvability of the system is not yet even known for the following "model problem"

$$-\Delta u_\nu^m + u_\nu^m |\nabla u^m|^2 = 0$$

(with nonhomogenous boundary condition).

In the scalar case, this equation is equivalent to Euler's equation of a variational integral

$$I(u) = \int_\Omega e^{-u^2} |\nabla u|^2 dx \ .$$

At the first glimpse it seems that $\inf I(u) = 0$ and that there is no minimum, but a more careful analysis shows that I has a regular minimum. (Study the variational inequality with the convex set

$$\mathscr{C}_L = \left\{ u \in g + H_o^1 \;\middle|\; -L \le u \le L \right\} .$$

The passage to the limit $L \to \infty$ can be justified. (Use testfunctions like $u - \epsilon_o e^{u^2}(u - g)G_\rho$), G_ρ = regularized Green function).

The open problem to pass to the limit in equation (1.13), under the condition (1.14) and (1.15) is of great importance for several fields in the theory of partial differential equations.

§2 Counterexamples concerning H^1-compactness of solutions of elliptic equations and systems

In this section we first present an example of a sequence (u_m) of weak solutions u_m of a *scalar* elliptic equation

$$(2.1) \qquad -\Delta u_m = F_m(x,\nabla u_m) + f_m(x)$$

such that u_m converges weakly, but not strongly in $H^{1,2}(\mathbb{R}^3)$ to the function $u = 0$ which does not satisfy the limiting equation

$$(2.2) \qquad -\Delta u = F(x,\nabla u) + f(x) \ .$$

Here the functions F_m will be measurable in the first argument x and C^∞ in the second argument; further the growth condition

(2.3) $$|F_m(x,\eta)| \le K|\eta|^2 \quad , \quad \eta \in \mathbb{R}^n \text{ , } m \in \mathbb{Z} \text{ ,}$$

will be satisfied with a uniform constant K. For fixed $z \in H^{1,2}$ our construction is such that

(2.4) $$F_m(\cdot,\nabla z) \to 0 \text{ almost everywhere } (m \to \infty) \text{ ,}$$

and

(2.5) $$f_m \to f \text{ , weakly in } L^\infty \text{ ,}$$

where f is equal to a constant $\neq 0$ on the cube $[0,1]^3$.
Obviously the functions u_m cannot be bounded uniformly since this would imply uniform C^α-regularity, which in turn implies H^1-compactness via the usual monotonicity argument. In fact each function u_m will be unbounded.

We now present the construction of the functions u_m above. We subdivide the unit cube $[0,1]^3 \subset \mathbb{R}^3$ by cubes with edge length $R = \frac{1}{m}$ and faces parallel to the coordinate axes. Let $\mathcal{S}_R$ be the set of such cubes, i.e. $\cup\{Q \mid Q \in \mathcal{S}_R\} = [0,1]^3$, $\operatorname{int} Q_1 \cap \operatorname{int} Q_2 = \phi$ for $Q_1, Q_2 \in \mathcal{S}_R$, $Q_1 \neq Q_2$. For each $Q \in \mathcal{S}_R$ we denote its center by x_Q. By $B_s(x_Q)$ (shorter: B_s) we denote the ball in $\mathbb{R}^3$ with center x_Q and radius s. We need the auxiliary function $w_R : \mathbb{R}^3 \to \mathbb{R}$ defined by

(2.6) $$w_R(x) = \begin{cases} \ln\frac{1}{|x|} - \ln\frac{1}{r} + 1 \text{ , } r = R^3 & \text{, for } x \in B_r \\ \frac{r}{|x|} \text{ , } r = R^3 & \text{, for } x \in B_R - B_r \\ \frac{1}{16} R^{-2}(5R^2 - x^2)^2 & \text{, for } x \in B_{\sqrt{5}R} - B_R \\ 0 & \text{, for } x \in \mathbb{R}^3 - B_{\sqrt{5}R} \end{cases} .$$

It is easy to see that $w_R \in C^1(\mathbb{R}^3 - 0)$ and $w_R \in H^{2,\infty}_{loc}(\mathbb{R}^3 - 0)$. For example, on the boundary ∂B_R of B_R we have with $w = w_R$,

$$w\big|_{R-0} = \frac{r}{R} = R^2 \quad , \quad w\big|_{R+0} = \frac{1}{16} R^{-2}(5R^2 - R^2)^2 = R^2$$

$$\nabla w(x)\Big|_{|x|=R-0} = -\frac{rx}{|x|^3} = -x$$

$$\nabla w(x)\Big|_{|x|=R+0} = -\frac{1}{4} R^3 x(5R^2 - x^2)\Big|_{|x|=R} = -x \text{ .}$$

We remark that the radius $\sqrt{5}\,R$ was chosen in order to make the rest of the coefficients simple, but still keep the "side condition" $w \in C^1$, $w \ge 0$. We further remark that $u \in H^{1,p}(\mathbb{R}^3)$ for all $p < 3$, $p \ge 1$.

By simple calculation, we obtain almost everywhere in $\mathbb{R}^3$

(2.7)
$$\Delta w_R = \begin{cases} -\dfrac{1}{|x|^2} & \text{for } x \in B_r \\ 0 & \text{for } x \in B_R - B_r \\ \frac{1}{4} R^{-2}(9x^2 - 15R^2) & \text{for } x \in B_{\sqrt{5}R} - B_R \\ 0 & \text{else} \end{cases} .$$

We now set $R = \frac{1}{6m}$, $m \in \mathbb{N}$, and

(2.8)
$$u_m = \Sigma_Q \, w_R(x - x_Q) \quad , \quad Q \in \mathcal{S}_R .$$

We state the most important properties of u_m in the following

<u>THEOREM 2.1:</u> *Let* $n = 3$ *and* u_m *be defined as in* (2.8). *Then* u_m , $m \in \mathbb{Z}$, *has the following properties*

(2.9) $u_m \to 0$ *weakly, but not strongly in* $H^{1,2}(\mathbb{R}^3)$

(2.10) $u_m \geq 0$ *a.e. in* $\mathbb{R}^n$, $u_m = 0$ *a.e. on* $\mathbb{R}^3 - [0,1]^3$

(2.11) $u_m \in H^{1,p}(\mathbb{R}^3) \cap H^{2,p/2}(\mathbb{R}^3) \cap H^{2,\infty}_{loc}(\mathbb{R}^3 - 0)$ *for all* $p \in [1,3[$

(2.12) $\Delta u_m = F_m(\cdot, \nabla u_m) + f_m$ *a.e. in* $\mathbb{R}^3$ *and also in the sense of distributions, where*

(2.13) $F_m(\cdot, z) = -\Sigma_Q \, \chi\big(B_{R^3}(x_Q)\big)|z|^2 \quad (Q \in \mathcal{S}_R)$, $R = \frac{1}{6m}$,

(2.14) $f_m = \Sigma_Q \big(\chi(B_{\sqrt{5}R}(x_Q)) - \chi(B_R(x_Q))\big) \, f^o_R(\cdot - x_Q)$
$(Q \in \mathcal{S}_R)$, $R = \frac{1}{6m}$, *with* $f^o_R(x) = \frac{1}{4} R^{-2}(9x^2 - 15R^2)$.

We have

(2.15) $F_m(.,z) \to 0$ *a.e. and in* L^1 *for every fixed function* $z \in L^2(\mathbb{R}^3, \mathbb{R}^3)$

(2.16) $-|z|^2 \leq F_m(.,z) \leq 0$

(2.17) $\|f_m\|_\infty \leq K$ *uniformly for* $m \to \infty$

(2.18) $f_m \to c_o \, \chi([0,1]^3)$ *weakly in* L^2 *for some constant* $c_o \neq 0$.

Here, and in what follows, we let $\chi(M)$ denote the characteristic function of the set M . Note that the weak limit $u = \lim u_m = 0$ does not satisfy the limiting equation $\Delta z = c_o$. It is simple to transform the equation for u_m into a related one with data converging *everywhere* (and not just weakly in L^∞): To see this solve $\Delta v_m = f_m$ and write the equation for $u_m - v_m$.

<u>Proof of Theorem 2.1:</u> Property (2.11) follows from the corresponding $H^{1,p}$- and $H^{2,q}$-inclusions of $\log \frac{1}{|x|}$ in $\mathbb{R}^3$ and the $H^{2,\infty}_{loc}$-inclusion

of w_R . (2.10) is obvious. For the weak convergence $u_m \to 0$ in H^1 we prove that (u_m) is uniformly bounded in H^1 as $m \to \infty$ and that $u_m \to 0$ almost everywhere. In fact, the uniform H^1-boundedness follows from the asymptotic behaviour:

$$\int_{B_{6R}} |\nabla w_R|^2 dx = O(R^3)$$

and

$$\int_{\mathbb{R}^3} |\nabla u_m|^2 dx = m^3 \int_{B_{6R}} |\nabla w_R|^2 dx = O(1) \quad , \quad R = \frac{1}{6m} .$$

The convergence $u_m \to 0$ a.e. follows since $w_R = O(R^2)$ on $B_{\sqrt{5}R} - B_R$, $w_R(x) \le \frac{r}{|x|} \le R$ on $B_R - B_{R^2}$, $r = R^3$, and $R = \frac{1}{6m}$. Thus $w_R = o(1)$ on $Q_R - B_{R^2}$ as $R \to 0$, and the exceptional set B_{R^2} has measure $O(R^6)$. Hence $u_m = o(1)$ except on a set of measure $O(m^3 \frac{1}{m^6})$, which is $o(1)$ as $m \to \infty$. This means that $u_m \to 0$ almost everywhere. (2.12), (2.13), (2.14) follow from the corresponding equations for w_R . The convergence $F_m(\cdot,z) \to 0$ a.e., for each $z \in L^2$, $(m \to \infty)$, follows since the support of F_m is contained in

$$\bigcup_Q B_r(x_Q) \quad , \quad (Q \in \mathcal{G}_R) \quad , \quad r = R^3 \quad , \quad R = \frac{1}{6m} ,$$

which has Lebesgue measure $O(\frac{1}{R^3} r^3) = O(R^6)$.

The corresponding L^1-convergence of $F_m(\cdot,z)$ follows from the absolute continuity of the Lebesgue integral. (2.16) is obvious, as is (2.17).

For disproving the *strong* convergence ∇u_m in L^2 $(m \to \infty)$ we show that

$$\int_e |\nabla u_m|^2 dx = O(1) \quad \text{as} \quad m \to \infty , \tag{2.19}$$

where

$$e = \bigcup_Q B_r(x_Q) \quad , \quad (Q \in \mathcal{G}_R) \quad , \quad r = R^3 \quad , \quad R = \frac{1}{6m} .$$

In fact

$$\int_e |\nabla u_m|^2 dx = m^3 \int_{B_r} |\nabla w_R|^2 dx$$

and

$$\int_{B_r} |\nabla w_R|^2 dx = \int_{B_r} \frac{1}{|x|^2} dx = 4\pi r = 4\pi R^3 = 4\pi/(6m)^3 ,$$

so we obtain (2.19). On the other hand, the Lebesgue measure of e is

$O(m^3|B_r|) = O(m^{-6}) = o(1)$ as $m \to \infty$. This implies that the absolute continuity of the integrals $\int |\nabla u_m|^2 dx$ is *not* uniform for $m \to \infty$, which disproves strong convergence.

We still need to show (2.18). From the definition of f_m it is clear that f_m converges weakly to a function which vanishes on the complement of $[0,1]^3$ and is equal to a constant $c_o \neq 0$ on $[0,1]^3$, where c_o is the mean value of the function

$$f_R^o(x) = \frac{1}{4} R^{-2}(9x^2 - 15R^2)\left[\chi(B_{\sqrt{5}R}) - \chi(B_R)\right]$$

taken over the cube with center 0 , edge length $6R$, and faces parallel to the axes. This gives

$$c_o = \frac{3}{4}\frac{1}{6^3} R^{-5} \int_R^{\sqrt{5}R} (3\xi^2 - 5R^2)\xi^2 \cdot 4\pi dx = \frac{3\pi}{6^3} R^{-5}\left[\frac{3}{5}\xi^5 - \frac{5}{3}R^2\xi^3\right]_R^{\sqrt{5}R}$$
$$= \frac{3\pi}{6^3}\left\{(\sqrt{5})^3\left[\frac{3}{5}(\sqrt{5})^2 - \frac{5}{3}\right] - \left[\frac{3}{5} - \frac{5}{3}\right]\right\} .$$

It follows that $c_o \neq 0$ since $\sqrt{5}$ is irrational. This completes the proof of Theorem 2.1.

From the example in Theorem 2.1 one can easily construct elliptic systems of two equations with solutions $v_m \in H^{1,2}(\mathbb{R}^3,\mathbb{R}^2)$ such that $\|v_m\|_\infty \leq K$ *uniformly* and $v_m \to v$ weakly but *not* strongly in $H^{1,2}$, and v does not satisfy the limiting equation. We simply take the functions u_m of Theorem 2.1 and define

$$v_m = (v_m^1, v_m^2) \quad , \quad v_m^1 = \sin u_m \quad , \quad v_m^2 = \cos u_m .$$

It can be shown easily that v_m satisfies the elliptic system

$$\Delta v_m^1 = \Delta u_m \cos u_m - |\nabla u_m|^2 \sin u_m = F_m(\cdot,\nabla u_m)v_m^2 - |\nabla v_m|^2 v_m^1 + f_m v_m^2$$
$$\Delta v_m^2 = -\Delta u_m \sin u_m - |\nabla u_m|^2 \cos u_m = -F_m(\cdot,\nabla u_m)v_m^1 - |\nabla v_m|^2 v_m^2 - f_m v_m^1$$

where F_m and f_m are defined in (2.13) and (2.14).

This yields

$$\text{(2.20)} \qquad \begin{cases} \Delta u_m^1 = a_m^1(\cdot,v_m,\nabla v_m) + f_m v_m^2 \\ \Delta v_m^2 = a_m^2(\cdot,v_m,\nabla v_m) - f_m v_m^1 \end{cases}$$

where

$$a_m^1(\cdot,y,z) = -|z|^2\left(y_1 + y_2 \Sigma_Q \chi(B_r(x_Q))\right) , \ (Q \in \mathcal{G}_R) , \ r = R^3 , \ R = \frac{1}{6m}$$
$$a_m^2(\cdot,y,z) = -|z|^2\left(y_2 - y_1 \Sigma_Q \chi(B_r(x_Q))\right) , \ (Q \in \mathcal{G}_R) , \ r = R^3 , \ R = \frac{1}{6m} .$$

Here $y = (y_1, y_2) \in \mathbb{R}^2$, $z \in \mathbb{R}^2 \times \mathbb{R}^3$.
for fixed $w \in H^{1,2}(\mathbb{R}^3, \mathbb{R}^2)$ we have both

$$a_m^1(\cdot, w, \nabla w) \to -|\nabla w|^2 w^1$$

and

$$a_m^2(\cdot, w, \nabla w) \to -|\nabla w|^2 w^2$$

in measure, as $m \to \infty$. The limiting equations for (2.20) are

$$\Delta v^1 = -|\nabla v|^2 v^1 + c_o v^2$$

$$\Delta v^2 = -|\nabla v|^2 v^2 + c_o v^1 .$$

But this is *violated* since $v_m^1 \to 0$, $v_m^2 \to 1$ weakly in $H^{1,2}$, and $c_o \neq 0$.

References

[1] Bensoussan, A., and Frehse, J.: Nonlinear elliptic systems in stochastic game theory. Journ. reine u. angew. Math. 350, 23-67 (1984)

[2] Frehse, J.: A refinement of Rellich's theorem, SFB-Preprint 664, 1984

[3] Frehse, J.: Existence and pertubation theorems for nonlinear elliptic systems. From: Nonlinear partial differential equations and their applications. College de France Seminar IV. Editors: H. Brézis-J.L. Lions-Pitman 1983

[4] Hildebrandt, S.: Quasilinear elliptic systems in diagonal form. From: Systems of nonlinear partial differential equations, 173-217. Editor: J.M. Ball. D. Reidel Publishing Company 1983

[5] Ladyženskaya, O.A., and Ural'ceva, N.N.: Linear and quasilinear elliptic equations. New York-London. Academic Press 1968. (English translation of the first Russian edition 1964)

QUASICONVEXITY, GROWTH CONDITIONS AND PARTIAL REGULARITY

Mariano Giaquinta

1. Introduction

Consider the variational integral

$$\mathcal{F}(u;\Omega) = \int_\Omega F(x,u,Du)\,dx \tag{1.1}$$

where Ω is an open set of $\mathbb{R}^n$, $n \geq 2$, $u(x) = (u^1(x), \ldots, u^N(x))$, $Du = \{D_\alpha u^i\}$ $\alpha = 1, \ldots, n$, $i = 1,\ldots,N$ stands for the gradient of u and $\mathcal{F}(x,u,p): \Omega \times \mathbb{R}^N \times \mathbb{R}^{nN} \to \mathbb{R}$ is a smooth function satisfying the growth condition

$$|F(x,u,p)| \leq c_o(1 + |p|^m) \qquad m \geq 2. \tag{G_o}$$

A *minimizer* of the functional $\mathcal{F}$ is a function $u \in H^{1,m}_{loc}(\Omega,\mathbb{R}^N)$ satisfying for every $\varphi \in H^{1,m}(\Omega,\mathbb{R}^N)$ with supp $\varphi \subset\subset \Omega$

$$\mathcal{F}(u;\ \mathrm{supp}\ \varphi) \leq \mathcal{F}(u + \varphi;\ \mathrm{supp}\ \varphi) \tag{1.2}$$

In the following I shall confine myself to consider vector valued minimizers, $N > 1$. As it is well known, see e.g. [6], in this case, we can expect only partial regularity of the minimizers, i.e. smoothness except possibly on a closed singular set, which in general is nonempty.

The regularity of minimizers of "differentiable" functionals and of weak solutions of related nonlinear elliptic systems has been intensively studied in the last twenty years; but the regularity of minimizers of "nondifferentiable" functionals of type (1.1) has been studied only recently, for an account of it I refer the reader to Giaquinta [6] [7].

The following theorem was proved in [9], in case $m = 2$, and then extended to the case $m \geq 2$ in [11].

Theorem 1.1: *Suppose that the integrand* $F(x,u,p)$ *in* (1.1) *satisfies the following assumptions:*

(S) $F(x,u,p)$ *is of class* C^2 *with respect to* p, *while* $(1+|p|^m)^{-1} F(x,u,p)$ *is Hölder-continuous in* (x,u) *uniformly with respect to* p.

(G) F <u>satisfies the following growth conditions</u>

(G_o) $|F(x,u,p) \leq c_o(1+|p|^m)$ $m \geq 2$

(G_1) $F(x,u,p) \geq c_1|p|^m$ $c_1 > 0$

(G_2) $|F_{pp}(x,u,p) \leq c_2(\mu^2 + |p|^2)^{\frac{m-2}{2}}$ $\mu \geq 0$

(E) F <u>is elliptic in the following sense</u>

$$F_{p_\alpha^i p_\beta^j}(x,u,p)\, \xi_i^\alpha \xi_j^\beta \geq \nu(\mu^2 + |p|^2)^{\frac{m-2}{2}} |\xi|^2 \quad \forall\, \xi \in \mathbb{R}^{nN};\ \nu > 0$$

(ND) <u>nondegeneracy assumption:</u> $\mu > 0$

<u>Let</u> $u \in H^{1,m}_{loc}(\Omega.\mathbb{R}^N)$ <u>be a minimizer of</u> $\mathcal{F}$ <u>in</u> (1.1). <u>Then there exists an open set</u> $\Omega_o \subset \Omega$ <u>where the first derivatives of</u> u <u>are locally Hölder-continuous; moreover the Lebesgue measure of the possible singular set</u> $\Omega - \Omega_o$ <u>is zero</u>.

Notice that we do not require that F be differentiable with respect to u, and even if it is, we do not require any information on the growth of $F_u(x,u,p)$. Therefore the first variation of $\mathcal{F}$, even in the direction of smooth functions φ, a priori does not exist. In this sense we say that $\mathcal{F}$ is not differentiable.

The method of the proof of theorem 1.1 is the so called 'direct approach' in [6]; it consists in comparing on balls $B_R(x_0)$ the minimizer u with the minimizers v (with $v = u$ on ∂B_R) of the functional $\int G(Du)dx$ where $G(p)$ is the second order Taylor polynominal of $F(x_0\ u_{x_0,R}, p)$ at $(Du)_{x_0,R}$, $f_{x_0,R}$ denoting the mean value of f in $B_R(x_0)$

$$f_{x_0,R} = \int_{B_R(x_0)} f\, dx = \frac{1}{|B_R|}\int_{B_R} f\, dx$$

This approach strongly relies on Caccioppoli type inequalities, which by means of a result on reverse Hölder inequalities due to Giaquinta and Modica [12], give rise to reverse Hölder inequalities for the gradient of u and for the gradient of u minus a first order polynomial.

The aim of this paper is to discuss to what extent the assumptions (S) (G) (E) (ND) of theorem 1.1 can be weakened. The smoothness assumption (S) is clearly optimal for what concerns the dependence in (x,u), while, although natural, it is not clear whether it can be weakened for what concerns the dependence in p. So in the rest of this paper _we shall always assume that_ (S) _holds_. Another assumption which, at least technically, seems difficult to avoid is the nondegeneracy hypothesis (ND). In the sequel we shall see where it interferes. So we shall restrict ourselves to discussing assumptions (G) and (E). Notice that, although explicitly stated as assumption, (G_1) essentially follows from (E).

Assumption (E) is the natural strengthening, in the context of the regularity theory, of the convexity condition of $F(x,u,p)$ with respect to p

$$F_{p_\alpha^i p_\beta^j}(x,u,p)\, \xi_i^\alpha \xi_j^\beta \ge 0 \qquad \forall \xi \in \mathbb{R}^{nN} .$$

As it is known, convexity of F with respect to p is a sufficient condition for the sequential weak lower semicontinuity of $\mathcal{F}$ in $H^{1,m}$ and therefore, together with the coercivity condition (G_2), for the existence of a minimizer (subject to given boundary conditions) for $\mathcal{F}$. But it is a necessary condition only in the scalar case $N = 1$. In 1952 Morrey showed essentially that a necessary and sufficient condition for the weak sequential lower semicontinuity of $\mathcal{F}$ is that F be quasiconvex (see [1] [15] for precise statements and improvements). F being quasiconvex means the following: _for all_ $x_0 \in \Omega$, $u_0 \in \mathbb{R}^N$, $p_0 \in \mathbb{R}^{nN}$ _and for all_ $\varphi \in H^{1,m}(\Omega,\mathbb{R}^N)$ _with_ $\operatorname{supp} \varphi \subset\subset \Omega$ _we have_

$$(1.3) \qquad \int_\Omega [F(x_0,u_0,p_0 + D\varphi) - F(x_0,u_0,p_0)]\, dx \ \ge \ 0$$

i. e. _the frozen functional_

$$\mathcal{F}^0(u;\Omega) = \int_\Omega F(x_0,u_0,Du)\, dx$$

has the (affine) linear functions as minimizers.

Notice that (1.3) is a kind of Jensen inequality. Quasiconvexity is strictly weaker than convexity if $N > 1$ while it reduces to convexity for $N = 1$. A typical function which is quasiconvex (but not convex) is roughly a convex function of the determinant of p (in case $N = n$).

As assumption (E) is the natural strengthening of convexity, the natural strengthening of quasiconvexity is the following:

(SQ) _strict quasiconvexity: for all_ $x_0 \in \Omega$, $u_0 \in \mathbb{R}^N$, $p_p \in \mathbb{R}^{nN}$ _and for all_ $\varphi \in C^1(\Omega,\mathbb{R}^N)$ _with_ supp $\subset\subset \Omega$ _we have_

$$\int_\Omega [F(x_0,u_0,p_0+ D\varphi) - F(x_0,u_0,p_0)]\, dx \geq \nu \int_\Omega [\gamma|D\varphi|^2+|D\varphi|^m]\, dx$$

$$\nu > 0 \quad .$$

It is not difficult to see that the nondegeneracy assumption amounts to the requirement that $\gamma > 0$ (i. e. $\gamma = 1$); in the following _whenever we assume_ (SQ) _we shall assume it with_ $\gamma = 1$.

At the end of 1984, Evans [3] (see also the related paper [4]) showed that if F does not depend explicitely on x, and u, is of class C^2 and satisfies (G_2) and (SQ), then a Caccioppoli inequality holds for a minimizer u minus any first-order polynomial. Then using the so-called blow-up method proved the following

Theorem 1.2: _Suppose_ $F = F(p)$ _is a_ C^2 _function satisfying_ (G_2) _and_ (SQ). _Then the conclusion of theorem 1.1 holds._

This theorem was extended independently in [13], using the direct approach, and in [5], still using the indirect approach, to the general case $F = F(x,u,p)$, but with an extra assumption. In fact when $F = F(p)$, (SQ) essentially implies a sort of control of $\mathcal{F}(u)$ from below; this is not true in the general case; so one needs a kind of control of $\mathcal{F}(u)$ from below as in (G_1): in [13] [5] exactly (G_1) was required. But soon M.C. Hong [14] noticed that actually in order to get Caccioppoli inequality (the only place where (G_1) was used) it suffices to require the following weaker condition: $(\tilde{G}_1)$ _there exists a continuous_ $\tilde{F}(p): \mathbb{R}^{nN} \to \mathbb{R}$ _which is quasiconvex at zero_, i. e.

$$\int_\Omega [\tilde{F}(D\varphi) - \tilde{F}(0)]\, dx \geq \tilde{\nu} \int_\Omega |D\varphi|^m\, dx \qquad \forall\varphi \in C^1_0(\Omega,\mathbb{R}^N) \qquad \tilde{\nu} > 0$$

satisfying

$$\tilde{F}(p) \leq F(x,u,p)$$

a condition which is clearly satisfied if $F = F(p)$ and F is strictly quasiconvex.

Notice that, for example, the integrand

$$F(u,p) := a(u)|p|^2 + L \det p$$

satisfies $(\tilde{G}_1)$ but not (G_1) for $a(u) \geq 1$.

Then in conclusion we have

Theorem 1.3: _Suppose that the integrand_ $F(x,u,p)$ _satisfies the assumptions_ (S) (G_0) $(\tilde{G}_1)$ (G_2) _and_ (SQ). _Then the conclusion of theorem 1.1 holds._

In this theorem, while assumption (G_2) would be quite natural if F were convex in p, it appears as too strong when F is only strictly quasiconvex. For example the integrand

$$|p|^2 + \sqrt{1 + (\det p)^2} \qquad n = N = 2$$

satisfies all the assumptions of theorem 1.3 except for (G_2).

A few months ago Acerbi and Fusco [2] were able to eleminate assumption (G_2), i. e. they were able to prove partial regularity without assuming any control on the second derivatives of F with respect to p. So finally we may state:

Theorem 1.4: _Suppose that the integrand_ $F(x,u,p)$ _satisfies the following assumptions:_

(G_0) $\quad |F(x,u,p)| \leq c_0(1 + |p|^m) \qquad m \geq 2$

(S) $\quad F(x,u,p)$ _is of class_ C^2 _with respect to_ p, $(1+|p|^m)^{-1}F(x,u,p)$ _is Hölder continuous in_ (x,u) _uniformly with repect to_ p

(SQ) $\quad F$ _is strictly quasi-convex:_ $\forall x_0, u_0, p_0 \quad \forall \varphi \in C_0^1(\Omega, \mathbb{R}^N)$

$$\int_\Omega [F(x_0,u_0,p_0 + D\varphi) - F(x_0,u_0,p_0)]dx \geq \nu \int_\Omega [|D\varphi|^2 + |D\varphi|^m]\, dx, \quad \nu > 0$$

$(\tilde{G}_1)$ _there is a continuous function_ $\tilde{F}(p)$ _which is quasiconvex at_ $p_0 = 0$, i. e.

$$\int_\Omega [\tilde{F}(D\varphi) - \tilde{F}(0)]\, dx \geq \tilde{\nu} \int_\Omega |D\varphi|^m\, dx \qquad \forall \varphi \in C_0^1(\Omega, \mathbb{R}^N) \qquad \tilde{\nu} > 0$$

<u>satisfying for all</u> p

$$\tilde{F}(p) \leq F(x,u,p)$$

<u>Then any minimizer of</u> $F(u)$ <u>has Hölder-continuous first derivatives except possibly on a singular closed set of measure zero.</u>

The proof of theorem 1.4 in [2] relies in principle on the blow-up technique, but it is quite complicated. It uses a very precise approximation lemma for $H^{1,m}$ functions combined with higher integrability of the gradient and with a variational principle of Ekeland.

In the sequel I shall give a simpler proof of theorem 1.4, exactely along the same lines of the proof of theorem 1.1, i. e. a direct proof which relies in a standard way on the higher integrability of the gradient, i. e. in the end on Caccioppoli type inequalities.

One final remark has to be added, and it is a remark on the growth of the first derivatives F_p with respect to p. Of course when we had assumptions (G_2), as a consequence, we also had

$$|F_p(x,u,p)| \leq c_3(1 + |p|^{m-1}) \tag{1.4}$$

but in theorem 1.4 no control of type (G_2) is assumed. <u>Still</u> (1.4) <u>holds</u>. In fact, see [16] [15]. Since F is quasiconvex, it follows that F, as function of each fixed component of p, is convex, so if $\Phi(p_\alpha^i)$ denotes the function $p_\alpha^i \to F(p)$ we have

$$\Phi'(p_\alpha^i) \lesseqgtr \frac{\Phi(p_\alpha^i + h) - \Phi(p_\alpha^i)}{h} \qquad h \gtrless 0$$

hence, choosing $h = \pm(1 + |p|)$, we derive at once, taking into account (G_0), inequality (1.4).

In the following therefore, when it is convenient, we shall assume freely that (1.4) holds and also that the following simple consequence of (1.4)

$$|F(x,u,p) - F(x,u,q)| \leq c_4(1 + |p|^{m-1} + |q|^{m-1})\, |p - q| \tag{1.5}$$

holds.

This work has been done while I was visiting the Mathematisches Institut der Universität Bonn under the support of the Sonderforschungsbereich 72, on leave from the University of Firenze. It naturally follows the lines presented in [6], which is an enlarged version of a series of lectures I gave in an analogous situation in 1980 -1981 at the University of Bonn. So it is a special pleasure that it could be included in this volume for the SFB 72.

Finally, it is a great pleasure to acknowledge the hospitality of the University of Bonn and of the SFB 72 on both occasions.

2. Caccioppoli's inequalities and higher integrability

In this section we shall discuss Caccioppoli's inequalities and their consequences, namely higher integrability of the gradient of a minimizer. For convenience we shall state and prove Caccioppoli's inequalities first in the case where the integrand F does not depend explicitly on x and u, then in the general case and finally we shall prove the so called higher integrability theorems we need in section 3.

Caccioppoli's inequalities in the case $F = F(p)$. Consider the variational integral

$$\mathcal{F}(u;\Omega) = \int_\Omega F(Du)\, dx \tag{2.1}$$

with integrand $F(p)$ of class C^2 satisfying the growth conditions

(G_0) $|F(p)| \leq c_0(\lambda_0 + |p|^2 + |p|^m)$

(G_3) $|F_p(p)| \leq c_0(\lambda_0 + |p| + |p|^{m-1})$

where c_0 and λ_0 are constants, $c_0 > 0$, $0 \leq \lambda_0 \leq 1$, and assume that F satisfies the strict quasiconvexity assumption (SQ), i. e.

$$\int_\Omega [F(p_0+D\varphi) - F(p_0)]dx \geq \nu \int_\Omega [|D\varphi|^2 + |D\varphi|^m]dx \quad \forall\varphi \in C_0^1(\Omega,\mathbb{R}^N).$$

Of course we may (and shall) assume that $F(0) = 0$. We have

Proposition 2.1: *Under the previous assumptions* (G_0) (G_3) (SQ), *if* $u \in H^{1,m}_{loc}(\Omega,\mathbb{R}^N)$ *is a minimizer for* $\mathcal{F}$ *in* (2.1), *then there is a constant* A *depending only on* n, N, m *and* c_0 *so that*

$$\int_{B_{\frac{r}{2}}} (|Du|^2+|Du|^m)\,dx \leq A\left\{\frac{1}{r^2}\int_{B_r}|u-u_0|^2dx + \frac{1}{r^m}\int_{B_r}|u-u_0|^m dx + \lambda_0 r^n\right\}$$

for all $B_r \subset\subset \Omega$ *and all* $u_0 \in \mathbb{R}^N$.

Proof: Compare [8] [9] [13] [2]. For $B_r \subset\subset \Omega$ and $\frac{r}{2} < t < s < r$ we choose $\eta \in C_0^1(B_s)$, $0 \leq \eta \leq 1$ with $\eta \equiv 1$ on B_t and $|D\eta| \leq 2/(s-t)$ and we set

$$\varphi = \eta(u-u_0) \qquad \psi = (1-\eta)(u-u_0).$$

Then in B_s

$$u - u_0 = \varphi + \psi \qquad Du = D\varphi + D\psi \quad ;$$

notice that $\psi \equiv 0$ in B_t.

From the minimality of u we deduce

$$\int_{B_s} F(Du)\,dx \leq \int_{B_s} F(Du-D\varphi)\,dx = \int_{B_s} F(D\psi) \tag{2.2}$$

and using this inequality together with the growth conditions (G_o) (G_3) and the strict quasiconvexity we obtain

$$\begin{aligned} \nu\int_{B_s}(|D\varphi|^2+|D\varphi|^m)\,dx &\leq \int_{B_s}F(D\varphi)\,dx = \int_{B_s}F(Du-D\varphi)\,dx = \\ &= \int_{B_s}F(Du)\,dx + \int_{B_s}[F(Du-D\psi) - F(Du)]\,dx \leq \qquad (2.3)\\ &\leq \int_{B_s}F(D\psi)\,dx + c_0\int_{B_s}[\lambda_0+|Du|+|D\psi|+(|Du|+|D\psi|)^{m-1}]\,|D\psi|\,dx \\ &\leq c_0\cdot c(m)\left\{\int_{B_s}(\lambda_0+|D\psi|^2+|D\psi|^m)\,dx + \int_{B_s-B_t}(|Du|^2+|Du|^m)\,dx\right\}. \end{aligned}$$

Since $\varphi = u - u_0$ on B_t and

$$|D\psi| \leq (1-\eta)\,|Du| + |D\eta|\,|u-u_o|$$

on B_s, we conclude from (2.3) that for c depending on c_o, n, m, ν we have

$$\begin{aligned} \int_{B_t}(|Du|^2+|Du|^m)\,dx \leq c\Big\{r^n &+ \int_{B_s-B_t}(|Du|^2+|Du|^m)\,dx + \\ &+ \frac{1}{(s-t)^2}\int_{B_r}|u-u_0|^2\,dx + \frac{1}{(s-t)^m}\int_{B_r}|u-u_o|^m\,dx\Big\}\ . \end{aligned}$$

We now fill the hole on the right hand side of (2.4), i. e. we sum c times the left hand side of (2.4) and we get for all $t < s$, $\frac{r}{2} < t < s < r$

$$\int_{B_t} (|Du|^2+|Du|^m)\,dx \le \frac{c}{1+c} \int_{B_s} (|Du|^2+|Du|^m)\,dx + \frac{c}{1+c} r^n \tag{2.5}$$
$$+ \frac{c}{(s-t)^2} \int_{B_r} |u-u_o|^2\,dx + \frac{c}{(s-t)^m} \int_{B_r} |u-u_o|^m\,dx \quad .$$

By means of the simple algebraic lemma 2.1 of [8], (2.5) gives the result at once.

q.e.d.

Remark 2.1: One easily sees that the previous proposition holds assuming strict quasiconvexity of F only at zero and for functions u that are not necessarily minimizers, but satisfy for all $B_R \subset\subset \Omega$

$$\int_{B_R} F(Du)\,dx \le Q_0 \int_{B_R} F(Du+D\varphi)\,dx +$$
$$+ Q_1 \int_{B_R} (\lambda_0 + |D(u+\varphi)|^2 + |D(u+\varphi)|^m)\,dx$$

compare also [14] [9].

As it is well known, in order to study the partial regularity of the derivatives of a minimizer, a second Caccioppoli inequalitiy is relevant, and precisely a "homogeneous Caccioppoli inequality" for u minus a first order polynominal.

In order to do that, for each fixed $p_0 \in \mathbb{R}^{nN}$ we consider the integrand

$$\bar{F}(p) := F(p_0 + p) - F(p_0) - F_p(p_0)p. \tag{2.6}$$

Lemma 2.1: We have

$$|\bar{F}(p)| \le c(p_0)\ [|p|^2+|p|^m]$$
$$|\bar{F}_p(p)| \le c(p_0)[|p|+|p|^{m-1}]\ .$$

Proof: Set $K(p_o) = \max\{F_{pp}(p) : |p| \le 1 + |p_o|\}$. Then we have for $|p| \le 1$ and a suitable $\theta \in (0,1)$

$$|\bar{F}(p)| = \frac{1}{2}\,|F_{pp}(p_0+\theta p)pp| \le \frac{1}{2}\,K(p_0)\,|p|^2$$

while for $|p| > 1$

$$\begin{aligned}|\bar{F}(p)| &\le c_0(\lambda_0+|p_0+p|^2+|p_0+p|^m) \le\\ &\le c(m,c_0)(\lambda_0+|p_0|^2+|p_0|^m+|p|^2+|p|^m) \le\\ &\le c(|p_0|)[|p|^2+|p|^m]\end{aligned}$$

and the first inequality is proved. The second one is proved analogously.

q.e.d.

Now we notice that

$\bar{F}$ is still quasiconvex at zero.

If u is a minimizer for $\mathcal{F}$ in (2.1), then $u - p_0x$ is a minimizer for

$$\bar{\mathcal{F}}(u,\Omega) := \int_\Omega \bar{F}(Du)\,dx.$$

Therefore from proposition 2.1, taking into account lemma 2.1 and remark 2.1, we conclude:

Theorem 2.1: <u>Let</u> F <u>be a strictly quasiconvex integrand satisfying</u> (G_0) (G_3) <u>and let</u> u <u>be a minimizer for</u> $\mathcal{F}$. <u>Then the following Caccioppoli's inequality holds: for all</u> $B_r \subset\subset \Omega$ <u>and for all first order polynomials</u> $P(x)$ <u>we have</u>

$$\int_{B_{\frac{r}{2}}} [|D(u-P)|^2 + |D(u-P)|^m)\,dx \le$$
$$\le A(|DP|)\left\{\frac{1}{r^2}\int_{B_r}|u-P|^2dx + \frac{1}{r^m}\int_{B_r}|u-P|^m\,dx\right\} \quad (2.7)$$

Notice that a control on the second derivatives F_{pp} of F would allow eliminating the dependence on $|DP|$ in the constant A. Compare [3] [13]. Notice moreover that the proof of this theorem would not work if we dropped the term $|D\varphi|^2$ on the right hand side of the inequality in (SQ); in fact due to lemma 2.1 in any case we would get on the right hand side a term of the type $|Du|^2$ which has no counterpart on the left hand side. Here one sees one of the difficulties in studying partial regularity without the non-degeneracy assumption. Instead, if we content ourselves with a Caccioppoli estimate (2.7) with $P = \text{const}$, then in SQ we can drop the term $|D\varphi|^2$ and one easily sees that the following Caccioppoli inequality follows:

$$\int_{B_{\frac{r}{2}}} |Du|^m dx \leq A \left\{ r^n + \frac{1}{r^m} \int_{B_r} |u-u_0|^m dx \right\} .$$

<u>Caccioppoli's inequalities in the general case.</u> Here the situation appears to be more complicated. Consider a general integrand $F(x,u,p)$ where F is of class C^1 with respect to p and

(G_0) $|F(x,u,p)| \leq c_0(\lambda_0+|p|^2+|p|^m)$

(G_3) $|F_p(x,u,p)| \leq c_0(\lambda_0+|p|+|p|^{m-1})$

$c_0 > 0$, $0 \leq \lambda_0 \leq 1$, and assume that F is strictly quasiconvex, i. e. (SQ) of the introduction holds.

A first difficulty is due to the fact that in order to use (SQ) we need to freeze the variable x, u, on the other hand, compare the proof of proposition 2.1, (2.2) does not hold for the frozen functional $\int F(x_0,u_0,Dv)dx$, since clearly u is not a minimizer for this functional. As mentioned in the introduction, this difficulty can be overcome, following M.C. Hong [15] by assuming that $(\tilde{G}_1)$ holds. Then we have, following the proof of proposition 2.1, compare with [14] [2].

<u>Proposition 2.2:</u> <u>Suppose</u> (G_0) (G_3) $(\tilde{G}_1)$ <u>hold</u>. <u>Suppose moreover that for all</u> x, u, p, q (1.5) <u>holds</u>. <u>Then for all</u> $B_r \subset\subset \Omega$ <u>and all</u> $u_0 \in \mathbb{R}^N$ <u>we have</u>

$$\int_{B_{\frac{r}{2}}(x_0)} |Du|^m dx \leq A \left\{ r^n + \frac{1}{r^m} \int_{B_r(x_0)} |u-u_0|^m dx \right\} . \tag{2.8}$$

As we remarked in the introduction (1.5) follows from the strict quasiconvexity of F plus (G_0). So in the statement of proposition 2.2 we simply have substituted the assumptions (SQ) (G_0) with the weaker condition (1.5) (G_0). We finally notice that this is the only point where we use $(\tilde{G}_1)$.

Proof: With the same notations and in the same way as in the proof of proposition 2.1, we have the following.
From the minimality of u we deduce

$$\int_{B_s} F(x,u,Du)\,dx \le \int_{B_s} F(x,u-\varphi,Du-D\varphi)\,dx =$$

(2.9)

$$= \int_{B_s} F(x,u_0+\psi,D\psi) = \int_{B_s\setminus B_t} F(x,u_0+\psi,D\psi) + \int_{B_t} F(x,u_0,0)\,dx$$

while from $(\tilde{G}_1)$ and (1.5)

$$\int_{B_s} [\tilde{\nu}|D\varphi|^m + \tilde{F}(0)]dx \le \int_{B_s} \tilde{F}(D\varphi) \le \int_{B_s} F(x,u,Du)dx =$$

$$= \int_{B_s} F(x,u,Du-D\psi) =$$

(2.10)

$$= \int_{B_s} F(x,u,Du)dx + \int_{B_s} [F(x,u,Du-D\psi)-F(x,u,Du)]dx \le$$

$$\le \int_{B_s} F(x,u,Du)dx + c\int_{B_s} (1+|Du|^{m-1}+|D\psi|^{m-1})|D\psi|dx.$$

Estimates (2.9) (2.10) together with (G_0) then give

$$\tilde{\nu}\int_{B_s} |D\varphi|^m dx \le \tilde{F}(0)|B_s| + c\int_{B_s} (1+|Du|^{m-1}+|D\psi|^{m-1})|D\psi|dx +$$

$$+ \int_{B_s} F(x,u,Du)\,dx \le$$

$$\le \tilde{F}(0)|B_s| + \int_{B_s-B_t} (1+|Du|^{m-1}+|D\psi|^{m-1})|D\psi| +$$

$$+ \int_{B_s\setminus B_t} F(x_0,u_0+\psi,D\psi) + c|B_t| \le$$

$$\le cr^n + c\int_{B_t-B_s} (1+|Du|^m+|D\psi|^m)\,dx$$

and in conclusion

$$\int_{B_t} |Du|^m dx \le c\left\{r^n + \int_{B_t-B_s} |Du|^m dx + \frac{1}{(s-t)^m}\int_{B_r} |u-u_0|^m dx\right\}.$$

The result then follows as in proposition 2.1.

q.e.d.

In the general case, theorem 2.1 takes a more comlicated form and we have to content ourselves with a nonhomogeneous Caccioppoli inequality, which fortunately suffices for our purposes.

From (S) we deduce

$$(2.11)\qquad |F(x,u,p)-F(y,v,p)| \le c(1+|p|^m)\omega(|x-y|+|u-v|) \;[1]$$

where ω is a bounded, concave nonnegative and increasing function satisfying

$$\omega(t) \le t^\sigma \quad .$$

Then we have, compare [13]

<u>Theorem 2.2: Let the integrand</u> $F(x,u,p)$ <u>satisfy</u> (S) (G_0) (G_3) (SQ) <u>and let</u> u <u>be a minimizer for</u> $\mathcal{F}$. <u>Then the following Caccioppoli inequality holds: for all</u> $B_r(x_0)$, <u>all</u> v_0 <u>and all first order polynomials</u> $P(x) = u_0 + p_0(x-x_0)$ <u>we have</u>

$$\int_{B_{\frac{r}{2}}(x_0)} [|D(u-P)|^2 + |D(u-P)|^m dx \le A(|DP|)\left\{\frac{1}{r^2}\int_{B_r(x_0)} |u-P|^2 dx + \right.$$

$$\left. + \frac{1}{r^m}\int_{B_r(x_0)} |u-P|^m dx + \int_{B_r(x_0)} (1+|Du|^m)\omega(|x-x_0|+|u-v_0|+|u-P|)\right\} .$$

<u>Proof:</u> For fixed x_0, u_0, p_0, v_0, set

$$\bar{F}(p) := F(x_0,v_0,p_0+p) - F(x_0,v_0,p_0) - F_p(x_0,v_0,p_0)\, p \quad .$$

Clearly $\bar{F}$ is strictly quasiconvex at zero, $\bar{F}(0) = 0$, and it satisfies the estimates of lemma 2.1 and

$$|\bar{F}(p+q)-\bar{F}(p)| \le c(p_0)(|p|+|q|+|p|^{m-1}+|q|^{m-1})|q| \; .$$

We now choose η as in the proofs of proposition 2.1 and 2.2 and set

$$\varphi = \eta(u-u_0-p_0(x-x_0)) \qquad \psi = (1-\eta)(u-u_0-p_0(x-x_0)) \quad ;$$

then obviously

$$u - u_0 - p_0(x-x_0) = \varphi + \psi \qquad Du - p_0 = D\varphi + D\psi \quad .$$

---

[1] actually uniformity does not follow from (S) (but it is not even needed); we might as well assume

$$|F(x,u,p) - F(y,v,p)| \le c(1+|p|^m)\omega(|y|,|v|;|x-y|+|u-v|)$$

but in order to simplify the writing, we confine ourselves to assuming (2.11).

From the strict quasiconvexity of $\bar{F}$ at zero we get

$$\tilde{\nu}\int_{B_s}(|D\varphi|^2+|D\varphi|^m)dx \le \int_{B_s}\bar{F}(D\varphi)dx = \int_{B_s}\bar{F}(Du-p_0-D\psi) =$$

(2.13)

$$= \int_{B_s}\bar{F}(Du-p_0)dx + \int_{B_s}[\bar{F}(Du-p_0-D_\psi)-\bar{F}(Du-p_0)]dx \le$$

$$\le \int_{B_s}\bar{F}(Du-p_0)dx + c(p_0)\int_{B_s-B_t}(|Du-p_0|+|Du-p_0|^{m-1}+|D\psi|+|D\psi|^{m-1})|D\psi| \;.$$

On the other hand

$$\int_{B_s}\bar{F}(Du-p_0)dx =$$

$$= \int_{B_s}F(x_0,v_0,Du) - \int_{B_s}F(x_0,v_0,p_0) - \int_{B_s}F_p(x_0,v_0,p_0)(Du-p_0) =$$

$$= \int_{B_s}F(x,u,Du)dx + \int_{B_s}[F(x_0,v_0,Du)-F(x,u,Du)]dx - \int_{B_s}F(x_0,v_0,p_0)$$

$$- \int_{B_s}F_p(x_0,v_0,p_0)(Du-p_0)$$

while

$$\int_{B_s}F(x,u,Du)dx \le \int_{B_s}F(x,u-\varphi,Du-D\varphi)dx =$$

$$= \int_{B_s}F(x,u_0+p_0(x-x_0)+\psi,p_0+D\psi)dx =$$

(2.15)

$$= \int_{B_s}F(x_0,v_0,p_0+D\psi) + \int_{B_s}[F(x,u_0+p_0(x-x_0)+\psi,p_0+D\psi) - F(x_0,v_0,p_0+D\psi]$$

$$= \int_{B_s}\bar{F}(D\psi) + \int_{B_s}F(x_0,v_0,p_0) + \int_{B_s}F_p(x_0,v_0,p_0)D\psi +$$

$$+ \int_{B_s}[F(x,u_0+p_0(x-x_0)+\psi,p_0+D\psi) - F(x_o,v_0,p_0+D\psi)] \;.$$

Putting together (2.13) (2.14) (2.15) we then obtain

$$\nu \int_{B_s} (|D\varphi|^2+|D\varphi|^m)dx \le \int_{B_s} \bar{F}(D\psi) + \int_{B_s} [F(x_0,v_0,Du)-F(x,u,Du)]$$

$$+ c \int_{B_s-B_t} (|Du-p_0)|^2+|Du-p_0|^m)dx + c \int_{B_s-B_t} (|D\psi|^2+|D\psi|^m)dx +$$

$$+ \int_{B_s} [F(x,u_0+p_0(x-x_0)+\psi,p_0+D\psi) - F(x_0,v_0,p_0+D\psi)]dx$$

i.e.

$$\int_{B_t} (|Du-p_0|^2+|Du-p_0|^m)dx \le c(p_0)\left\{ \int_{B_s\setminus B_t} (|Du-p_0|^2+|Du-p_0|^m)dx + \right.$$

$$+ \frac{1}{r^2}\int_{B_r} |u-P|^2dx + \frac{1}{r^m}\int_{B_r} |u-P|^m dx +$$

(2.16)

$$+ \int_{B_r} \omega(|x-x_0|+|u-u_0|)[|Du|^2+|Du|^m]\ dx +$$

$$\left. + \int_{B_r} \omega(|x-x_0|+|u-v_0|+|u-P|)(|p_0|^m+|Du|^m dx \right\} .$$

From (2.16) the result follows as previously (Compare the proof of proposition 2.2).

q.e.d.

<u>Higher integrability of the gradient</u>. We recall the following result due to Giaquinta-Modica [12], see also [6] chap. V, on reverse Hölder inequalities with increasing supports.

<u>Theorem 2.3:</u> <u>Let</u> Ω <u>be a bounded open set in</u> $\mathbb{R}^n$. <u>Suppose we have</u>

$$\left(\fint_{B_{\frac{R}{2}}} |g|^q dx\right)^{\frac{1}{q}} \le b\left(\fint_{B_R} |g|^r dx\right)^{\frac{1}{r}} + \left(\fint_{Q_R} |f|^q dx\right)^{\frac{1}{q}}$$

<u>for all</u> $Q_R \subset\subset \Omega$, <u>where</u> $g \in L^q(\Omega)$, $f \in L^s(\Omega)$, $0 < r < q < s < +\infty$. <u>Then there exists a positive</u> $\epsilon = \epsilon(n,q,r,s,b)$ <u>so that</u> $g \in L^{q+\epsilon}_{loc}(\Omega)$. <u>Moreover for any</u> $\Omega' \subset\subset \Omega$ <u>we have</u>

$$\left(\fint_{\Omega'} |g|^{q+\epsilon}dx\right)^{\frac{1}{q+\epsilon}} \le c\left(\fint_{\Omega}|g|^q dx\right)^{\frac{1}{q}} + c\left(\fint_{\Omega}|f|^{q+\epsilon}dx\right)^{\frac{1}{q+\epsilon}}$$

where c is a constant depending on n, q, r, s, b and on $|\Omega|/\text{dist}(\Omega',\partial\Omega)^n$ and $|\Omega|/|\Omega'|$.

As it is known, compare e. g. [6], Caccioppoli's inequality in conjunction with theorem 2.3 give rise to higher integrability of the gradient and to reverse Hölder inequalities. More precisely we get, as consequences of theorem 2.1 and 2.2, compare [14]

Theorem 2.4: *Let u be a minimizer of*

$$\int F(Du)\,dx$$

where the integrand F satisfies the hypothesis of theorem 1.4. Then there exist $\delta > 0$ and $c > 0$ such that

$$\left[\fint_{B_{\frac{R}{2}}(y_0)} \left(|Du-p_0|^2+|Du-p_0|^m\right)^{1+\delta} dx\right]^{\frac{1}{1+\delta}} \le$$

$$\le c \fint_{B_R(y_0)} (|Du-p_0|^2+|Du-p_0|^m)\,dx$$

holds for all $B_R(y_0) \subset\subset \Omega$ and all $p_o \in \mathbb{R}^{nN}$.

Theorem 2.5: *Let u be a minimizer of*

$$\int F(x,u,Du)\,dx$$

where the integrand F satisfies the hypotheses of theorem 1.4. Then there exist $\delta > 0$, $\gamma > 0$, $c > 0$ and a nondecreasing function $h(t)$ so that

$$\left[\fint_{B_{\frac{R}{2}}(y_0)} (|Du-p_0|^2+|Du-p_0|^m)^{1+\delta} dx\right]^{\frac{1}{1+\delta}}$$

$$\le c \fint_{B_R(y_0)} (|Du-p_0|^2+|Du-p_0|^m)\,dx +$$

$$+ R^\gamma h\left[|u_{x_0,R}|+|p_0|+\left(\fint_{B_R(y_0)} |Du-p_0|^m dx\right)^{\frac{1}{m}}\right]$$

holds for all $B_R(y_0) \subset\subset \Omega$ and all $p_0 \in \mathbb{R}^{nN}$.

Notice that the constant c in theorems 2.4 and 2.5 depends on $|p_0|$. As already noticed, in the case we have a control on the second derivatives F_{pp} of F, as in [13], then c is independent of $|p_0|$.

The proof of theorems 2.4, 2.5 is exactly the same as the analogous one in [13] for the case c independent of p_0; for the reader's convenience we shall repeat it.

In (2.7) and (2.12.) we choose $u_0 = u_{x_0,r}$, and we recall the well-known Sobolev-Poincare inequality: <u>for any</u> $q \geq \frac{n}{n-1}$, <u>set</u>

$$q_* = \frac{nq}{n+q}$$

<u>then there is a constant</u> $c = c(n,q)$ <u>so that</u>

$$\int_{B_r(x_0)} |u-u_{x_0,r}-p_0(x-x_0)|^q dx \leq c(n,q)\left[\int_{B_r(x_0)} |Du-p_0|^{q_*} dx\right]^{\frac{q}{q_*}} .$$

We estimate the first two integrals on the right hand sind of (2.7), (2.12) by

$$\frac{c}{r^2}\left[\int_{B_r} |Du-p_0|^{2_*}dx\right]^{\frac{2}{2_*}} + \frac{c}{r^m}\left[\int_{B_r} |Du-p_0|^{m_*}dx\right]^{\frac{m}{m_*}}$$

and then, noticing that $m_* \leq 2_* m$, by a simple use of Hölder inequality, by

$$\frac{c}{r^2}\left[\int_{B_r} (|Du-p_0|^2+|Du-p_0|^m)^{\frac{2_*}{2}} dx\right]^{\frac{2}{2_*}} .$$

Therefore we have

$$\fint_{B_{\frac{r}{2}}(x_0)} (|Du-p_0|^2+|Du-p_0|^m)dx \leq c_1\left[\fint_{B_r} (|Du-p_0|^2+|Du-p_0|^m)^{\frac{2_*}{2}} dx\right]^{\frac{2}{2_*}} +$$

$$+ c_2 \fint_{B_r} (1+|Du|^m)\omega(|x-x_0|+|u-v_0|+|u-u_0-p_0(x-x_0)|)\, dx$$

with $c_0 = 0$ under the hypotheses of theorem 2.4 This implies at once theorem 2.4, taking into account theorem 2.3.

The proof of theorem 2.5 needs two steps. First we notice taht from proposition 2.2 it follows that $1 + |Du|^m \in L^{1+\epsilon}_{loc}(\Omega)$ for some positive ϵ and that for all $B_\rho \subset\subset \Omega$

$$\left[\fint_{B_{\frac{\rho}{2}}} (1+|Du|^m)^{1+\epsilon}dx\right]^{\frac{1}{1+\epsilon}} \leq c \fint_{B_\rho} (1+|Du|^m)dx . \tag{2.18}$$

Set now

$$\alpha(x) = \min\{1,|u(x)-v_0|^\sigma\}$$

$$\beta(x) = \min\{1,(|x-x_0|+|p_0||x-x_0|+|u_0-u_0|)^s\}$$

then

$$\fint_{B_r} (1+|Du|^m)\omega(|x-x_0|+|u-v_0|+|u-u_0-p_0(x-x_0)|)dx \le$$

$$\le c \fint_{B_r} \alpha(x)(1+|Du|^m)dx + c \fint_{B_r} \beta(x)(1+|Du|^m)dx .$$

We estimate the last integral, using (2.18) and Poincare inequality as follows:

$$\fint_{B_r} \beta(x)(1+|Du|^m)dx \le c\left[\fint_{B_r} (1+|Du|^m)^{1+\epsilon}dx\right]^{\frac{1}{1+\epsilon}}\left[\fint_{B_r} \beta(x)dx\right]^{\frac{\epsilon}{1+\epsilon}} \le$$

$$\le c \int_{B_{2r}} (1+|Du|^m)dx \left[(1+|p_0|)r + \fint_{B_r} |u-u_0|dx\right]^{\frac{\delta\epsilon}{1+\epsilon}} \le$$

$$\text{(2.19)} \qquad \le c \fint_{B_{2r}} (1+|Du|^m)dx \cdot c(p_0)\, r^{\frac{\delta\epsilon}{1+\epsilon}} \left[\fint_{B_{2r}} (1+|Du|^m)dx\right]^{\frac{1}{m}\frac{\epsilon\delta}{1+\epsilon}} =$$

$$= c(p_0)\, r^{\frac{\delta\epsilon}{1+\epsilon}} \left[\fint_{B_{2r}} (1+|Du|^m)dx\right]^{1+\frac{1}{m}\frac{\epsilon\delta}{1+\epsilon}} \le$$

$$\le c(p_0)\, r^{\frac{\delta\epsilon}{1+\epsilon}} \int_{B_{2r}} (1+|Du|^m)^{1+\frac{1}{m}\frac{\epsilon\delta}{1+\epsilon}} dx$$

i. e. for $m\gamma = \frac{\delta\epsilon}{1+\epsilon}$

$$\fint_{B_r} \beta(x)(1+|Du|^m)dx \le c(p_0)r^{m\gamma} \fint_{B_{2r}} (1+|Du|^m)^{1+\gamma} dx$$

where γ is as near to zero as we like.

For $B_{R_0}(y_0)$ fixed, set now

$$g(x) = [|Du-p_0|^2+|Du-p_0|^m]^{\frac{2_*}{2}}$$

$$f(x) = [\alpha(x)(1+|Du|^m)+c(p_0)R_0^{m\gamma}(1+|Du|^m)^{1+\gamma}]^{\frac{2_*}{2}}$$

then $f(x) \in L^s$ for some $s > \frac{2}{2_*}$, provided γ is sufficiently small and inequality (2.17) can be written for all $B_r(x_0) \subset\subset B_{R_0}(y_0)$ as

$$\left[\fint_{B_{\frac{r}{2}}} g^{\frac{2}{2_*}}dx\right]^{\frac{2_*}{2}} \le c \fint_{B_{2r}} g\, dx + \left[\fint_{B_{2r}} f^{\frac{2}{2_*}} dx\right]^{\frac{2_*}{2}} .$$

Applying theorem (2.3) we conclude that there is an exponent $q > \frac{2}{2_*}$ so that

$$\left[\int\!\!\!\!\!-_{B_{R_0}} g^q \, dx\right]^{\frac{1}{q}} \le c\left[\int\!\!\!\!\!-_{B_{R_0}} g^{\frac{2}{2_*}} dx\right]^{\frac{2_*}{2}} + \left[\int\!\!\!\!\!-_{B_{R_0}} f^q \, dx\right]^{\frac{1}{q}}$$

if q is suffiently near to $\frac{2}{2_*}$, we can again estimate

$$\left[\int\!\!\!\!\!-_{B_R} f^q \, dx\right]^{\frac{1}{q}}$$

choosing now $v_0 = u_{x_0,2R}$, as in (2.9) and the result follows at once.

3. Partial regularity

In this section we prove theorem 1.4. We proceed as in [13]. Theorem 1.4 follows in a standard way, see [6] pp. 197-199, from the following proposition.

Proposition 3.1: *Set*

$$\Phi(x_0,R) = \int\!\!\!\!\!-_{B_R(x_0,R)} [|Du-(Du)_R|^2 + |Du-(Du)_R|^m] \, dx .$$

Then for every $x_0 \in \Omega$, $\epsilon > 0$, *and for every* ρ, R, $0 < \rho < R < \text{dist}(x_0,\partial\Omega)$ *we have*

$$(3.1)\quad \Phi(x_0,\rho) \le A\left\{\left[\left(\frac{\rho}{R}\right)^2 + \epsilon + \left(\frac{R}{\rho}\right)^n \chi(x_0,R)\right]\Phi(x_0,R) + R^{\sigma}H(x_0,R)\right\}$$

where $\gamma > 0$, $A = A(|u_{x_0,R}| + |(Du)_{x_0,R}|)$ *and*

$$\chi(x_0,R) = \gamma\left[\frac{1}{\epsilon}(1+|u_{x_0,R}|+|(Du)_{x_0,R}|) + \Phi(x_0,R)^{\frac{1}{m}},\ \Phi(x_0,R)\right]$$

$$H(x_0,R) = \begin{cases} 0 & \text{if } F \text{ is independent of } x,u \\ \tilde{H}\left[|u_{x_0,R}|+|(Du)_{x_0,R}|+\Phi(x_0,R)^{\frac{1}{m}}\right] & \text{otherwise} \end{cases}$$

$\gamma(t,s)$ *being an increasing function in* t *going to zero as* s *goes to zero uniformly for* t *in a bounded set, and* $\tilde{H}(t)$ *and* $A(t)$ *are increasing functions of* t.

Let us first show how theorem 1.4 follows from the previous proposition.

Inequality (3.1) can be written, for $0 < \tau < 1$, as

$$\Phi(x_0,\tau R) \le A\tau^2\{1+\epsilon\tau^{-2}+\chi(x_o,R)\tau^{-n-2}\}\Phi(x_0,R)+\tau^{-n}H(x_0,R)R^{\sigma} \quad .$$

For

$$|u_{x_0,R}| + |(Du)_{x_0,R}| \le M_1$$

we choose α with $\sigma < \alpha < 2$ and then we fix τ in such a way that

$$2A(M_1)\tau^{2-\alpha} = 1 \quad .$$

Then we fix ϵ so that

$$\epsilon\tau^{-2} < \frac{1}{2} \quad .$$

Since we have

$$\chi(x_0,R) < \frac{1}{2}\tau^{n+2}$$

provided $\Phi(x_0,R)$ is less than some $\epsilon_1(M_1)$, setting

$$H_0 = \tau^{-n}A(M_1)H(M_1+\epsilon_1)$$

we have: <u>if for some</u> R

$$|u_{x_0,R}| + |(Du)_{x_0,R}| < M_1 \quad \text{and} \quad \Phi(x_0,R) < \epsilon_1$$

<u>then</u>

$$\Phi(x_0,\tau R) \le \tau^{\alpha}\Phi(x_0,R) + H_0R^{\sigma} \quad .$$

On the other hand we have for every k

$$|u_{x_0,\tau^kR}| + |(Du)_{x_0,\tau^kR})$$
$$\le |u_{x_0,R}| + |(Du)_{x_o,R}| + H_1(n,N,\tau)\sum_{s=0}^{k}\Phi(x_0,\tau^sR)^{\frac{1}{m}} \ .$$

Therefore by induction we get

$$\Phi(x_0,\tau^kR) \le \tau^{k\alpha}\Phi(x_0,R) + H_0(\tau^{k-1}R)^{\sigma}\sum_{s=0}^{\infty}(\tau^{\alpha-\sigma})^s$$
$$\le \Phi(x_0,R) + H_0\frac{R^{\sigma}}{\tau^{\sigma}-\tau^{\alpha}}\tau^{k\sigma} < \epsilon_1 \quad \epsilon_1^{\frac{1}{m}} + |u_{x_0,\tau^kR}| + |(Du)_{x_0,\tau^kR}|$$
$$\le \epsilon_1^{\frac{1}{m}} + |u_{x_0,R}| + |(Du)_{x_0,R}| + H_1(\frac{\epsilon_1}{2})^{\frac{1}{m}}\sum_{s=0}^{k}(\tau^{\frac{\sigma}{m}})^s \le$$
$$\le |u_{x_o,R}| + |(Du)_{x_0,R}| + H_1(\frac{\epsilon_1}{2})^{\frac{1}{m}}\frac{1}{1-\tau^{\sigma/m}} + \epsilon_1^{\frac{1}{m}} \le M_1$$

if

$$R < R_0 \quad , \quad \Phi(x_0,R) < \frac{\epsilon_1}{2} \quad , \quad |u_{x_0,R}| + |(Du)_{x_0,R}| < \frac{M_1}{2}$$

and ϵ_1 and R_0 are chosen in such a way that also

$$\epsilon_1^{\frac{1}{m}} + H_1\left(\frac{\epsilon_1}{2}\right)^{\frac{1}{m}} \frac{1}{1-\tau^{\sigma/m}} < M_1$$

$$H_0 \frac{R^{\epsilon}}{\tau^{\sigma}-\tau^{\alpha}} < \frac{\epsilon_1}{2} \qquad \text{for } R < R_0 \quad .$$

Then we can conclude: _for any_ M_1 _if_

$$(3.2) \qquad |u_{x_0,R}| + |(Du)_{x_0,R}| < \frac{M_1}{2} \qquad \Phi(x_0,R) < \frac{\epsilon_1}{2}$$

for some $R < R_0(M_1)$, _then_

$$\Phi(x_0,\tau^k R) \leq \epsilon_1 \tau^{k\sigma}$$

and hence for every $\rho < R_0$

$$(3.3) \qquad \Phi(x_0,\rho) \leq \text{const}\left(\frac{\rho}{R_0}\right)^{\sigma} \quad .$$

Now, since $\Phi(x_0,R)$ and $|u_{x_0,R}| + |(Du)_{x_0,R}|$ are continuous functions of x_0, if (3.2) holds for a point $x_0 \in \Omega$, then there is a ball $B_r(x_0)$ such that (3.2) (and therefore (3.3)) holds for every $x \in B_r(x_0)$. This implies that the derivatives of u are Hölder continuous in an open set Ω_0. see [6], and obviously we have

$$\Omega - \Omega_0 \subset \left\{x: \liminf_{R\to 0} \int_{B_R(x)} \left[|Du-(Du)_{x,R}|^2+|Du-(Du)_{x,R}|^m\right]dy > 0\right\}$$
$$\cup \left\{x: \sup_R \left(|u_{x,R}|+|(Du)_{x,R}|\right) = +\infty\right\}$$

in particular $\text{meas}(\Omega-\Omega_0) = 0$.

Now let us prove proposition 3.1.

Fix a point $x_0 \in \Omega$ and a radius $R < \text{dist}(x_0,\partial\Omega)$, set

$$u_0 = u_{x_0,R}$$
$$p_0 = (Du)_{x_0,R}$$

and denote by $F^0(p)$ the frozen integrand

$$F^0(p) = F(x_0,u_0,p)$$

of course we have the Taylor expansion

$$F^0(p) = F^0(p_0) + F^0_{p_0}(p_0)(p-p_0) +$$
$$+ \int_0^1 (1-t)F^0_{pp}(p_0+t(p-p_0))(p-p_0)(p-p_0)dt \ .$$

Define the integrand

$$G(p) = F^0(p_0) + F^0_p(p_0)(p-p_0) + \tfrac{1}{2}F^0_{pp}(p_0)(p-p_0)(p-p_0)$$

and notice that for

$$|p-p_0| \le B$$

we have

$$|F^0(p)-G(p)| \le \eta(|p_0|+B;|p-p_0|)|p-p_0|^2 \tag{3.4}$$

where $\eta(t,s)$ is the modulus of continuity of $F^0_{pp}(p)$, that we may assume concave in s.
Since the strict quasiconvexity implies (see [17],theor. 4.4.1) that

$$\lambda(|u_0|+|p_0|)|\xi|^2|\eta|^2 \le F^0_{p^i_\alpha p^j_\beta}(p_0)\xi^\alpha\xi^\beta\eta_i\eta_j \le \Lambda(|u_0|+|p_0|)|\xi|^2|\eta|^2$$

the elementary Hilbert space theory together with Gårding inequality ensures the existence and uniqueness of a solution $v \in H^{1,2}(B_{\frac{R}{2}},\mathbb{R}^N)$ of

$$\begin{cases} \displaystyle\int_{B_{\frac{R}{2}}(x_0)} G(Dv)dx \to \min \\ v-u \in H^1_0(B_{\frac{R}{2}}(x_0),\mathbb{R}^N) \end{cases} .$$

Note that v is the solution of the Dirichlet boundary value problem

$$\begin{cases} F^0_{p^i_\alpha p^j_\beta}(p_0)D_\alpha D_\beta v^i = 0 & j = 1,\dots,N \quad \text{in} \quad B_{\frac{R}{2}}(x_0) \\ v - u = 0 & \text{on} \quad \partial B_{\frac{R}{2}}(x_0) \end{cases} .$$

Therefore we have, see e. g. [6] Chap. III, for all $\pi \in \mathbb{R}^{nN}$ and any $\rho < \frac{R}{2}$

$$(3.5) \qquad \int_{B_{\frac{R}{2}}(x_0)} |Dv-\pi|^2 dx \le c \int_{B_{\frac{R}{2}}(x_0)} |Du-\pi|^2 dx$$

$$(3.6) \qquad \fint_{B_\rho(x_0)} |Dv|^2 dx \le c \fint_{B_{\frac{R}{2}}(x_0)} |Dv|^2 dx$$

$$(3.7) \qquad \fint_{B_\rho(x_0)} |Dv-(Dv)_{x_0,\rho}|^2 dx \le c\left(\frac{\rho}{R}\right)^2 \fint_{B_{\frac{R}{2}}(x_0)} |Dv-(Dv)_{x_0,\frac{R}{2}}|^2 dx .$$

Moreover from the L^p-theory for elliptic systems we deduce that if $u \in H^{1,p}(B_{\frac{R}{2}}(x_0),\mathbb{R}^N)$, $p > 2$, then $v \in H^{1,p}(B_{\frac{R}{2}},\mathbb{R}^N)$ and

$$(3.8) \qquad \int_{B_{\frac{R}{2}}(x_0)} |Dv-\pi|^p dx \le c \int_{B_{\frac{R}{2}}(x_0)} |Du-\pi|^p dx$$

$$(3.9) \qquad \fint_{B_\rho(x_0)} |Dv-(Dv)_{x_0,\rho}|^p dx \le c\left(\frac{\rho}{R}\right)^p \fint_{B_R(x_0)} |Dv-(Dv)_{x_0,\frac{R}{2}}|^p dx .$$

Notice that the constant c appearing in (3.5) ... (3.9) does depend on u_0 and p_0; from now on constants c can depend on u_0, p_0.

From (3.7) (3.9) we deduce

$$(3.10) \qquad \rho^n \Phi(x_0,\rho) \le c\left(\frac{\rho}{R}\right)^{n+2} R^n \Phi(x_0,R) + c \int_{B_{\frac{R}{2}}} [|Du-Dv|^2+|Du-Dv|^m] dx$$

and we only need to estimate the last integral.
Using the strict quasiconvexity we have, writing

$$w = u - v$$

$$\nu \int_{B_{\frac{R}{2}}} (|Dw|^2+|Dw|^m) dx \le \int_{B_{\frac{R}{2}}} [F^0(p_0+Dw)-F^0(p_0)] dx =$$

$$= \int_{B_{\frac{R}{2}}} [F^0(p_0+Dw)-G(p_0+Dw)] dx + \frac{1}{2} \int_{B_{\frac{R}{2}}} F^0_{pp}(p_0) Dw\, Dw =$$

$$= (I) + \frac{1}{2} \int_{B_{\frac{R}{2}}} F^0_{pp}(p_0) Dw\, Dw \quad .$$

On the other hand we have

$$\frac{1}{2} \int_{B_{\frac{R}{2}}} F^0_{pp}(p_0)\ Dw\ Dw\ dx = \int_{B_{\frac{R}{2}}(x_0)} [G(Du)-G(Dv)]dx =$$

$$= \int_{B_{\frac{R}{2}}} [G(Du)-F^0(Du)]dx + \int_{B_{\frac{R}{2}}} [F^0(Du)-F(x,u,Du)]dx +$$

$$+ \int_{B_{\frac{R}{2}}} [F(x,u,Du)-F(x,v,Dv)]dx + \int_{B_{\frac{R}{2}}} [F(x,v,Dv)-F^0(Dv)]dx +$$

$$+ \int_{B_{\frac{R}{2}}} [F^0(Dv)-G(Dv)]dx = (II) + (III) + (IV) + (V) + (VI) \ .$$

Notice that $(IV) \leq 0$, since u is a minimizer. Therefore we conclude:

$$(3.11) \qquad \int_{B_{\frac{R}{2}}(x_0)} [|Dw|^2+|Dw|^m]dx \leq c|(I)+(II)+(III)+(V)+(VI)| \ .$$

In case F does not depend on x and u also III and V are zero. So we distinguish two situations

a) $F = F(p)$. We need only to estimate (I), (II), (VI). Let us estimate (II), i. e.

$$\left| \int_{B_{\frac{R}{2}}(x_0)} [G(Du)-F^0(Du)]dx \right| \ .$$

We have using (3.11) and G_0 and setting

$$E_{\frac{R}{2}} = \{x \in B_{\frac{R}{2}}: |Du(x)-p_0| \geq B\}$$

for some B to be fixed later

$$\int_{B_{\frac{R}{2}}} |G(Du)-F^0(Du)|dx = \int_{B_{\frac{R}{2}}-E_{\frac{R}{2}}} \ldots dx + \int_{E_{\frac{R}{2}}} \ldots dx \le$$

$$\le \int_{B_{\frac{R}{2}}} \eta(|p_0|+B;|Du-p_0|)|Du-p_0|^2 dx + c\int_{E_{\frac{R}{2}}} (|Du|^2+|Du|^m+1)dx \le$$

$$(3.12)\quad \le \int_{B_{\frac{R}{2}}} \eta(|p_0|+B;|Du-p_0|)|Du-p_0|^2 dx + c\int_{E_{\frac{R}{2}}} (|Du-p_0|^m+|Du-p_0|^2)dx + c|E_{\frac{R}{2}}| \quad .$$

Using the L^p-estimate in theorem 2.1 we get

$$\int_{B_{\frac{R}{2}}} \eta(|p_0|+B;|Du-p_0|)|Du-p_0|^2 \le$$

$$(3.13)\quad \le \left[\int |Du-p_0|^{2(1+\epsilon)}dx\right]^{\frac{1}{1+\epsilon}} \left[\int \eta(|p_0|+B;|Du-p_0|)dx\right]$$

$$\le \int_{B_R} \left[|Du-p_0|^2+|Du-p_0|^m\right]dx\ \eta\left[|p_0|+B;\ \fint_{B_R(x_0)} |Du-p_0|dx\right] \quad .$$

On the other hand

$$\int_{E_{\frac{R}{2}}} (|Du-p_0|^2+|Du-p_0|^m)dx \ge (B^2+B^m)|E_{\frac{R}{2}}|$$

i. e.

$$(3.14)\qquad |E_{\frac{R}{2}}| \le \frac{1}{B^2+B^m}\int_{B_R} (|Du-p_0|^2+|Du-p_0|^m)dx$$

hence

$$\int_{E_{\frac{R}{2}}} (|Du-p_0|^2+|Du-p_0|^m)dx \le \left[\int_{B_{\frac{R}{2}}} (|Du-p_0|^2+|Du-p_0|^m)^{1+\delta} dx\right]^{\frac{1}{1+\delta}} |E_R|^{\frac{\delta}{1+\delta}}$$

$$(3.15)\qquad \le c\int_{B_R} (|Du-p_0|^2+|Du-p_0|^m)dx \left(\frac{|E_{R/2}|}{|B_R|}\right)^{\frac{\delta}{1+\delta}} \le$$

$$\leq c\left[\frac{1}{B^2+B^m}\right]^{\frac{\delta}{1+\delta}}\left[\fint_{B_R} (|Du-p_0|^2+|Du-p_0|^m)\,dx\right]^{\frac{\delta}{1+\delta}} \int_{B_R} (|Du-p_0|^2+|Du-p_0|^m)\,dx.$$

From (3.12) ... (3.15) we then conclude choosing for all $\epsilon > 0$

$$B^2 + B^m \geq \frac{1}{\epsilon}[1+c(p_0)]$$

with

$$\text{(3.16)} \qquad |(II)| \leq [\chi(x_0,R)+\epsilon] \int_{B_R(x_0)} (|Du-p_0|^2+|Du-p_0|^m)\,dx \,.$$

Exactly in the same way, taking into account (3.8) (3.10), we see that (I) and (VI) are estimated by the right hand side of (3.16). This concludes the proof in the case $F = F(p)$.

b) The general case $F = F(x,u,p)$. (I) (III) (VI) can be estimated as before, using theorem 2.2 instead of theorem 2.1. One easily sees that the extra term that appears is exactly of the type

$$R^{n+\sigma}H(x_0,R) \,.$$

So in order to conclude the proof, it suffices to estimate III and V. This is done using the L^p-estimates in propositions 2.1, 2.2 and assumption (S), see e. g. [13].

References

[1] Acerbi, E.; Fusco, N.: Semicontinuity problems in the calculus of variations. Arch. Rat. Mech. Anal. 86 (1984) 125-145.

[2] Acerbi, E.; Fusco, N.: A regularity theorem for minimizers of quasiconvex integrals. Preprint 1986.

[3] Evans, L.C.: Quasiconvexity and partial regularity in the calculus of variations. Preprint 1984.

[4] Evans, L.C.; Gariepy, R.F.: Blow-up, compactness and partial regularity in the calculus of variations. Preprint 1986.

[5] Fusco, N.; Hutchinson, J.: $C^{1,\alpha}$ partial regularity of functions minimizing quasiconvex integrals. manuscripta math. 54 (1985) 121-143.

[6] Giaquinta, M.: Multiple integrals in the calculus of variations and nonlinear elliptic systems. Annals of Math. Studies 105, Princeton University Press, Princeton 1983.

[7] Giaquinta, M.: The problem of the regularity of minimizers. Address of the International Congress of Mathematics, Berkeley 1986.

[8] Giaquinta, M.; Giusti, E.: On the regularity of minima of variational integrals. Acta Math. 148 (1982) 31-46.

[9] Giaquinta, M.; Giusti, E.: Differentiability of minima of nondifferentiable functionals. Inventiones Math. 72 (1983) 285-298.

[10] Giaquinta, M.; Giusti, E.: Quasi-minima. Ann. Inst. H. Poincare, Analyse non lineaire, 1 (1984) 79-107.

[11] Giaquinta, M.; Ivert, P.-A.: Partial regularity of minima of variational integrals. Preprint FIM, ETH Zürich 1984.

[12] Giaquinta, M.; Modica, G.: Regularity results for some classes of higher order nonlinear elliptic systems. J. reine angew. Math. 311/312 (1979) 145-169.

[13] Giaquinta, M.; Modica, G.: Partial regularity of minimizers of quasiconvex integrals. Ann. Inst. H. Poincare, Analyse non lineaire 3 (1986) 185-208.

[14] Hong, M.C.: Existence and partial regularity in the calculus of variations. Preprint 1986.

[15] Marcellini, P.: Approximation of quasiconvex functions and lower seimcontinuity of multiple integrals. manusripta math. 51 (1985) 1-28.

[16] Morrey, C.B. jr.: Multiple integrals in the calculus of variations. Springer Verlag, New York 1966.

Mariano Giaquinta
Istituto di Matematica Applicata "G. Sansone"
Facolta di Ingegneria
Universita di Firenze
Via S. Marta, 2

I-50139 Firenze

THE MONOTONICITY FORMULA IN GEOMETRIC MEASURE THEORY, AND AN APPLICATION TO A PARTIALLY FREE BOUNDARY PROBLEM

Michael Grüter

1. Introduction

In this paper we want to describe a method which originated in Geometric Measure Theory (GMT) and which in recent years has been successfully applied to the classical regularity theory for two-dimensional variational problems. The key word is "Monotonicity of area ratios for minimal surfaces", and in section 2 we sketch the arguments from GMT and the relation to the classical theory of minimal surfaces.

For its role in GMT let me only refer to [FH], [AW1], [MS], and [SL].

The method we shall describe which uses the monotonicity formula to prove regularity results was first introduced in [GM1] and [GM2].

It was shown how to prove regularity of weak two-dimensional surfaces of bounded mean curvature (with finite area) in Riemannian manifolds. Using a simple trick even the regularity question for two-dimensional stationary harmonic mappings could be settled, see [SR], [GM3], [GM4]. In [GHN1] and [DG2] the method was extended to cover the regularity problem for stationary minimal surfaces near a free boundary.

In this situation one is looking at two-dimensional parametrized minimal surfaces whose boundary is partly prescribed and partly allowed to vary freely in a supporting surface. For an exact formulation of the problem see section 3. Before that corresponding results for area minimizing (in contrast to just stationary) surfaces had been obtained by H. Lewy [LH] and W. Jäger [JW].

More general problems for stationary surfaces involving a free boun dary were later treated in [GHN2], [DG4], [JJ1], [JJ2], [DU], [YR] by using the same idea.

Since regularity near the fixed part of the boundary is understood well enough, see e.g. [DG1] and the references given there, it remains to investigate the points where the fixed boundary meets the supporting surface. The best one can hope for in the general case is Hölder continuity of the solution if some lift-off condition is satisfied. Only if the Jordan curve (prescribing the fixed boundary values) and the supporting surface intersect orthogonally one can prove higher regularity. This is easily seen by reflecting the minimal surface which meets the supporting surface at a right angle. The resulting surface may be considered as a solution to a Plateau problem, and the corresponding regularity theory applies.

In the second part of the paper we show that the method of [GHN1] can be used to prove continuity up to the corners without any lift-off condition. Actually, the same result already follows from an estimate for the length of the free boundary, given in [HN]. The additional assumption (made there) about the absence of branch points of odd order in the free boundary could later be removed, see [DG5]. However, it seems to be worthwile and more satisfactory to have a direct argument for the continuity in the corners. Furthermore, the proof may serve as an illustration of the general method.

In some sense this part of the paper may be considered as a continuation of and supplement to [GHN1]. For completeness we show in section 4 how one can obtain Hölder continuity in case the Jordan arc is not tangential to the supporting surface.

Let me finally thank the Sonderforschungsbereich 72 for continuous and generous support during the past years.

2. <u>The monotonicity formula for minimal surfaces</u>

Here, we want to give two proofs of the monotonicity formula which is used in GMT as a fundamental tool in the regularity theory for n-dimensional surfaces with controlled mean curvature. We conclude this section by sketching the argument - involving the monotonicity formula - that leads to regularity results for classical surfaces (minimal surfaces, surfaces of bounded mean curvature, and stationary harmonic mappings).

Since we do not want to give an introduction to GMT we formulate everything in the language of differential geometry, i.e. we consider the following situation.

Suppose that $M \subset \mathbb{R}^{n+k}$ is an n-dimensional C^1-submanifold that is *area-minimizing* in U, $U \subset \mathbb{R}^{n+k}$ an open set such that $\partial M \cap U = \emptyset$. This means that for every n-dimensional C^1-submanifold N satisfying

$$\text{clos}\,[(M \sim N) \cup (N \sim M)] \subset K$$

for some compact subset K of U, we have ($\mathcal{H}^n$ denotes n-dimensional Hausdorff-measure)

$$\mathcal{H}^n(M \cap K) \leq \mathcal{H}^n(N \cap K) \quad .$$

(*) *Then the function*

$$r \mapsto \mathcal{H}^n(M \cap B_r(p))r^{-n}$$

is increasing on $(0,R)$ *for every* $p \in M$ *and* $R > 0$ *such that* $B_R(p) \subset U$.

This is easily seen as follows:
Fo almost all $r \in (0,R)$ consider the (n-1)-dimensional compact submanifold M_r (without boundary) defined by

$$M_r = M \cap \partial B_r(p) \quad .$$

Without loss of generality we may assume $p = 0$. We now define (for fixed r as above) N_r as the cone over M_r with vertex 0, i.e.

$$N_r = \{\lambda x\colon x \in M_r,\ 0 \leq \lambda \leq 1\} \quad .$$

Since $\partial N_r = M_r$ we get

$$\mathcal{H}^n(M \cap B_r) \leq \mathcal{H}^n(N_r \cap B_r) \quad . \tag{1}$$

Actually, one has to approximate N_r (extended by M outside of B_r) by C^1-submanifolds. But this is obviously possible while keeping inequality (1). A simple calculation shows that

$$\mathcal{H}^n(N_r \cap B_r) = \frac{r}{n}\,\mathcal{H}^{n-1}(M_r) \quad . \tag{2}$$

Setting $\phi(r) = \mathcal{H}^n(M \cap B_r)$ and noting that ϕ is increasing, hence differentiable almost everywhere, with $\phi'(r) = \mathcal{H}^{n-1}(M_r)$ we may combine (1) and (2) to get

$$n\phi(r) \leq r\phi'(r) \tag{3}$$

for almost every $r \in (0,R)$.
Multiplying (3) by r^{-n-1} we write the corresponding differential inequality as

$$\frac{d}{dr}\,(r^{-n}\phi(r)) \geq 0 \quad .$$

Using the montonicity of ϕ we get (*).

In fact, the same result holds for *stationary* (or *minimal*) submanifolds. The proof obviously has to be different from the one given above and will involve the *first variation formula*. The argument is as follows. Again, we assume that $M \subset \mathbb{R}^{n+k}$ is an n-dimensional C^1-submanifold such that $\mathcal{H}^n(M \cap K) < \infty$ for each compact subset $K \subset \mathbb{R}^{n+k}$. We call M *stationary* (with respect to n-area) in $U \subset \mathbb{R}^{n+k}$, U open such that $\partial M \cap U = \phi$, if

(4) $$\frac{d}{dt} \mathcal{H}^n(\Phi_t(M \cap U))\Big|_{t=0} = 0$$

holds for every $\Phi\colon (-1,1) \times \mathbb{R}^{n+k} \to \mathbb{R}^{n+k}$ such that the following is true:

(i) Φ is continuously differentiable

(ii) $\Phi_t := \Phi(t,\cdot)$ is a diffeomorphism for each $t \in (-1,1)$, $\Phi_0 = \mathrm{Id}$,

(iii) there is some compact set $K \subset U$ with

$$\Phi_t|_{\mathbb{R}^{n+k} \sim K} = \mathrm{Id}|_{\mathbb{R}^{n+k} \sim K}\,,$$

$$\Phi_t(K) \subset K \qquad , \; t \in (-1,1)\,.$$

To calculate the derivative in (4) we introduce the initial velocity vector field X defined by

(5) $$X(x) = \frac{d}{dt}\, \Phi(t,x)\Big|_{t=0}\,.$$

Let us remark that, given any smooth vector field on $\mathbb{R}^{n+k}$ with compact support in U, there always exists Φ as above such that (5) holds.

An easy calculation shows that (4) is equivalent to

(6) $$\int_M \mathrm{div}_M X(x)\, d\mathcal{H}^n(x) = 0$$

for each $X \in C^1_c(U,\mathbb{R}^{n+k})$.

Here, the tangential divergence $\mathrm{div}_M X$ is defined by

$$\mathrm{div}_M X(x) = \sum_{i=1}^{n} D_{\tau_i} X(x) \cdot \tau_i$$

where $\{\tau_1,\ldots,\tau_n\}$ is any orthonormal basis for $T_x M$ and $D_{\tau_i} X$ denotes the usual directional derivative of X in the direction τ_i.

Thus, if M is C^2 the classical divergence theorem tells us that stationarity is equivalent to the fact that the mean curvature vector

$\underline{\underline{H}}$ of M vanishes identically (in U), where

$$\underline{\underline{H}}(x) = \sum_{i=1}^{n} B(\tau_i, \tau_i) \quad ,$$

B the second fundamental form of M.

This is the classical definition of a minimal submanifold. In fact, one can show that each minimal submanifold of class C^2 is locally area-minimizing.

For C^1-submanifolds M which are stationary in U and $p \in U$, $B_R(p) \subset U$, we have the following *monotonicity identity*

$$(**) \qquad \begin{aligned} \rho^{-n}\mathcal{H}^n(M \cap B_\rho(p)) &= \sigma^{-n}\mathcal{H}^n(M \cap B_\sigma(p)) + \\ &+ \int_{M \cap (B_\rho(p) \sim B_\sigma(p))} \frac{|(x-p)^\perp|^2}{|x-p|^{n+2}} \, d\mathcal{H}^n(x) \quad , \end{aligned}$$

for each $0 < \sigma < \rho < R$ where $(x-p)^\perp$ denotes the component of $x - p$ normal to T_xM.

From (**) we easily get the monotonicity in (*). To prove (**) one makes an appropriate choice of X in the first variation formula (6). For the convenience of the reader we repeat the argument.

We let $r = |x-p|$ and choose $X(x) = \gamma(r)(x-p)$, where $\gamma \in C^1(\mathbb{R})$, $\operatorname{spt} \gamma \subset (-\infty, R)$, and $\gamma \equiv$ const. near 0. To calculate $\operatorname{div}_M X$ we use the fact that

$$\begin{aligned} \operatorname{div}_M X(x) &= \sum_{i=1}^{n+k} e_i \cdot P_{T_xM}(\operatorname{grad} X^i(x)) = \\ &= \sum_{i,j=1}^{n+k} P_{ij}(x) \, D_j X^i(x) \quad , \end{aligned}$$

where $\{e_i\}_{i=1,\dots,n+k}$ is the canonical basis of $\mathbb{R}^{n+k}$ and (P_{ij}) is the matrix of P_{T_xM} (= orthogonal projection of $\mathbb{R}^{n+k}$ onto T_xM) relative to this basis. We get

$$\operatorname{div}_M X(x) = \gamma(r) \sum_{j=1}^{n+k} P_{jj}(x) + r\gamma'(r) \sum_{i,j=1}^{n+k} P_{ij}(x) \, \frac{(x^i-p^i)}{r} \, \frac{(x^j-p^j)}{r} \; .$$

From $\sum_{i=1}^{n+k} P_{ii} = n$ and

$$\sum_{i,j=1}^{n+k} P_{ij} \frac{(x^i-p^i)}{r} \frac{(x^j-p^j)}{r} = 1 - \left|\frac{(x-p)^{\perp}}{r}\right|^2$$

we conclude

$$(7) \qquad \int_{M\cap U} n\gamma(r) + r\gamma'(r)\,d\mathscr{H}^n(x) = \int_{M\cap U} r\gamma'(r)\left|\frac{(x-p)^{\perp}}{r}\right|^2 d\mathscr{H}^n(x).$$

Now let $\varepsilon \in (0,1)$, $\rho \in (0,R)$, and consider $\varphi = \varphi_{\epsilon} \in C^1(\mathbb{R})$ such that φ is decreasing, $\varphi(t) = 0$ for $t > 1$, and $\varphi = 1$ for $t \le 1-\varepsilon$. For $\gamma(r) := \varphi(r/\rho)$ we have

$$r\gamma'(r) = -\rho \frac{\partial}{\partial\rho} \varphi(r/\rho) \quad .$$

Thus (7) leads to

$$(8) \qquad nI(\rho) - \rho\, I'(\rho) \ge 0 \quad ,$$

where $I(\rho) = \int_{M\cap U} \phi(r/\rho)\,d\mathscr{H}^n(x)$.

Letting $\varepsilon \downarrow 0$ and integrating (8) we derive (*). If we do not discard the (non-negative) right hand side of (7) we instead get (**).

A similar result holds if we assume that the mean curvature $\underline{\underline{H}}$ of M is bounded in U, say $|\underline{\underline{H}}| \le \Lambda$. In this case one shows that the function

$$(***) \qquad \rho \mapsto e^{\Lambda\rho}\,\rho^{-n}\mathscr{H}^n(M\cap B_\rho(p))$$

is non-decreasing on $(0,R)$.

Before we are going to describe the use of the monotonicity formula for the classical theory of (minimal) surfaces let me finally mention one of the main open problems related to it.

It was shown by W.K. Allard [AW2] that if one can prove a monotonicity result for a parametric elliptic integrand using similar arguments as for area then it follows that the integrand already is the area integrand.

Therefore, to prove such a result for more general integrands one has to give a completely new proof of the monotonicity formula.

In fact, even to prove a lower bound for mass ratios of surfaces which are stationary with respect to a general parametric elliptic integrand is one of the outstanding problems in GMT.

For a different approach to regularity see the recent paper [AW3].

Let us now briefly describe how one uses the monotonicity formula in order to show continuity of two-dimensional parametric minimal surfaces. For more details the reader is referred to the papers mentioned in the introduction and to the special case treated in the next section.

Let $\Omega \subset \mathbb{R}^2$ be open and let $x \in C^1(\Omega,\mathbb{R}^3)$ be a minimal surface with locally finite area. Pick $w_0 \in \Omega$ and $r > 0$ such that $B_r(w_0) \subset \Omega$ and

$$(9) \qquad [\operatorname*{osc}_{\partial B^2_{r'}(w_0)} x]^2 \leq 2C \operatorname{Area}\left[x(B^2_r(w_0))\right] = C \int_{B^2_r(w_0)} |\nabla x|^2 \, dw$$

holds for some $r' \in [\frac{r}{2},r]$. This is possible because of the Courant-Lebesgue-Lemma.

Since the area is locally finite we may assume that r has been chosen small enough so that the oscillation of the boundary values of x on $\partial B^2_{r'}(w_0)$ is as small as we want.

Now, if $w_1 \in B^2_{r'}(w_0)$ and if $x_1 = x(w_1)$ were far away from these bounday values, say

$$(10) \qquad \inf_{|w-w_0|=r'} |x_1-x(w)| > R \quad ,$$

then the monotonicity would imply

$$R^{-2}\operatorname{Area}\left[B^3_R(x_1) \cap x(B^2_{r'}(w_0))\right] \geq \rho^{-2}\operatorname{Area}\left[B^3_\rho(x_1) \cap x(B^2_{r'}(w_0))\right]$$

for $0 < \rho < R$. But it is well known that the $\limsup_{\rho \downarrow 0}$ of the right hand side is at least π.

Thus, we conclude

$$\pi R^2 \leq \operatorname{Area}\left[B^3_R(x_1) \cap x(B^2_{r'}(w_0))\right]$$

which implies

$$2\pi R^2 \leq \int_{B^2_{r'}(w_0)} |\nabla x|^2 \, dw \quad .$$

But this means that R cannot be too big (in (10)), i.e.

$$(11) \qquad \inf_{|w-w_0|=r'} |x_1-x(w)|^2 \leq \frac{1}{2\pi} \int_{B^2_{r'}(w_0)} |\nabla x|^2 \, dw \quad .$$

Now from (9) and (11) we get for some constant C

$$(12) \qquad \left[\operatorname*{osc}_{B^2_{r'}(w_0)} x\right]^2 \leq C \int_{B^2_r(w_0)} |\nabla x|^2 \, dw \quad .$$

Hence we may choose a sequence $r_k \downarrow 0$ such that (9) is satisfied (for r'_k) and deduce from (12) the continuity of x in w_0.

The assumption that $x \in C^1(\Omega,\mathbb{R}^3)$ was only made to avoid technical complications. The advantage of the method described above is that it works for general *weak* two-dimensional surfaces of finite area and bounded mean curvature in Riemannian manifolds. See the introduction for references and generalizations. The geometric idea behind it is very simple:

If a surface has bounded mean curvature, passes through the center of a ball, and has no boundary inside this ball, then it has to use a certain amount of area to leave the ball.

Thus, the oscillation on small disks is comparable to the area of the image of the disk.

By looking at thin cylinders one sees that the bound for the mean curvature has to enter the estimate, see (***).

3. Continuity near the corners.

In this section we are going to show how the arguments of [GHN1] can be used to obtain everywhere-continuity of any solution to our free boundary problem, even in the corners.

Let us first recall the setting of the problem. Note that, as in [GHN1], the arguments work in much more general situations.

We assume that $\Gamma \subset \mathbb{R}^3$ is a regular Jordan arc of class C^1 such that its two end points P_1, P_2 lie in a given two-dimensional surface $S \subset \mathbb{R}^3$ and, moreover, $\Gamma \cap S = \{P_1,P_2\}$.

We consider mappings defined on subsets of $\mathbb{R}^2$ which will often be identified with $\mathbb{C}$. For $w \in \mathbb{R}^2$ we write $w = (u,v) = u + iv$. Let

$$B = \{w \in \mathbb{R}^2 : |w| < 1,\ v > 0\}$$

be the open semi-disk,

$$C = \{w \in \mathbb{R}^2 : |w| = 1,\ v \geq 0\}$$

the closed circular arc, and

$$I = \{w \in \mathbb{R}^2 : |w| < 1,\ v = 0\}$$

the open interval.
Thus

$$\partial B = C \cup I \quad .$$

The class of *admissible surfaces* is defined as follows:

$$\mathscr{C}(S,\Gamma) = \left\{ x \in H_2^1(B,\mathbb{R}^3) : x \text{ is bounded by the configuration } \langle \Gamma, S \rangle \right\}.$$

Here, $H_2^1(B,\mathbb{R}^3)$ is the Sobolev space of $\mathbb{R}^3$-valued L^2-functions on B such that their first derivatives also belong to $L^2(B)$.

The statement "x is bounded by the configuration $\langle \Gamma,S \rangle$" is supposed to mean the following.

Denote by x_C and x_I the L^2-traces of x on C and I, respectively. We assume that x_C maps C continuously and weakly monotonic onto Γ such that

$$x_C(-1) = P_1 \quad , \quad x_C(1) = P_2 \quad .$$

Furthermore, x_I has to satisfy

$$x_I(w) \in S \quad \text{for a.e.} \quad w \in I \quad .$$

For $x \in H_2^1(B,\mathbb{R}^3)$ and measurable $\Omega \subset B$ we consider the Dirichlet integral

$$D_\Omega(x) = \int_\Omega |\nabla x|^2 \, du\, dv \quad ,$$

$$D(x) = D_B(x) \quad .$$

We call $x: B \to \mathbb{R}^3$ a *minimal surface* if it is real analytic and satisfies

$$\Delta x = 0 \quad \text{(harmonic)}$$

as well as the conformality relations

$$|x_u|^2 - |x_v|^2 = 0 = x_u \cdot x_v$$

on B, and $x \not\equiv$ const..

An *admissible variation* of a surface $x \in \mathscr{C}$ is a family of surfaces $x_\epsilon \in \mathscr{C}$, $|\epsilon| < \epsilon_0$, $\epsilon_0 > 0$, where $\{x_\epsilon\}_{|\epsilon|<\epsilon_0}$ is of one of the following types:

1. $$x_\epsilon = x \circ \tau_\epsilon$$

where $\{\tau_\epsilon\}_{|\epsilon|<\epsilon_0}$ is a family of diffeomorphisms of $\bar{B}$ such that

$\tau_0 = \mathrm{id}_{\bar{B}}$ and that $\tau(w,\epsilon) = \tau_\epsilon(w)$ is of class C^1 on $\bar{B} \times (-\epsilon_0, \epsilon_0)$.

2. $$x_\epsilon(w) = x(w) + \epsilon\Psi(w,\epsilon)$$

where $D(\Psi(\cdot,\epsilon)) \leq C$ for $|\epsilon| < \epsilon_0$ and $\Psi(\cdot,\epsilon) \to \Phi(\cdot)$ as $\epsilon \to 0$ a.e. on B, for some $\Phi \in H_2^1(B,\mathbb{R}^3)$.

A surface $x \in \mathscr{C}$ is called *stationary* in $\mathscr{C}$ if

$$\lim_{\epsilon\to 0} \frac{1}{\epsilon} [D(x_\epsilon)-D(x)] = 0$$

holds for all admissible variations of x.

It is well known that a nonconstant stationary $x \in \mathscr{C}$ is a minimal surface, and in [DG1] it was shown that

$$x \in C^{0,\alpha}(B \cup (\mathrm{int}\ C), \mathbb{R}^3)$$

for every $\alpha \in (0,1)$.

To investigate the behavior of a stationary surface near the free part of the boundary we make the following assumptions on the supporting surface S.

<u>**ASSUMPTION (A)**</u>

$S \subset \mathbb{R}^3$ is a two-dimensional submanifold. There are numbers $\rho_0 > 0$, $K_0 \geq 0$, $K \geq 1$ such that the following holds.

For each $f \in S$ there is a neighborhood U of f in $\mathbb{R}^3$ and a C^2-diffeomorphism $h: \mathbb{R}^3 \to \mathbb{R}^3$ such that

$$h^{-1}(f) = 0$$

$$h^{-1}(U) = \{y \in \mathbb{R}^3: |y| < \rho_0\}$$

and, furthermore,

$$h^{-1}(S \cap U) = \{y \in \mathbb{R}^3: |y| < \rho_0,\ y^3 = 0\} \quad .$$

If we set

$$g_{ij}(y) = \delta_{kl}\, h^k_{y^i}(y) h^l_{y^j}(y)$$

then

$$K^{-1}|\xi|^2 \leq g_{ij}(y)\xi^i\xi^j \leq K|\xi|^2$$

for all ξ, $y \in \mathbb{R}^3$, as well as

$$\sup_{i,j,k} \left|\frac{\partial g_{ij}}{\partial y^k}(y)\right| \leq K_0 \quad .$$

Note that each compact C^2-submanifold of $\mathbb{R}^3$ automatically satisfies (A). For non-compact S the assumption (A) imposes a certain uniformity condition on the metric $ds^2 = g_{ij}(y)dy^i dy^j$ at infinity, which implies that S is complete with respect to the induced metric of $\mathbb{R}^3$.

If S satisfies (A) we shall call it a *strict* C^2-*surface* in $\mathbb{R}^3$.

We have the following result.

Theorem 1: Suppose that S is a strict C^2-surface and that $x \in \mathscr{C}$ is stationary. Then

$$x \in C^0(\bar{B},\mathbb{R}^3) \quad .$$

Remarks:

(i) For higher regularity near the corners we refer to Theorem 2 in the next section.

(ii) Regularity of the free boundary away from P_1, P_2 was shown in [GHN1], [DG2]; we have

$$x \in C^{1,\alpha}(B \cup I,\mathbb{R}^3)$$

for every $\alpha \in (0,1)$. Of course, we get higher regularity (away from the corners) of the free boundary if we strengthen the hypotheses on S.

(iii) As was mentioned in the introduction the same result follows from [HN].

The proof of Theorem 1 will follow from the crucial Lemma below (analogous to Lemma 4 in [GHN1]). Before we can state it we need some more notation. Let $e \in \{-1,1\}$ be one of the corners of B. For $0 < r < 1$ we set

$$S_r(e) = \{w \in B: |w-e| < r\} \quad ,$$

$$C'_r(e) = \{w \in B: |w-e| = r\} \quad ,$$

$$C''_r(e) = \{w \in \mathbb{R}^2: |w-e| < r\} \cap C \quad ,$$

and finally

$$C_r = C'_r \cup C''_r \quad .$$

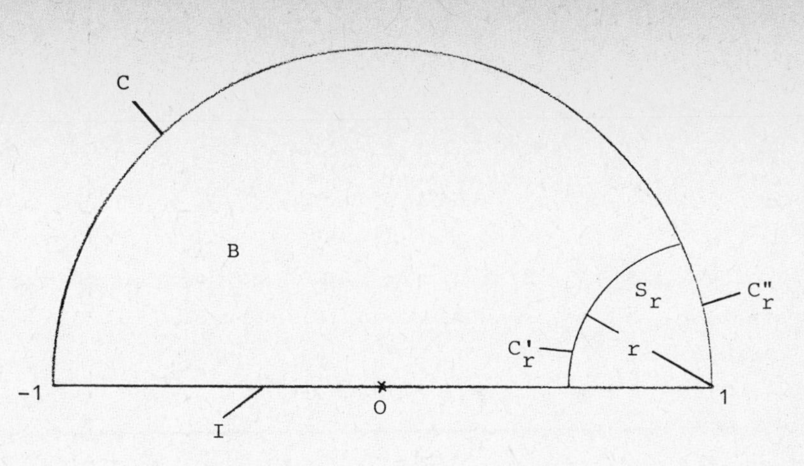

Lemma: Let S be a strict C^2-surface with constants ρ_0, K_0, K and assume that $x \in \mathscr{C}$ is stationary.

There is a number $K_* = K_*(\rho_0, K_0, K) > 0$ such that the following is true.

If $e \in \{-1,1\}$, $w^* \in S_r$, $0 < R < \rho_1 = \rho_0\sqrt{K}$, and if

$$\inf_{w \in C_r} |x(w) - x(w^*)| > R$$

then

$$R^2 \leq K_*^2 D_{S_r}(x) \quad . \tag{*}$$

Proof: Let $w^* \in S_r$, $0 < R < \rho_1$, $x^* = x(w^*)$, and $\delta(x^*) = \mathrm{dist}(S, x^*)$. First, suppose $\delta(x^*) > 0$. Define $d = (2K)^{-1} \leq \frac{1}{2}$, $R^* = \min\{\delta(x^*), d^2R\}$. Using the standard variation, c.f. section 2 and [GM1], [GM2], [GHN1], [GHN2], for $0 < \rho < R^*$ we get

$$2\pi \leq R^{*-2} \int_{S_r \cap K_{R^*}(x^*)} |\nabla x|^2 \, du \, dv \quad ,$$

where $K_{R^*}(x^*) = \{w \in B: |x(w) - x^*| < R^*\}$. Thus, if $R^* = d^2R$ we have shown that (*) is true with

$$K_* = (d^2 \sqrt{2\pi})^{-1} \quad .$$

If, however, $R^* = \delta(x^*) < d^2 R$ we get

$$(1) \qquad 2\pi \le \frac{1}{\delta(x^*)^2} \int_{S_r \cap K_{\delta(x^*)}(x^*)} |\nabla x|^2 \, du\, dv \quad .$$

Now choose $f \in S$ such that $\delta(x^*) = |x^* - f|$ and consider the coordinate transformation h as in assumption (A). For $d^{-1}\delta(x^*) < \rho < dR$ define the variation

$$x_\epsilon(w) = h(y(w) + \epsilon\eta(w))$$

where

$$y(w) = h^{-1}(x(w)) \quad ,$$

and

$$\eta(w) = \begin{cases} 0 & , \ w \in B \sim \bar{S}_r \\ \lambda(\rho - \|y(w)\|)\, y(w) & , \ w \in \bar{S}_r \end{cases} \quad .$$

Here,

$$\|y(w)\|^2 = g_{ij}(y(w))\, y^i(w)\, y^j(w)$$

and $\lambda \in C^1(\mathbb{R},\mathbb{R})$ has the properties ($\tau > 0$ some small number)

$$\lambda' \ge 0,$$
$$\lambda(t) = 0 \quad \text{for} \quad t \le 0 \quad ,$$
$$\lambda(t) = 1 \quad \text{for} \quad t \ge \tau \quad .$$

We have to show that this variation is admissible, i. e.

$$\text{(i)} \qquad \eta \in H_2^1(B, \mathbb{R}^3) \quad \text{with} \quad \eta\big|_{C_r''} \equiv 0 \ ,$$

$$\text{(ii)} \qquad x_\epsilon : I \to S \quad \mathcal{L}^1\text{-a.e.} \quad .$$

For $w \in C_r$ we get

$$\begin{aligned} \|y(w)\| &\ge K^{-1}|x(w) - f| \ge K^{-1}\{|x(w) - x^*| - \delta(x^*)\} \\ &> K^{-1}\{R - d^2 R\} = R(1 - d^2)K^{-1} = \\ &= dR\, 2(1 - d^2) > dR \quad . \end{aligned}$$

Thus, for $\rho < dR$ and $w \in C_r$ we have $\|y(w)\| > \rho$. But this implies $\eta(w) = 0$ and therefore $x_\epsilon(w) = x(w)$. To check (ii) one has to perform a similar calculation (c.f. the proof of Lemma 4 in [GHN1]). Using the above variation for $d^{-1}\delta(x^*) < \rho < dR$ we may conclude

$$(2) \qquad \frac{1}{\delta(x^*)^2} \int_{S_r \cap K_{2\delta(x^*)}(f)} |\nabla x|^2 \, du\, dv \le \frac{C(R)}{d^4} \frac{1}{R^2} \int_{S_r} |\nabla x|^2 \, du\, dv$$

with $C(R) = (1 + \frac{1}{2} R K_0 K) e^{RK_0 K}$.

Taking (1) into account (2) now implies $(K_{\delta(x^*)}(x^*) \subset K_{2\delta(x^*)}(f))$

$$R^2 \leq \frac{C(R)}{2\pi d^4} \int_{S_r} |\nabla x|^2 \, du \, dv \, .$$

This shows that (*) is true with

$$K_* = d^{-2} \sqrt{C(\rho_1)/2\pi} \quad .$$

Next, assume $\delta(x^*) = 0$ $(x^* = f)$. In this case we get - using the variation defined before -

$$(3) \qquad \frac{1}{\rho^2} \int_{S_r \cap K_{\rho/K}(x^*)} |\nabla x|^2 \, du \, dv \leq \frac{C(R)}{d^2} \frac{1}{R^2} \int_{S_r} |\nabla x|^2 \, du \, dv$$

for $0 < \rho < dR$. Letting $\rho \downarrow 0$ and noting that

$$\limsup_{\rho \downarrow 0} \frac{1}{\rho^2} \int_{S_r \cap K_{\rho/K}(x^*)} |\nabla x|^2 \, du \, dv \geq 8\pi \, d^2$$

(c.f. Lemma 1 in [GHN1] and observe that a corresponding estimate was already used in the derivation of (1)) we have established (*) with the same constant as before. □

We now give the

<u>Proof</u> of Theorem 1: Let $0 < \epsilon < \rho_1$ and $e \in \{-1,1\}$. Since $D(x) < \infty$ we can find $R_0 > 0$ such that (K_* as in the previous Lemma)

$$(1) \qquad K_*^2 \, D_{S_{R_0}}(x) < \epsilon^2 \quad .$$

Applying the previous Lemma we get for $0 < r \leq R_0$

$$(2) \qquad \sup_{w^* \in S_r} \{\inf_{w \in C_r} |x(w) - x(w^*)|\} \leq \epsilon \quad .$$

We may additionally assume that R_0 has been chosen so small that

$$(3) \qquad \operatorname{osc}_{C''_{R_0}} x \leq \epsilon \quad .$$

Note, that x_C is continuous by assumption. By the well known Courant-Lebesgue-Lemma there is $r \in [R_0/2, R_0]$ such that (because of (1))

$$(4) \qquad (\operatorname{osc}_{C'_r} x)^2 \leq \frac{\pi}{\log 2} D_{S_{R_0}}(x) < (K')^2 \, \epsilon^2$$

with $K'^2 = \dfrac{\pi}{K_*^2 \log 2}$.

Collecting the estimates (1) - (4) we arrive at

$$\operatorname*{osc}_{S_r} x \lesssim 2\epsilon + \epsilon + K'\epsilon = (3+K')\epsilon \quad .$$

This implies the continuity of x at e. □

4. Higher regularity.

To improve Theorem 1 we have to impose some condition on the arc Γ and the supporting surface S near the intersections P_1, P_2.

We shall assume the following

Lift-off Condition:

The arc Γ and the surface S intersect transversely at P_1 and P_2, i.e. the angle between the tangent to Γ and the tangent plane to S at P_i $(i = 1,2)$ is positive.

Remark: For a discussion of the various but essentially equivalent conditions on the behavior of Γ and S near P_i the reader is referred to [NJ; §499, §511] and [HN].

Let us only mention the so called chord-arc condition: There exist $\epsilon, c > 0$ such that any two points $x_1, x_2 \in \Gamma \cup S$ with $|x_1-x_2| < \epsilon$ can be connected by a rectifiable arc contained in $\Gamma \cup S$ whose length is bounded by $(1+c)|x_1-x_2|$.

Theorem 2: Suppose that S is a strict C^2-surface, that S and Γ satisfy the lift-off condition, and that $x \in \mathscr{C}$ is stationary. Then

$$x \text{ is Hölder continuous on } \bar{B} \quad .$$

Remarks: (i) The Hölder exponent will of course depend on the angle between Γ and S. In general it is not possible to achieve higher (at least C^1) regularity of x near the corners, even if Γ and S are analytic, unless Γ and S intersect at a right angle. The reason is that x is a conformal map and has to preserve angles.

(ii) Although the assertion of Theorem 2 is more or less standard once one has continuity (guaranteed by Theorem 1) for the convenience of the reader we chose to indicate how this result can be derived from well known facts.

<u>Proof</u> of Theorem 2: Since x is already continuous on $\bar{B}$ we may extend it by reflection to get a function

$$y \in C^0(\bar{D}_\delta) \cap H_2^1(D_\delta,\mathbb{R}^3)$$

for some $\delta > 0$, where $D_\delta = \{w \in \mathbb{R}^2: |w| < 1, |v| < \delta\}$. It is easy to verify that y satisfies a system of the form

$$\Delta y(w) = f(w)$$

with $f \in L^1(D_\delta,\mathbb{R}^3)$, $|f| \leq c|\nabla y|^2$ for some $c \geq 0$. Furthermore y is almost conform on D_δ in the sense that

$$\max \{|y_u \cdot y_v|, ||y_u|^2 - |y_v|^2|\} \leq c(|v|)|y_u||y_v|$$

for some function c satisfying $\lim_{r\downarrow 0} c(r) = 0$. For details we refer to Hilfssatz 3 in [DG2].

Next, observe that the boundary values of y near $e \in \{-1,1\}$ are contained in two C^1-Jordan arcs meeting at an angle $\alpha \in (0,\pi]$.

Thus, we are able to apply (a modified version of) Satz 1 in [DG3]. One only has to convince oneself that the almost conformality of y near e is sufficient to show Hölder continuity of y near e. In particular $x|_{D_\delta} = y|_{B\cap D_\delta}$ is Hölder continuous near e.

<u>Remark:</u> The proof of Satz 1 in [DG3] actually involves another reflection. □

References

[AW1] Allard, W.K.: On the first variation of a varifold; Ann. of Math. <u>95</u> (1972), 417-491.

[AW2] Allard, W.K.: A characterization of the area integrand; Symposia Math. XIV (1974), 429-444.

[AW3] Allard, W.K.: An integrability theorem and a regularity theorem for surfaces whose first variation with respect to a parametric elliptic integrand is controlled. Proc. Symposia in Pure Math. <u>44</u> (1986), 1-28.

[DG1] Dziuk, G.: Das Verhalten von Flächen beschränkter mittlerer Krümmung an C^1-Randkurven; Nachr. Akad. Wiss. Göttingen, II. Math.-Phys. Kl. (1979), 21-28.

[DG2] Dziuk, G.: Über die Stetigkeit teilweise freier Minimalflächen; man. math. <u>36</u> (1981), 241-251.

[DG3] Dziuk, G.: Über quasilineare elliptische Systeme mit isothermen Parametern an Ecken der Randkurve; Analysis **1** (1981), 63-81.

[DG4] Dziuk, G.: C^2-regularity for partially free minimal surfaces; Math. Z. **189** (1985), 71-79.

[DG5] Dziuk, G.: On the length of the free boundary of a minimal surface; Control and Cybernetics **14** (1985), 161-170.

[DU] Dierkes, U.: A geometric maximum principle for surfaces of prescribed mean curvature in Riemannian manifolds; Preprint.

[FH] Federer, H.: Geometric Measure Theory; Springer Verlag, 1969.

[GM1] Grüter, M.: Über die Regularität schwacher Lösungen des Systems $\Delta x = 2H(x)x_u \wedge x_v$; Dissertation, Düsseldorf 1979.

[GM2] Grüter, M.: Regularity of weak H-surfaces; J. Reine Angew. Math. **329** (1981), 1-15.

[GM3] Grüter, M.: Conformally invariant variational integrals and the removability of isolated singularities; man. math. **47** (1984), 85-104.

[GM4] Grüter, M.: Eine Bemerkung zur Regularität stationärer Punkte von konform invarianten Variationsintegralen; man. math. **55** (1986), 451-453.

[GHN1] Grüter, M., Hildebrandt, S., and Nitsche, J.C.C.: On the boundary behavior of minimal surfaces with a free boundary which are not minima of the area; man. math **35** (1981), 387-410.

[GHN2] Grüter, M., Hildebrandt, S., and Nitsche, J.C.C.: Regularity for stationary surfaces of constant mean curvature with free boundaries; Acta math. **156** (1986), 119-152.

[HN] Hildebrandt, S., and Nitsche, J.C.C.: Geometric properties of minimal surfaces with free boundaries; Math. Z. **184** (1983), 497-509.

[JJ1] Jost, J.: On the regularity of minimal surfaces with free boundaries in Riemannian manifolds; man. math. **56** (1986), 279-291.

[JJ2] Jost, J.: Continuity of minimal surfaces with piecewise smooth free boundaries; Math. Ann. **276** (1987), 599-614.

[JW] Jäger, W.: Behavior of minimal surfaces with free boundaries; Comm. Pure Appl. Math. **23** (1970), 803-818.

[LH] Lewy, H.: On minimal surfaces with free boundary; Comm. Pure Appl. Math. **4** (1951), 1-13.

[MS] Michael, J.H., and Simon, L.: Sobolev and mean-value inequalities on generalized submanifolds of $\mathbb{R}^n$; Comm. Pure Appl. Math. **26** (1973), 361-379.

[NJ] Nitsche, J.C.C.: Vorlesungen über Minimalflächen; Springer-Verlag, 1975.

[SL] Simon, L.: Survey lectures on minimal submanifolds; Ann. of Math. Stud. 103 (1983), 3-52.

[SR] Schoen, R.: Analytic aspects of the harmonic map problem; Seminar on Nonlinear Partial Differential Equations, 321-358, Springer-Verlag, 1984.

[YR] Ye, R.: Regularity of a minimal surface at its free boundary; Preprint.

ISOPERIMETRIC PROBLEMS HAVING CONTINUA OF SOLUTIONS

Robert Gulliver

The isoperimetric problem asks for the hypersurface of smallest area which bounds a given volume v_o. In Euclidean $\mathbb{R}^{n+1}$, the unique solution - up to translations - is the sphere of the appropriate radius. The isometry group of $\mathbb{R}^{n+1}$ generates an (n+1)-parameter family of solutions, all of which are congruent to each other. A similar situation occurs in any symmetric space.

We are interested in finding a continuum, or family depending continuously on one or more real variables, of embedded stationary solutions which are not congruent. An embedded compact hypersurface Σ will be stationary for the isoperimetric problem in a Riemannian manifold M^{n+1} if and only if it bounds the prescribed volume v_o and has constant mean curvature; this follows from the first-variation formula (equation (3) below). According to a theorem of Aleksandrov, when M is the Euclidean space (or, by analogous means, the hyperbolic space) Σ must be standard sphere ([1]). The family we seek can therefore exist only for less well-understood Riemannian metrics.

Theorem: *There exists a Riemannian manifold* M *of any dimension* $n + 1 \geq 2$ *, in which the compact embedded hypersurfaces bounding volume* $v_o = 1$ *and having constant mean curvature include a one-parameter family* $\{\Sigma_t\}$ *with distinct mean curvatures.*

Note that the hypersurfaces found in the theorem cannot be congruent, since their mean curvatures are different. This latter property also has consequences for the so-called Lagrange multiplier method. Namely, one may find hypersurfaces of constant mean curvature H as stationary solutions of a second variational problem, in which the functional $A(\Sigma) + n\,H\,V(D(\Sigma))$ is considered. Here $D(\Sigma)$ is a domain with boundary Σ ; V denotes (n+1)-dimensional volume; and A is the n-dimensional measure, which we also call "area". In this problem, no constraint is assumed, and the constant H is prescribed in advance. By contrast, in the isoperimetric problem, the resulting value $H = H(v_o)$ of constant mean curvature may be difficult to determine in advance. The two problems may be seen to be equivalent once it is known that the function $H(v_o)$ is strictly monotone. But according to our theorem, this function is no better than a relation which may assume an entire interval of values for a single number v_o. In particular, our example may be relevant to attempts to generalize Gerhardt's construction of constant-mean-curvature foliations of certain Lorentzian manifolds ([2]).

This paper was stimulated by our recent collaboration with Stefan Hildebrandt ([3]), in which a variety of boundary-value problems were considered. In each case, an example was found having an interesting continuum of solutions. In particular, we have exploited the idea which is implicit in sections 3 and 4 of [3], that one should first find a plausible family of submanifolds, and then construct the problem of which each is a solution.

We gratefully acknowledge the support of the Sonderforschungsbereich 72 at the University of Bonn, and of the Max-Planck-Instiutut für Mathematik.

1. Deformation of metrics according to a foliation

It is easy to show that any codimension-one foliation has leaves of constant mean curvature in some metric. For our purpose, however, we need more control over the resulting metric. The following lemma will be applied with the Euclidean metric playing the role of ds^2.

Lemma 1: *Let* $\{\Sigma_t : t_2 < t < t_3\}$ *be an oriented codimension-one foliation of the Riemannian manifold* (M^{n+1}, ds^2), *with unit normal vector field* ν. *Suppose a second Riemannian metric* $d\tilde{s}^2$ *is introduced, so that* ν *remains normal to the leaves* Σ_t. *Write* $\psi(x)$ *for the* $d\tilde{s}^2$*-length of* ν, *and write the* $d\tilde{s}^2$*-area form of* Σ_t *as* $d\tilde{A}_{\Sigma_t} = (\varphi(x))^n dA_{\Sigma_t}$. *Then the* $d\tilde{s}^2$*-mean curvature* $\tilde{H}$ *of* Σ_t *satisfies*

$$\psi\, \tilde{H} = H - d(\log \varphi)(\nu) . \tag{1}$$

Proof: We shall compute the mean curvature of the leaf Σ_o by variational means. Choose a test function $\eta : \Sigma_o \longrightarrow \mathbb{R}$ having compact support, and let a family of hypersurfaces F_t be chosen starting from

$F_o = \Sigma_o$, such that the distance from F_o to F_t along the integral curves of υ through x equals $t\eta(x) + O(t^2)$, asymptotically as $t \longrightarrow 0$.

We first show that

$$(2) \qquad d\tilde{A}_{F_t} = \left[(\varphi(x))^n + O(t^2)\right] dA_{F_t} .$$

Let $\theta(x,t)$ be the ds^2-angle between υ and the unit normal vector to F_t. Observe that $\theta(x,t) = O(t)$ as $t \longrightarrow 0$. If $\{e_1,\ldots,e_n\}$ is the local ds^2-orthonormal basis for $T\Sigma_t$ such that $e_2,\ldots,e_n$ are also tangent to F_t, then

$$\left\{e_1 \cos\theta + \upsilon \sin\theta \ , \ e_2,\ldots,e_n\right\}$$

is a ds^2-orthonormal basis for TF_t. Since $d\tilde{A}_{\Sigma_t} = \varphi^n dA_{\Sigma_t}$ by hypothesis, we have $\widetilde{\|e_1 \wedge \ldots \wedge e_n\|} = \varphi^n$. It follows that

$$\begin{aligned} &\widetilde{\|(e_1 \cos\theta + \upsilon \sin\theta) \wedge e_2 \wedge \ldots \wedge e_n\|}^2 \\ &= \cos^2\theta \widetilde{\|e_1 \wedge \ldots \wedge e_n\|}^2 + \sin^2\theta \widetilde{\|\upsilon \wedge e_2 \wedge \ldots \wedge e_n\|}^2 , \end{aligned}$$

since υ is $d\tilde{s}^2$-orthogonal to Σ_t. But this expression equals φ^{2n} plus terms of order $O(t^2)$, and formula (2) follows.

Now the first-variation formula for area is

$$(3) \qquad \left.\frac{d}{dt}\right|_{t=0} \tilde{A}(F_t) = -\int_{F_o} n\, \tilde{H}(F_o,x)\tilde{\eta}(x)\, d\tilde{A}(x) ,$$

where $\tilde{\eta} = \psi\eta$ is the $d\tilde{s}^2$-length of the variation vector field $\eta\upsilon$. Thus using equation (2), we see that

$$\begin{aligned} \int_{F_o} n\, \tilde{H}\, \psi\eta\, \varphi^n dA &= \int_{F_o} n\, \tilde{H}\, \tilde{\eta}\, d\tilde{A} \\ &= -\left.\frac{d}{dt}\right|_{t=0} \int_{F_o} \left(\varphi^n + O(t^2)\right) dA \\ &= \int_{F_o} n\left[d(\log\varphi)(\upsilon) - H\right]\eta\varphi^n dA . \end{aligned}$$

Since η is arbitrary, formula (1) follows. Q.e.d.

<u>Remark 1:</u> It is also possible, although more difficult, to give a direct derivation of the formula (1) for the mean curvature using moving frames.

It should be apparent from formula (1) that the mean curvature of the foliation $\{\Sigma_1\}$ is most conveniently controlled by choosing φ to be a constant. In particular, this choice yields stronger results than, for example, a conformal deformation $(\varphi = \psi)$.

<u>Corollary 1:</u> *Suppose that the family* $\{\Sigma_t\}$ *of hypersurfaces of a Riemannian manifold* (M, ds^2) *covers a subset* Ω *of* M *, and forms a foliation of* $\Omega \setminus \Gamma$ *for some closed subset* Γ *of* M *. Assume that* Σ_t *has constant mean curvature* $H(\Sigma_t, x) = h(t)$ *for all* x *in a neighborhood* U *of* Γ *, and that* $H(\Sigma_t, x) > 0$ *everywhere. Let the metric* $d\tilde{s}^2$ *be defined on* $\Omega \cup U$ *by*

$$d\tilde{s}^2 := ds^2\Big|_{T\Sigma_t} + \psi^2\, ds^2\Big|_{N\Sigma_t}$$

where $\psi(x) := H(\Sigma_t, x)/h(t)$ *. Then in the metric* $d\tilde{s}^2$ *,* Σ_t *has constant mean curvature* $h(t)$ *.*

<u>Proof:</u> We apply formula (1) with $\varphi \equiv 1$. On U , we have $\psi \equiv 1$, so that $d\tilde{s}^2 = ds^2$ and $\tilde{H}(\Sigma_t, x) = H(\Sigma_t, x) = h(t)$. Thus ψ is well defined, smooth and positive on $\Omega \cup U$. The conclusion now follows from (1). Q.e.d.

2. <u>Construction of the example</u>

The proof of our theorem requires us to construct hypersurfaces which bound constant volume. However, the method of Lemma 1 allows only direct control of area. Fortunately, for hypersurfaces of constant mean curvature, constant volume and constant area are equivalent.

<u>Lemma 2:</u> *Suppose that* $\{\Sigma_t\}$ *is a family of immersed compact hypersurfaces of a Riemannian manifold* M *, and that* Σ_t *is the boundary of a region* D_t *with multiplicities. Suppose that each* Σ_t

has constant mean curvature $h(t) \neq 0$ *, and that its area* $A(\Sigma_t)$ *is constant. Then the volume* $V(D_t)$ *is constant.*

Proof: If $\eta : \Sigma_t \longrightarrow \mathbb{R}$ is the normal component of the variation vector field, then the first variation of volume is

$$\frac{d}{dt} V(D_t) = \int_{\Sigma_t} \eta \, dA .$$

Meanwhile, as in equation (3), the first variation of area is

$$\frac{d}{dt} A(\Sigma_t) = - \int_{\Sigma_t} n\, H(\Sigma_t, x)\eta(x)dA(x) = -n\, h(t) \frac{d}{dt} V(D_t) .$$

Since area is assumed constant and $h(t) \neq 0$, we conclude that volume is constant. Q.e.d.

Remark 2: The hypothesis $h(t) \neq 0$ is necessary, as may be seen from examples in which M is locally the Riemannian product $\Sigma_o \times [-1,1]$.

In order to construct our example, we shall find a smooth family of strictly convex closed hypersurfaces Σ_t in $\mathbb{R}^{n+1}$, whose Euclidean area is constant. We may observe that such a family cannot be a foliation, while Lemma 1 is valid only for foliations. Therefore, as in Corollary 1, we shall choose $d\tilde{s}^2 = ds^2$ on a neighborhood U of the singular set Γ of $\{\Sigma_t\}$. This requires that each hypersurface Σ_t must have constant Euclidean mean curvature on U . We shall choose Σ_t to be a piece of a sphere near Γ .

Choose values $r > \epsilon > 0$, and let Γ be the $(n-1)$-sphere $\left\{(x^o,x') \in \mathbb{R} \times \mathbb{R}^n = \mathbb{R}^{n+1} : x^o = -\epsilon \ , \ |x'| = r\right\}$.

Lemma 3: *There is a smooth family* $\{\Sigma_t\}$ *of hypersurfaces passing through* Γ *, each having constant Euclidean mean curvature* $h(t)$ *on a neighborhood* U *of* Γ *, forming a foliation except at* Γ *, and with constant Euclidean area.*

Proof: Without loss of generality, we may choose ϵ to be small, by translating the hypersurfaces Σ_t .

For each $t > 0$ we shall first choose Σ_t so that it agrees in the half-space $\{x^0 \leq 0\}$ with the sphere passing through Γ and centered at $(t,0)$. The radius $R = R(t)$ of this sphere satisfies

$$R(t)^2 = r^2 + (t+\epsilon)^2 , \tag{4}$$

and its Euclidean mean curvature is $h(t) := 1/R(t)$. Note that $dR/dt > 0$.

We next extend the above spherical cap to form a closed, convex, Lipschitz-continuous hypersurface Σ_t having area independent of t . For simplicity, we may choose Σ_t to agree in the half-space $\{x^0 \geq 0\}$ with the sphere of radius $\rho(t)$ and centered at $(\gamma(t),0)$, where $\gamma > 0$ and

$$\rho(t)^2 - \gamma(t)^2 = \rho_1(t)^2 := r^2 + 2\epsilon t + \epsilon^2 . \tag{5}$$

Note that Σ_t meets $\{x^0 = 0\}$ in the (n-1)-sphere of radius $\rho_1(t)$. Write α_{n-1} for the measure of the unit (n-1)-sphere in $\mathbb{R}^n$. We may compute

$$\frac{1}{\alpha_{n-1}} A(\Sigma_t) = \int_0^{\rho_1} (R^2-s^2)^{1/2} Rs^{n-1}ds + \left(\int_{\rho_1}^{\rho} + \int_0^{\rho}\right)(\rho^2-s^2)^{1/2}\rho s^{n-1}ds . \tag{6}$$

Observe that for fixed s , the integrand of the first integral is a decreasing function of R . More precisely, we find

$$\frac{1}{\alpha_{n-1}} \frac{d}{dt} A\left(\Sigma_t \cap \{x^0 \leq 0\}\right) \leq -\delta_n\, t\, r^{n-2} , \tag{7}$$

where $\delta_n > 0$ is independent of $\epsilon \in (0,r)$ and uniform for $0 < b_o \leq t/r \leq b_1$. Meanwhile, the last term of equation (6) is an increasing function of $\rho > \rho_1$, if ρ_1 is held constant. Now choose t_o satisfying $rb_o < t_o < rb_1$, and choose a value for $\rho(t_o)$ with $\rho_1(t_o) < \rho(t_o) < R(t_o)$. These choices determine Σ_{t_o} ; let a_o be its area. For all other $t \in (rb_o, rb_1)$ we define $\rho(t)$ by the condition that

$$A(\Sigma_t) = a_o , \tag{8}$$

and let $\gamma(t) > 0$ be determined by equation (5).

We shall now show that $\left\{\Sigma_t : rb_o \leq t \leq rb_1\right\}$ is a Lipschitz foliation in the half-space $\{x^o > -\epsilon\}$, for small values of ϵ. As t tends to a value t_1, we may compute the derivative at $t = t_1$ of the distance to $\Sigma_t \cap \{x^o \geq 0\}$ along various radial lines from $(\gamma(t_1),0)$. This derivative assumes its minimum either along the x^o-axis, where it equals $d\big(\rho(t) + \gamma(t)\big)/dt$, or at $\Sigma_t \cap \{x^o = 0\}$, where it equals $(\rho_1/\rho)d\rho_1/dt = \epsilon/\rho > 0$. That is, the foliation property will follow from positivity of

$$\frac{d}{dt}\big(\rho(t) + \gamma(t)\big) = \big((\gamma+\rho)d\rho/dt - \epsilon\big)/\gamma\ , \tag{9}$$

where we have used the derivative of equation (5). Now for $\epsilon = 0$, we have $\rho_1(t) = r$ independent of t, and also

$$\frac{d}{dt} A\left(\Sigma_t \cap \{x^o \leq 0\}\right) < 0$$

by inequality (7). Equation (8) now implies that $A\left(\Sigma_t \cap \{x^o > 0\}\right)$, which we know to be an increasing function of ρ, is also a strictly increasing function of t; therefore $d\rho/dt > 0$. But this implies that $d(\rho+\gamma)/dt > 0$ by equation (9), for $\epsilon = 0$. By continuity, we have $d(\rho+\gamma)/dt > 0$ for sufficiently small $\epsilon > 0$.

Finally, the inequality $\rho(t) < R(t)$ remains true for t in an appropriate closed interval about t_o. Geometrically, this means that Σ_t has an interior angle less than π at $\{x^o = 0\}$. We now smooth Σ_t, in some canonical way, inside the region $\left\{-\epsilon/2 < x^o < \epsilon/2\right\}$, so that its area is decreased by a constant value $\delta > 0$. For example, the smoothing may be done by convolution with one of a family of positive mollifiers on S^n, where Σ_t is represented as a graph in central projection. For δ sufficiently small, the resulting family of smooth hypersurfaces is a foliation except at Γ, and Σ_t has constant mean curvature $h(t)$ on $U := \left\{x^o < -\epsilon/2\right\}$. Q.e.d.

We may now summarize the proof of the theorem stated in the introduction. From Lemma 3, we find a smooth family (with uniform estimates) of strictly convex hypersurfaces Σ_t in $\mathbb{R}^{n+1}$, each containing Γ, forming a foliation of an open set $\Omega \setminus \Gamma$, having constant Euclidean mean curvature $h(t)$ on a neighborhood U of Γ, and such that $A(\Sigma_t)$ has a constant value a_1. Let D_t be the open set bounded by Σ_t. As in Corollary 1, we write $\psi(x) := H(\Sigma_t,x)/h(t)$ and define a new Riemannian metric on $\Omega \cup U$:

$$d\tilde{s}^2 = ds^2\Big|_{T\Sigma_t} + \psi^2\, ds^2\Big|_{N\Sigma_t} .$$

Then Σ_t has constant $d\tilde{s}^2$-mean curvature $\tilde{H}(\Sigma_t,x) = h(t)$ in the new metric. We may extend $d\tilde{s}^2$ arbitrarily to all of $\mathbb{R}^{n+1}$. Note that $\tilde{A}(\Sigma_t) \equiv a_1$. It now follows from Lemma 2, applied to $\mathbb{R}^{n+1}$ with the metric $d\tilde{s}^2$, that the volume $\tilde{V}(D_t)$ is constant. By homothetic rescaling, we may assume $\tilde{V}(D_t) \equiv 1$. The theorem is proved.

Remark 3: It will be obvious to the reader that the family of hypersurfaces $\{\Sigma_t\}$ may be perturbed in many independent ways so that it still satisfies the conclusions of Lemma 3 for some closed set Γ. It appears likely that few of these families will lead to isometric Riemannian structures on $\mathbb{R}^{n+1}$. Intuitively, in other words, one expects that the class of metrics having the properties stated in the theorem has infinite dimension and infinite codimension among all Riemannian metrics.

Remark 4: It is not clear to us whether the extension of $d\tilde{s}^2$ to all of $\mathbb{R}^{n+1}$ may be chosen so that the specific family of hypersurfaces we have constructed will have minimum area among hypersurfaces bounding volume one. It may be noted, however, that the variation vector field has nodal domains which are consistent with a minimizing condition subject to one constraint.

References

1. Aleksandrov, A.D.: Uniqueness theorems for surfaces in the large, V. Vestnik, Leningrad Univ. 13, 19 5-8 (1958). Amer. Math. Soc. Transl. 21, 412-416.

2. Gerhardt, C.: H-surfaces in Lorentzian manifolds. Comm. Math. Phys. 89, 523-553 (1983).

3. Gulliver, R. and S. Hildebrandt: Boundary configurations spanning continua of minimal surfaces. Manuscripta Math. 54, 323-347 (1986).

HARMONIC MAPS - ANALYTIC THEORY AND GEOMETRIC SIGNIFICANCE

Jürgen Jost

In this paper, the analytic and geometric aspects of the theory of harmonic maps are discussed. We are not aiming at completeness (and refer to [EL1], [EL2], [J4], [J5] for some of the topics not treated here), nor do we give detailed proofs. Instead, we represent the main lines of development of the theory and explain them through examples.

We shall start with the analytic foundations and, building upon this, treat existence, regularity, and uniqueness of harmonic maps; in the course of that, already some geometric applications will occur. The rôle of harmonic maps in Riemannian and Kählerian geometry will then be discussed somewhat more systematically in the subsequent paragraphs. The last paragraph finally is devoted to a novel development of Teichmüller theory via harmonic maps.

There are two main motivations for studying harmonic maps. The first one is that the occuring partial differential equations constitute the prototype of a whole class of nonlinear elliptic systems that are important for many applications inside and outside of pure mathematics (we refer to [G] at this point). Often, insights gained from studying harmonic maps can be generalized to these elliptic systems.

For investigating harmonic maps, their geometric meaning is often very helpful, and this geometric significance also constitutes the second motivation for looking at harmonic maps.

Since we are passing from the analytic to the geometric aspects, most of the geometric motivation for harmonic maps will become clear only towards the end of this survey.

Many of the results covered here were stimulated by the SFB 72. Since my own work was supported over many years by this institution, it is a great pleasure for me to be able to give a survey of these achievement for the occasion of the present book.

In particular, I want to thank Professor Dr. Stefan Hildebrandt who was the director of the SFB 72 over many years and to whom I personally own the generous support of my research.

1. Let us first exhibit the notation used in this paper:
X and Y are Riemannian manifolds of dimension n and N, resp. . X may or may not have a boundary, whereas Y is without boundary and complete; usually both X and Y are compact. In local coordinates $(x^1,\dots,x^n)$ and $(u^1,\dots,u^N)$, resp., the metric tensors are given by $(\gamma_{\alpha\beta})$ and (g_{ij}), resp.. Let $(\gamma^{\alpha\beta}) := (\gamma_{\alpha\beta})^{-1}$. $\gamma := \det(\gamma_{\alpha\beta})$;
$\Gamma^i_{jk} := \frac{1}{2} g^{il}(g_{jl,k} + g_{kl,j} - g_{jk,l})$ (Christoffel symbols of Y).
Here, as in the sequel, we use the standard summation convention.

Let $f := X \to Y$ be a map, at first of class C^1 for simplicity. $< , >$ is the scalar product in the bundle $T^*X \otimes f^{-1}TY$.
We define the energy density as $(df = \frac{\partial f^i}{\partial x^\alpha} dx^\alpha \otimes \frac{\partial}{\partial f^i})$

$$e(f) := \frac{1}{2} < df, df > = \frac{1}{2} \gamma^{\alpha\beta}(x)\, g_{ij}(f(x)) \frac{\partial f^i}{\partial x^\alpha} \frac{\partial f^j}{\partial x^\beta} \tag{1.1}$$

in local coordinates and the energy of f is

$$E(f) := \int_X e(f)\, dX. \tag{1.2}$$

Herewith, the variational problem of finding critical points of the energy is posed. If $f \in C^2$, $E(f) < \infty$, is a critical point of E, f has to satisfy the corresponding Euler-Lagrange equations, namely

$$\Delta_X f^i + \gamma^{\alpha\beta} \Gamma^i_{jk}(f(x)) \frac{\partial f^i}{\partial x^\alpha} \frac{\partial f^k}{\partial x^\beta} = 0 \quad (i = 1,\dots,N) \tag{1.3}$$

in local coordinates, where

$$\Delta_X f^i := \frac{1}{\sqrt{\gamma}} \frac{\partial}{\partial x^\alpha} (\gamma^{\alpha\beta} \sqrt{\gamma} \frac{\partial f^i}{\partial x^\beta})$$

is the Laplace-Beltrami operator of X.

In invariant notation, (1.3) becomes

(1.4) $$\tau(f) := \text{trace } \nabla \, df = 0,$$

where ∇ here is the covariant derivative in $T^*X \otimes f^{-1}TY$. The covariant derivative in T^*X yields in (1.3) the Laplace-Beltrami operator of X, whereas the nonlinear term in (1.3) comes from the derivative in $f^{-1}TY$.

A Riemannian structure of Y different from Euclidean structure thus leads to the nonlinearity of the system (1.3). Incidentally, there is another nonlinearity, namely of a global topological type, because Y in general is not homeomorphic to $\mathbb{R}^N$ so that (1.3) cannot be written globally. This problem can be overcome, however, by a different point of view. Because of Nash's imbedding theorem, Y can be isometrically imbedded into some Euclidian space $\mathbb{R}^l$. Therefore, we can consider f as a map from X into $\mathbb{R}^l$, satisfying the nonlinear constraint

$$f(X) \in Y \quad \text{for all } x \in X ;$$

(1.3) then becomes

(1.5) $$\Delta f - D^2\pi(f)\,(df,df) = 0$$

where $\pi : \mathbb{R}^l \to Y$ is nearest point projection, and $D^2\pi$ is the second fundamental form of π (cf. [J5]; section 1.3 for details).

From this point of view, it also becomes clear in which function space one should look for critical points of E, in particular minima. Since we can always localize the problem in the domain, we can assume w.l.o.g. for the moment that X is compact. We then define the Sobolev space

$$W^{1,2}(X,Y) := \{f \in W^{1,2}(X,\mathbb{R}^l) : f(x) \in Y \text{ for almost all } x \in X\}.$$

We have $W^{1,2}(X,\mathbb{R}^l) = H^{1,2}(X,\mathbb{R}^l)$ whence every map of finite energy can be approximated by C^∞-maps. In general, however, these approximating maps need not have their image in Y, as in general the equality $W^{1,2}(X,Y) = H^{1,2}(X,Y)$ does not hold (see [SU2]).

A map $f \in C^2(X,Y)$ satisfying (1.3) is called harmonic. A critical point of E, however need not be of class C^2, and therefore in general only satisfies a weak form of (1.3) which we now want to discuss.

Variations of f can be described by sections φ of $f^{-1}TY$; in symbols $\varphi \in \Gamma_o(X,f^{-1}TY)$ where the subscript o denotes compact support. We put

$$f_t(x) := \exp_{f(x)} t\varphi(x). \tag{1.6}$$

We have $f_o = f$, and f_t is a variation of f.

If f is a critical point of E,

$$\frac{d}{dt} E(f_t)|_{t=0} = 0, \tag{1.7}$$

yielding

$$\int < df, d\varphi > = 0 \tag{1.8}$$

(If $\varphi = \varphi^i(x) \frac{\partial}{\partial f^i}$,

$$d\varphi = \nabla_{\frac{\partial}{\partial x^\alpha}} (\varphi^i \frac{\partial}{\partial f^i})\, dx^\alpha = \frac{\partial \varphi^i}{\partial x^\alpha} \frac{\partial}{\partial f^i} + \varphi^i \Gamma^k_{ij} \frac{\partial f^i}{\partial x^\alpha} \frac{\partial}{\partial f^k}.) \tag{1.9}$$

In order that (1.8) is welldefined, we need

$$\varphi \in L^\infty \cap W_o^{1,2}(X, f^{-1}TY)$$

meaning that φ is bounded and satisfies

$$\int < d\varphi,\, d\varphi > < \infty$$

In local coordinates, (1.8) becomes

$$\int \{\gamma^{\alpha\beta} \frac{\partial f^i}{\partial x^\alpha} \frac{\partial \psi^i}{\partial x^\beta} - \gamma^{\alpha\beta}\, \Gamma^i_{jk} \frac{\partial f^j}{\partial x^\alpha} \frac{\partial f^k}{\partial x^\beta}\} \sqrt{\gamma}\, dx = 0, \tag{1.10}$$

with $\psi^i = g_{ij}\, \varphi^j \in W_o^{1,2} \cap L^\infty$.

The task of regularity theory now is to show that a weak solution (i.e. a $W^{1,2}$-solution of (1.8)) is regular, i.e. of class C^2. (1.6) suggests a useful type of test vectors, namely

$$\varphi(x) = \eta(x)\; k(f(x)) \tag{1.11}$$

where $\eta \in Lip_o(X,\mathbb{R})$ typically is a cut-off function ($\eta \equiv 1$ on $B(x_o,R) := \{x \in X: d(x,x_o) \le R\}$, $\eta \equiv 0$ on $X \backslash B(x_o,2R)$, and linearly decaying in between, i.e. $|\nabla\eta| \le \frac{1}{R}$), whereas $k(f)$ is a regular function of f. Particularly useful are test vectors of the form

$$k = \nabla a \tag{1.12}$$

where a is a real valued function of $f(x)$. If a is strictly convex, then one gets in (1.8) the positive term

$$\begin{aligned} < df,\, d(\nabla a(f)) > &= D^2 a(df,df) \\ &\ge \lambda < df,df > \end{aligned} \tag{1.13}$$

where λ is a lower bound for the smallest eigenvalue of the Hessian $D^2 a$.

Thus taking as a test vector

$$\varphi(x) = \eta^2(x)\; (\nabla a)(f(x)),$$

where η is a cut-off function as discussed above (note that one needs η^2 instead of η in order to deduce (1.15) below), (1.8) becomes

(1.14) $$\int_{B(x_o,2R)} (\eta^2(x)\, D^2a(df,df) + < df, d\eta^2 \otimes \nabla a >) = 0$$

Using (1.13), noting $\nabla\eta \equiv 0$ on $B(x_o,R)$, one gets

$$\lambda \int_{B(x_o,2R)} \eta^2 < df,df > \leq \frac{2}{R} \left(\int_{B(x_o,2R)} \eta^2 < df,df >\right)^{1/2} \left(\int_{B(x_o,2R)\setminus B(x_o,R)} |\nabla a(f)|\right)^{1/2},$$

thus

(1.15) $$\int_{B(x_o,R)} < df,df > \leq \frac{4}{\lambda^2 R^2} \int_{B(x_o,2R)\setminus B(x_o,R)} |(\nabla a)(f)|^2$$

Therefore, a higher order term, namely the energy of f on $B(x_o,R)$, is estimated by a lower order term.

In such a situation, one can often use a suitable imbedding inequality between function spaces and start an iteration procedure. For the sake of explanation, let us assume that $a(p) = \frac{1}{2} d^2(p,p_o)$, $p_o \in Y$ fixed, is strictly convex on $f(X)$.

Then $(\nabla a)(f(x)) = d(f(x),p_o)$. For simplicity, we also assume

(1.16) $$⨍_{B(x_o,2R)\setminus B(x_o,R)} d(f(x),p_o) = 0$$

(Here, ⨍ is the average integral). By Poincaré's inequality, in this case

$$\int_{B(x_o,2R)\setminus B(x_o,R)} d^2(f(x),p_o) \leq c_1 R^2 \int_{B(x_o,2R)\setminus B(x_o,R)} < df,df >$$

Using this in (1.15), one gets

(1.17) $$\int_{B(x_o,R)} < df,df > \leq \frac{c_2}{\lambda^2} \int_{B(x_o,2R)\setminus B(x_o,R)} < df,df >$$

Putting for abbreviation $E(x_o,R) := \int_{B(x_o,R)} < df,df >$, $\gamma := \frac{c_2}{\lambda^2}$,

we get by adding $\gamma\, E(x_o,R)$ on both sides of (1.17) ("hole filling" in the sense of Widman)

(1.18) $$E(x_o,R) \leq \frac{\gamma}{\gamma+1} E(x_o,2R)$$

Iteratively, one gets with $\alpha := \log_2 (\frac{\gamma+1}{\gamma})$ for $r<R$

(1.19) $$E(x_o,r) \leq 2^\alpha \left(\frac{r}{R}\right)^\alpha E(x_o,R).$$

If this holds for every x_o, a theorem of Morrey ("Dirichlet growth theorem") yields Hölder continuity of f. Of course, (1.16) is not necessarily satisfied, but the above scheme of proof, developped by Hildebrandt-Widman ([HW1], [HW2]) can nevertheless be used with a more complicated choice of test functions to prove Hölder continuity of f.

Once Hölder continuity is proved, one can show $f \in C^2$ in a known manner (cf. however, in this regard also § 3).

Decisive for the success of the previous argument was the existence of strictly convex functions on f(X). At first sight, the necessity of this assumption may appear as an artefact of the argument, but it turns out that without this assumption, there do exist singular solutions of (1.8). The map

$$\frac{x}{|x|} : \mathbb{R}^n\backslash\{0\} \to S^{n-1} := \{x \in \mathbb{R}^n: |x| = 1\}$$

can be considered as a harmonic map; namely, in case $X = \mathbb{R}^n$, $Y = S^{n-1}$, (1.3) becomes

$$\Delta f + f\,|df|^2 = 0, \tag{1.20}$$

and $f(x) = \frac{x}{|x|}$ is a solution of (1.20) on $\mathbb{R}^n\backslash\{0\}$.

Moreover, $f \in W^{1,2}(B(0,1),S^{n-1})$ for $n \geq 3$, and it therefore has finite energy in the unit ball and can be extended to a weak solution of (1.20) on B(0,1). Obviously, $f(x) = \frac{x}{|x|}$ is discontinuous at 0, and thus not regular. Of course, in this case, already the topology of the setting prevents the continuity of f, but if one embeds S^{n-1} as an equator sphere into S^n ($i: S^{n-1} \to S^n$), then $i \circ f : B(0,1) \to S^n$ still is weakly harmonic. Now, there is no topological obstruction for continuity anymore, and we see that the discontinuity has a geometric reason. We recall that there exists no strictly convex function on the equator sphere of S^n.

Incidentally, one can also show directly that for a harmonic map $f : X \to Y$ and a function $a : Y \to \mathbb{R}$, one has

$$\Delta(a \circ f) = D^2 a(df,df) \tag{1.21}$$

If a is convex, $a \circ f$ is subharmonic. If X is compact and $\partial X = \phi$, it follows that $a \circ f$ is constant, and if a is strictly convex, f itself is constant. Thus, we see that there is no nontrivial harmonic map $f: X \to Y$, X compact, $\partial X = \phi$, with a strictly convex function on f(X).

In the example above, $i: S^{n-1} \to S^n$ is also harmonic, and there is no strictly convex function on $i(S^{n-1})$. More generally, each closed geodesic in a Riemannian manifold Y, considered as a map

$$\gamma : S^1 \to Y$$

(with parametrization proportional to arclenght), is harmonic. Therefore, unless $\gamma \equiv$ const., there cannot exist any strictly convex function on $\gamma(S^1)$.

Later on (§ 3), we shall see that nontrivial harmonic maps

$$h : S^n \to Y$$

are precisely the obstruction for regularity theory.

We want to consider the above reasoning again for the form (1.10) of the weak equations, in order to gain an insight into the difficulties of regularity theory from a slightly different viewpoint. We choose as testvector in (1.10)

$$\psi^i = \eta^2 f^i(x)$$ [1]

and obtain

$$(1.22)\qquad \int \{ \gamma^{\alpha\beta}\, \eta^2 \Big(\frac{\partial f^i}{\partial x^\alpha}\frac{\partial f^i}{\partial x^\alpha} - f^i\, \Gamma^i_{jk}\, \frac{\partial f^j}{\partial x^\alpha}\frac{\partial f^k}{\partial x^\beta}\Big) + 2\,\gamma^{\alpha\beta}\, \eta\, \frac{\partial f^i}{\partial x^\alpha}\, f^i\, \frac{\partial \eta}{\partial x^\beta} \}\sqrt{\gamma}\, dx = 0$$

If now

$$(1.23)\qquad \gamma^{\alpha\beta}\Big(\frac{\partial f^i}{\partial x^\alpha}\frac{\partial f^j}{\partial x^\beta} - f^i\, \Gamma^i_{jk}\, \frac{\partial f^j}{\partial x^\alpha}\frac{\partial f^k}{\partial x^\beta}\Big) \geq \lambda\, |df|^2$$

with $\lambda > 0$, we can again deduce an inequality of the type

$$(1.24)\qquad \int_{B(x_o,R)} |df|^2 \leq \frac{c_3}{\lambda^2 R^2} \int_{B(x_o,2R)\setminus B(x_o,R)} |f|^2 \quad ,$$

and one can then draw the same consequences as from (1.15) above. From the assumption (1.23) we see that the nonlinearity in (1.3), being of the same weight as $\Delta f \cdot f$, namely quadratic in df, may destroy the regularizing properties of the Laplace-Beltrami operator Δ_X, unless this nonlinearity is a priori controlled by restrictive assumptions. One has to assume that $f^i(x)\ \Gamma^i_{jk}(f(x))$ is small or has at least the right sign. It turns out that this is equivalent to the existence of strictly convex functions on f(X), because of the fact that one can choose local coordinates adapted to a strictly convex function.

2.

First of all, we have the old theorem attributed to Hilbert and Birkhoff

<u>Theorem 2.1:</u>

Let Y be a compact Riemannian manifold.

Then every free homotopy class of closed loops in Y contains a closed geodesic. Also, any two (not necessarily distinct) points in Y can be connected by a geodesic curve in a prescribed homotopy class.

---

[1] this is equivalent to the above choice $\varphi = \eta^2\ (\nabla a)(f)$, $a = \frac{1}{2}\, d^2\ (f(x), p_o)$, and if as local coordinates we choose normal coordinates centered at p_o.

These geodesics are obtained by minimizing the energy among all maps $S^1 \to Y$ of $[0,1] \to Y$ with prescribed end points, resp., in the given homotopy class.

It is also not difficult to produce unstable geodesics by saddle point constructions, for example to prove that every metric g on S^2 admits a closed geodesic. In order to achieve this, one just considers the class H of all topologically nontrivial maps

$$h: S^1 \times [0,1] \to (S^2, g)$$

with $h(\cdot,0) = \text{const.}$, $h(\cdot,1) = \text{const.}$, and shows that there exists a closed geodesic $f: S^1 \to (S^2,g)$ with

$$E(f) = \inf_{h \in H} \sup_{t \in [0,1]} E(h(\cdot,t)).$$

The essential difficulty of the theory of closed geodesics consists of showing that geodesics obtained by different saddle point constructions are geometrically different and not just multiple coverings of each other. The theorem of Lyusternik and Schnirelman showing that there are always at least three distinct closed geodesics without self-intersections on (S^2,g) is an example. For a more general account, we refer to [Kl].

If one then poses the general existence problem for harmonic maps, namely to find a harmonic map $f: X \to Y$ in a prescribed homotopy class, with prescribed boundary values in case $\partial X \neq \phi$, one might believe that this could be solved in an analogous fashion, by first minimizing the energy among all maps in the given class, and then showing that the infimum of energy is realized by a smooth, hence harmonic map.

In the light of the insights of § 1, we can already see, however, that carrying out this program will encounter serious difficulties. First of all, the class in which one would naturally minimize E is the Sobolev space $W^{1,2}(X,Y)$, and for $n = \dim X \geq 2$, maps in $W^{1,2}(X,Y)$ are not necessarily continuous. In particular, the topology of $W^{1,2}(X.Y)$ does not necessarily respect homotopy classes. Thus, even if one manages to find a minimizing sequence that converges to a continuous minimum of energy, this convergence will only take place in $W^{1,2}(X,Y)$ (in general, it will be weak $W^{1,2}$ convergence), and the limit of the minimizing sequence, i.e. the map obtained, may lie in a different homotopy class.

In addition, in general, one cannot expect that the limit of a minimizing sequence, although a weak solution of (1.3), is continuous, and therefore, it may not even make sense to say that such a map is in a homotopy class.

The full difficulties occur only for $n \geq 3$, but already for $n = 2$ a direct analogue of Theorem 2.1 fails to hold. The following example of Lemaire ([L]) is instructive:

Theorem 2.2:

Each harmonic map $u : D \to S^2$ with constant boundary values is constant (here $D := \{z \in \mathbb{R}^2 : |z| \leq 1\}$ is the unit disk). In particular, no non-trivial homotopy class contains a harmonic map.

Proof:

As u has constant boundary values,

$$\frac{d}{dt} E(u \circ \sigma_t)\big|_{t=0} = 0, \tag{2.1}$$

where $\sigma_t: D \to D$ is a family of diffeomorphisms with $\sigma_o = \mathrm{id}$, depending differentiably on t. One concludes that u is conformal and hence, since constant on the boundary, constant itself.

q.e.d.

We now want to consider what happens if one chooses a minimizing sequence $(u_n)_{n \in \mathbb{N}}$ for the energy of maps $u_n : D \to S^2$ of degree 1 with constant boundary values. First of all,

$$E(u_n) \geq 4\pi \quad (= \text{Area } (S^2)), \tag{2.2}$$

since the u_n are surjective. This follows from an elementary computation; equality in (2.2) could only occur for a conformal map; however, no u_n can be conformal. On the other hand

$$\inf_{n \in \mathbb{N}} E(u_n) = 4\pi \tag{2.3}$$

for a minimizing sequence. To see this, let $p_o := u_n(\partial D) \in S^2$. If $\delta > 0$ is sufficiently small, we can map $\{z \in D: \frac{1}{2} \leq |z| \leq 1\}$ with energy $< \epsilon$ onto $B(p_o,\delta)$, $\{|z| = \frac{1}{2}\}$ going to $\partial B(p_o,\delta)$, and $\{|z| \leq \frac{1}{2}\}$ can be mapped conformally onto $S^2 \backslash B(p_o,\delta)$; of course the maps on $\{|z| = \frac{1}{2}\}$ can be chosen in such a way that the resulting maps $D \to S^2$ are continuous. Thus, we have constructed a map of energy $< 4\pi + \epsilon$, proving (2.3). Projecting S^2 stereographically onto $\mathbb{C} \cup \{\infty\}$, with $p_o = \infty$, $S^2 \backslash B(p_o,\delta)$ is identified with $\{|z| \leq N\}$, $N \to \infty$ as $\delta \to 0$. The above conformal map from $\{|z| \leq \frac{1}{2}\}$ onto $S^2 \backslash B(p_o,\delta)$ is then given by $z \to 2Nz$. The preimage of $\{|w| \leq K\}$ $(0 < K < \infty)$ under this map is $\{|z| \leq \frac{K}{2N}\}$, and hence shrinks to the origin as $N \to \infty$. Also, we see that the limit map is constant. One possible interpretation of this phenomenon is that in the limit we obtain a (surjective) harmonic map of a single point, here the origin, onto S^2 which accounts for the loss of energy (0 instead of 4π). If one performs rescalings in the domain and considers $ND := \{|z| \leq N\}$, one obtains in the limit a (conformal) harmonic map $\mathbb{C} \to S^2$ which can be extended to a surjective conformal map $S^2 \to S^2$. In this interpretation, the limit of a minimizing sequence is a constant map $D \to S^2$ plus a harmonic map $S^2 \to S^2$ of energy 4π.

It turns out that for $n = 2$ singularities can only develop in the manner just described; indeed, we have the following general theorem

Theorem 2.3:
Let Σ be a compact surface [2]*, Y a compact Riemannian manifold, B a parameter domain, for simplicity without boundary, e.g. $B = S^d$, $H \in [\Sigma \times B, Y]$ a homotopy class*

$$m := \inf_{h \in H} \sup_{t \in B} E(h(\cdot,t)). \tag{2.4}$$

Then there exist a harmonic map $u_o : \Sigma \to Y$ and possibly (non constant) harmonic maps $u_1, \ldots, u_k : S^2 \to Y$ with

$$\sum_{i=0}^{k} E(u_i) = m. \tag{2.5}$$

Furthermore, there exist a sequence of map $h_n(\cdot,t_n) : \Sigma \to Y$, $t_n \in B$, with

$$E(h_n(\cdot,t_n)) \to m \tag{2.6}$$

$$h_n \to u_o \quad \textit{(weak convergence in } W^{1,2}) \tag{2.7}$$

and points $z_1, \ldots, z_m \in \Sigma$ and sequences $(\lambda_{i,n})_{n \in \mathbb{N}}$ of real numbers $(i=1,\ldots,m)$ with $\lambda_{i,n} \to \infty$ as $n \to \infty$ and

$$h_n(\lambda_{i,n}(z - z_i), t_n) \to u_i(z) : \mathbb{C} \cup \{\infty\} \to Y, \tag{2.8}$$

where $\lambda_{i,n}(z - z_i)$ is a rescaling as in the above example.

---

[2] In this paragraph, a surface is a two-dimensional Riemannian manifold; it turns out, however, that for $n = 2$ the definition of a harmonic map is independent of the choice of a special metric on Σ and depends only on the conformal structure of Σ; thus, Σ can just be a Riemann surface.

Such a result was first proved by Sacks-Uhlenbeck ([SkU]), obtaining however instead of (2.5) only

$$\sum_{i=0}^{k} E(u_i) \le m. \tag{2.9}$$

In the present form, the result is due to the author ([J8]), using a different method. Let us also mention that Struwe ([St]) using still another method, namely the heat flow method to be discussed below, was able to reproduce the results of [SkU], again however obtaining only (2.9) and not (2.5).

Analogous results hold for the Dirichlet problem, i.e. where $\partial\Sigma \neq \phi$ and boundary values are prescribed on $\partial\Sigma$. Let us draw some consequences of Theorem 2.3:

Corollary 2.1:

Let Y be a compact Riemannian manifold with $\pi_2(Y) = 0$, Σ *a compact surface. Then every continuous map* $g : \Sigma \to Y$ *is homotopic to an energy minimizing harmonic map f. If* $\pi_2(Y) \neq 0$, *then there exists at least a harmonic map f with*

$$f_\# = g_\#, \tag{2.10}$$

where $g_\# : \pi_1(\Sigma) \to \pi_1(Y)$ *is the induced map of the fundamental groups.*

This is the fundamental existence result of Lemaire ([L]) and Sacks-Uhlenbeck ([SkU]). Other proofs were given by Schoen-Yau ([SY2]) and the author (cf. [J2]).

Corollary 2.2:

i) *For a homotopy class* α *of maps* $\Sigma \to Y$, *we let* $a_\#$ *be the induced maps of the fundamental groups,*

$$E_\alpha := \inf \{E(g) : g \in \alpha\} \tag{2.11}$$

Moreover, for $\pi : \pi_1(\Sigma) \to \pi_1(Y)$

$$E_\pi := \inf \{E(g) : g_\# = \pi\} \tag{2.12}$$

Finally, let

$$E_o := \inf \{E(h) : h : S^2 \to Y \text{ harmonic}, h \neq \text{const.}\} \tag{2.13}$$

Then every homotopy class α *with*

$$E_\alpha < E_{\alpha_\#} + E_o \tag{2.14}$$

contains a harmonic map.

ii) *If* $\partial\Sigma \neq \phi$, $\varphi : \partial\Sigma \to Y$ *given, for a homotopy class* β *of maps* $g : \Sigma \to Y$ *with* $g_{|\partial\Sigma} = \varphi$, *let*

(2.15)
$$E_\beta := \inf \{E(g) : g \in \beta\}$$

and

(2.16)
$$E_\varphi := \inf \{E(g) : g_{|\partial\Sigma} = \varphi\}$$

If

(2.17)
$$E_\beta < E_\varphi + E_o,$$

β contains a harmonic map (with boundary values φ).

Special cases (beyond those of Corollary 2.1) where (2.14) or (2.17) can be verified are found in [BC], [J3], [J6]. In the first two papers, for example it is proved that any nonconstant $\varphi : \partial D \to S^2$ admits at least two harmonic extensions $D \to S^2$ (compare with Theorem 2.2!).

Corollary 2.3 ([SkU], with refinements in [MY], [SiY]):
Let α be a homotopy class of maps $S^2 \to Y$. Then there exist energy minimizing (in their homotopy class) harmonic maps

$$u_i : S^2 \to Y \quad (i=1,\dots k)$$

with

(2.18)
$$\alpha = \sum_{i=1}^{k} [u_i] \text{ in } \pi_2(Y)$$

([u] := homotopy class of u)
and

(2.19)
$$\sum_{i=1}^{k} E(u_i) = E_\alpha.$$

Corollary 2.4 ([SkU]):
If for some $k \geq 2$, $\pi_k(Y) \neq 0$ (Y compact), then there exists a nontrivial harmonic map $n : S^2 \to Y$.

We also want to state the following result of [JS]:

Theorem 2.4:
Let Σ_1, Σ_2 be compact surfaces (with Riemannian metrics), $\varphi : \Sigma_1 \to \Sigma_2$ a diffeomorphism. Then there exists a harmonic diffeomorphism homotopic to φ.

Of course, the fact that φ is homotopic to some harmonic map already follows from Corollary 2.2. The fact that one can find such a harmonic map that is a diffeomorphism itself is much deeper. The proof uses a-priori estimates from below for the functional determinant of univalent harmonic maps, going back to the work of E. Heinz ([Hz]; cf. also

[JK] and [J8]). One minimizes the energy in the class of all diffeomorphisms homotopic to φ and shows via a replacement argument based on [J1] that the minimum even in this restriced class is still harmonic and moreover a diffeomorphism. (In case where Σ_2 has nonpositive curvature, it had been shown earlier by Schoen-Yau ([SY1]) and Sampson ([Sa1]) that the harmonic map of Corollary 2.1 is a diffeomorphism if homotopic to a diffeomorphism.)

Let us point out that neither in dimension $n = 1$ (theorem 2.1) nor for $n = 2$ (theorem 2.3) one needs any further geometric assumptions on the image Y, althrough the geometry of Y enters into the nonlinearity of the equations (1.3).

This is different for $n \geq 3$ where one does indeed need additional geometric assumptions in order to show the existence of a harmonic map. We first state the theorem of Eells-Sampson ([ES])

Theorem 2.5:

Let X and Y be compact Riemannian manifold, and let Y have nonpositive sectional curvature. Then any continuous map $g : X \to Y$ *is homotopic to some harmonic map.*

Hamilton ([Hm]) proved a corresponding result for the Dirichlet problem. If one also wants to admit positive image curvature, one has to impose a size restiction on the image. We have the theorem of Hildebrandt-Kaul-Widman ([HKW 3])

Theorem 2.6:

Let $\varphi : \partial X \to Y$ *be continuous, with*

$$\varphi(\partial X) \subset B(p,r)$$

for some $p \in Y$ *with*

$$r < \min\left(\frac{\pi}{2\kappa}, i(p)\right) \tag{2.20}$$

where κ^2 *is an upper bound for the sectional curvature of* Y *and* $i(p)$ *is the injectivity radius of* p. *Then there exists a harmonic map*

$$f : X \to B(p,r)$$

with

$$f_{|\partial X} = \varphi.$$

In case $Y = S^N$ (2.20) means that $\varphi(\partial X)$ is contained in an open hemisphere. Jäger-Kaul ([JäK 2]), Baldes ([Ba]), Schoen-Uhlenbeck ([SU 3]), and Giaquinta-Soucek ([GS]) then investigated in special cases what happens if (2.20) is violated. Their results show that one can neither expect existence nor regularity of harmonic maps. Note that in § 1 we already discussed the singular map $\frac{x}{|x|} : B(0,1) \to S^{n-1}$.

Condition (2.20) means that $d^2(\cdot,p)$ is strictly convex on $B(p,r)$, and in § 1, we discussed the importance of strictly convex functions for regularity theory. From the results of Schoen-Uhlenbeck ([SU 1], [SU 2]) one can deduce the following generalization of Theorem 2.6 (cf. also [J2]):

Theorem 2.7:
Let Y' be a compact part of Y with strictly convex boundary $\partial Y'$, and suppose there exists a strictly convex function on Y'. If X is compact, and if $\varphi : \partial X \to Y'$ has a finite energy extension $\varphi: X \to Y'$, there exists a harmonic map

$$f : X \to Y'$$

with

$$f_{|\partial X} = \varphi.$$

For the proof of theorem 2.5, Eells-Sampson did not use variational methods but rather studied the associated parabolic problem

$$f(x,t) : X \times [0,\infty) \to Y$$

$$\tau(f) = \frac{\partial f(x,t)}{\partial t} \quad \text{for } t > 0 \quad (\text{cf. } (1.4)) \tag{2.21}$$

$$f(x,0) = g(x)$$

where $g : X \to Y$ is some given continuous map. It is then shown that, if Y has nonpositive curvature, a solution $f(x,t)$ of (2.21) exists for all $t \geq 0$, and that, as $t \to \infty$, $f(\cdot,t)$ converges to a harmonic map homotopic to g (cf. also [Ht]). The estimates necessary for this procedure are based on the formula

$$\begin{aligned} \Delta_X e(f(x,t)) - \frac{\partial e(f(x,t))}{\partial t} &= |\nabla df(x,t)|^2 + \langle df \cdot Ric^X(e_\alpha), df \cdot e_\alpha \rangle \\ &- \langle R^Y(df \cdot e_\alpha, df \cdot e_\beta) df \cdot e_\alpha, df \cdot e_\beta \rangle \end{aligned} \tag{2.22}$$

where Ric^X is the Ricci tensor of X, R^Y is the curvature tensor of Y, $(e_\alpha)_{\alpha=1,\dots,n}$ is an orthonormal frame on X. Thus, if $R^Y \leq 0$, (2.22) implies

$$\Delta_X e(f) - \frac{\partial e(f)}{\partial t} \geq -c\, e(f). \tag{2.23}$$

The strong maximum principle then yields pointwise bounds for $e(f(x,t))$ in terms of $E(f(\cdot,t))$.

On the other hand, for a solution of (2.21), $E(f(\cdot,t))$ is a decreasing function of t, because one computes

$$(2.24) \qquad \frac{d}{dt} E(f(\cdot,t)) = - \int_X < \frac{\partial f}{\partial t}, \frac{\partial f}{\partial t} > dx.$$

This easily implies the necessary estimates for the above scheme. The behaviour of (2.21) without the assumption $R^Y \leq 0$ is not yet understood, except for special cases. Struwe ([St]) could use (2.21) in case dim M = 2 in order to reproduce the result of Sacks-Uhlenbeck discussed above. It is not ruled out, however, that even for n = 2 a solution of (2.21) might develop singularities for finite t.

The proof of Theorem 2.6 is based on a-priori estimates for weak solutions of (1.10) (cf. also [HJW]). A weak solution can be obtained as the minimum of E under prescribed boundary conditions. One can also directly obtain a smooth solution via a degree argument (cf. [HKW 2]). Although the methods for a-priori estimates work for arbitrary weak solutions of (1.10) (satisfying the constraint (2.20)), not necessarily minima, it turns out a posteriori that - as a consequence of a uniqueness result to be discussed in § 4 - under the constraint (2.20) any solution already is a minimum.

Finally, the proof of Theorem 2.7 is based on methods that only work for energy minima. These will be discussed in the next paragraph.

3.

The regularity theory for energy minimizing maps has been developped by Schoen-Uhlenbeck in [SU1], [SU2]. Analogous results in a somewhat different context were obtained by Giaquinta-Giusti ([GG1], [GG2], [G]; cf. also [JM] for boundary regularity). For a detailed account of those latter results we refer to Giaquinta's article in the present volume. Here, we shall only discuss the results of Schoen-Uhlenbeck.

Theorem 3.1:

Let X,Y be Riemannian manifolds, X compact and possibly with boundary ∂X*; in the latter case, let* $\varphi \in C^{2,\alpha}(\partial X, Y)$ *be given. Let* $f \in W^{1,2}(X,Y)$ *be energy minimizing, with* $f_{|\partial X} = \varphi$ *in case* $\partial X \neq \phi$*. Then there exists a set* $S \subset \overset{\circ}{X}$ *of Hausdorff dimension* $\leq n-3$ $(n = \dim X)$ *with the property that f is regular on* $X \setminus S$*. In particular, f is regular in a neighborhood of* ∂X*. If* $n = 3$*, S is discrete.*

Thus an energy minimizing map is regular outside a set of codimension at least 3. Baldes ([Ba]) showed that the estimate on the codimension cannot be improved in general.

The singularities of f can be characterized as follows

Theorem 3.2:
Under the assumption of theorem 3.1, let $z \in S$. *Then there exists a sequence* $(\lambda_n)_{n\in N} \subset \mathbb{R}^+$, $\lambda_n \to 0$ *for which the maps* $f_n \in W^{1,2}(B(0,1),Y)$, $f_n(x) := f(\exp_z \lambda_n x)$ *converge strongly in* $W^{1,2}$ *towards some map* $g \in W^{1,2}(B(0,1),Y)$ *which is of the form* $g(x) = w(\frac{x}{|x|})$ *where* $w \in W^{1,2}(S^{n-1},Y)$ *is a nonconstant harmonic map.*

Thus, "blowing up" the singularity of the energy minimizing map f leads to a singular map of the type discussed in § 1. Conversely it is a consequence of Theorem 3.2 that f is regular if there is no nonconstant harmonic map $S^{n-1} \to Y$. This leads to Theorem 2.7, remembering that we already saw in § 1 that the existence of strictly convex functions on the image excludes the existence of nontrivial harmonic maps from a compact manifold without boundary. Theorem 2.5 can also be deduced if one uses the result of White ([Wh]) that the infimum of energy is obtained in a given homotopy class if all homotopy groups $\pi_k(Y)$, $k \geq 2$, vanish.

For simplifications and extensions of the work of Schoen-Uhlenbeck, we also refer to [HKL], [HL], and [Lu].

We now want to discuss a-priori estimates for harmonic maps. Having geometric applications in mind, such estimates should only depend on intrinsic geometric quantities of the manifolds involved, but not on the choice of local coordinates.

Theorem 3.3 ([HJW]):
Let $B(p,r) \subset Y$

(3.1) $\quad r < \min\,(i(p), \frac{\pi}{2\kappa})$ (notations as in theorem 2.6).

Let $D(0,2d) := \{x \in \mathbb{R}^n : x < 2d\}$ *be a chart on X; assume for the metric tensor of X w.r.t. this chart*

(3.2) $$\lambda(\xi)^2 \leq \gamma_{\alpha\beta}(x)\, \xi^\alpha \xi^\beta \leq \mu\, |\xi|^2 \quad (0 < \lambda \leq \mu)$$

for all $x \in D(0,2d)$, $\xi \in \mathbb{R}^n$.
If $f : (D(0,2d), (\gamma_{\alpha\beta}))$ *is harmonic, we have for all* $x,y \in D(0,d)$

(3.3) $$d(f(x),f(y)) \leq \frac{c}{d^\beta}\, d(x,y)^\beta$$

for some $\beta \in (0,1)$ *where the constant* c *depends only on* n, dim Y, *curvature bounds* $-\omega^2 \leq K \leq \kappa^2$ *on* $B(p,r)$, r, λ, μ, *but not on* d.

As a corollary, one gets the following Liouville type theorem

Corollary 3.1:

Suppose X is homogeneously regular, meaning that X is diffeomorphic to $\mathbb{R}^n$ *and that in such a global coordinate chart the metric tensor satisfies*

(3.4) $$\lambda|\xi|^2 \leq \gamma_{\alpha\beta}(x)\, \xi^\alpha \xi^\beta \leq \mu\, |\xi|^2 \qquad 0 < \lambda \leq \mu$$

for all $x \in \mathbb{R}^n$, $\xi \in \mathbb{R}^n$.

If $f : X \to Y$ *is harmonic with*

(3.5) $$f(X) \subset B(p,r), \quad r < \min\left(\frac{\pi}{2\kappa}, i(p)\right),$$

then f is constant.

The importance of Corollary 3.1 rests on the fact that in combination with the following theorem of Ruh-Vilms it allows to deduce Bernstein type theorems for minimal submanifolds of $\mathbb{R}^n$.

Theorem 3.4 ([RV]):

Let $F : X \to \mathbb{R}^{n+p}$ *be an immersion of class* C^3 $(n=\dim X)$. *Then the Gauss map*

$$G : F(M) \to G(n,p)$$

is harmonic, if and only if F immerses M with parallel mean curvature vector, for example if F(M) is a minimal submanifold of $\mathbb{R}^{n+p}$. *Here,* $G(n,p)$ *is the Grassmann manifold of oriented n-dimensional linear subspaces of* $\mathbb{R}^{n+p}$, *equiped with the Riemannian metric induced by the symmetric structure.*

The Bernstein theorem of [HJW], a consequence of Corollary 3.1 and Theorem 3.4, then is

Theorem 3.5:

Let $F : \mathbb{R}^n \to \mathbb{R}^{n+p}$ *be an immersion of class* C^3, $X := F(\mathbb{R}^n)$ *be minimal, or, more generally, have parallel mean curvature vector. Suppose there exists some oriented n-dimensional plane* P_o *and some* α_o *with*

(3.6) $$\alpha_o > \cos^m\left(\frac{\pi}{2\kappa\sqrt{m}}\right), \quad m = \min(n,p),$$

$$\kappa^2 := \begin{cases} 1 & \text{for } m = 1 \\ 2 & \text{for } m \geq 2 \end{cases}$$

and

(3.7) $$\langle P, P_o \rangle \geq \alpha_o$$

for all oriented tangent planes P of X.

Finally, let the metric

(3.8) $$\gamma_{\alpha\beta}(x) := F_{x^\alpha}(x)\, F_{x^\beta}(x)$$

on X be uniformly equivalent to the Euclidean one in the sense of (3.4). Then X is an affine linear subspace of $\mathbb{R}^{n+p}$.

Theorem 3.5 in particular applies to minimal graphs whose tangent planes are restricted by (3.7).

Higher order a-priori estimates were obtained by Giaquinta-Hildebrandt ([GH]), Sperner([Sp]), and Choi ([Ci]). The most comprehensive results are due to Jost-Karcher.

Theorem 3.6 ([JK]):
Let $f : X \to Y$ *be a continuous harmonic map. Then the* $C^{2,\alpha}$ *norm of f* $(0 < \alpha < 1)$ *is bounded by some constant depending only on* α, *curvature bounds for X and Y, lower bounds for the injectivity radius of X and Y, and on the modulus of continuity of f (cf. Theorem 3.3 for estimates of the latter).*

The results of [JK] depend on results about harmonic coordinates, i.e. coordinates whose components are harmonic functions:

Theorem 3.7 ([JK]) :
Let M be a Riemannian manifold of dimension n; assume for the sectional curvature K of M

(3.9) $$|K| \le \Lambda^2.$$

Let $p \in M$. *There exists* $R_o > 0$, *depending on n, the injectivity radius i(p), and* Λ *((3.9) actually only has to hold on* $B(p,R_o)$*), with the property that for each* $R \le R_o$ *there exist harmonic coordinates on B(p,R), the corresponding metric tensor g of which satisfies*

(3.10) $$|dg(x)| \le \frac{c_1\, \Lambda^2\, R^2}{d(x,\partial B(p,R))} \qquad \text{for } x \in B(p,R)$$

with $c_1 = c_1(n,\Lambda R_o)$, *and*

(3.11) $$|dg|_{C^\alpha(B(p,(1-\delta)\mathbb{R}))} \le \frac{c_2}{\delta^2}\, \Lambda^2\, R^2 \qquad (0 < \alpha < 1)$$

with $c_2 = c_2(\alpha, \Lambda R_o, n)$.

In particular, the Hölder norms of the corresponding Christoffel symbols are bounded in terms of ΛR_o *and n.*

Theorem 3.7 has been applied by Peters in [P1] (and independently by Greene-Wu ([GWu]) to show

Theorem 3.8:

Let $0 < \alpha < 1$. Let $\{M^k\}_{k \in N}$ be a sequence of compact n-dimensional Riemannian manifolds with bounded diameters

(3.12) $$\operatorname{diam} M^k \le d,$$

volumina

(3.13) $$\operatorname{vol}(M^k) \ge V > 0,$$

and sectional curvatures

(3.14) $$|K(M^k)| \le \Lambda^2.$$

Then a subsequence converges w.r.t. the Lipschitz topology to some compact C^∞-manifold M with a metric $g \in C^{1,\alpha}$.

The proof uses suitable coverings by balls and convergence properties of the corresponding harmonic coordinates. Peters' Theorem is the optimal version of a compactness result for Riemannian manifolds. Originally, Gromov ([Gv]) had sketched a proof of a weaker variant by different methods. Carrying out this proof in detail encountered some difficulties, however.

Peters ([P2]) also showed that Theorem 3.8 (and also Theorems 3.6 and 3.7) no longer hold for $\alpha = 1$.

4.

Hartmann showed

Theorem 4.1 ([Ht]):

Let Y have nonpositive curvature, X be compact, and let $f_1, f_2 : X \to Y$ be homotopic harmonic maps. For fixed $x \in X$, let $f(x,s)$ be the unique geodesic connection between $f_1(x)$ and $f_2(x)$ in the homotopy class of curves determined by the homotopy between f_1 and f_2; let $s \in [0,1]$ be proportional to arclength. Then, for each $s \in [0,1]$, $f(\cdot,s) : X \to Y$ is a harmonic map with

(4.1) $$E(f(\cdot,s)) = E(f_1) = E(f_2)$$

and the length of the geodesic $f(x,\cdot)$ is independent of x. Thus, any two homotopic harmonic maps can be connected by a parallel family of harmonic maps of equal energy. If the curvature of Y is strictly negative, then a harmonic map $f : X \to Y$ is unique in its homotopy class, unless f is constant or maps X onto a closed geodesic. In the latter case, nonuniqueness can only occur by reparametrization, i.e. rotation of this geodesic.

Hamilton ([Ht]) proved a corresponding result for the Dirichlet problem. Schoen-Yau ([SY3]) proved that Theorem 4.1 remains true if X is complete with finite volume, but not necessarily compact.

Jäger-Kaul proved

Theorem 4.2 ([JäK1]):

Let X be compact with boundary ∂X, $f_1, f_2 : X \to Y$ *harmonic maps of class* $C^0(X,Y) \cap \overset{\circ}{C^2}(X,Y)$. *Suppose*

(4.2) $$f_i(X) \subset B(p,r) \quad \textit{for some } p \in Y$$

with

(4.3) $$r < \min\left(i(p), \frac{\pi}{2\kappa}\right) \quad \textit{(notations as in theorem 2.6).}$$

If

$$f_1|_{\partial X} = f_2|_{\partial X},$$

then

$$f_1 \equiv f_2 \quad .$$

In particular, under the assumptions of Theorem 2.5 or 2.6, harmonic maps are always energy minimizing. Likewise, a posteriori the estimates of Theorem 3.3 apply only to energy minima although the technique of proof does not use this assumption.

Thus, under the assumptions of Theorems 2.5 and 2.6, no unstable harmonic maps exist, and indeed for topological reasons, one would not expect the existence of unstable harmonic maps in those situations.

If G_1 and G_2 operate as groups of isometries on X and Y, resp. and if $f : X \to Y$ is harmonic, the maps

$$f \circ g \text{ and } \gamma \circ f$$

are also harmonic for $g \in G_1$, $\gamma \in G_2$.

Thus, Schoen-Yau ([SY3]) could use theorem 4.1 to find restrictions on group actions on manifolds. (If G_1 is compact and operates on a compact manifold X, we can average the metric of X to find a Riemannian metric on X for which G_1 operates by isometries.)

Under the assumption of Theorem 2.7, uniqueness does not hold as is shown by geodesic arcs on a paraboloid of revolution. Anyway, most examples of nonuniqueness of harmonic maps come from geodesics or analogous constructions. Therefore, one would expect stronger uniqueness results for harmonic homotopy equivalences between compact manifolds; at present, this is unknown.

5.

The applicability of harmonic maps in geometry rests on the following three principles

1) Maps that are distinguished as being canonical w.r.t. some geometric structure are harmonic.
2) Within a given class, a harmonic map is distinguished as satisfying a differential equation.
3) In many cases, one can show the existence of a harmonic map in a given class - cf. § 2.

These principles are put into force in the following way:

One seeks a "canonical" map, i.e. a map with a certain geometric property, which, if it exists at all, has to be harmonic by 1). In order to find such a map, conversely one first shows the existence of a harmonic map, i.e. 3). One then shows that under suitable geometric assumptions the harmonic map must have additional geometric properties; this is usually achieved by skillfully exploiting the differential equation according to 2). Then one shows that the harmonic map already is of the required canonical type.

In § 6, we shall see an example of such a kind of argument. 3) was already treated in § 2. Examples for 1) are, among others (we met several of them already in the preceding §§).

- Isometries of Riemannian manifolds (cf. § 4)
- Harmonic functions on Riemannian manifolds (cf. § 3; the result of Theorem 3.5 is interesting for this case, too)
- Geodesics parametrized proportionally to arclength as harmonic maps $S^1 \to Y$ (cf. § 2)
- Minimal immersions and parametric minimal surfaces; Gauss maps of minimal submanifolds of some $\mathbb{R}^{n+p}$ (cf. Theorem 3.4)
- Holomorphic maps between Kähler manifolds (cf. § 6 below), in particular conformal maps between surfaces
- Hopf maps $S^3 \to S^2$, $S^7 \to S^4$, $S^{15} \to S^8$

We now want to discuss a simple example for 2).
As in (2.22), one has for a harmonic map $f : X \to Y$

$$\Delta_X e(f) = |\Delta df|^2 + \langle df\, Ric^X(e_\alpha), df\cdot e_\alpha \rangle - \langle R^Y(df\cdot e_\alpha, df\cdot e_\beta)\, df\cdot e_\alpha, df\cdot e_\beta \rangle \tag{5.1}$$

with the same notations as in (2.22).

Integration over X yields ($\int \Delta g = 0$, for $g \in C^2(X,\mathbb{R})$, X compact)

Corollary 5.1 ([ES]):

Let $f : X \to Y$ *be harmonic,* X *compact,* $\partial X = \phi$, $\mathrm{Ric}^X \geq 0$, *and let* Y *have nonpositive sectional curvature.*

Then f *is totally geodesic (i.e.* $\nabla\, df \equiv 0$*) and has constant energy density* e(f). *If* $\mathrm{Ric}^X > 0$ *for at least one point in* X, f *is constant. If the sectional curvature of* Y *is negative,* f(X) *is onedimensional, hence a point or a closed geodesic.*

Theorem 2.5 and Corollary 5.1 directly imply the wellknown result that a nonpositively curved compact Riemannian manifold Y is a $K(\pi_1(Y), 1)$ space.

Corollary 5.2:

Let Y *be a nonpositively curved compact Riemannian manifold. Then*

$$\pi_k(Y) = 0 \quad \text{for } k \geq 2. \tag{5.2}$$

Proof:

Let $g : S^k \to Y$ be continuous. By Theorem 2.5, g is homotopic to a harmonic map $f : S^k \to Y$. By Corollary 5.1, f is constant. Hence g is homotopic to a constant map, and $\pi_k(Y) = 0$ for $k \geq 2$.

q.e.d.

In a similar way, we deduce Preissmann's theorem

Corollary 5.3:

If Y *is a negatively curved compact Riemannian manifold, then every abelian subgroup of* $\pi_1(Y)$ *is infinite cyclic.*

Proof:

Let a, b be commuting elements of $\pi_1(Y)$. The homotopy between ab and ba can be considered as a continuous map $g: T^2 \to Y$, where T^2 is a two-dimensional flat torus. By Theorem 2.5 g is homotopic to a harmonic map $f : T^2 \to Y$. By Corollary 5.1, $f(T^2)$ is contained in a closed geodesic. Hence both a and b are homotopic to some multiple of this geodesic. We did not preserve the base point in this argument, but one easily convinces oneself that this does not affect the reasoning. Finally, Theorem 4.1 implies that a multiply covered nontrivial closed geodesic is not homotopic to zero. Thus, $\pi_1(Y)$ does not contain elements of finite order (cf. Theorem 2.1 showing that each element of $\pi_1(Y)$ can be represented by a closed geodesic; in the present case also Theorem 2.5 applies).

Similarly, cf. [J7], one can also show a more general result of Gromoll-Wolf ([GW]) and Lawson-Yau ([LY]):

Theorem 5.1:
Let Y be a compact Riemannian manifold of nonpositive sectional curvature. Suppose that $\Gamma := \pi_1(Y)$ *contains a solvable subgroup* Γ_o. *Then Y contains an isometric immersion of a compact flat manifold X with* $\pi_1(X) = \Gamma_o$. *In particular,* Γ_o *is a Bieberbach group.*

6.
As already mentioned, holomorphic mappings between Kähler manifolds are harmonic. If conversely, one seeks holomorphic maps, one can try (by the scheme of § 5) to show first the existence of a harmonic map and then that - under additional assumptions - this harmonic map is actually holomorphic.

According to our previous considerations, such an assumption would consist in an negativity condition on the image curvature. If one wants to compare a Kähler manifold X with a negatively curved Kähler manifold Y, it turns out, however, that formula (5.1) is not very well suited for this purpose. Namely, one would have to require that X has nonnegative Ricci curvature, whereas naturally one would rather like to treat the case of an X which is negatively curved, too.

Therefore, let us consider the derivation of (5.1). In local coordinates (cf. (1.1))

$$e(f) = \frac{1}{2}\,\gamma^{\alpha\beta}(x)\;g_{ij}(f(x))\;\frac{\partial f^i}{\partial x^\alpha}\,\frac{\partial f^j}{\partial x^\beta} \tag{6.1}$$

If one computes $\Delta e(f)$, no third derivatives occur because f is harmonic. Second derivatives of g_{ij} yield the term with R^Y, whereas the Ricci term comes from the second derivatives of $\gamma^{\alpha\beta}$. Thus, this term will disappear if one does not contract as in (6.1) with $\gamma^{\alpha\beta}$ but rather looks at a 2-form (instead of the function e(f)) formed by first derivatives of f.

At least in the Kähler context this leads to a useful formula for a harmonic map $f : X \to Y$, discovered by Siu ([Si1])

$$\begin{aligned}\partial\bar\partial(g_{i\bar j}\;\bar\partial f^i\wedge\partial\overline{f^j}) &= R_{i\bar j k\bar l}\;\bar\partial f^i\wedge\partial\overline{f^j}\wedge\partial f^k\wedge\overline{\partial f^l}\\ &\quad - g_{i\bar j}\;D\,\bar\partial f^i\wedge\bar D\;\partial\overline{f^j}\end{aligned} \tag{6.2}$$

with

$$\bar\partial f^i := \frac{\partial f^i}{\partial z^\alpha}\, \overline{dz^\alpha}$$

$$D\bar\partial f^i := \partial\bar\partial f^i + \Gamma^i_{jk}\, \partial f^j \wedge \bar\partial f^k$$

$$(R_{i\bar j k\bar l}) := \text{curvature tensor of } Y$$

Let ω be the Kähler form of X, $n := \dim_{\mathbb{C}} X$. One then gets (the Kähler condition is $\bar\partial\omega = 0 = \partial\omega$)

$$\text{(6.3)} \qquad \partial\bar\partial\, (g_{i\bar j}\, \bar\partial f^i \wedge \overline{\partial f^j} \wedge \omega^{n-2})$$

$$= R_{i\bar j k\bar l}\, \bar\partial f^i \wedge \overline{\partial f^j} \wedge \partial f^k \wedge \overline{\partial f^l} \wedge \omega^{n-2} - g_{i\bar j}\, D\, \bar\partial f^i \wedge \bar D\, \partial f^j \wedge \omega^{n-2}$$

Integrating (6.3) over X (compact, $\partial X = \phi$), the integral over the left hand side and hence also the one over the right hand side vanishes. On the other hand, since f is harmonic

$$\text{(6.4)} \qquad g_{i\bar j}\, D\, \bar\partial f^i \wedge \bar D\, \partial f^j \wedge \omega^{n-2} = a\omega^n$$

with a nonnegative function a.

Thus, one only has to check under which conditions

$$\text{(6.5)} \qquad R_{i\bar j k\bar l}\, \bar\partial f^i \wedge \overline{\partial f^j} \wedge \partial f^k \wedge \overline{\partial f^l} \wedge \omega^{n-2} = b\, \omega^n$$

with $b \leq 0$, and furthermore, which consequences one can draw from the vanishing of (6.4) and (6.5).

This was carried out in [Si1]; the most important special case of his result is

Theorem 6.1:
Let D be an irreducible hermitian symmetric bounded domain in $\mathbb{C}^N$, $N \geq 2$, Y a compact quotient of D by a fix point free discrete group of automorphisms. Let X be a compact Kähler manifold which is homotopically equivalent to Y. Then X and Y are (±) biholomorphically equivalent.

First of all, Y is nonpositively curved. Therefore, a homotopy equivalence $g : X \to Y$ by Theorem 2.5 is homotopic to a harmonic homotopy equivalence $f : X \to Y$. Furthermore, one can show that (6.5) indeed holds with a nonpositive b (Y is called "strongly seminegatively curved"). This is a stronger condition than the nonpositivity of the sectional curvature, as the arguments of the curvature tensor in (6.3) have fewer symmetries than in (6.1), as we did not contract with the inverse metric tensor of the domain. In any case, the curvature tensor of Y as in Theorem 6.1 does satisfy this stronger requirement. Hence both (6.4) and (6.5) vanish identically. The vanishing of (6.5) means

$$D\bar{\partial} f^i \equiv 0 \quad \text{for } i = 1,\dots \dim Y,$$

Since f is a homotopy equivalence and $\dim_{\mathbb{R}} X \geq 4$, df has real rank at least Y on an open subset of X, and algebraic considerations then show that in this case the vanishing of (6.4) implies that ∂f or $\bar{\partial} f$ vanishes. Since f is harmonic, one can then conclude by unique continuation that $\partial f \equiv 0$ or $\bar{\partial} f \equiv 0$ in X. Thus, f is ± holomorphic, and, as a homotopy equivalence, therefore also bijective.

So much for a sketch of Siu's argument.

Theorem 6.1 generalizes Mostow's rigidity theorem. Its importance lies in the fact that, in contrast to Mostow's theorem, one need not require that X is locally symmetric, but only that it is Kählerian.

On the other hand, Mostow's rigidity theorem also holds for locally symmetric spaces of finite volume, not necessarily compact. For spaces of rank 1 this was proved by Prasad ([Pr]), whereas the case $n \geq 2$ is one of the celebrated results of Margulis ([Mg]). This raises the question, whether theorem 6.1 remains true in this more general situation. This was investigated by Jost-Yau ([JY3], [JY4]); one has

<u>Theorem 6.2:</u>

Let D be an irreducible hermitian symmetric bounded domain in $\mathbb{C}^N$, $N \geq 2$; let $Y = D/\Gamma$ (Γ a discrete group of automorphisms of D) have finite volume (w.r.t. to the induced (Kähler) metric of D).

Let X have a finite cover X' that is a Kähler manifold. Assume that X' is quasiprojective, i.e. that X' has a projective algebraic compactification $\bar{X}$, for which $\bar{X}\backslash X'$ is an algebraic subvariety.

In case rank $D \geq 2$, we also assume that the complex codimension of $S := \bar{X}\backslash X'$ in $\bar{X}$ is at least 3. If X is properly homotopically equivalent to Y (X then is a quotient of a contractible Kähler manifold by a group isomorphic to Γ), then X is ± biholomorphically equivalent to Y.

It is still unknown whether the assumption $\operatorname{codim}_{\mathbb{C}} S \geq 3$ (in case rank $D \geq 2$) is really necessary. In any case, it is satisfied for the Satake compactification of all spaces $Y = D/\Gamma$, rank $D \geq 2$, i.e. all spaces to which Margulis' result applies, with the sole exception of quotients of the Siegel upper half plane of degree 2. In this case

$\dim_{\mathbb{C}} D = 3$, and quotients of finite volume have a compactifying divisor of dimension 1, hence of codimension 2.

The assumption is needed in the proof, in order to construct a homotopy equivalence of finite energy between X and Y. This being achieved, one can also find a harmonic homotopy equivalence f of finite energy, and one can apply the arguments of the proof of Theorem 6.1 to f. (The noncompactness requires some additional aguments.)

Let us also state the following result of [JY4]):

<u>Theorem 6.3:</u>
Let M_p be the moduli space of Riemann surface of genus p, i.e. $M_p = T_p/\Gamma_p$ (T_p = Teichmüller space, Γ_p = modular group), $p \geq 2$.
Let X satisfy the assumptions of Theorem 6.2 i.e. let X be quasiprojective with a compactifying divisor of codimension at least 3.
If X is properly homotopically equivalent to M_p, then also ± biholomorphically equivalent. Thus, X is also a quotient of Teichmüller space by the modular group.

The assumption of a compactification by a divisor of codimension at least 3 is satisfied for M_p, if $p \geq 3$. If $p = 2$, the codimension is again 2.

Finally, we want to discuss the case $D = H^N$ ($H: = \{z \in \mathbb{C}: z = + i\, y, y > 0\}$). If Y is a quotient of H^N, and $f : X \to Y$ is a harmonic map from a Kähler manifold X, one can again deduce the vanishing of (6.4) and (6.5). The vanishing of (6.5), however, no longer implies that f is ± holomorphic. Nevertheless

<u>Lemma 6.1:</u> ([JY1]):
Let X be a Kähler manifold, $Y = H^N/\Gamma$, ($\Gamma \subset \mathrm{Aut}(H^N)$), $f: X \to Y$ a harmonic map of finite energy. If $\tilde{f} = (\tilde{f}^1,\ldots,\tilde{f}^N) : \tilde{X} \to H^N$ is the lift of f to universal covers, $z_o \in X$, $\partial\tilde{f}^i(z_o) \neq 0$ (for some $i \in \{1,\ldots,N\}$), then a neighborhood of z_o is holomorphically fibered by level surfaces of $\tilde{f}^i$.

Although the map f in general need not be holomorphic, we see that the level surfaces of the components of f are unions of analytic hypersurfaces. The work of Jost-Yau ([JY1], [JY2], [JY3]) and Mok ([M1], [M2]) then yields

$\dim_{\mathbb{C}} D = 3$, and quotients of finite volume have a compactifying divisor of dimension 1, hence of codimension 2.

The assumption is needed in the proof, in order to construct a homotopy equivalence of finite energy between X and Y. This being achieved, one can also find a harmonic homotopy equivalence f of finite energy, and one can apply the arguments of the proof of Theorem 6.1 to f. (The noncompactness requires some additional aguments.)

Let us also state the following result of [JY4]):

<u>Theorem 6.3:</u>
Let M_p be the moduli space of Riemann surface of genus p, i.e. $M_p = T_p/\Gamma_p$ (T_p = Teichmüller space, Γ_p = modular group), $p \geq 2$.
Let X satisfy the assumptions of Theorem 6.2 i.e. let X be quasiprojective with a compactifying divisor of codimension at least 3.
If X is properly homotopically equivalent to M_p, then also $\pm$ biholomorphically equivalent. Thus, X is also a quotient of Teichmüller space by the modular group.

The assumption of a compactification by a divisor of codimension at least 3 is satisfied for M_p, if $p \geq 3$. If $p = 2$, the codimension is again 2.

Finally, we want to discuss the case $D = H^N$ ($H := \{z \in \mathbb{C} : z = + i y, y > 0\}$). If Y is a quotient of H^N, and $f : X \to Y$ is a harmonic map from a Kähler manifold X, one can again deduce the vanishing of (6.4) and (6.5). The vanishing of (6.5), however, no longer implies that f is $\pm$ holomorphic. Nevertheless

<u>Lemma 6.1:</u> ([JY1]):
Let X be a Kähler manifold, $Y = H^N/\Gamma$, ($\Gamma \subset \mathrm{Aut}(H^N)$), $f: X \to Y$ a harmonic map of finite energy. If $\tilde{f} = (\tilde{f}^1, \ldots, \tilde{f}^N) : \tilde{X} \to H^N$ is the lift of f to universal covers, $z_o \in X$, $\partial \tilde{f}^i(z_o) \neq 0$ (for some $i \in \{1, \ldots, N\}$), then a neighborhood of z_o is holomorphically fibered by level surfaces of $\tilde{f}^i$.

Although the map f in general need not be holomorphic, we see that the level surfaces of the components of f are unions of analytic hypersurfaces. The work of Jost-Yau ([JY1], [JY2], [JY3]) and Mok ([M1], [M2]) then yields

Theorem 6.4:

i) *Let* $Y = H^N/\Gamma$ (Γ *a discrete subgroup of* Aut (H^N)) *be compact*, X *a compact Kähler manifold homotopically equivalent to* Y. *Then the universal cover of* X *is* H^N, *and if* Γ *is irreducible*, X *is* ± *biholomorphically equivalent to* Y.

ii) *Let* $Y = H^N/\Gamma$ *have finite volume*, Γ *irreducible*, X *quasiprojective and properly homotopically equivalent to* Y. *Then* X *is* ± *biholomorphically equivalent to* Y.

If Γ is not irreducible, Y may be a product of Riemann surfaces. As these admit deformations, in this case a harmonic homotopy equivalence need not be holomorphic. Theorem 6.4.i) however tells us that deformations of Y can only arise through deformations of the individual factors.

In similar vein, some restrictions on harmonic maps from a Kähler manifold into a Riemannian manifold were obtained by Sampson ([Sa2]).

7.

In Theorem 6.3, we already saw an application of harmonic maps to Teichmüller theory. Since the elements of the moduli space M_p are Riemann surfaces of genus p, it is also natural to use harmonic maps between surfaces for studying the structure of Teichmüller space. This was carried out by Wolf ([W]) and the author ([J8]).

For this, let S be a compact topological surface (i.e. a two-dimensional manifold) of genus p. In order to exclude trivial cases, let $p \geq 2$.

For each conformal structure g on S there exists a unique hyperbolic metric which we shall denote by the same letter g. Let us recall the relevant results of § 2 on harmonic maps between surfaces:

Lemma 7.1:

Let (S,g), (S,γ) *be hyperbolic structures on* S. *Then there exists a unique harmonic map*

$$u : (S,g) \to (S,\gamma)$$

homotopic to the identity of S.

u *is a diffeomorphism.*

$$\rho^2(u)\, u_z\, \bar{u}_z\, dz^2 \tag{7.1}$$

is a holomorphic quadratic differential on S, *where* $\rho^2(u)\, du\, d\bar{u}$ *is a local representation of the hyperbolic metric* γ, *and* z *is a conformal parameter on* (S,g).

Lemma 7.1 in particular yields a map

$$q_g : T_p \to Q(g)$$

of Teichmüller space T_p into the space Q(g) of holomorphic quadratic differentials on (S,g). One then shows ([W], [J8]) that q_g is bijective and that the transformations $q_{g_1} \circ (q_{g_2})^{-1}$ are diffeomorphisms. Since

$$\dim_{\mathbb{C}} Q(g) = 3p - 3$$

by Riemann-Roch, one deduces "Teichmüller's theorem" that T_p is diffeomorphic to $\mathbb{R}^{6p-6}$.

Although the map q_g itself is not holomorphic, studying its properties more closely one can nevertheless obtain the canonical complex structure of T_p. Moreover, looking at the variations of the energy of the harmonic map u induced by variations of g and γ, one can also recover the Weil-Petersson metric of T_p and compute its curvature (this was first with a different method done by Tromba ([T]), and also by Wolpert ([Wp]), and, in a more general context by Siu ([Si])). Finally, Wolf ([W]) showed that by adding the sphere at infinity to Q(g), one recovers Thurston's compactification of T_p.

Thus, one can obtain all the basic structures of T_p by using harmonic maps rather than quasiconformal maps as in previous investigations of T_p. Compared to quasiconformal maps, harmonic maps have the advantage that they are defined by an integral instead of a pointwise variational principle and therefore have better regularity properties.

References

[Ba] Baldes, A., Stability and uniqueness properties of the equator map from a ball into an ellipsoid, Mz 185 (1984),505/516

[BC] Brezis, H., and J.M. Coron, Large solutions for harmonic maps in two dimensions, Comm. Math. Phy. 92, 1983, 203-215

[Ci] Choi, H.J., On the Liouville theorem for harmonic maps, Preprint

[EL1] Eells, J., and L. Lemaire, A report on harmonic maps, Bull London Math. Soc. 10 (1978), 1-68

[El2] Eells, J., and L. Lemaire, Selected Topics in Harmonic Maps, CBMS Regional Conf. Ser. 50 (1983)

[ES] Eells, J., and J.H. Sampson, Harmonic Mappings of Riemannian Manifolds, Am. J. Math. 86 (1964), 109-160

[EW] Eells, J., and J.C. Wood, Restrictions on Harmonic Maps of Surfaces, Top. 15 (1976), 263-266

[G] Giaquinta, M., Multiple integrals in the calculus of variations and nonlinear elliptic systems, Ann. Math. Studies, Princeton, 1983

[GG1] Giaquinta, M., and Giusti, E., On the regularity of the minima of variational integrals, Acta Math. 148 (1982), 31-46

[GG2] Giaquinta, M., and Guisti, E., The singular set of the minima of certain quadratic functionals, Ann. Sc. Norm. Sup. Pisa 11 (1984), 45-55

[GH] Giaquinta, M., and S. Hildebrandt, A priori estimates for harmonic mappings, J. Reine Angew. Math. 336 (1982), 124-164

[GS] Giaquinta, M., and J. Soucek, Harmonic maps into a hemisphere,

[GWu] Greene, R., and H.-H. Wu, Lipschitz convergence of Riemannian manifolds

[GW] Gromoll, D. and J. Wolf, Some relations between the metric structure and the algebraic structure of the fundamental group in manifolds of nonpositive curvature, Bull. AMS 77 (1971), 545-552

[Gv] Gromov, M., Structures métriques pour les variétés riemanniennes, redigeés par J. Lafontaine u. P. Pansu, Cedic-Fernand Nathan, Paris, 1981

[Hm] Hamilton, R., Harmonic maps of manifolds with boundary, L.N.M. 471, Springer, Berlin, Heidelberg, New York, 1975

[HKL] Hardt, R., D. Kinderlehrer, and F.-H. Lin, Existence and partial regularity of static liquid crystal configurations

[HL] Hardt, R., and F.-H. Lin, Mappings minimizing the L^p norm of the gradient

[Ht] Hartmann, P., On homotopic harmonic maps, Can. J. Math. 19 (1967) 673-687

[Hz] Heinz, E., Zur Abschätzung der Funktionaldeterminante bei einer Klasse topologischer Abbildungen, Nachr. Akad. Wiss. Gött. (1968), 183-197

[HJW] Hildebrandt, S., J. Jost, and K.-O. Widman, Harmonic Mappings and Minimal Submanifolds, Inv. math. 62 (1980), 269-298

[HKW1] Hildebrandt, S., H. Kaul, and K.-O. Widman, Harmonic Mappings into Riemannian Manifolds with Non-positive Sectional curvature, Math. Scand. 37 (1975), 257-263

[HKW2] Hildebrandt, S., H. Kaul and K.-O. Widman, Dirichlet's Boundary Value Problem for Harmonic Mappings of Riemannian Manifolds, M. Z. 147 (1976), 225-236

[HKW3] Hildebrandt, S., H. Kaul, and K.-O. Widman, An Existence Theorem for Harmonic Mappings of Riemannian Manifolds Acta Math. 138 (1977) 1-16

[HW1] Hildebrandt, S., and K.-O. Widman, Some Regularity Results for Quasilinear Elliptic Systems of Second Order, Math. Z. 142 (1975), 67-86

[HW2] Hildebrandt, S., and K.-O. Widman, On the Hölder Continuity of Weak Solutions of Quasilinear Elliptic Systems of Second Order, Ann. Sc. N. Sup. Pisa IV (1977), 145-178

[JäK1] Jäger, W., and H. Kaul, Uniqueness and stability of harmonic maps and their Jacobi fields, Man. Math. 28 (1979), 269-291

[JäK2] Jäger, W., and H. Kaul, Rotationally symmetric harmonic maps from a ball into a sphere and the regularity problem for weak solutions of elliptic systems, J. Reine Angew. Math. 343 (1983), 146-161

[J1] Jost, J., Univalency of Harmonic Mappings between Surfaces, Journ. Reine Angew. Math. 324, 141-153

[J2] Jost, J., Existence Proofs for Harmonic Mappings with the Help of a Maximum Principle, M. Z. 184 (1983), 489-496

[J3] Jost, J., The Dirichlet problem for harmonic maps from a surface with boundary onto a 2-sphere with non-constant boundary values, J.Diff. Geom. 19 (1984), 393-401

[J4] Jost, J., Harmonic maps between surfaces, Springer Lecture Notes in Math., 1062 (1984)

[J5] Jost, J., Harmonic mappings between Riemannian manifolds, Proc. CMA, Vol 4, ANU-Press, Canberra, 1984

[J6] Jost, J., On the existence of Harmonic maps from a surface into the real projective plane, Comp. Math. 59 (1986), 15-19

[J7] Jost, J., Nonlinear methods in complex geometry, Birkhäuser (Reihe: DMV-Seminare) to appear

[J8] Jost, J., Two-dimensional geometric variational problems

[JK] Jost, J., and H. Karcher, Geometrische Methoden zur Gewinnung von a-priori-Schranken für harmonische Abbildungen, Man. Math. 40 (1982), 27-77

[JM] Jost, J., and M. Meier, Boundary regularity for minima of certain quadratic functionals, Math. Ann. 262 (1983), 549-561

[JS] Jost, J., and R. Schoen, On the existence of harmonic diffeomorphisms between surfaces, Inv. Math. 66 (1982), 353-359

[JY1] Jost, J., and S.T. Yau, Harmonic mappings and Kähler manifolds, Math. Ann. 262 (1983), 145-166

[JY2] Jost, J., and S.T. Yau, A strong rigidity theorem for a certain class of compact complex analytic surfaces, Math. Ann. 271 (1985), 143-152

[JY3] Jost, J., and S.T. Yau, The strong rigidity of locally symmetrics complex manifolds of rank one and finite volume, Math. Ann. 275 (1986), 291-304

[JY4] Jost, J. and S.T. Yau, On the rigidity of certain discrete groups and algebraic varieties, Math. Ann., to appear

[Kl] Klingenberg, W., Lectures on closed geodesic, Springer, Berlin etc., 1978

[LY] Lawson, B., and S.T. Yau, Compact manifolds of nonpositive curvature, J. Diff. Geom. 7 (1972), 211-228

[L] Lemaire, L., Applications harmoniques de surfaces Riemanniennes, J. Diff. Geom. 13 (1978), 51-78

[Lu] Luckhaus, St., Partial Hölder continuity for minima of certain energies among maps into Riemannian manifold

[Mg] Margulis, G., Discrete groups of motions of manifolds of nonpositive curvature (in Russian), Proc. Int. Cong. Math., Vancouver, 21-34 (1974)

[MY] Meeks, W., and S.T. Yau, Topology of three dimensional manifolds and the embedding problems in minimal surface theory, Ann. Math. 112, 441-483 (1982)

[M1] Mok, N., The holomorphic or antiholomorphic character of harmonic maps into irreducible compact quotients of polydisks, Math. Ann. 272 (1985), 197-216

[M2] Mok, N., La rigidité forte des quotients compacts des polydisques en terme des groupes fondamentaux,

[Mo] Mostow, G., Strong rigidity of locally symmetric spaces, Ann. Math. Stud. 78, Princeton, 1973

[P1] Peters, St., Konvergenz Riemannscher Mannigfaltigkeiten, Dissertation, Bonn, 1986

[P2] Peters, St., On the sharpness of a $C^{1,\alpha}$-result for limit metrics,

[Pr] Prasad, G., Strong rigidity of Q-rank 1 lattices, Inv. math. 21 (1973), 255-286

[RV] Ruh, E.A., and J. Vilms, The Tension Field of the Gauss Map, Trans. A.M.S. 149 (1970), 569-573

[SkU] Sacks, J., and K. Uhlenbeck, The Existence of Minimal Immersions of 2-Spheres, Ann. Math. 113 (1981), 1-24

[Sa1] Sampson, J., Some Properties and Applications of Harmonic Mappings, Ann. Sc. Ec. Sup. 11 (1978), 211-228

[Sa2] Sampson, J., Applications of harmonic maps to Kähler geometry, Contemp. Math. 49 (1986), 125-134

[SU1] Schoen, R., and K. Uhlenbeck, A regularity theory for harmonic maps, J. Diff. Geom. 17 (1982), 307-335

[SU2] Schoen, R., and K. Uhlenbeck, Boundary regularity and miscellaneous results on harmonic maps, J. Diff. Geom. 18 (1983), 253-268

[SU3] Schoen, R., and K. Uhlenbeck, Regularity of minimizing harmonic maps into the sphere, Inv. math. 78 (1984), 89-100

[SY1] Schoen, R., and S.T. Yau, On univalent harmonic maps between surfaces, Inv. Math. 44 (1978), 265-278

[SY2] Schoen, R., and S.T. Yau, Existence of incompressible minimal surfaces and the topology of three dimensional manifolds with non-negative scalar curvature, Ann. Math. 110 (1979), 127-142

[SY3] Schoen, R., and S.T. Yau, Compact group actions and the topology of manifolds with non-negative curvature Top. 18 (1979), 361-380

[Si1] Siu, Y.T., The complex analyticity of harmonic maps and the strong rigidity of compact Kähler manifolds, Ann. Math. 112 (1980), 73-111

[Si2] Siu, Y.T., Curvature of the Weil-Petersson metric in the moduli space of compact Kähler-Einstein manifolds of negative first Chern class

[SiY] Siu, Y.T., and S.T. Yau, Compact Kähler manifolds of positive bisectional curvature. Inv. math. 59, 189-204 (1980)

[Sp] Sperner, E., A priori gradient estimates for harmonic mappings, SFB 72 - Preprint 513, Bonn, 1982

[St] Struwe, M., On the evolution of harmonic mappings of Riemann surfaces, Comm. Math. Helv. 60 (1985), 558-581

[T] Tromba, A., On a natural algebraic affine connection on the space of almost complex structures and curvature of Teichmüller space with respect to its Weil-Petersson metric, manuscripta math.

[Wh] White, B., Homotopy classes in Sobolev spaces of mappings

[W] Wolf, M., The Teichmüller theory of harmonic maps, Stanford, 1986

[Wp] Wolpert, S., Chern forms and the Riemann tensor for the moduli space of curves, Inv. Math.

ASYMPTOTIC BEHAVIOR OF SOLUTIONS OF SOME QUASILINEAR ELLIPTIC SYSTEMS IN EXTERIOR DOMAINS

Michael Meier

Introduction

Quasilinear elliptic systems of diagonal form[1)]

$$(0.1) \qquad - D_\alpha (a^{\alpha\beta}(x)\, D_\beta\, u^i) = F^i(x,u,\nabla u) \qquad (i=1,\ldots,N),$$

whose nonlinearity $F = (F^1,\ldots,F^N)$ grows quadratically with respect to the derivatives ∇u, have been of considerable interest for more than a decade. Examples of (0.1) arise, for instance, in physics and in differential geometry. We only mention that the local coordinate representations of a harmonic map between Riemannian manifolds satisfy equations of this type. For further examples, the reader is referred to the surveys [20] and [21] by Hildebrandt.

On the one hand, it was found that solutions u of (0.1) may exhibit a behavior drastically different from that of solutions h of the corresponding linear equation

$$(0.2) \qquad - D_\alpha (a^{\alpha\beta}(x)\, D_\beta h) = 0.$$

In particular, bounded weak solutions of (0.1) need not be continuous ([9], [14]), and neither Liouville theorems nor a priori bounds for the Hölder norm are generally valid for classical solutions ([16], [28]). The celebrated function $u(x) = x|x|^{-1}$ exemplifies the first fact for the simple system $-\Delta u = u|\nabla u|^2$, $n \geq 3$.

On the other hand, various properties of solutions u and h of (0.1) and (0.2) turned out to agree when the supremum of $|u|$ is sufficiently small (and sometimes even if only a one-sided condition for $u \cdot F$ is satisfied). For instance, we mention the regularity results and a priori estimates due to Wiegner and to Hildebrandt-Widman, as well as Liouville theorems, maximum principles, and removable singularity results by various authors. See, e.g., [4], [11], [14]-[19], [23], [24], [28], [30]-[32], [43]-[46]; a more detailed discussion of the literature is given in [21] and [22].

In the present paper, we wish to point out another close similarity

---

1) Here and in the sequel, $x = (x^1,\ldots,x^n)$, $n \geq 2$, $u = u(x)$, and the summation convention with respect to α, $\beta = 1,\ldots,n$ is adopted.

between (0.1) and (0.2), concerning the asymptotic behavior of solutions in an exterior domain. Specifically, we shall derive some analogues of a well known theorem asserting that any bounded solution $h = h(x)$ of a uniformly elliptic equation (0.2) in $\{|x| > 1\}$ tends to a limit as $x \to \infty$ (cf. [12], [33], [37]).

We have chosen this topic for several reasons. First of all, it provides the opportunity to describe, in a new context, some of the more recent ideas introduced by Caffarelli [4] and the author [31], [32] in the study of elliptic systems (0.1) and of harmonic maps, the main tool being Moser's Harnack inequality for weak supersolutions to (0.2). Actually, we shall present some modifications and refinements here that yield results not yet previously established. Moreover, it is possible to include, without any additional effort, certain classes of *non-uniformly elliptic systems*. Although we only treat some simple kinds of degeneracy, our theorems turn out to be applicable, for example, to harmonic maps on arbitrary radially symmetric graphs. We found it to be beyond the scope of this report, however, to incorporate appropriate results from [6], [41] concerning weighted coefficients of a rather general nature.
Finally, we shall demonstrate that our limit theorems can be used effectively to deduce Liouville type theorems under hypotheses much weaker than needed for the standard approach via a priori estimates combined with a blow-down argument [23], [18].

Our paper is organized as follows.
Section 1 serves as an introduction to the problem. Here we discuss the main results in the simple case of a scalar equation and describe possible extensions and applications. Also the assumptions made in the sequel are motivated and stated.
Section 2 deals with bounded solutions of diagonal elliptic systems (0.1), while in Section 3 harmonic maps of Riemannian manifolds are considered. In addition to a limit theorem, we further include an application to *minimal submanifolds* of higher codimension, some generalization of a *Liouville theorem* due to Hildebrandt-Jost-Widman [18], as well as an estimate for the *growth of unbounded harmonic maps* into manifolds of non-positive sectional curvature.

It is a pleasure to acknowledge the continuous support of the Sonderforschungsbereich 72. Moreover, I wish to express my gratitude to Professor Stefan Hildebrandt for his helpful interest and advice over many years.

1. Scalar equations

As was observed by Gilbarg and Serrin [12], Harnack's inequality for elliptic equations can be used to study the behavior of solutions in the neighborhood of infinity and near isolated singular points. Later on, Moser [33] and Serrin [37] established Harnack type inequalities under very general structural assumptions, whence the results of [12] also extended to a rather broad class of equations. In the linear case, the following theorem was obtained, cf. [33, Theorem 5], [37].

THEOREM 1.1.

Let $a^{\alpha\beta}(x)$ *be bounded measurable coefficients in* $\Omega_1 := \{x \in \mathbb{R}^n : |x| > 1\}$. *Suppose also that the equation*

$$-D_\alpha(a^{\alpha\beta}(x)\, D_\beta u) = 0 \tag{1.1}$$

is uniformly elliptic.

If $u \in H^1_{2,loc}(\Omega_1)$ *is a bounded weak solution of* (1.1) *in* Ω_1 *then* $\lim_{x\to\infty} u(x) = \bar{u}_\infty$ *exists.*

REMARK.

In addition, it was noticed in [33, p. 589, footnote] that, when $n > 2$,

$$|u(x) - \bar{u}_\infty| = O(|x|^{2-n}) \text{ as } |x| \to \infty. \tag{1.2}$$

For a proof of (1.2), cf. [38, p.84][2]. There Serrin further showed that the expression $|u(x) - \bar{u}_\infty|$ can be estimated from below by a positive multiple of $|x|^{2-n}$ provided that the strong maximum principle at infinity is violated. We emphasize that Theorem 1.1 does *not* constitute a removable singularity result in this case. The problem of determining the rate of convergence had been studied previously by Finn and Gilbarg [8, Theorem 1] in connection with compressible, irrotational, subsonic fluid flow whose velocity components satisfy equations of the form (1.1); the authors deduced an improved decay rate for $u(x)-\bar{u}_\infty$ from the vanishing of a certain outflow integral. We do not want to give a more detailed discussion of this question here but refer to [39], in particular Theorems 5 through 8, for further information.

Theorem 1.1 immediately applies to *bounded harmonic functions* on subdomains G of Riemannian manifolds, provided one can introduce coordinates $x \in \Omega_1$ in G with respect to which the eigenvalues of the metric tensor are uniformly bounded away from 0 and ∞. An important example is furnished by the graph G of any uniformly Lipschitz

---

[2] See also the end of Section 2 below.

continuous scalar (or vector-valued) function on Ω_1.

The following application of Theorem 1.1 to *non-parametric minimal surfaces* of codimension one was already given by Moser [33, Theorem 6].

COROLLARY 1.2.

Let f be a solution of the minimal surface equation

$$(1.3) \qquad D_\alpha\left(\frac{D_\alpha f}{\sqrt{1+|\nabla f|^2}}\right) = 0 \text{ in } \Omega_1 = \{x \in \mathbb{R}^n : |x| > 1\}.$$

If $|\nabla f|$ *is bounded in* Ω_1, *then* $\nabla f(x)$ *approaches a limit as* $x \to \infty$.

We remark that the boundedness of ∇f *follows* from (1.3) if $n = 2$, by virtue of a well known result due to Bers [2]. This was recently generalized to dimensions $3 \leq n \leq 7$ by Simon [40]. Incidentally, for $n \geq 3$, Corollary 1.2 yields the existence of a tangent hyperplane to the graph of f at infinity. An extension of Corollary 1.2 to the case of higher codimensions will be presented in Section 3.

Next, we shall address the question whether Theorem 1.1 also holds for certain *non-uniformly elliptic equations*.

By modifying a counter-example due to Finn [7, p. 394] it can be seen that the answer is negative in general. In fact, let $\nu = n-1 \geq 1$, $\tau > 0$, $r = |x|$ and $u(x) = (1 + r^{-\tau}) \dfrac{x_1}{r}$,

$a^{\alpha\beta}(x) = \delta^{\alpha\beta} r^{(1+\tau)(\tau-\nu)/\tau} \cdot \exp(-\nu\tau^{-2} r^\tau)$.

Then (1.1) holds in Ω_1. Moreover, u and the coefficients $a^{\alpha\beta}$ are bounded, while the ellipticity constant tends to zero as $x \to \infty$.

On the other hand, an inspection of the proof of Theorem 1.1 shows that uniform ellipticity of (1.1) need only be assumed to hold on annular regions of the form, e.g., $\{x \in \mathbb{R}^n : R < |x| < 4R\}$ which we will denote by $(R|4R)$ in the sequel. Indeed, it is a sequence of domains $(R_k|4R_k)$, $R_k \to \infty$, on which Harnack's inequality is applied, while the ordinary maximum principle can be used to treat the regions inbetween (see the proof of Theorem 1.3 below). We therefore consider a class of admissible coefficients $a^{\alpha\beta}$ with the property

$$|a^{\alpha\beta}(x)| \leq w(x),$$

$$\text{(A)} \qquad Q(x,\xi) := a^{\alpha\beta}(x)\, \xi_\alpha \xi_\beta \geq \lambda\, w(x)\, |\xi|^2$$

$$\text{for every } \xi \in \mathbb{R}^n \text{ and a.e. } x \in \Omega_1.$$

Here, λ is a positive constant, while the measurable weight w is positive and finite a.e. and satisfies

(W) $\qquad \sup_{(R|4R)} w \le \lambda^{-1} \inf_{(R|4R)} w \qquad$ for all $R > 1$.

Although part of our subsequent discussion would require the validity of (W) only for some sequence $R = R_k \to \infty$, we prefer to impose the uniform condition above.

Note that all sums, products, and quotients of weights with the property (W) again satisfy (W), after suitable redefinition of λ. Simple examples are given by the weights $w(x) = |x|^\tau(\log 2|x|)^\sigma$ with σ, $\tau \in \mathbb{R}$, etc. We also remark that equation (1.1) in Ω_1 with coefficients satisfying (A) for $w(x) = |x|^\tau$, $\tau > 2-n$ (resp. $\tau < 2-n$), can be transformed into a uniformly elliptic equation in Ω_1 (resp. in a punctured ball) by means of a simple coordinate transformation of the independent variables x (cf. [6], [1]). The "limiting" weight $|x|^{2-n}$ naturally occurs, e.g., when considering the Laplace-Beltrami operator of a cylinder (see Example 1.4). Finally we notice that all of our results concerning the behavior of solutions at infinity immediately imply *corresponding theorems near isolated point singularities*, the concept of admissibility on punctured balls being defined analogous to (W).
Let us now state an extension of Theorem 1.1.

<u>THEOREM 1.3.</u>
Assume that the coefficients $a^{\alpha\beta}(x)$ *satisfy* (A), (W), *and let* $u \in H^1_{2,loc}(\Omega_1,\mathbb{R})$ *be a bounded weak solution of*

$$(1.4) \qquad - D_\alpha(a^{\alpha\beta} D_\beta u) = F(x,u,\nabla u) \text{ in } \Omega_1,$$

where F satisfies the estimate $|F(x,u,\nabla u)| \le a\, Q(x,\nabla u)$ *a.e. on* Ω_1 *with a positive constant* a. *Then* $\lim_{x\to\infty} u(x)$ *exists.*

Because of the simplicity and elegance of the argument, we include a sketch of the proof, following an idea due to Gilbarg and Serrin [12]. W.l.o.g. it may be assumed that $\lim_{R\to\infty} \inf_{|x|>R} u(x) = 0$. Fixing an arbitrary $\epsilon > 0$, one thus finds an increasing and unbounded sequence of radii $R_k > 1$ such that $u(x) \ge -\epsilon$ for $|x| > R_1$ and $\inf_{(2R_k|3R_k)} u \le \epsilon$. We now apply Trudinger's Harnack inequality [42, Theorem 1.1] to the nonnegative function $u + \epsilon$ on the domains $(R_k|4R_k)$ where (1.4) is uniformly elliptic. This yields that $\sup_{(2R_k|3R_k)} u \le C\,\epsilon$ with a constant C independent of k. Finally, an easy computation shows that $-D_\alpha(\tilde{a}^{\alpha\beta} D_\beta u) \leq 0$ in Ω_1, where $\tilde{a}^{\alpha\beta} = a^{\alpha\beta}\cdot\exp(au)$. One can thus apply the maximum principle on all regions $(2R_1|2R_k)$ and deduces that $u(x) \leq C\epsilon$ for $|x| > 2R_1$. Since ϵ was chosen arbitrarily, it follows that $\lim_{R\to\infty} \sup_{|x|>R} u \leq 0$, and the assertion is proved.

In the sequel, we present an application of Theorem 1.3 to *harmonic functions on Riemannian manifolds*. Specifically, we restrict ourselves to the case of a manifold $G \subset \mathbb{R}^{n+m}$ which is given as the graph of a function $f = (f^1,\dots,f^k) \in C^2$, $f = f(x)$, defined either for $|x| > 1$ or for $0 < |x| < 1$. Let h be bounded and harmonic on G. As is shown by hyperbolic graphs over $\mathbb{R}^2$ - first constructed by Ossermann [34] - the expression $h(x,f(x))$ need not tend to a limit as $x \to \infty$ (resp. $x \to 0$). On the other hand, one infers from Theorem 1.3 that a limit does always exist if f is radially symmetric.

EXAMPLE 1.4.

Let $f(\bar{x}) = \underline{f}(|\bar{x}|)$ for all $\bar{x}$ in the domain of f. Then one can introduce new coordinates $x \in \Omega_1$ on the graph G of f with respect to which the coefficients of the Laplace-Beltrami operator satisfy the conditions (A), (W); moreover, $x \to \infty$ iff $\bar{x} \to \infty$ (resp. $\bar{x} \to 0$).
In fact, consider the solution $g(\rho)$ of the initial value problem $+g'(\rho) = \rho^{-1}\{1 + [\underline{f}'(g(\rho))]^2\}^{-1/2} g(\rho)$, $\rho \geq 1$, $g(1) = 1$.
(In case that f is defined on the punctured ball, the sign in the differential equation must be reversed). For $x \in \Omega_1$ let $\bar{x} = g(|x|)x/|x|$; then with respect to the coordinate chart $x \to (\bar{x}, f(\bar{x}))$ on G, the metric tensor is given by the identity matrix times $g(r)^2/r^2$, where $r = |x|$. The differential equation implies that g is monotone increasing and satisfies $g(2\rho) \leq 2g(\rho)$ for $\rho \geq 1$. Hence, the above assertion about the coefficients $\delta^{\alpha\beta}(g(r)/r)^{n-2}$ of the Laplace-Beltrami operator follows. Note that for the cylinder $\partial\Omega_1 \times \mathbb{R}^+ \subset \mathbb{R}^{n+1}$ one readily constructs a coordinate chart with respect to which the coefficients are given by $\delta^{\alpha\beta}|x|^{2-n}$.

REMARK 1.5.

In case of non-constant weights $w(x)$, the limit in Theorem 1.3 need not be approached Hölder continuously, in contrast to (1.2). To see this, let $\tau > 0$, $r = |x|$, $w(x) = r^{2-n}(\log 2r)^{\tau+1}$. Then the function $u(x) = (\log 2r)^{-\tau}$ is a solution of the equation

$$(1.5)\qquad -D_\alpha(w(x)\, D_\alpha u) = 0 \quad \text{in } \Omega_1.$$

Incidentally, when $n \geq 3$ and u is interpreted as a harmonic function with respect to an appropriate metric on Ω_1, the example shows that there does not exist a uniform least rate of approach of the kind $O(\|x\|^{-\sigma})$, $\sigma > 0$, in terms of the geodesic distance $\|x\|$ from x to a fixed point of Ω_1.

At the end of this section, we want to state a "*gap theorem*" for unbounded solutions of (1.1). A very precise result of this kind was proved by Serrin and Weinberger [39, Theorem 1]. Here we only present a weaker version concerning solutions on all of $\mathbb{R}^n$, which allows us to include the case of $\mathbb{R}^2$ as well as that of non-uniformly elliptic equations.

THEOREM 1.6.

Let u be a non-constant weak solution of the linear equation (1.1) *in* $\mathbb{R}^n$ *with coefficients satisfying* (A),(W). *Suppose also that* (1.1) *is uniformly elliptic for* $|x| \le 1$. *Then there exist constants* $\sigma = \sigma(n,\lambda) > 0$, c, $r_o > 0$ *such that*

$$\underset{|x|=r}{\operatorname{osc}}\ u \ge c r^{\sigma} \quad \text{for } r \ge r_o. \tag{1.6}$$

The proof is the same as that of Theorem 4 in [33] and rests upon Harnack's inequality applied to annular regions. Our assumptions within the unit ball are only required to guarantee that $\sup_{|x|=r} u$ (resp. $\inf_{|x|=r} u$) is monotone increasing (resp. decreasing) in r.

We remark that Theorem 1.6 is closely related to interior Hölder estimates for solutions of (1.1) defined on bounded domains. The optimal exponent σ was recently computed by Lin and Ni [27] in the special case of uniformly elliptic Laplace-Beltrami operators on $\mathbb{R}^2$; if the coefficients $a^{\alpha\beta}$ satisfy (A) in $\mathbb{R}^2$ with $w \equiv 1$, then $\sigma = \lambda^{-1/2}$.

A result analogous to Theorem 1.6 will be proved in Section 3 for harmonic maps into manifolds of non-positive sectional curvature.

It should be noted that, when Theorem 1.6 is applied to a harmonic function u on the manifold $\mathbb{R}^n$ with an appropriate metric, the quantity $\underset{\|x\|=\rho}{\operatorname{osc}}\ u$ (where $\|x\|$ denotes geodesic distance, as in 1.5) can sometimes be shown to grow faster than any power of ρ. In particular, for manifolds with a cylindrical end this growth is of exponential order in ρ, as a consequence of (1.6).

2. Diagonal elliptic systems with quadratic growth

In the following, we shall extend Theorem 1.3 to solutions $u = (u^1,..,u^N)$ of diagonal elliptic systems with quadratic growth,

$$(2.1)\qquad -D_\alpha(a^{\alpha\beta}(x)\, D_\beta\, u^i) = F^i(x,u,\nabla u) \qquad (i=1,..,N)$$

in $\Omega_1 = \{x \in \mathbb{R}^n : |x| > 1\}$, imposing a smallness condition on $\sup_{\Omega_1} |u|$.
Except for the more general kind of ellipticity admitted here, our hypotheses completely correspond to the ones of existing Liouville type theorems and regularity results already mentioned in the Introduction.

Throughout this section, the coefficients $a^{\alpha\beta}$ are required to satisfy (A),(W), and furthermore it is assumed that

$$(2.2)\qquad |F(x,u,\nabla u)| \le a\, Q(x,\nabla u),$$

$$(2.3)\qquad u\cdot F(x,u,\nabla u) \le a^*\, Q(x,\nabla u)$$

a.e. in Ω_1. Here $F = (F^1,\ldots,F^N)$, Q denotes the quadratic form defined in (A), while $a > 0$ and a^* are real constants.

THEOREM 2.1.
Let $u \in H^1_{2,loc}(\Omega_1,\mathbb{R}^N)$ *be a bounded weak solution of* (2.1) *in* Ω_1 *with the property that*

$$(2.4)\qquad a^* + a \sup_{\Omega_1} |u| < 2.$$

Then $\lim_{x\to\infty} u(x) = \bar{u}_\infty$ *exists. Moreover,*

$$(2.5)\qquad \int_{|x|>2r} |\nabla u|^2 |x|^{2-n}\, dx = O(\omega^2(r)) \quad \text{as } r \to \infty,$$

where $\omega(r) = \sup_{|x|=r} |u(x) - u_\infty|$.

In case that $w \equiv 1$,

$$(2.6)\qquad \omega(r) = \begin{cases} O(r^{-\sigma}) & \text{if } n = 2, \\ O(r^{2-n}) & \text{if } n \ge 3, \end{cases}$$

$$(2.7)\qquad \int_{|x|>r} |\nabla u|^2\, dx = \begin{cases} O(r^{-2\sigma}) & \text{if } n = 2, \\ O(r^{2-n}) & \text{if } n \ge 3 \end{cases}$$

as $r \to \infty$, σ *being a positive number.*

REMARKS.

- The existence of a limit of u can also be obtained if additional inhomogeneous terms of the form $b(x)w(x)$ are admitted in the estimates (2.2), (2.3), provided, e.g., $b(x) = o(|x|^{-2})$ as $|x| \to \infty$. This follows by a straightforward generalization of our proof below.
- Examples like the one in Remark 1.5 show that (2.5) is essentially optimal.
- The last part of the proof will indicate how $\omega(r)$ could be estimated for non-constant weights w, using adapted capacity measures which involve w.

- Due to the coordinate invariant formulation of (2.2)-(2.4) and the generality of the weights considered, Theorem 2.1 contains some considerable extensions and improvements of a result by Baldes [1].
- One can use Theorem 2.1 to deduce some generalization of a Liouville theorem by Hildebrandt and Widman [16] for a class of non-uniformly elliptic coefficients. We shall describe this simple argument in Section 3, in the context of harmonic maps.
- We note that weak solutions on a domain Ω with non-compact boundary converge, under similar hypotheses as above, to a limit at infinity, provided the boundary values do and $\limsup_{R\to\infty} R^{-n}\text{meas}\,\{x \in \mathbb{R}^n - \Omega : R < |x| < 2R\} > 0$. The proof follows from a suitable boundary version of the Harnack inequality in Lemma 2.2 by adapting the argument subsequent to Lemma 2.4, cf. also [32, § 3].

The proof of Theorem 2.1 rests upon a modification of a method introduced in [32], [31, pp. 711-713] to obtain simple proofs of continuity and Liouville type theorems for solutions of quasilinear systems (2.1), as well as for harmonic maps; cf. also [22, pp. 26-30] for a review. Our investigation in [32] was originally inspired by Caffarelli's elegant approach to the regularity problem, cf. [4].
The starting point of the argument is Moser's Harnack inequality for weak supersolutions of the operator $L = D_\alpha(a^{\alpha\beta}\, D_\beta)$. In the present context we need a version with solid balls replaced by annular domains $(R|R') = \{x \in \mathbb{R}^n : R < |x| < R'\}$. In terms of subsolutions, the basic inequality reads as follows.

LEMMA 2.2.
Let v *be a weak solution of* $-Lv \leq 0$ *in* $(R|8R)$. *There exists a constant* $\delta = \delta(n,\lambda) > 0$ *such that*

$$\sup_{(2R|4R)} v \leq (1 - \delta) \sup_{(R|8R)} v + \delta\, \bar{v}(R) \tag{2.8}$$

where $\bar{v}(R)$ *denotes the average of* v *on* $(2R|4R)$.

This assertion is proved by applying the Harnack inequality [13, Theorem 8.18] to the weak supersolution $\{\sup_{(R|8R)} v\} - v$ on a suitable covering of $(2R|4R)$ by balls.
A simple computation (cf. [22, p. 27]) shows the following

LEMMA 2.3.
Let u *be a weak solution of* (2.1) *in some domain* Ω, *and suppose that* (2.3) *holds with* $a^* < 1$. *Then, for any* $\zeta \in \mathbb{R}^n$ *with norm* $|\zeta| \leq (1-a^*)/a$,

the function $v = |u - \zeta|^2$ *satisfies* $-Lv \leq 0$ *in* Ω.

We now consider a bounded subsolution v of L in Ω_1. While v need not approach a limit at infinity[3)], it is important to note that the average $\bar{v}(R)$ as well as both suprema in (2.8) converge to $\lim_{R\to\infty} \sup_{|x|>R} v$ as $R \to \infty$. In fact, the maximum principle implies that $\sup_{|x|=R} v$ is monotone with respect to R for large R, and hence (2.8) yields the convergence of $\bar{v}(R)$. Applying this argument to the functions v from Lemma 2.3, we obtain

LEMMA 2.4.
Let u *satisfy the assumptions of Theorem* 2.1 *with* (2.4) *replaced by the weaker condition that* $a^* < 1$. *Then the averages of* u *on* $(R|2R)$ *tend to a limit* $\bar{u}_\infty \in \mathbb{R}^N$ *as* $R \to \infty$. *Moreover,*

$$R^{-n} \int_{(R|2R)} |u - \bar{u}_\infty|^2 \, dx \to 0 \quad \text{as } R \to \infty. \tag{2.9}$$

Actually, the proof of (2.9) requires a small additional consideration, choosing the direction of ζ in Lemma 2.3 in such a way that $\lim_{R\to\infty} \{ \sup_{|x|>R} |u - \zeta| - \sup_{|x|>R} |u| \} = |\zeta|$. We omit the details and refer to [31, p. 712] or [22, p. 29] for an analogous reasoning.

Let us now proceed to the *proof of Theorem 2.1*. For $0 \leq t \leq 1$ we set $u_t = u - t\bar{u}_\infty$, $M_t = \lim_{R\to\infty} \sup_{|x|>R} |u_t|$ and define the closed set $I = \{t \in [0,1]: M_t \leq (1-t) M_o\}$ containing 0. Here $\bar{u}_\infty$ is the vector from Lemma 2.4, whence $|\bar{u}_\infty| = M_o$.

Suppose that $t' \in I$, $t' < 1$. By virtue of (2.4) one then finds a number t, $t' < t < 1$, and a radius $K > 1$ with the property that $a^* + t|\bar{u}_\infty| + a \sup_{|x|>K} |u_t| < 2$. Thus, the function $v = |u_t|^2$ satisfies $-Lv \leq 0$ for $|x| > K$. By our discussion preceding Lemma 2.4, and by (2.9),

$$M_t^2 = \lim_{R\to\infty} \bar{v}(R) \leq (1 - t)^2 |\bar{u}_\infty|^2,$$

whence $t \in I$. From this consideration it follows that $1 \in I$, and the convergence of u(x) to $\bar{u}_\infty$ is proved. (An inspection shows that the condition (W) has hitherto been used only for the proof of Lemma 2.2.

---

3) An example in the uniformly elliptic case can easily be constructed using the function from (3.1) of [28].

Hence the result just established is valid in case of any other weight w for which an analogue of Lemma 2.2 holds.)

The integrals in (2.5), (2.6) can easily be estimated by using the following straightforward modification of a result due to Giaquinta and Giusti [10, pp. 43-44] whose proof also rests upon the weak Harnack inequality for supersolutions.

<u>LEMMA 2.5.</u>

Let u be a weak solution of (2.1) *in* $(R|8R)$, *and suppose that* (2.3) *holds with* $a^* < 1$. *Then*

$$\int_{(2R|4R)} |\nabla u|^2 \, dx \leq C R^{n-2} \{ \sup_{(R|8R)} |u|^2 - \sup_{(2R|4R)} |u|^2 \}, \tag{2.10}$$

where the constant C *depends on* n, λ *and* a^*.

In fact, if $r > 1$ is sufficiently large and $R \geq r$, we may apply Lemma 2.5 to the function $u_1 = u - \bar{u}_\infty$ instead of u. Since $-L|u_1|^2 \leq 0$ for $|x| > r$, the maximum principle yields that the quantity $\omega(R)$ is monotone decreasing with respect to R. Hence,

$$\int_{(2R|4R)} |\nabla u|^2 \, dx \leq C R^{n-2} \{\omega^2(R) - \omega^2(2R)\} \quad \text{for } R > r. \tag{2.11}$$

The assertion (2.5) now follows by summation of (2.11) for $R = R_k = 2^k r$, $k = 1,2,\ldots$. Note that (2.7) can be derived from (2.6) in an analogous fashion since $R^{n-2}\{\omega(2R) + \omega(R)\}$ is bounded in this case.
Thus, it only remains to prove (2.6) when $w \equiv 1$. Actually, the case $n = 2$ is somewhat exceptional and will therefore first be outlined separately. By means of the change of coordinates $x \to x|x|^{-2}$, the system (2.1) on Ω_1 is transformed into a corresponding one (also uniformly elliptic since $n = 2$) on the punctured unit disk. Lemma 2 of [31] together with a regularity result due to Wiegner [46] now implies that the transformed solution extends to a Hölder continuous solution in the whole disk, even if instead of (2.4) one merely requires that $a^* < 1$ in (2.3). Hence Theorem 2.1 can be improved in this special case. (Under the stronger assumption (2.4), several different methods yield Hölder continuity, cf. [43], [15], [4], [32]).
When $n \geq 3$, on the other hand, the singularity at infinity need not be removable. We then deduce (2.6) from the corresponding result for solutions of linear equations. Recalling that $u_1 = u - \bar{u}_\infty$ and that $\omega(R) \to 0$ as $R \to \infty$, let us first state two facts upon which the subsequent argument is based.

(2.12) *There exists a constant* $K > 1$ *such that* $-\bar{L}|u_1| := -D_\alpha(\bar{a}^{\alpha\beta}(x)\, D_\beta|u_1|) \leq 0$ *for* $|x| > K$, *where* $\bar{a}^{\alpha\beta} = a^{\alpha\beta}\exp(K|u_1|)$.

(2.13) *If* $w \equiv 1$, *one finds a function* h *with the properties that* $\bar{L}h = 0$ *for* $|x| > K$, $h(x) = 1$ *for* $|x| = K$, $h(x) = O(|x|^{2-n})$ *as* $x \to \infty$.

Postponing the proof of (2.12), (2.13) for a moment, we derive (2.6) as follows. With a suitable choice of $C > 0$, the function $v = |u_1| - Ch$ satisfies $v(x) \leq 0$ for $|x| = K$. Moreover, $-\bar{L}v \leq 0$ for $|x| > K$ and $v(x) \to 0$ as $x \to \infty$. Hence the maximum principle yields that $v(x) \leq 0$ for $|x| > K$, which combined with (2.13) implies (2.6).

The assertion (2.12) is valid for any weight w and can be deduced from the estimate (2.2) alone since $\omega(R) \to 0$ as $R \to \infty$. This follows by testing the system (2.1) with the vector $u_1|u_1|^{-1}\exp(K|u_1|)\varphi$, where φ is any non-negative C^∞-function with compact support in $\{|x| > K\}$. In the resulting integral inequality, one uses the observation that $Q(x,\nabla|u_1|) \leq Q(x,\nabla u_1)$ to be able to estimate a term of the form $\varphi Q(x,\nabla u_1)\{1 - a|u_1|\} - \varphi Q(x,\nabla|u_1|)\{1 - K|u_1|\}$ from below by zero for K sufficiently large. Actually, the computation is a refined version of the one presented in [31, Lemma 3].

Finally, we should like to include a sketch of the proof of (2.13), although this is a well known result which can be extracted, for instance, from Serrin's work [38, p. 84]. There (2.13) appears as a consequence of more subtle investigations on the precise asymptotic behavior of solutions. A simple argument from [38], however, suffices to prove (2.13) and also indicates how the corresponding decay rate may be estimated in case of more general weights w.
First of all, the solution h is easily constructed as the weak limit of approximating solutions h_R of $Lh_R = 0$ on $(K|R)$ taking boundary values 1 and 0 for $|x| = K$ and $|x| = R$, respectively. Uniform bounds on the L_2-norms of ∇h_R, and hence on the Sobolev norms of the (naturally extended) functions h_R, are immediate. As a result, Moser's theorem 1.1 implies that the limit function $h = h(x) \geq 0$ thus obtained converges to *zero* as $x \to \infty$. It only remains to give the required estimate on the rate of convergence. By applying the Harnack inequality to h on annular

regions, we see that we need only estimate the quantity $m(r) = \min_{|x|=r} h$ (which decreases with respect to r). To this end, let $r > K$ be fixed and $m = m(r)$. We consider the twofold truncation $\underline{h}$ of $h - m/2$ by $m/2$ from above and 0 from below. Note that $\underline{h}$ has bounded support and $\underline{h} \equiv m/2$ on the ball of radius r. Since $\bar{L}h = 0$ in $\{|x| > K\}$ and $\nabla\underline{h} = \nabla h$ on the set where $\underline{h} \neq 0$, one infers easily that, with some constant c independent of r and m,

$$cm = \int a^{\alpha\beta} D_\alpha \underline{h} D_\beta \underline{h} dx$$

which can be estimated below by a fixed positive multiple of $\int |\nabla\underline{h}|^2 w(x) dx$. In the case $w \equiv 1$ considered here, a lower bound for the latter integral is given by $m^2/4$ times the ordinary capacity of the ball of radius r (cf. [37, Lemma 10]). We have thus shown that $m(r) = O(r^{2-n})$ as $r \to \infty$. This concludes the proof of Theorem 2.1.

3. Harmonic maps and minimal submanifolds

In this section, we shall establish an analogue of Theorem 2.1 for harmonic maps of Riemannian manifolds with image in a regular geodesic ball. Moreover, applications to a Liouville theorem and to minimal submanifolds of higher codimension are presented, and a "gap theorem" for harmonic maps into manifolds of non-positive sectional curvature will be proved. Related problems concerning existence, regularity, a priori estimates, Liouville and Bernstein theorems, as well as removable singularities have been treated in a number of papers, e.g. [17]-[19], [11], [31], [32]. For a detailed discussion and for additional references, see [19], [22], and the companion article by Jost in this volume.

We shall henceforth consider an n-dimensional Riemannian manifold X_1 of class C^1, globally described by coordinates $x \in \Omega_1 = \{x \in \mathbb{R}^n : |x| > 1\}$ with respect to which the coefficients $a^{\alpha\beta}(x)$ of the Laplace-Beltrami operator of X_1 satisfy (A),(W).[4)] In terms of the coefficients $\gamma_{\alpha\beta}(x)$ of the Riemannian metric, $a^{\alpha\beta} = \sqrt{\gamma}\, \gamma^{\alpha\beta}$, using standard notation. The points of X_1 and Ω_1 will always be identified and denoted by x in the sequel, while $|x|$ stands for the *Euclidean norm*.
Moreover, let Y be a complete N-dimensional Riemannian manifold of class C^3 and without boundary.
We first state our results, postponing the outline of their proofs to the end of the section.

---

4) Cf., e.g., Example 1.4.

THEOREM 3.1.

Let $U : X_1 \to Y$ *be a (weakly) harmonic map. Suppose that* $U(X_1)$ *is contained in a closed geodesic ball* $B_M(P)$ *of* Y *which does not meet the cut locus of its center* P *and for which*

$$M < \pi/(2\sqrt{\kappa}),$$

where $\kappa \geq 0$ *denotes an upper bound for the sectional curvature of* Y *on* $B_M(P)$. *Then* $U(x)$ *tends to a limit as* $|x| \to \infty$.

By a theorem due to Ruh and Vilms [36], the previous result can be applied to certain minimal submanifolds of Euclidean space. Here we restrict ourselves to the non-parametric case. By combining Theorem 3.1 with computations of [18] (cf. also [31]), one readily obtains the following generalization of Corollary 1.2 to graphs of higher codimension.

COROLLARY 3.2.

Let $f = (f^1, \ldots, f^k) \in C^3(\Omega_1, \mathbb{R}^k)$ *be a solution of the minimal surface system in* Ω_1 *(cf. [35], [18]), i.e., suppose that the graph of* f *is a minimal submanifold of* $\mathbb{R}^{n+k}$ *(or, more generally, has parallel mean curvature field). Assume also that there exists a constant* b *with*

$$(3.1)\quad b < \cos^{-m}(\pi/(2\sqrt{\kappa}\, m)),\ m = \min(n,k),\ \kappa = \begin{cases} 1 \text{ if } m = 1, \\ 2 \text{ if } m > 1, \end{cases}$$

such that

$$(3.2)\qquad \det\Big(\delta_{\alpha\beta} + \sum_{j=1}^{k} D_\alpha f^j\, D_\beta f^j\Big) \leq b^2 \text{ for all } x \in \Omega_1.$$

Then $\nabla f(x)$ *tends to a limit as* $x \to \infty$.

Examples due to Lawson and Osserman [26] show that a quantitative assumption of the kind (3.2) is necessary, although it is not clear in which cases the bound for b in (3.1) is sharp.

Next we wish to point out that an approach to Liouville theorems via Theorems 2.1 and 3.1 yields interesting generalizations of results due to Ivert [23] and to Hildebrandt-Jost-Widman [18] which were originally deduced from a priori estimates by means of a blow-down argument. As was noted by the author [29] in a similar context, sometimes the validity of a Liouville theorem is not affected by relaxing certain assumptions in a compact subset of the domain. The same phenomenon occurs here (in contrast to [5]). Furthermore, the weighted ellipticity condition (A) allows us to include various examples other than "simple manifolds" X considered in [18].

COROLLARY 3.3.

Let X *be an n-dimensional Riemannian manifold containing* X_1*, and suppose that the complement of* $\{x \in X_1 : |x| > R\}$ *in* X *is compact with boundary* $\{x \in X_1 : |x| = R\}$ *for* $R > 1$. *Let* $U: X \to B_M(P)$ *be a harmonic map,* $B_M(P)$ *satisfying the hypotheses of Theorem 3.1.*

Then U *is a constant map.*

In case when $\kappa = 0$, Corollary 3.3 can be complemented by the following "gap theorem".

THEOREM 3.4.

Let X *be as in Corollary 3.3, while* Y *is simply connected and has non-positive sectional curvature. If* $U: X \to Y$ *is a non-constant harmonic map, there exist numbers* $r_o > 1$, $c > 0$, $\sigma = \sigma(n,\lambda) > 0$ *such that*

$$\omega(r) \geq c\, r^{\sigma} \quad \textit{for } r \geq r_o, \tag{3.3}$$

where $\omega(r)$ *stands for the infimum of all radii of geodesic balls in* Y *containing* $\{U(x) : |x| = r\}$.

Theorem 3.4 can be viewed as counterpart to interior a priori estimates for the Hölder norm of harmonic maps due to Giaquinta and Hildebrandt [11]. An alternative method for deriving these bounds was outlined in [32], the main ideas of which also appear in the proof below. Yet the emphasis is somewhat different in the present context. Here we require estimates *without assuming a bound on the size of* U, while some technical simplifications arise from the curvature assumption. Note also that the exponent σ in (3.3) does not depend on the dimension N of Y and that no lower bound for the sectional curvature need be imposed.

Let us now turn to the *proofs of Theorem 3.1 and Corollary 3.3.* By $u = u(x)$ we denote a representation of U with respect to normal coordinates $u = (u^1,..,u^N)$ centered at P. It follows from some differential geometric estimates in [18, Lemma 1] that Lemma 2.4 is applicable to the solution u of the harmonic map system, whence one finds a limit $\bar{u}_\infty$ of u in the average sense (2.9). Let $\bar{P}$ denote the point in $B_M(P)$ with the coordinates $\bar{u}_\infty$, and let M_o stand for the distance from $\bar{P}$ to P. Finally, for $0 \leq t \leq 1$, $u_t = u_t(x)$ denotes a normal coordinate representation of U centered at that point on the geodesic joining P and $\bar{P}$ in $B_M(P)$ which is at distance tM_o from P.

One can now adapt the argument following Lemma 2.4 in the proof of

Theorem 2.1, defining $M_t = \lim_{R\to\infty} \sup_{|x|>R} |u_t|$ and $I = \{t \in [0,1] : M_t \leq (1-t)\, M_o\}$. Indeed, if $t' \in I$ and $t' < 1$, one finds a number t, $t' < t < 1$, and a radius $K > 1$ such that $|u_t(x)|^2 < \pi/(2\sqrt{\kappa})$ for $|x| > K$. By the differential geometric estimates in [18, Lemma 1], $|u_t|^2$ is subharmonic for $|x| > K$ (with respect to the metric on X_1). Hence M_t^2 equals the limit of the averages of $|u_t|^2$ on $(R|2R)$ as $R \to \infty$. To see that this limit is bounded by $(1-t)^2 M_o^2$, we may use (2.9) together with the estimates $|u_t| \leq (1-t)\, M_o + |u_1|$ and $|u_1| \leq C|u-\bar{u}_\infty|$ for some constant C.

Thus $t \in I$, and the preceding consideration shows that $1 \in I$, whence Theorem 3.1 is established. For a similar reasoning, cf. also [31, pp. 712 f.], [32], [22].

In order to prove Corollary 3.3, we let $\underline{M}_t = \sup_X |u_t|$ $(0 \leq t \leq 1)$. From the first part of the argument above it follows that, if $t' < 1$ and $\underline{M}_{t'} = M_{t'}$, then $|u_t|^2$ is subharmonic on X for some $t > t'$. By the structure of X, the maximum principle now implies that $\underline{M}_t = M_t$ (and $\underline{M}_o = M_o$). Thus finally $\underline{M}_1 = M_1 = 0$, concluding the proof of Corollary 3.3.

It should be noted here that the asymptotic estimates (2.5) - (2.7) of Theorem 2.1 carry over immediately to the coordinate representation u_1 of U in Theorem 3.1.

At the end of this section, we sketch the *proof of Theorem* 3.4. Let u be the same coordinate representation of U as before, where $P \in Y$ will be chosen later. Since $\kappa = 0$ the function $v = |u|^2$ is subharmonic *on* X, whence one infers from the properties of X and from the maximum principle that the quantity

$$p(r) = \sup_{|x|=r} |u|^2, \quad r > 1,$$

increases with respect to r.

By *iterating the estimate* (2.8) *of Lemma* 2.2, we thus find a positive integer $J = J(n,\lambda) \geq 8$ of the form 2^k such that

$$p(4r) \leq 1/8\; p(Jr) + C\, \bar{v}_r \quad \text{for } r > 1, \tag{3.4}$$

where $\bar{v}_r$ denotes the average of v on $T_r = (2r|Jr/2)$, while C stands for a generic constant depending only on n and λ. In fact, we have crudely estimated a specific convex combination of averages of v on annular subdomains of T_r by $C\,\bar{v}_r$ in (3.4). It is worth noting that in the present context our previous iteration argument can be replaced by an alternative proof; employing standard local L_∞-bounds for the subharmonic function $|u|^2$ (or alternatively $|u|$) one can deduce (3.4) even with $J = 16$ and with the term $1/8\ p(Jr)$ omitted (cf. [13, Theorem 8.17] and [31, Lemma 3]). On the other hand, the iteration method provides some more precise quantitative information (as indicated above) which is crucial when establishing a priori estimates in the case $\kappa > 0$ (cf. [32]).

For the remainder of the proof, let $r > 1$ be fixed.

First one chooses P to be the Riemannian center of mass of the restriction of the map U to T_r, since in this case the average on T_r of the coordinate representation u vanishes (cf. [3, 8.1.3]). The average $\bar{v}_r$ in (3.4) can thus be estimated by the Poincaré inequality.

Moreover, since P lies in each geodesic ball containing $\{U(x): x \in T_r\}$, the maximum principle yields that $p(Jr) \leq 4\ \omega^2(Jr)$. Obviously $\omega^2(4r) \leq p(4r)$, whence we finally conclude from (3.4) that

$$\omega^2(4r) \leq 1/2\ \omega^2(Jr) + C\ r^{2-n} \int_{T_r} |\nabla u|^2\ dx. \tag{3.5}$$

Now let $g_{ij}(u)$ denote the coefficients of the metric of Y and consider the quantity $E(u) = a^{\alpha\beta}\ g_{ij}(u)\ D_\alpha u^i D_\beta u^j$ representing the energy density of the map U, up to a factor $2\sqrt{\gamma}$ (cf. [11, p. 133]). By our assumptions (A), (W), and since $\kappa = 0$, we have $|\nabla u|^2 \leq C\ \Lambda_r\ E(u)$ on T_r with the normalization constant $\Lambda_r = \sup_{T_r} w^{-1}$, and thus

$$2\omega^2(4r) \leq \omega^2(Jr) + C\ \Lambda_r\ r^{2-n} \int_{T_r} E(u)\ dx. \tag{3.6}$$

From now on, let $u = u(x)$ again denote a normal coordinate representation of U with respect to an *arbitrary center* $P \in Y$. A simple invariance argument shows that (3.6) continues to be valid for any such u. Moreover, we may apply Lemma 2.5 to u, noting that $a^* = 0$ for the harmonic map system under consideration. Actually, one obtains a slight (yet important) improvement in the present case, since the integrand $|\nabla u|^2$ in the estimate (2.10) can be replaced by $\Lambda_r\ E(u)$ for $r \leq R \leq Jr/8$. This follows by an inspection of the proof of [11, Proposition 1]. The resulting improved inequality is now iterated with respect to R, and one arrives at

$$\Lambda_r\ r^{2-n} \int_{T_r} E(u)\ dx \leq C\ \{p(Jr) - p(4r)\}. \tag{3.7}$$

Finally we combine (3.6) and (3.7) and take the infimum with respect to $P \in Y$, thus obtaining that

$$2\omega^2(4r) \le \omega^2(Jr) + C\,\{\omega^2(Jr) - \omega^2(4r)\}.$$

Since $r > 1$ is arbitrary and $C + 1 < C + 2$, the assertion of Theorem 3.4 follows at once.

References

[1] Baldes, A., *Degenerate elliptic operators, diagonal systems and variational integrals*; manuscripta math. 55, 467-486 (1986).

[2] Bers, L., *Isolated singularities of minimal surfaces*; Ann. of Math. (2) 53, 364-386 (1951).

[3] Buser, P., and H. Karcher, *Gromov's almost flat manifolds*; Astérisque 81 (1981).

[4] Caffarelli, L.A., *Regularity theorems for weak solutions of some nonlinear systems*; Comm. Pure Appl. Math. 35, 833-838 (1982).

[5] Donelly, H., *Bounded harmonic functions and positive Ricci curvature*; Math. Z. 191, 559-565 (1986).

[6] Fabes, E.B., C.E. Kenig, and R.P. Serapioni, *The local regularity of solutions of degenerate elliptic equations*; Comm. P.D.E. 7 (1), 77-116 (1982).

[7] Finn, R., *Isolated singularities of solutions to non-linear partial differential equations*; Trans. Amer. Math. Soc. 75, 385-404 (1953).

[8] Finn, R., and D. Gilbarg, *Three-dimensional subsonic flows, and asymptotic estimates for elliptic partial differential equations*; Acta Math. 98 265-296 (1957).

[9] Frehse, J., *A discontinuous solution of a mildly nonlinear elliptic system*; Math. Z. 134, 229-230 (1973).

[10] Giaquinta, M., and E. Giusti, *On the regularity of the minima of variational integrals*; Acta Math. 148, 31-45 (1982).

[11] Giaquinta, M., and S. Hildebrandt, *A priori estimates for harmonic mappings*; J. reine angew. Math. 336, 124-164 (1982).

[12] Gilbarg, D., and J. Serrin, *On isolated singularities of solutions of second order elliptic differential equations*; J. Analyse Math. 4, 309-340 (1955-1956).

[13] Gilbarg, D., and N. S. Trudinger, *Elliptic partial differential equations of second order*; Springer, Berlin/Heidelberg/-New York (1977).

[14] Hildebrandt, S., and K.-O. Widman, *Some regularity results for quasilinear elliptic systems of second order*; Math. Z. 142, 67-86 (1975).

[15] Hildebrandt, S., and K.-O. Widman, *On the Hölder continuity of weak solutions of quasilinear elliptic systems of second order*; Ann. Scuola Norm. Sup. Pisa (IV), 4, 145-178 (1977).

[16] Hildebrandt, S., and K.-O. Widman, *Sätze vom Liouvilleschen Typ für quasilineare elliptische Gleichungen und Systeme*; Nachrichten der Akad. Wiss. Göttingen, II. Math.-Phys. Klasse, Nr. 4, 41-59 (1979).

[17] Hildebrandt, S., H. Kaul, and K.-O. Widman, *An existence theory for harmonic mappings of Riemannian manifolds*; Acta Math. 138, 1-16 (1977).

[18] Hildebrandt, S., J. Jost, and K.-O. Widman, *Harmonic mappings and minimal submanifolds*; Inventiones math. 62, 269-298 (1980).

[19] Hildebrandt, S., *Liouville theorems for harmonic mappings and an approach to Bernstein theorems*; in "Seminar on Diff. Geometry", ed. S.-T. Yau, Annals of Math. Studies 102, 107-131, Princeton Univ. Press, 1982.

[20] Hildebrandt, S., *Nonlinear elliptic systems and harmonic mappings*; Proc. of the 1980 Beijing Symposium of Diff. Geom. and Diff. Equ., Vol. 1, pp. 481-615, Science Press, Beijing, China 1982.

[21] Hildebrandt, S., *Quasilinear elliptic systems in diagonal form*; in "Systems of nonlinear partial differential equations", ed. J.M. Ball, Reidel, Dordrecht/Boston/Lancester (1983).

[22] Hildebrandt, S., *Harmonic mappings of Riemannian manifolds*; in "Harmonic mappings and minimal immersions", ed. E. Giusti, Springer Lecture Notes in Math. 1161, 1-117 (1985).

[23] Ivert, P.-A., *On quasilinear elliptic systems of diagonal form*; Math. Z. 170, 283-286 (1980).

[24] Jäger, W., *Ein Maximumsprinzip für ein System nichtlinearer Differentialgleichungen*; Nachrichten der Akad. Wiss. Göttingen 11, 157-164 (1976).

[25] Karp, L., *Subharmonic functions on real and complex manifolds*; Math. Z. 179, 535-554 (1982).

[26] Lawson, H.B., and R. Osserman, *Non-existence, non-uniqueness and irregularity of solutions to the minimal surface system*; Acta Math. 139, 1-17 (1977).

[27] Lin, F.H., and W.M. Ni, *On the least growth of harmonic functions and the boundary behavior of Riemann mappings*; Comm. P.D.E. 10(7), 767-786 (1985).

[28] Meier, M., *Liouville theorems for nonlinear elliptic equations and systems*; manuscripta math. 29, 207-228 (1979).

[29] Meier, M., *Liouville theorems for nondiagonal elliptic systems in arbitrary dimensions*; Math. Z. 176, 123-133 (1981).

[30] Meier, M., *Removable singularities for weak solutions of quasilinear elliptic systems*; J. reine angew. Math. 344,87-101 (1983).

[31] Meier, M., *Removable singularities of harmonic maps and an application to minimal submanifolds;* Indiana Univ. Math. J. 35, 705-726 (1986).

[32] Meier, M., *On quasilinear elliptic systems with quadratic growth;* Preprint 1984.

[33] Moser, J., *On Harnack's theorem for elliptic differential equations;* Comm. Pure Appl. Math. 14, 577-591 (1961).

[34] Osserman, R., *A hyperbolic surface in 3-space;* Proc. Amer. Math. Soc. 7, 54-58 (1956).

[35] Osserman, R., *Minimal varieties;* Bull. Amer. Math. Soc. 75, 1092-1120 (1969).

[36] Ruh, E.A., and J. Vilms, *The tension field of the Gauss map;* Trans. Amer. Math. Soc. 149, 569-573 (1970).

[37] Serrin, J., *Local behavior of solutions of quasilinear equations;* Acta Math. 111, 247-302 (1964).

[38] Serrin, J., *Singularities of solutions of nonlinear equations;* Proc. Symp. Appl. Math. Amer. Math. Soc. 17, 68-88 (1965).

[39] Serrin, J., and H.F. Weinberger, *Isolated singularities of solutions of linear elliptic equations;* Amer. J. Math. 88, 258-272 (1966).

[40] Simon, L., *Asymptotic behaviour of minimal graphs over exterior domains;* Ann. Inst. Henri Poincaré, Analyse non linéaire 4, 231-242 (1987).

[41] Stredulinski, E.W., *Weighted inequalities and degenerate partial differential equations;* Springer Lecture Notes in Math. 1074 (1984).

[42] Trudinger, N.S., *On Harnack type inequalities and their application to quasilinear elliptic equations;* Comm. Pure Appl. Math. 20, 721-747 (1967).

[43] Wiegner, M., *Ein optimaler Regularitätssatz für schwache Lösungen gewisser elliptischer Systeme;* Math. Z. 147, 21-28 (1976).

[44] Wiegner, M., *A-priori Schranken für Lösungen gewisser elliptischer Systeme;* manuscripta math. 18, 279-297 (1976).

[45] Wiegner, M., *Das Existenz- und Regularitätsproblem bei Systemen nichtlinearer elliptischer Differentialgleichungen;* Habilitationsschrift, Bochum (1977).

[46] Wiegner, M., *On two-dimensional elliptic systems with a one-sided condition;* Math. Z. 178, 493-500 (1981).

DECOMPOSITION THEOREMS AND THEIR APPLICATION TO NON-LINEAR ELECTRO- AND MAGNETO-STATIC BOUDNARY VALUE PROBLEMS

A. Milani
Departimento di Matematica, Università di Torino
Italia

R. Picard
Department of Mathematical Sciences,
The University of Wisconsin-Milwaukee, USA

0. Introduction

Hodge-Kodaira type decompositions and their applications in linear electro- and magneto-statics are well-known. As it will be shown in this paper they also can be applied to solve the standard static boundary value problems in fairly general non-linear media.

We shall attempt to demonstrate the applicability of some variants of these decompositions to the solution of the following boundary value problems:

$$(0.1)\quad \begin{cases} \left.\begin{array}{l} \operatorname{curl} E = J\,, \\ \operatorname{div} D = Q\,, \end{array}\right\} \text{ in a domain } G \text{ of } \mathbb{R}^3 \\ n \times E = 0 \text{ on } \partial G\,, \\ \text{and} \\ \left.\begin{array}{l} \operatorname{curl} H = K\,, \\ \operatorname{div} B = R\,, \end{array}\right\} \text{ in } G \subset \mathbb{R}^3 \\ n \cdot B = 0 \text{ on } \partial G\,, \end{cases}$$

where E,D denote the electric field and the displacement current, H,B are the magnetic field and magnetic induction respectively. J is called current density, Q is the charge density. K,R denote the analogous magnetic quantities which can be thought of as introduced by a known magnetic flow of B through the surface ∂G.

We shall assume that ∂G is compact. Therefore we will essentially have to distinguish two cases: the interior and the exterior domain case.

The connection between E,D and H,B respectively is described by the following constitutive relations

$$(0.2) \qquad D = \epsilon(E) \quad , \quad B = \mu(H) \ ,$$

where ϵ, μ shall be specified later.

We shall employ the Hilbert space approach known from the linear theory to obtain corresponding results for the non-linear case. For this purpose we have to specify the meaning of (0.1),(0.2) suitably. Due to refinements in the formulation of the problem we shall obtain results that lead to improvements of known results even if restricted to the linear case. This refers in particular to the handling of the occurence of non-trivial homogeneous solutions of (0.1),(0.2) in the linear case.

In the first section we shall develop the main tools for the solution theory that is given in section 2. The following section will also provide the necessary terminology to formulate (0.1),(0.2) in a more precise way.

1. Decomposition Theorems

To introduce suitable Hilbert spaces to interpret (0.1) we shall use a definition scheme already employed in [Pi6]. Let a be one of the differential expressions 'curl, grad, div' . Then we introduce the Hilbert spaces

$$(1.1) \qquad \begin{aligned} H(a,G) &:= \left\{ \phi \in L_2(G) \mid a\phi \in L_2(G) \right\} \\ \mathring{H}(a,G) &:= \left\{ \phi \in H(a,G) \mid (a\phi,\psi) = (\phi, a^*\psi) \quad \text{for all} \quad \psi \in H(a^*,G) \right\} , \end{aligned}$$

where a^* denotes the formal adjoint of a with respect to the inner product $(\cdot,\cdot)$ of the space of real-valued, measurable, square integrable functions or vector fields $L_2(G)$ defined on a region $G \subset \mathbb{R}^3$. In order to simplify the notation we do not indicate the number of components (1 or 3) of the elements of these L_2-spaces. This information will be clear from the context. The norm of $L_2(G)$ will be denoted by $\|\cdot\|$. The inner product of $H(a,G)$, $\mathring{H}(a,G)$ is of course the graph inner product

$$(\cdot,\cdot) + (a\cdot,a\cdot) \ .$$

Apparently, we have

$$(1.2) \qquad \mathring{H}(a,G) \subset H(a,G) \ ,$$

i.e. $\mathring{H}(a,G)$ is a subspace of $H(a,G)$. For convenience we also introduce the subspaces

$$H_o(a,G) = \left\{\phi \in H(a,G) \mid a\phi = 0\right\},$$

(1.3)

$$\mathring{H}_o(a,G) = \left\{\phi \in \mathring{H}(a,G) \mid a\phi = 0\right\},$$

and

(1.4)
$$\mathring{H}_o(a,b,G) = \left\{\phi \in \mathring{H}_o(a,G) \mid b\phi = 0\right\} = \mathring{H}_o(a,G) \cap H_o(b,G) .$$

Here b is one of the formal differential operators 'div,curl' . The definition (1.4) will only be used for a either 'curl' , or 'div' and $a \neq b$. With the same convention we introduce the Hilbert space

(1.5)
$$\mathring{H}(a,b,G) = \mathring{H}(a,G) \cap H(b,G) ,$$

with inner product

$$(\cdot,\cdot) + (a\cdot,a\cdot) + (b\cdot,b\cdot) .$$

We shall usually skip the reference to the region G if it is clear from the context.

Using the language developed so far we may describe the set where we shall seek for solutions in the following way:

(1.6)
$$E \in \mathring{H}(\mathrm{curl}) \quad , \quad D \in H(\mathrm{div}) \quad , \quad H \in H(\mathrm{curl}) \quad , \quad B \in \mathring{H}(\mathrm{div}) .$$

Here the first and the last statement generalizes not only the existence of curl E and div B in the sense of distributional vector fields in $L_2(G)$, but also the boundary conditions $n\times E = 0$, $n\cdot B = 0$ on the boundary ∂G of G . In the following we shall occasionally use the classical notation '$n\times E = 0$, $n\cdot B = 0$' etc. in a symbolic way to refer to the respective boundary condition in a more readily understandable way. It should be kept in mind, however, that strictly speaking expressions like $n\times E, n\cdot B$ are meaningless, since the generalizations of the boundary conditions used here need no restriction on the regularity of the boundary.

We note that the Hilbert spaces

$$\mathring{H}_o(\mathrm{curl},\mathrm{div}) \quad \text{and} \quad \mathring{H}_o(\mathrm{div},\mathrm{curl})$$

are the spaces of harmonic vector fields with vanishing tangential or normal component on the boundary ∂G (in the above sense).

We shall assume that

(1.7) ∂G is compact and separates $\mathbb{R}^3$ into non-empty open domains.

As a consequence of (1.7) we have

$$\text{(1.8)} \quad \text{and} \quad \begin{aligned} H_o(\mathrm{grad},G) &= \mathrm{span}\left(\{\chi_i(G)\}_i\right) \\ \mathring{H}_o(\mathrm{grad},G) &= \{0\} \;. \end{aligned}$$

The functions $\chi_i(G)$ in (1.8) are the characteristic functions of the bounded, connected components of G. We note in particular that $H(\mathrm{grad},G)$, $\mathring{H}(\mathrm{grad},G)$ are the usual Sobolev spaces $W_1(G)$, $\mathring{W}_1(G)$ (see e.g. [Le], p.10,11). A simple application of the projection theorem for Hilbert spaces leads to the following Lemma (compare [Pi2]).

<u>Lemma 1:</u> *We have the following orthogonal decompositions of the (vectorial) space* $L_2(G)$:

$$\begin{aligned} L_2(G) &= \overline{\mathrm{grad}\,\mathring{H}(\mathrm{grad})} \oplus \mathring{H}_o(\mathrm{curl},\mathrm{div}) \oplus \overline{\mathrm{curl}\,H(\mathrm{curl})} \\ &= \overline{\mathrm{grad}\,\mathring{H}(\mathrm{grad})} \oplus H_o(\mathrm{div}) \\ &= \mathring{H}_o(\mathrm{curl}) \oplus \overline{\mathrm{curl}\,H(\mathrm{curl})} \end{aligned}$$

(1.9) *and*

$$\begin{aligned} L_2(G) &= \overline{\mathrm{grad}\,H(\mathrm{grad})} \oplus \mathring{H}_o(\mathrm{div},\mathrm{curl}) \oplus \overline{\mathrm{curl}\,\mathring{H}(\mathrm{curl})} \\ &= \overline{\mathrm{grad}\,H(\mathrm{grad})} \oplus \mathring{H}_o(\mathrm{div}) \\ &= H_o(\mathrm{curl}) \oplus \overline{\mathrm{curl}\,\mathring{H}(\mathrm{curl})} \;. \end{aligned}$$

As an immediate consequence of Lemma 1 we have the following decomposition result.

<u>Lemma 2:</u> *The Hilbert spaces* $H(\mathrm{curl})$, $H(\mathrm{div})$ *allow the following representations as direct sums:*

$$\begin{aligned} \mathring{H}(\mathrm{curl}) &= \overline{\mathrm{grad}\,\mathring{H}(\mathrm{grad})} \oplus \mathring{H}_o(\mathrm{curl},\mathrm{div}) \oplus \overline{\mathrm{curl}\,H(\mathrm{curl})} \cap \mathring{H}(\mathrm{curl}) \\ &= \overline{\mathrm{grad}\,\mathring{H}(\mathrm{grad})} \oplus H_o(\mathrm{div}) \cap \mathring{H}(\mathrm{curl}) \\ &= \mathring{H}_o(\mathrm{curl}) \oplus \overline{\mathrm{curl}\,H(\mathrm{curl})} \cap \mathring{H}(\mathrm{curl}) \;, \end{aligned}$$

$$\mathring{H}(\mathrm{div}) = \overline{\mathrm{grad}\, H(\mathrm{grad})} \cap \mathring{H}(\mathrm{div}) \oplus \mathring{H}_0(\mathrm{div},\mathrm{curl}) \oplus \overline{\mathrm{curl}\, \mathring{H}(\mathrm{curl})}$$

$$= \overline{\mathrm{grad}\, H(\mathrm{grad})} \cap \mathring{H}(\mathrm{div}) \oplus \mathring{H}_0(\mathrm{div})$$

$$= H_0(\mathrm{curl}) \cap \mathring{H}(\mathrm{div}) \oplus \overline{\mathrm{curl}\, \mathring{H}(\mathrm{curl})} .$$

In these decompositions $\mathring{H}(\mathrm{curl})$, $\mathring{H}(\mathrm{div})$ *can be replaced by* $H(\mathrm{curl})$, $H(\mathrm{div})$ *respectively, and the direct sum representations remain valid.*
Moreover, we have

$$H(\mathrm{div}) = \overline{\mathrm{grad}\, \mathring{H}(\mathrm{grad})} \cap H(\mathrm{div}) \oplus \mathring{H}_0(\mathrm{curl},\mathrm{div}) \oplus \overline{\mathrm{curl}\, H(\mathrm{curl})}$$

$$= \overline{\mathrm{grad}\, \mathring{H}(\mathrm{grad})} \cap H(\mathrm{div}) \oplus H_0(\mathrm{div})$$

$$= \mathring{H}_0(\mathrm{curl}) \cap H(\mathrm{div}) \oplus \overline{\mathrm{curl}\, H(\mathrm{curl})} ,$$

$$H(\mathrm{curl}) = \overline{\mathrm{grad}\, H(\mathrm{grad})} \oplus \mathring{H}_0(\mathrm{div},\mathrm{curl}) \oplus \overline{\mathrm{curl}\, \mathring{H}(\mathrm{curl})} \cap H(\mathrm{curl})$$

$$= \overline{\mathrm{grad}\, H(\mathrm{grad})} \oplus \mathring{H}_0(\mathrm{div}) \cap H(\mathrm{curl})$$

$$= H_0(\mathrm{curl}) \oplus \overline{\mathrm{curl}\, \mathring{H}(\mathrm{curl})} \cap H(\mathrm{curl}) .$$

Lemma 2 implies the following result.

<u>Lemma 3:</u> *We have*

$$\mathrm{curl}\, H(\mathrm{curl}) = \mathrm{curl}\left(\overline{\mathrm{curl}\, \mathring{H}(\mathrm{curl})} \cap H(\mathrm{curl})\right) ,$$

$$\mathrm{curl}\, \mathring{H}(\mathrm{curl}) = \mathrm{curl}\left(\overline{\mathrm{curl}\, H(\mathrm{curl})} \cap \mathring{H}(\mathrm{curl})\right) ,$$

$$\mathrm{div}\ \ H(\mathrm{div}) = \mathrm{div}\left(\overline{\mathrm{grad}\, \mathring{H}(\mathrm{grad})} \cap H(\mathrm{div})\right) ,$$

$$\mathrm{div}\ \ H(\mathrm{div}) = \mathrm{div}\left(\overline{\mathrm{grad}\, H(\mathrm{grad})} \cap \mathring{H}(\mathrm{div})\right) .$$

<u>Proof:</u> The statements follow directly from Lemma 2 by differentiation. □

Another consequence of Lemma 2 is the following density result.

<u>Lemma 4:</u> *The set*

$H(\mathrm{curl}) \cap \overline{\mathrm{curl}\, \mathring{H}(\mathrm{curl})}$ *is dense in the subspace* $\overline{\mathrm{curl}\, \mathring{H}(\mathrm{curl})}$.

Similarly, we have

$\mathring{H}(\mathrm{curl}) \cap \overline{\mathrm{curl}\, H(\mathrm{curl})}$ *is dense in the subspace* $\overline{\mathrm{curl}\, H(\mathrm{curl})}$.

Proof: Noting that $\mathring{H}(\mathrm{curl})$ is dense in $L_2(G)$, we only have to apply the decompositions of Lemma 2. The respective projections are continuous (in $L_2(G)$) and therefore the desired result follows.

□

To proceed in the description of the intended approach we need more assumptions on the boundary of G . In order to give an easily accessible assumption, we shall require that

(1.10) G is a Lipschitz transformation of a domain with smooth boundary.

Let χ_K denote the characteristic function of $K \subset \mathbb{R}^3$. According to [Pi4] our assumption on G guarantees the following compactness results:

$$\begin{aligned} &\chi_K : \mathring{H}(\mathrm{curl},\mathrm{div}) \longrightarrow L_2(G) , \\ &\chi_K : \mathring{H}(\mathrm{div},\mathrm{curl}) \longrightarrow L_2(G) , \end{aligned} \qquad (1.11)$$

and

$$\begin{aligned} &\chi_K : \mathring{H}(\mathrm{grad}) \longrightarrow L_2(G) , \\ &\chi_K : H(\mathrm{grad}) \longrightarrow L_2(G) , \end{aligned} \qquad (1.12)$$

are compact mappings for any bounded domain $K \subset G$.

In case G is bounded (1.11),(1.12) imply the compact imbedding results

$$\begin{aligned} &\mathring{H}(\mathrm{curl},\mathrm{div}) \to\to L_2(G) , \\ &\mathring{H}(\mathrm{div},\mathrm{curl}) \to\to L_2(G) , \\ &\mathring{H}(\mathrm{grad}) \to\to L_2(G) , \\ &H(\mathrm{grad}) \to\to L_2(G) . \end{aligned} \qquad (1.13)$$

If

$$L_2^{loc}(G) = \left\{\phi \,|\, \phi \text{ regular distribution s.th. } \chi_K\phi \in L_2(G) \text{ for any measurable } K \subset \mathbb{R}^3\right\} ,$$

denotes the topological vector space of measurable, locally square integrable functions or vector fields respectively, we obtain from (1.11),(1.12) the local imbedding result:

$$\text{(1.14)} \quad \begin{aligned} \mathring{H}(\mathrm{curl},\mathrm{div}) &\to\to L_2^{loc}(G) , \\ \mathring{H}(\mathrm{div},\mathrm{curl}) &\to\to L_2^{loc}(G) , \\ \mathring{H}(\mathrm{grad}) &\to\to L_2^{loc}(G) , \\ H(\mathrm{grad}) &\to\to L_2^{loc}(G) . \end{aligned}$$

Assumption (1.10) also yields the existence of particular representers of a base of the absolute and relative cohomology groups of G. These will be used to formulate suitable conditions to force a unique solution of the respective boundary value problem. For the smooth case the construction is elementary and can be performed in a way shown in [Pi5], (note the modification in the unbounded case given in [Pi6]). According to the results in [Pi4] the essential properties (see below) are Lipschitz invariant and can therefore be transferred to G. We shall not give the details here, but rather rephrase the result of this construction in a way suited for our purposes ('$\perp$' denotes orthogonality, as upper index it denotes the orthocomplement):

($\mathfrak{B}_e$) There exists a finite set $\mathfrak{B}_e \subset \mathring{H}_o(\mathrm{curl})$ that is linearly independent modulo elements in the closure of $\mathrm{grad}\, \mathring{H}(\mathrm{grad})$ with the property:

$$\mathring{H}_o(\mathrm{curl},\mathrm{div}) \cap \mathfrak{B}_e^{\perp} = \{0\} .$$

($\mathfrak{B}_m$) There exists a finite set $\mathfrak{B}_m \subset \mathring{H}_o(\mathrm{div})$ that is linearly independent modulo elements in the closure of $\mathrm{curl}\, \mathring{H}(\mathrm{curl})$ with the property:

$$\mathring{H}_o(\mathrm{div},\mathrm{curl}) \cap \mathfrak{B}_m^{\perp} = \{0\} .$$

We note here that by their construction elements of $\mathfrak{B}_e, \mathfrak{B}_m$ have compact support in G. The number of elements in $\mathfrak{B}_e$ is given by the number β_1 of connected components of the boundary ∂G minus 1 . In the unbounded case the point at ∞ has to be counted as a connected component of the boundary. The cardinality of $\mathring{\mathscr{B}}_m$ is determined by the number β_2 of handles of domain G . Alternatively to $\mathring{\mathscr{B}}_m$ the following set can be used:

$(\mathscr{B}_m)$ There exists a finite set $\mathscr{B}_m \subset H_o(\mathrm{curl})$ that is linearly independent modulo elements in the closure of $\mathrm{grad}\, H(\mathrm{grad})$ with the property:

$$\mathring{H}_o(\mathrm{div},\mathrm{curl}) \cap \mathscr{B}_m^{\perp} = \{0\} .$$

Elements in $\mathscr{B}_m$ have bounded support in G . A similar replacement for $\mathring{\mathscr{B}}_e$ can be found analogously if G is bounded:

$(\mathscr{B}_e)$ There exists a finite set $\mathscr{B}_e \subset H_0(\mathrm{div})$ that is linearly indpendent modulo elements in the closure of $\mathrm{curl}\, H(\mathrm{curl})$ with the property:

$$\mathring{H}_o(\mathrm{curl},\mathrm{div}) \cap \mathscr{B}_e^{\perp} = \{0\} .$$

In the case where G is unbounded a different construction has to be used - as has been pointed out in [Pi6]:

$(\mathscr{B}_e)'$ There exists a finite set of continuous functionals $\mathscr{B}_e$ defined on

$$D := \left\{\phi \in L_2(\mathbb{R}^3) \mid \phi = 0 \text{ in } \mathbb{R}^3\setminus G\right\} \cup \overline{\mathrm{curl}\, H(\mathrm{curl},\mathbb{R}^3)}$$

as a subspace of $L_2(\mathbb{R}^3)$, such that any $\Lambda \in \mathscr{B}_e$ annihilates elements in $\mathrm{grad}\, \mathring{H}(\mathrm{grad},G)$ (regarded as element of $L_2(\mathbb{R}^3)$ via continuation by zero) i.e.

(1.15) $$\mathrm{grad}\, \mathring{H}(\mathrm{grad},G) \subset \mathscr{B}_e^a := \left\{\phi \in D \mid \Lambda\phi = 0 \text{ for all } \Lambda \in \mathscr{B}_e\right\} .$$

Moreover, we have

(1.16) $$\overline{\mathrm{curl}\, H(\mathrm{curl},\mathbb{R}^3)} \subset \mathscr{B}_e^a .$$

Finall, $\mathscr{B}_e$ is complete in the sense that

$$\mathring{H}_o(\mathrm{curl},\mathrm{div}) \cap \mathscr{B}_e^a = \{0\} .$$

Apparently, relation (1.15) generalizes the concept of - in G - divergence free fields. Inclusion (1.16) formulates that in a certain sense 'curl Λ' is zero in $\mathbb{R}^3$ for $\Lambda \in \mathfrak{B}_e$. The sense in which those functionals are continuous is considered in [Pi6], but is irrelevant for our purposes.

As an immediate consequence of the above existence statements we obtain a slight modification of Lemma 1.

Lemma 5: *We have the following representations of the (vectorial) space* $L_2(G)$ *as a direct sum*

$$(1.17)\qquad L_2(G) = \overline{\text{grad } \mathring{H}(\text{grad})} \oplus \text{span}(\mathring{\mathfrak{B}}_e) \oplus \overline{\text{curl } H(\text{curl})} ,$$

$$(1.18)\qquad L_2(G) = \overline{\text{grad } H(\text{grad})} \oplus \text{span}(\mathring{\mathfrak{B}}_m) \oplus \overline{\text{curl } \mathring{H}(\text{curl})} ,$$

$$(1.19)\qquad L_2(G) = \overline{\text{grad } H(\text{grad})} \oplus \text{span}(\mathfrak{B}_m) \oplus \overline{\text{curl } \mathring{H}(\text{curl})} .$$

If G *is bounded we also have the decomposition*

$$(1.20)\qquad L_2(G) = \overline{\text{grad } \mathring{H}(\text{grad})} \oplus \text{span}(\mathfrak{B}_e) \oplus \overline{\text{curl } H(\text{curl})} .$$

If G *is unbounded (1.20) may be modified to*

$$(1.21)\qquad L_2(G) \cap \mathfrak{B}_e^a = \overline{\text{grad } \mathring{H}(\text{grad})} \oplus \overline{\text{curl } H(\text{curl})} .$$

The second and the third term in (1.17) and (1.19) as well as the first and the second term in (1.18) and (1.20) are orthogonal.

Moreover, we have

$$\mathring{H}_o(\text{curl}) = \overline{\text{grad } \mathring{H}(\text{grad})} \oplus \text{span}(\mathring{\mathfrak{B}}_e)$$

$$\mathring{H}_o(\text{div}) = \text{span}(\mathring{\mathfrak{B}}_m) \oplus \overline{\text{curl } \mathring{H}(\text{curl})}$$

$$H_o(\text{curl}) = \overline{\text{grad } H(\text{grad})} \oplus \text{span}(\mathfrak{B}_m) .$$

If G *is bounded we also have*

$$H_o(\text{div}) = \text{span}(\mathfrak{B}_e) \oplus \overline{\text{curl } H(\text{curl})} .$$

Proof: Since the reasoning for (1.18) is similar we only show (1.17). The orthogonality statement of the lemma follows immediately from the definition of $\mathfrak{B}_e$ and Lemma 1. For any $f \in L_2(G)$ we have according to Lemma 1:

$$f = f_1 + f_2 + f_3 \quad \text{for some} \quad f_1 \in \overline{\text{grad}\, \mathring{H}(\text{grad})}\ ,\ f_2 \in \mathring{H}_o(\text{curl},\text{div})\ ,$$
$$\text{and} \quad f_3 \in \overline{\text{curl}\, H(\text{curl})}\ .$$

By the properties of $\mathfrak{B}_e$ we have that $f_2 \in \mathring{H}_o(\text{curl},\text{div})$ can be written in a unique way as a sum of two terms h_1 and h_2 with

$$h_1 \in \overline{\text{grad}\, \mathring{H}(\text{grad})} \quad \text{and} \quad h_2 \in \text{span}(\mathfrak{B}_e)\ .$$

Indeed, any $\phi^k \in \mathfrak{B}_e$ can be written as

$$\phi^k = \phi_1^k + \phi_2^k \quad , \quad \phi_1^k \in \overline{\text{grad}\, \mathring{H}(\text{grad})}\ ,\ \phi_2^k \in \mathring{H}_o(\text{curl},\text{div})\ .$$

The set $\{\phi_2^k\}_k$ is linearly independent by the linear independence of the ϕ^k, $k = 1,2,\ldots,\beta_1$, modulo the closure of $\text{grad}\, \mathring{H}(\text{grad})$. Moreover, it is a base of $\mathring{H}_o(\text{curl},\text{div})$, since any $h \in \mathring{H}_o(\text{curl},\text{div})$ orthogonal to all ϕ_2^k, $k = 1,2,\ldots\beta_1$, is also orthogonal to $\mathfrak{B}_e$ and therefore equal to zero.
Consequently,

$$f = (f_1+h_1) + h_2 + f_3\ ,$$

which is the desired representation. Uniqueness is clear since $f \equiv 0$ implies first $f_3 = 0$. Then, since $(\phi_2^k,h_2) = 0$ for all k it follows that $h_2 = 0$, and thus $f_1 + f_2 = 0$. The remaining decomposition results follow by comparison with Lemma 1.

□

Because of the technical differences between the interior and exterior domain case we shall now separate the discussion of these two cases.

1.1 G bounded

For a bounded domain we have - as indicated in (1.13) - a compact imbedding result that allows to formulate Friedrich's type inequalities.

Lemma 6: *There is a uniform constant* C *such that*

$$\|E\| \leq C\Big(\|\text{curl } E\| + \|\text{div } E\| + \sum_{\phi \in \mathring{\mathcal{B}}_e} |(\phi,E)|\Big), \tag{1.22}$$

$$\|E\| \leq C\Big(\|\text{curl } E\| + \|\text{div } E\| + \sum_{\phi \in \mathcal{B}_e} |(\phi,E)|\Big), \tag{1.23}$$

$$\|H\| \leq C\Big(\|\text{curl } H\| + \|\text{div } H\| + \sum_{\phi \in \mathring{\mathcal{B}}_e} |(\phi,H)|\Big), \tag{1.24}$$

$$\|H\| \leq C\Big(\|\text{curl } H\| + \|\text{div } H\| + \sum_{\phi \in \mathcal{B}_e} |(\phi,H)|\Big). \tag{1.25}$$

for all $E \in \mathring{H}(\text{curl},\text{div})$ *and* $H \in \mathring{H}(\text{div},\text{curl})$.

Proof: The four estimates can be shown by a standard contradiction argument making use of the compact imbedding result (1.13) (see [Fr], [Pil]).

□

Due to Poincaré's inequality we know that $\text{grad } H(\text{grad})$ and $\text{grad } \mathring{H}(\text{grad})$ are closed subspaces of $L_2(G)$. By a similar reasoning we obtain

Lemma 7: *The following spaces are closed subspaces of* $L_2(G)$:

$$\begin{aligned} \text{curl } H(\text{curl}) &= \text{curl } \big(\text{curl } H(\text{curl}) \cap H(\text{curl})\big), \\ \text{curl } \mathring{H}(\text{curl}) &= \text{curl } \big(\text{curl } H(\text{curl}) \cap \mathring{H}(\text{curl})\big), \\ \text{div } H(\text{div}) &= \text{div } \big(\text{grad } H(\text{grad}) \cap H(\text{div})\big), \\ \text{div } \mathring{H}(\text{div}) &= \text{div } \big(\text{grad } H(\text{grad}) \cap \mathring{H}(\text{div})\big). \end{aligned} \tag{1.26}$$

Proof: The closedness for the spaces on the right-hand side follows using Lemma 6. Applying the previous lemma to sequences $(\text{curl } \phi_n)_n$ in

$$\begin{aligned} &\text{curl}\big(\overline{\text{curl } \mathring{H}(\text{curl})} \cap H(\text{curl})\big), \\ &\text{curl}\big(\text{curl } H(\text{curl}) \cap \mathring{H}(\text{curl})\big), \text{ respectively,} \end{aligned} \tag{1.27}$$

yields the convergence of $(\phi_n)_n$ in $L_2(G)$. Note that $\operatorname{curl} H(\operatorname{curl}) \perp \mathcal{B}_e$ and $\operatorname{curl} \mathring{H}(\operatorname{curl}) \perp \mathcal{B}_m$. By the closedness of the differential operators involved and taking Lemma 3 into account this shows the closedness of the spaces in (1.27) and also that the closure bar can be omitted. Thus, the first two statements are shown to be valid. The closedness of the other two spaces follows in the same way.

□

1.2 G unbounded

In order to handle this case we first have to extend the spaces introduced so far. The asymptotic behaviour of the fundamental solution of Maxwell's equation suggests to consider analogous Hilbert spaces as in the bounded case weighted however with a weight function $\rho = (1+|\cdot|)^{-1}$, where $|\cdot|$ denotes the Euclidean norm in $\mathbb{R}^3$. We first define

$$L_2^{\sim}(G) = \left\{\phi \in L_2^{loc}(G) \mid \rho\phi \in L_2(G)\right\} \tag{1.28}$$

with inner product

$$(\cdot,\cdot)^{\sim} = (\rho\cdot,\rho\cdot) . \tag{1.29}$$

The corresponding norm will be denoted by $\|\cdot\|^{\sim}$. Based on (1.28), (1.29) we define

$$H^{\sim}(a,G) := \left\{\phi \in L_2^{\sim}(G) \mid a\phi \in L_2(G)\right\} ,$$
$$\mathring{H}^{\sim}(a,G) := \left\{\phi \in H^{\sim}(a,G) \mid (a\phi,\psi) = (\phi,a^*\psi) \text{ for all } \psi \in H(a^*,G) \text{ with bounded support}\right\} .$$

Here a is again one of the formal differential operators 'curl,div,grad' and a^* its formal adjoint. In analogy to (1.5) we define the Hilbert space

$$\mathring{H}^{\sim}(a,b,G) = \mathring{H}^{\sim}(a,G) \cap H^{\sim}(b,G) . \tag{1.30}$$

We shall use analogous conventions with respect to these weighted spaces as we did for the unweighted spaces.
We note in particular that elements with bounded support are dense in these weighted spaces. Thus, we have

$$\begin{array}{lll} \mathring{H}(a,G) & \text{dense in} & \mathring{H}^{\sim}(a,G) , \\ H(a,G) & \text{dense in} & H^{\sim}(a,G) , \\ \mathring{H}(a,b,G) & \text{dense in} & \mathring{H}^{\sim}(a,b,G) . \end{array} \tag{1.31}$$

For the case that G is unbounded we have to modify some of our previous results in a suitable way. First we note the following variant of Friedrichs' inequality.

<u>Lemma 8:</u> *There is a uniform constant* C *such that*

$$(1.32) \qquad \|E\|^{\sim} \leq C\Big(\|\mathrm{curl}\ E\| + \|\mathrm{div}\ E\| + \sum_{\phi\in\mathfrak{B}_e} |(\phi,E)|\Big)$$

$$(1.33) \qquad \|H\|^{\sim} \leq C\Big(\|\mathrm{curl}\ H\| + \|\mathrm{div}\ H\| + \sum_{\phi\in\mathfrak{B}_e} |(\phi,H)|\Big)$$

for all $E \in \mathring{H}^{\sim}(\mathrm{curl},\mathrm{div})$ *and* $H \in \mathring{H}^{\sim}(\mathrm{div},\mathrm{curl})$.

<u>Proof:</u> Since the reasoning is analogous in both cases we shall only consider inequality (1.32).

Let ζ be a C_∞-function with compact support in $\mathbb{R}^3$ such that $\zeta = 1$ in a ball B containing ∂G and the supports of all elements of $\mathfrak{B}_e$. For any vector field $E \in \mathring{H}^{\sim}(\mathrm{curl},\mathrm{div})$ we have

$$E = \zeta E + (1-\zeta)E .$$

Without loss of generality we may also assume that E is a smooth vector field on $\mathrm{supp}(1-\zeta)$ and has bounded support in G . Applying Lemma 6 to ζE yields

$$(1.34) \qquad \|\zeta E|^{\sim} \leq C\Big(\|\mathrm{curl}\ E\| + \|\mathrm{div}\ E\| + \sum_{\phi\in\mathfrak{B}_e} |(\phi,E)| + \|\chi_Z E\|\Big) ,$$

for some constant C (all norms and inner products are with respect to G). Here Z denotes a compact set containing the support of $\mathrm{grad}\ \zeta$. For the term $(1-\zeta)E$ we have the following Poincaré type inequality in $\mathbb{R}^3$

$$(1.35) \qquad \Big((1-\zeta)E,(1-\zeta)E\Big)^{\sim} \leq C \sum_{i=1}^{n} \Big\|\partial_i(1-\zeta)E\Big\|^2 = C\Big(\Big\|\mathrm{curl}(1-\zeta)E\Big\|^2 + \Big\|\mathrm{div}(1-\zeta)E\Big\|^2\Big) .$$

Taking (1.34),(1.35) together we get

$$(1.36) \qquad \|E\|^{\sim} \leq C\Big(\|\mathrm{curl}\ E\| + \|\mathrm{div}\ E\| + \sum_{\phi\in\mathfrak{B}_e} |(\phi,E)| + \|\chi_Z E\|\Big) .$$

Here again all norms and inner products refer to the domain G . Estimate (1.32) now follows again by a standard contradiction argument based on (1.14) (compare [Pil]).

□

Again referring to weighted Poincaré type inequalities we know that

$$\text{(1.37)}\qquad \begin{aligned} \overline{\operatorname{grad} H(\operatorname{grad})} &= \operatorname{grad} \tilde{H}(\operatorname{grad}) \,, \\ \overline{\operatorname{grad} \mathring{H}(\operatorname{grad})} &= \operatorname{grad} \tilde{\mathring{H}}(\operatorname{grad}) \,. \end{aligned}$$

In a similar way the following closedness result (' $\overline{\qquad}^{\sim}$ ' means closure in $\tilde{L}_2(G)$) follows from Lemma 8.

<u>Lemma 9:</u> *The sets*

$$\operatorname{curl} \tilde{H}(\operatorname{curl}) \quad , \quad \operatorname{curl} \tilde{\mathring{H}}(\operatorname{curl})$$

are closed subspaces of $L_2(G)$.

Moreover, we have

$$\overline{\operatorname{curl} H(\operatorname{curl})} = \operatorname{curl} \tilde{H}(\operatorname{curl}) = \operatorname{curl}\left(\tilde{H}(\operatorname{curl}) \cap \overline{\operatorname{curl} \mathring{H}(\operatorname{curl})}^{\sim}\right) ,$$

$$\overline{\operatorname{curl} \mathring{H}(\operatorname{curl})} = \operatorname{curl} \tilde{\mathring{H}}(\operatorname{curl}) = \operatorname{curl}\left(\tilde{\mathring{H}}(\operatorname{curl}) \cap \overline{\operatorname{curl} H(\operatorname{curl})}^{\sim}\right) .$$

<u>Proof:</u> The result follows immediately from Lemma 3, Lemma 4 and Lemma 8. Note that

$$\mathcal{B}_m \perp \overline{\operatorname{curl} \mathring{H}(\operatorname{curl})} \quad \text{and} \quad \mathring{\mathcal{B}}_e \perp \overline{\operatorname{curl} H(\operatorname{curl})} \,.$$

□

Finally we note the following density lemma that corresponds to Lemma 4.

<u>Lemma 10:</u> *The set*

$$\tilde{H}(\operatorname{curl}) \cap \overline{\operatorname{curl} \mathring{H}(\operatorname{curl})}^{\sim} \quad \textit{is dense in the subspace} \quad \overline{\operatorname{curl} \mathring{H}(\operatorname{curl})}^{\sim}$$

of $\tilde{L}_2(G)$.

Similarly, we have

$\overset{\circ}{H}^{\sim}(\text{curl}) \cap \overline{\text{curl } H(\text{curl})}^{\sim}$ *is dense in the subspace* $\overline{\text{curl } H(\text{curl})}^{\sim}$ *of* $L_2^{\sim}(G)$.

Proof: The result follows from the observation that any u in the $L_2^{\sim}(G)$-closure of $\text{curl } \overset{\circ}{H}(\text{curl})$ can be approximated by smooth elements of the form $\text{curl } \phi_n$ of $\overset{\circ}{H}(\text{div})$, n positive integer, such that ϕ_n has bounded support and

$$\text{curl } \phi_n \longrightarrow u \text{ in } L_2^{\sim}(G) \quad \text{as} \quad n \longrightarrow \infty .$$

Thus, the first statement follows. A similar reasoning by using the techniques of [MS] shows the second density statement.

□

We have now prepared the necessary tools to handle the non-linear boundary value problems of electro-magnetic theory. The approach will be via the theory of monotone operators (see e.g. [Br],[Li]) and is developed in the next section.

2. Application to Boundary Value Problems of Electro- and Magneto-Statics

In order to solve problem (0.1),(0.2) we shall reformulate it as an abstract equation of the form

$$A(u) = F , \tag{2.1}$$

where u is to be found in a certain Hilbert space V continuously and densely imbedded into another Hilbert space H (with norm $\|\cdot\|$ and inner product $(\cdot,\cdot)$), F is a given right-hand side in the normed dual V' of V and $A : V \longrightarrow V'$ is a monotone operator. Thus, by the usual identification of H with its dual, we have a Gelfand triple

$$V \longrightarrow H \longrightarrow V' . \tag{2.2}$$

The operator A is usually given by a continuous form $\alpha : V \times V \longrightarrow \mathbb{R}$ via the relations

$$\langle A(u),v\rangle = \alpha(u,v) \quad \text{for all} \quad v \in V$$

(2.3) and

$$u \in D(A) = \left\{ u \in V \mid \alpha(u,\cdot) \ \text{ is a continuous functional on } \ V \right\}.$$

Consequently, (2.1) may be written in variational form

$$\alpha(u,v) = \langle F,v\rangle \quad \text{for all} \quad v \in V \ . \tag{2.4}$$

We recall that solvability of problems (2.1),(2.4) is assured by the following proposition.

<u>Proposition:</u> *Let* A *(or* α *) be a strongly monotone and coercive operator (or form), i.e. there is a constant* $\ell > 0$ *such that*

$$\langle A(u) - A(v) \ , \ u - v\rangle \geq \ell\|u-v\|^2 \quad (\text{or} \quad \alpha(u,u-v) - \alpha(v,u-v) \geq \ell\|u-v\|^2) \ ,$$

and

$$\langle A(u),u\rangle \geq \ell\|u\|^2 \quad , \quad (\text{or} \quad \alpha(u,u) \geq \ell\|u\|^2) \ ,$$

for all $u,v \in D(A)$ $(u,v \in V)$ *. Then there exists a unique solution of* (2.1) *(or* (2.4)*).*

In order to satisfy the assumption of this proposition we assume:

<u>Assumption:</u> ϵ,μ and also their inverses are Lipschitz continuous and strongly monotone mappings from $L_2(G)$ to $L_2(G)$.

2.1 <u>The Electro-Static Boundary Value Problem</u>

We want to find a solution $E \in H(\text{curl})$, $D \in H(\text{div})$ of the system

$$\left.\begin{array}{l} \text{curl } E = J \ , \\ \text{div } D = Q \ , \end{array}\right. \quad \text{in} \quad G \subset \mathbb{R}^3 \tag{2.5}$$

such that

$$n \times E = 0 \quad \text{on} \quad \partial G \ , \tag{2.6}$$

in the generalized sense (i.e. $E \in \mathring{H}(\text{curl})$) and

$$D = \epsilon(E) \ . \tag{2.7}$$

As necessary solvability conditions we obtain immediately:

(2.8) $\quad J \in \mathring{H}_o(\mathrm{div}) \quad , \quad J \perp \mathfrak{B}_m \; , \; Q \in L_2(G) \; .$

In order to obtain uniqueness we have to impose a finite number of additional conditions:

(2.9) $\quad (D,\phi) = c_\phi \quad \text{for all} \quad \phi \in \mathring{\mathfrak{B}}_e \; ,$

where $c_\phi, \phi \in \mathring{\mathfrak{B}}_e$, are given real numbers.

We shall see that conditions (2.8) are also sufficient to obtain a unique solution of problem (2.5),(2.6),(2.7),(2.9).

2.1.1 G bounded

In order to find a solution in the case of a bounded domain we decompose D according to Lemma 1 and take (1.26) into account

(2.10) $\quad D = \mathrm{grad}\;\theta_o + h_o + \mathrm{curl}\;\theta \; ,$

where $\theta_o \in \mathring{H}(\mathrm{grad})$, $h_o \in \mathring{H}_o(\mathrm{curl},\mathrm{div})$ and $\theta \in H(\mathrm{curl})$. Since $h_o + \mathrm{curl}\;\theta$ is in $H_o(\mathrm{div})$ the function θ_o is determined as the unique solution of the Dirichlet problem

(2.11) $\quad \mathrm{div}\;\mathrm{grad}\;\theta_o = \mathrm{div}\;D = Q \; .$

Moreover, since according to (2.9)

$$c_\phi \quad \text{is known for all} \quad \phi \in \mathring{\mathfrak{B}}_e \; ,$$

and

$$\mathrm{curl}\;H(\mathrm{curl}) \perp \mathring{\mathfrak{B}}_e \; ,$$

we have

(2.12) $\quad (h_o,\phi) = c_\phi - (\mathrm{grad}\;\theta_o,\phi) \quad \text{for all} \quad \phi \in \mathring{\mathfrak{B}}_e \; .$

As in the proof of Lemma 5 we see that (2.12) determines h_o uniquely. Thus it remains to determine θ . From (2.5) we get

$$(2.13) \qquad \operatorname{curl} \epsilon^{-1}\big(\operatorname{curl} \theta + (h_o + \operatorname{grad} \theta_o)\big) = J \ .$$

Without loss of generality we may assume (see (1.26)) that

$$(2.14) \qquad \theta \in H := \operatorname{curl} \mathring{H}(\operatorname{curl}) \ .$$

The density of V in H follows according to Lemma 4. In order to solve the problem (2.13),(2.14) we define

$$\alpha(u,v) = \big(\epsilon^{-1}(\operatorname{curl} u + \kappa_o) - \epsilon^{-1}(\kappa_o), \operatorname{curl} v\big) ,$$

where $\kappa_o = h_o + \operatorname{grad} \theta_o$, for $u,v \in V = H(\operatorname{curl}) \cap \operatorname{curl} \mathring{H}(\operatorname{curl})$. Due to the assumption on ϵ we have that α is strongly monotone and coercive so that the Proposition can be applied. I.e. we get a unique solution $\theta \in V$ of

$$(2.15) \qquad \alpha(\theta,v) = (J,v) - \big(\epsilon^{-1}(\kappa_o), \operatorname{curl} v\big) \quad \text{for all} \quad v \in V \ .$$

Note that the right-hand side of (2.15) is indeed a continuous functional in V' by (1.25). A solution of (2.15) apparently also satisfies

$$\big(\epsilon^{-1}(\operatorname{curl} u + \kappa_o), \operatorname{curl} v\big) = (J,v) ,$$

which in turn implies (2.13) and also

$$n \times E = n \times \epsilon^{-1}(D) \equiv n \times \epsilon^{-1}(\operatorname{curl} \theta + \kappa_o) = 0 \quad \text{on} \quad \partial G ,$$

if (2.8) is satisfied. This shows existence. But uniqueness of a solution of (2.5),(2.6),(2.7),(2.9) is clear since in fact this problem is equivalent to (2.11),(2.12),(2.13) and the latter is equivalent to its variational formulation (2.15). Thus, we obtain our first theorem.

Theorem 1: *The interior boundary value problem of electro-statics (2.5),(2.6),(2.7),(2.9) has a unique solution* E,D *iff conditions (2.8) are satisfied.*

The additional condition (2.9) which is formulated with respect to D may be replaced by a conditon for E :

$$(2.16) \qquad (E,\phi) = c_\phi \quad \text{for all} \quad \phi \in \mathcal{B}_e \ .$$

Then by a similar reasoning, starting with a decomposition of E rather than D, we are led to solving the problem

$$\text{div}\, \epsilon(\text{grad}\, \theta + \lambda_o) = Q \quad , \quad \theta \in \overset{\circ}{H}(\text{grad}) \, , \tag{2.17}$$

where λ_o is known from the data J and c_ϕ for $\phi \in \mathfrak{B}_e$. Solving (2.17) by the method of monotone operators results in our second theorem.

<u>Theorem 2:</u> *The interior boundary value problem of electro-statics* (2.5),(2.6),(2.7),(2.16) *has a unique solution* E,D *iff conditions* (2.8) *are satisfied.*

<u>Remark:</u> *The existence part of the proof of* Theorem 1 *involves the a-priori estimate*

$$\begin{aligned} \ell \|\text{curl } u\|^2 &\leq \left(\epsilon^{-1}(\text{curl } u + \kappa_o) - \epsilon^{-1}(\kappa_o) \, , (\text{curl } u + \kappa_o) - \kappa_o\right) \\ &= a(u,u) = (J,u) - \left(\epsilon^{-1}(\kappa_o) \, , \text{curl } u\right) \\ &\leq \|J\| \, \|u\| + \|\epsilon^{-1}(\kappa_o)\| \, \|\text{curl } u\| \end{aligned}$$

(by Lemma 6)

$$\leq C\|J\| \, \|\text{curl } u\| + \|\epsilon^{-1}(\kappa_o)\| \, \|\text{curl } u\| \, .$$

In other words

$$\ell \|\text{curl } u\| \leq \left(C\|J\| + \|\epsilon^{-1}(\kappa_o)\|\right) . \tag{2.18}$$

Such a constant $\ell > 0$ *exists by the monotonicity assumption on* ϵ^{-1}. *If the assumptions on* ϵ, μ *are satisfied only locally, then we obtain a local result provided the data* J,Q,c_ϕ *are sufficiently small for* $\phi \in \mathfrak{B}_e$ *, and* $\epsilon(0) = 0$, $\mu(0) = 0$ *(compare the 'Non-Regular Theorem' in* [Si]). *The 'smallness' has to be determined such that* (2.18) *prevents* curl u *from getting out of the domain of validity of the assumptions on* ϵ, μ . *A similar remark applies to* Theorem 2 *and the following theorems. Note in particular that the coorespondence between data and solutions is one-to-one. The basic idea of reducing the electro-static problem in the way described has also been used by* D. Graffi, see [Gr], p.11-14.

2.1.2 G unbounded

The considerations in this case are very similar to the case of a bounded domain. In fact there are only minor changes like using Lemma 10 instead of Lemma 4. Generally speaking at several stages the unweighted Hilbert spaces have to be replaced by the corresponding weighted variants. Since those modifications are fairly obvious we shall only indicate the reasoning. Eventually, we are led to the analogue of problem (2.13),(2.14) and (2.17) respectively. Restricting our attention to the first problem we have to find

$$\theta \in V := H^{\sim}(\mathrm{curl}) \cap \overline{\mathrm{curl}\ \mathring{H}(\mathrm{curl})}^{\sim} \longrightarrow H := \overline{\mathrm{curl}\ \mathring{H}(\mathrm{curl})}^{\sim} ,$$

such that

$$\mathrm{curl}\ \varepsilon^{-1}(\mathrm{curl}\ \theta + \kappa_o) = J \ . \tag{2.19}$$

Here κ_o is again determined uniquely by Q and the numbers c_ϕ for $\phi \in \mathring{\mathcal{B}}_e$, if we assume that

$$\rho^{-1}Q \in L_2(G) \ . \tag{2.20}$$

Problem (2.19) can be solved in the same way as (2.13) if we assume in addition to (2.8) that

$$\rho^{-1}J \in L_2(G) \ . \tag{2.21}$$

Thus we obtain the following result:

<u>Theorem 3:</u> *The exterior boundary value problem of electro-statics* (2.5),(2.6),(2.7),(2.9) *has a unique solution* E,D *if conditions* (2.8),(2.20),(2.21) *are satisfied. This statement remains valid if* (2.9) *is replaced by*

$$\lambda(E) = c_\lambda \quad \text{for} \quad \lambda \in \mathcal{B}'_e \ . \tag{2.22}$$

<u>Remark:</u> *An 'iff'-condition can be formulated as well. Since we are looking for* L_2*-solutions* E,D *and* curl $\mathring{H}$(curl) , div H(div) *are not closed, such a condition, however, has the undesirable form*

$$J \in \mathrm{curl}\ \mathring{H}(\mathrm{curl}) \quad , \quad Q \in \mathrm{div}\ H(\mathrm{div}) \ . \tag{2.23}$$

2.2 The Magneto-Static Boundary Value Problem

The techniques to solve the magneto-static case are very similar to

those employed in the electro-static case. The underlying similarity can be displayed by using differential forms. Here, however, we prefer to formulate the problems in a vector analytical language which in turn forces us to consider all cases as separate. Since the modifications are only minor and quite obvious, we can, however, be very brief.
The system of equations to be considered is

$$\text{(2.24)} \qquad \begin{aligned} &\operatorname{curl} H = K , \\ &\operatorname{div} B = R , \end{aligned} \quad \text{in } G , \qquad n \cdot B = 0 \text{ on } \partial G .$$

Here

$$\text{(2.25)} \qquad B = \mu(H) .$$

Since we are looking for solutions $H \in H(\text{curl})$, $B \in H(\text{div})$ we have as necessary conditions for solvability of (2.24),(2.25)

$$\text{(2.26)} \qquad K \in H_o(\text{div}) \quad , \quad R \in L_2(G) \quad , \quad R \perp H_o(\text{grad}) .$$

In (2.26) the last condition is implied by the boundary condition for B. We also have to have

$$\text{(2.27)} \qquad K \perp \mathfrak{H}_e .$$

Uniqueness of solutions to (2.24),(2.25) can be guaranteed by the additional conditions

$$\text{(2.29)} \qquad (H,\phi) = c_\phi \quad \text{for} \quad \phi \in \mathfrak{H}_m ,$$

where $c_\phi, \phi \in \mathfrak{H}_m$, are given real numbers. Alternatively we may require the following conditions on B

$$\text{(2.30)} \qquad (B,\phi) = c_\phi \quad \text{for} \quad \phi \in \mathfrak{H}_m .$$

2.2.1 G bounded

In case of condition (2.29) we use the decomposition

$$\text{(2.31)} \qquad H = \operatorname{grad} \theta + h_o + \operatorname{curl} \theta_o ,$$

where $h_o \in \mathring{H}_o(\mathrm{div},\mathrm{curl})$ and $\theta_o \in \mathrm{curl}\,\mathring{H}(\mathrm{curl})$ are uniquely determined by the data c_ϕ for $\phi \in \mathring{\mathcal{B}}_m$ and K. Thus, θ can be uniquely determined as the solution of the Neumann problem

(2.32) $$\mathrm{div}\,\mu\bigl(\mathrm{grad}\,\theta + (h_o + \mathrm{curl}\,\theta_o)\bigr) = R .$$

Thus, equivalence of the magneto-static boundary value problem and the variational formulation of (2.32) can be shown. If assuming conditions of the type (2.30) we base our reasoning on decomposing B in the same way

(2.33) $$B = \mathrm{grad}\,\theta_o + h_o + \mathrm{curl}\,\theta ,$$

where now $\theta \in \mathring{H}(\mathrm{curl}) \cap \mathrm{curl}\,H(\mathrm{curl})$ has to be determined by solving

(2.34) $$\mathrm{curl}\,\mu^{-1}(\mathrm{grad}\,\theta_o + h_o + \mathrm{curl}\,\theta) = K ,$$

with boundary condition

(2.35) $$n\times\theta = 0 \quad \text{on} \quad \partial G .$$

This way we can again establish equivalence. Summarizing we have

Theorem 4: *The interior boundary value problem of magneto-statics* (2.24),(2.25),(2.29) *has a unique solution* H,B *iff conditions* (2.26),(2.27) *are satisfied. This statement remains valid if* (2.29) *is replaced by* (2.30).

2.2.2 G unbounded

Dealing with the unbounded case involves similar modifications as discussed in Section 2.1.2. Taking into account that (2.27) has to be replaced by (2.28) and that in the unbounded case a reasonable additional requirement for K,R is

(2.36) $$\rho^{-1}K, \rho^{-1}R \in L_2(G) ,$$

we obtain

Theorem 5: *The exterior boundary value problem of magneto-statics* (2.24),(2.25),(2.29) *has a unique solution* H,B *if conditions*

(2.26),(2.28),(2.36) are satisfied. This statement remains valid if (2.29) is replaced by (2.30).

Remark: *The comments given in previous remarks apply similarly. In particular, an 'iff'-condition is provided by*

$$(2.37) \qquad K \in \text{curl } H(\text{curl}) \quad , \quad R \in \text{div } \mathring{H}(\text{div}) .$$

Acknowledgements

The present work has been started during a stay of the second author at the Dipartimento di Matematica, Università di Torino, sponsored by the C.N.R., Italia. We gratefully acknowledge the support by the C.N.R. as well as by the SFB 72 at the University of Bonn where the work has been completed. In particular we would like to express our gratitude to the respective institutions for their kind hospitality.

References

[Br] H.Brezis: Operateurs Maximaux Monotones. North Holland Publ., Amsterdam (1973)

[Fr] K.O. Friedrichs: Differential Forms on Riemannian Manifolds. Comm. Pure Appl. Math. 8, 551-590 (1955)

[Gr] D. Graffi: Nonlinear partial differential equations in physical problems. Pitman, Boston, London, Melbourne (1980)

[Le] R. Leis: Initial Boundary Value Problems in Mathematical Physics. B.G. Teubner, Stuttgart, and J. Wiley & Sons Ltd., Chichester (1986)

[Li] J.L. Lions: Quelques méthodes de résolution des problèmes aux limites non-linéaires. Dunod-Gauthiers-Villars, Paris (1969)

[MS] N.G. Meyers & J. Serrin: H = W. Proc. Nat. Acad. Sci. USA 51, 1055-1056 (1964)

[Pi1] R. Picard: Randwertaufgaben der verallgemeinerten Potentialtheorie. Math. Meth. Appl. Sci. 3, 218-228 (1981)

[Pi2] R. Picard: On Boundary Value Problems of Electro- and Magnetostatics. Proc. Roy. Soc. Edin. 92A, 165-174 (1982)

[Pi3] R. Picard: Ein Hodge-Satz für Mannigfaltigkeiten mit nichtglattem Rand. Math. Meth. Appl. Sci. 5, 153-161 (1983)

[Pi4] R. Picard: An Elementary Proof for a Compact Imbedding Result in Generalized Electromagnetic Theory. Math. Z. 187, 151-161 (1984)

[Pi5] R. Picard: On the Low Frequency Asymptotics in Electromagnetic Theory. J. für Reine Angew. Math. 354, 50-73 (1985)

[Pi6] R. Picard: On the Low Frequency Asymptotics in Acoustics. Math. Meth. Appl. Sci. 8, 436-450 (1986)

[Si] L.M. Sibner & R.J. Sibner: A Non-Linear Hodge-DeRham Theorem. Acta Math. 125, 57-73 (1970)

INITIAL BOUNDARY VALUE PROBLEMS IN THERMOELASTICITY

Reinhard Racke
Institut für Angewandte Mathematik der Universität Bonn
Wegelerstr. 10, D-5300 Bonn 1

0. Introduction

We consider a three-dimensional body B, the elastic and thermal behaviour of which is described by the equations of balance of linear momentum and balance of energy (cf. [Ca]):

(0.1) $$\rho X_{tt} = \text{Div } S + \rho b ,$$

(0.2) $$\rho T\eta_t = -\text{Div } q + \rho r ,$$

where we use the following notation:

X,S,b,T,η,q and r are understood to be functions of $x = (x^1,x^2,x^3)$ and of time t, $0 \leq t < \infty$. X^i, $i = 1,2,3$, are the coordinates at time t of that material point of B which has coordinates (x^1,x^2,x^3) when B is in a fixed undistorted reference configuration R. The subindex t denotes a derivative with respect to t, Div the divergence with regard to x. Also we have

ρ : material density in R
S : Piola-Kirchhoff stress tensor
b : specific extrinsic body force
T : temperature
η : specific entropy
q : heat flux
r : specific extrinsic heat supply.

Moreover we set

F : deformation gradient: $F_{ij} = \partial X^i/\partial x^j$
ψ : specific Helmholtz free energy.

In thermoelasticity the constitutive relations are

$$(0.3)\qquad \begin{cases} \psi = \psi(F,T) \\ \eta = -(\partial\psi/\partial T)'(F,T) \quad (\text{': transposition}) \\ S = \rho(\partial\psi/\partial F)(F,T) \end{cases}$$

especially one has from the Clausius-Duhem inequality:

$$(0.4)\qquad q\cdot\nabla T \leq 0 \quad (\nabla\text{: x-gradient}) .$$

Additionally one has *initial conditions* at time $t = 0$ and *boundary conditions*.

The structure of (0.1), (0.2) shows the coupling of a hyperbolic part (elasticity equations) with a parabolic one (heat equation). This coupling is responsible for the arising difficulties and the special effects in the behaviour of solutions.

In the following section 1 we describe the results on the linearized equations having been obtained with support of the SFB 72. In section 1.1 we rewrite the linearized system as an evolution equation of order one in t; in section 1.2 the asymptotic behaviour of the solution as $t \to \infty$ for inhomogeneous, anisotropic media in bounded and in unbounded domains is analyzed; in section 1.3 a spectral theorem is given in the case of a homogeneous, isotropic medium filling $\mathbb{R}^3$, and in section 1.4 aspects of a possible scattering theory are discussed. The results of section 1 apart from those of 1.4 are contained in [Le1-3], [Ra1-2].

In section 2 we give a survey of results for the nonlinear equations, mostly for one-dimensional models. These results are essentially taken from [D&H], [Ra4], [Sl], [Zh], [Z&S].

1.1 The linearized equations as system of order one in t

Let $U := X - x$ be the displacement vector and let $\theta := T - T_o$ be the temperature difference with respect to a fixed reference temperature T_o. Then one obtains for small values of $|\nabla U|$, $|\nabla U_t|$, $|\theta|$, $|\theta_t|$, $|\nabla\theta|$ the system (cf. [Ca], [Le3], [Ra1]):

$$(1.1)\qquad PU_{tt} - D'SDU + D'\Gamma\theta = b ,$$

$$(1.2)\qquad c\cdot\theta_t - \nabla'L\nabla\theta + T_o\Gamma'DU_t = r .$$

Here $P = P(x)$ denotes the mass density matrix and $S = S(x)$ a positive definite 6×6-matrix containing the elastic moduli. $\Gamma = \Gamma(x) = (\gamma_1,\ldots,\gamma_6)'(x)$ consists of the coefficients of the stress temperature tensor, $c = c(x)$ denotes the specific heat, $L = L(x)$ denotes the heat conductivity tensor and D abbreviates the following differential symbol

$$D = \begin{bmatrix} \partial_1 & 0 & 0 \\ 0 & \partial_2 & 0 \\ 0 & 0 & \partial_3 \\ 0 & \partial_3 & \partial_2 \\ \partial_3 & 0 & \partial_1 \\ \partial_2 & \partial_1 & 0 \end{bmatrix} , \quad \partial_j = \partial/\partial x^j .$$

Initial conditions are given by

$$(1.3) \quad U(0,x) = U_o(x) \quad , \quad U_t(0,x) = U_1(x) \quad , \quad \theta(0,x) = \theta_o(x) ,$$

where $x \in G$; $G \subset \mathbb{R}^3$, open, describes the reference configuration R .

As *boundary conditions* we choose for simplicity the Dirichlet boundary conditions:

$$(1.4) \qquad U\big|_{\partial G} = 0 \quad , \quad \theta\big|_{\partial G} = 0 ,$$

that is we consider a model with rigidly clamped boundary and constant temperature at the boundary. Without loss of generality we assume $T_o = 1$.

Results on existence and uniqueness of solutions (U,θ) were obtained in [Ku] for homogeneous, isotropic media and bounded domains with smooth boundary using the integral equation method. The first comprehensive results for inhomogeneous, anisotropic media were given in [Da1] for bounded domains using Hilbert space methods. In [Da1] the asymptotic behaviour of the solutions is described too.

We choose an approach via an evolution equation which admits an elegant description of existence and uniqueness results in bounded and in unbounded domains and for inhomogeneous and anisotropic media. Moreover the asymptotic behaviour of the solutions can be derived easily.

For that purpose we rewrite the system (1.1), (1.2) as a first order system of order one in t. For simplicity we assume both the body forces b and the heat supply r to be zero.

Let $V = (V^1,V^2,V^3)' := (SDU,U_t,\theta)'$. Then one obtains

$$V_t + AV = 0 \quad , \quad V(0) = V^o \, , \tag{1.5}$$

as a system in the Hilbert space $\mathcal{H} = \left(L_2(G)\right)^{10}$ with inner product $\langle V,W\rangle := (V,QW)$, where $(\cdot,\cdot)$ denotes the usual inner product in $\left(L_2(G)\right)^{10}$ and Q denotes the following weight matrix

$$Q := \begin{bmatrix} S^{-1} & 0 & 0 \\ 0 & P & 0 \\ 0 & 0 & c \end{bmatrix} .$$

Let

$$M := -\begin{bmatrix} 0 & -D & 0 \\ -D' & 0 & D'\Gamma \\ 0 & \Gamma'D & -\nabla'L\nabla \end{bmatrix} .$$

Then the differential operator A is given by

$$D(A) := \left\{ V \in \mathcal{H} \mid V^2 \in \left(W_o^{1,2}(G)\right)^3 , \; V^3 \in W_o^{1,2}(G) \; , \; MV \in \mathcal{H} \right\}$$

$$AV := Q^{-1}MV$$

($W_o^{1,2}(G)$ is the usual Sobolev space generalizing the Dirichlet boundary conditions, cf. [Ad].)

We are looking for a *solution* $V \in C_o\left(\mathbb{R}_o^+, D(A)\right) \cap C_1(\mathbb{R}_o^+,\mathcal{H})$ of (1.5) with $V^o := (SDU_o,U_1,\theta_o)' \in D(A)$.

One shows that $-A$ generates a contraction semigroup and thus obtains in the general case (bounded or unbounded domain, inhomogeneous and anisotropic medium):

<u>Theorem 1.1:</u> *There is a unique solution of* (1.5).

The characteristic coupling of the hyperbolic part with the parabolic one, which appears again in the structure of the symbol M, prevents A from being selfadjoint but still allows A to generate a semigroup.

Since the null space of A reduces A we consider in the following the restriction of A to the orthogonal complement of the null space but keep the same notation $A, \mathcal{H}, \ldots$.

1.2 The asymptotic behaviour of the solution

One obtains that $\nabla V^3(t)$ tends to zero in $\left(L_2(G)\right)^3$ as $t \to \infty$, which implies for bounded domains the convergence of $V^3(t)$ as well - a typical behaviour of solutions of the heat equation.

The general asymptotic behaviour is described as follows:

Let

$$J := \left\{ W \in D(A) \mid \forall t \geq 0 : \left\| H(t)W \right\| = \left\| H^*(t)W \right\| = \|W\| \right\} ,$$

where $\left\{H(t)\right\}_{t\geq 0}$ denotes the semigroup generated by A, $*$ denotes the adjoint, and $\|\cdot\|$ symbolizes the norm in $\mathcal{H}$. One has the orthogonal decomposition of the Hilbert space $D(A)$ (inner product $\langle\cdot,\cdot\rangle + \langle A\cdot, A\cdot\rangle$):

$$D(A) = J + J^{\perp} . \tag{1.6}$$

<u>Theorem 1.2:</u> *Let* $V^o \in D(A)$, $V^o = V_1^o + V_2^o$ *according to* (1.6), $V(t) := H(t)V^o$ *the solution of* (1.5). *Then one has*

(i) $\forall t \geq 0 : \left\|H(t)V_1^o\right\| = \left\|H^*(t)V_1^o\right\| = \|V_1^o\|$,

(ii) $\forall R > 0 : \lim\limits_{t\to\infty} \left\|V(t) - H(t)V_1^o\right\|_{G_R} = 0$,

especially: $V(t) - H(t)V_1^o$ *converges to zero weakly.*

Here $G_R := G \cap B(0,R)$ denotes the part of G lying in the ball of radius R with centre 0, and $\|\cdot\|_{G_R}$ denotes the induced norm in G_R.

Remark:
For bounded domains (ii) expresses the strong convergence of the solution to an "oscillating" part described in J. This oscillating part is characterized by the fact that J is the closure (in $D(A)$) of the eigenfunctions of A to purely imaginary eigenvalues. Examples for the (non-)existence of purely imaginary eigenvalues, which is related to the shape of the domain G, are cited in [Le3], [Ra1] (for domains $G \subset \mathbb{R}^2$ where the above results accordingly hold).

The reduction onto J may be interpreted as a reduction onto the hyperbolic part with certain side conditions. Thus one can show by using results on the equations of linear elasticity (cf. [Le3], [Mo]) that in exterior domains (domains with bounded complement) one has "local energy decay" if the medium is homogeneous and isotropic outside a fixed ball $B(0,R_o)$ and if the principle of unique continuation holds for the operator $D'SD + \lambda$, $\lambda > 0$ (cf. [Le3]).

In general one obtains the following theorem:

Theorem 1.3: *Let G be a domain with bounded complement and $V^o \in D(A)$ with V_1^o (according to (1.6)) belonging to the absolutely continuous subspace $\mathcal{H}_{ac}(iA|\mathcal{H}_J)$ of the selfadjoint operator $iA|\mathcal{H}_J$ which acts on the closure $\mathcal{H}_J$ of J in $\mathcal{H}$. Then one has for $V(t) := H(t)V^o$:*

$$(1.7) \qquad \forall R > 0 : \lim_{t\to\infty} \|V(t)\|_{G_R} = 0 .$$

The local energy decay expressed in Theorem 1.3 will become important again in section 1.4.

1.3 The Cauchy problem for homogeneous, isotropic media

In the case $G = \mathbb{R}^3$ filled with a homogeneous and isotropic medium we want to give detailed descriptions of the spectrum of A, of an expansion into generalized eigenfunctions and of the asymptotic behaviour. As usual the means shall be the Fourier transform. The operator A is now denoted by A_o and means the operator *before* reducing it to the orthogonal complement of the null space. It may be represented as

$$A_o = \begin{bmatrix} S & 0 & 0 \\ 0 & Id & 0 \\ 0 & 0 & 1 \end{bmatrix} \begin{bmatrix} 0 & -D & 0 \\ -D' & 0 & \gamma\nabla \\ 0 & \gamma\nabla' & -\Delta \end{bmatrix} \quad , \quad \gamma \in \mathbb{R}\setminus\{0\} .$$

(Id: identity on $\mathbb{C}^3$; $\gamma = 0$ would correspond to a decoupling of the elasticity equation and the heat equation.) Moreover

$$S = \begin{bmatrix} 2a+b & b & b & 0 & 0 & 0 \\ b & 2a+b & b & 0 & 0 & 0 \\ b & b & 2a+b & 0 & 0 & 0 \\ 0 & 0 & 0 & a & 0 & 0 \\ 0 & 0 & 0 & 0 & a & 0 \\ 0 & 0 & 0 & 0 & 0 & a \end{bmatrix} \quad \text{with} \quad a > 0 \ , \ 2a+3b > 0$$

$\hat{A}_o(p)$, $p \in \mathbb{R}^3$, denotes the symbol of A_o . Then we obtain for $\xi \in \mathbb{C}$ the following determinant $\det\left(\hat{A}_o(p) - \xi\right)$:

$$\det\left(\hat{A}_o(p) - \xi\right) = -\xi^3\left(-\xi^3 + \xi^2|p|^2 - \xi(2a + b + \gamma^2)|p|^2 + (2a + b)|p|^4\right)(\xi^2 + a|p|^2)^2 .$$

If we abbreviate the second factor by $\Delta_o(\xi,p,2a+b,\gamma)$ we obtain for the spectrum $\Lambda_o := \sigma(A_o)$:

<u>Theorem 1.4:</u> (i) $\Lambda_o = \Lambda_o^1 \cup \Lambda_o^2$, *where*

$$\Lambda_o^1 := \left\{\xi \in \mathbb{C} \mid \exists p \in \mathbb{R}^3 : \Delta_o(\xi,p,2a+b,\gamma) = 0\right\} ,$$
$$\Lambda_o^2 := i\mathbb{R}$$

(ii) $P_\sigma(A_o) = \{0\}$ *(point spectrum)*

(iii) $R_\sigma(A_o) = \phi$ *(residual spectrum)*

(iv) Λ_o^1 *consists of the positive real axis* $(\beta \equiv \xi_2 = 0)$ *and two branches which are symmetric with respect to the* $(\alpha \equiv)$ ξ_1*-axis.*

Typical curves for $2a + b = 1$ and several values of γ are drawn below (from [Le3]):

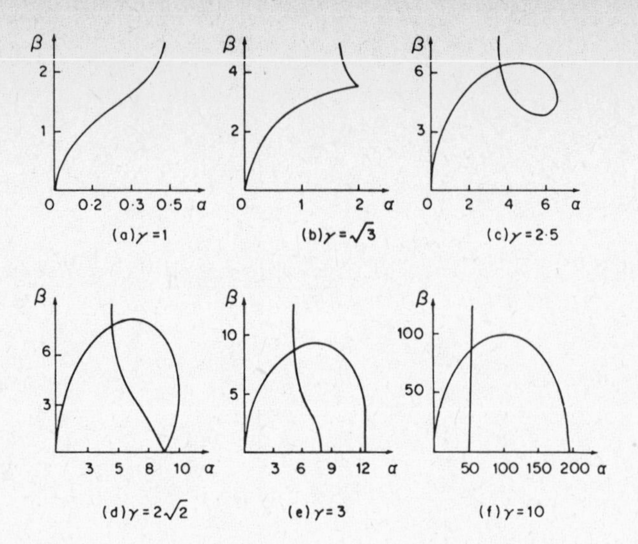

In the following we assume

$$\frac{\gamma^2}{2a+b} < 1 \quad \text{(physically motivated, cf. [Ra2]).} \tag{1.8}$$

We obtain generalized eigenfunctions for A (A^*) by setting

$$\Phi_m^{(\prime)}(x,p) := -(2\pi)^{-3/2}\, e^{-ipx}\, P_m^{(*)}(p) \quad , \quad x,p \in \mathbb{R}^3 \ ,$$
$$m = 0,1,\ldots,5 \ .$$

Here $P_m(p)$ denotes the projection onto the eigenspace for the eigenvalue $\xi_m(p)$ of $\hat{A}_o(p)$, $m = 0,1,\ldots,5$; $\xi_o(p) = 0$ is a triple eigenvalue, $\xi_4(p) = ai|p|$ and $\xi_5(p) = -ai|p|$ are double eigenvalues; brackets refer to A^*.

We have $\mathrm{Id} = \sum_{m=o}^{5} P_m(p)$, $\hat{A}_o(p)P_m(p) = \xi_m(p)P_m(p)$, the projections not being necessarily orthogonal. We obtain:

$$A^{(*)}\Phi_m^{(\prime)}(\cdot,p) = \overset{(\overline{\qquad})}{\xi_m(p)}\ \Phi_m^{(\prime)}(\cdot,p) \ , \tag{1.9}$$

(A to be understood as a formal differential symbol).

If $\Phi_m^{(')}$ denotes the following integral transform

$$\left(\Phi_m^{(')} V\right)(p) := \int_{\mathbb{R}^3} \Phi_m^{(')}(x,p) V(x)\,dx = -P_m^{(*)}(p)\hat{V}(p)$$

(where the head " ^ " symbolizes the Fourier transform), we obtain for $V,W \in Y := \left\{V \in \mathscr{H} \mid \hat{V} \text{ is continuous and has compact support in } \mathbb{R}^3 \setminus \{0\}\right\}$

<u>Parseval's equation</u>

$$\text{(1.10)} \qquad \langle V,W\rangle = \langle \pi_o V,W\rangle + \sum_{m=1}^{5} \langle \Phi_m V, \Phi_m' W\rangle ,$$

where π_o denotes the orthogonal projection onto the null space of A_o.

The following expansion theorem holds:

<u>Theorem 1.5:</u> *Let* $V \in Y$, $\pi_m := \Phi_m'^{*} \Phi_m$, $m = 1,\ldots,5$. *Then*

$$V = \sum_{m=0}^{5} \pi_m V \quad \textit{and} \quad A_o V = \sum_{m=1}^{5} \Phi_m'^{*} \xi_m(\cdot)\Phi_m V .$$

For the solution $V = V(t)$ of the time dependent problem (1.5) we obtain for $V^o \in Y$:

<u>Theorem 1.6:</u> $$V(t) = \pi_o V^o + \sum_{m=1}^{5} \Phi_m'^{*} e^{-\xi_m(\cdot)t} \Phi_m V^o .$$

The asymptotic behaviour then can be described precisely in analogy to Theorem 1.2. The role of the purely imaginary eigenvalues of A (cf. the remark following Theorem 1.2) is played by those of the symbol $\hat{A}_o(p)$. This is expressed in the next Theorem.

<u>Theorem 1.7:</u> *Let* $V^o \in Y$, $V(t) = H(t)V^o$. *Then the following holds:*

(i) $$\forall t \geq 0 : \left\|\pi_o V^o + \sum_{m=4}^{5} \Phi_m'^{*} e^{-\xi_m(\cdot)t} \Phi_m V^o\right\|^2 = \|\pi_o V^o\|^2 + \|\pi_4 V^o\|^2 + \|\pi_5 V^o\|^2 ,$$

(ii) $\lim_{t\to\infty} \left\| V(t) - \left(\pi_o V^o + \sum_{m=4}^{5} \Phi_m'^{*} e^{-\xi_m(\cdot)t} \Phi_m V^o\right) \right\| = 0$,

(iii) $\lim_{t\to\infty} \| V(t) \| = 0 \Leftrightarrow \forall m = 0,4,5 : P_m(\cdot)\hat{V}^o(\cdot) = o$.

The results in the Theorems 1.5, 1.6 and 1.7 are extensions of the considerations in [G&S] for the formally selfadjoint Maxwell's equations, where a dissipative boundary condition in the halfspace is studied, which makes the problem nonselfadjoint. Here the nonselfadjointness of our problem is already contained in the (formal) differential operator. In [D&G] the Cauchy problem is studied too, and the asymptotic behaviour of the mechanical and the thermal energy is investigated.

1.4 On the scattering theory for exterior domains

The local energy decay (Theorem 1.3) suggests that a wave operator exists for this non-selfadjoint problem. We assume that G is a domain with bounded complement, that the medium is homogeneous and isotropic outside a fixed ball of radius $R_o > 0$ around zero and that the principle of unique continuation holds for the operator $D'SD + \lambda$, $\lambda > 0$ (which is satisfied for example for isotropic media, see [Le3]). These assumptions guarantee that the wave operators for the elasticity equations exist and hence are natural requirements, because the reduction onto the space J described in section 1.2 represents a reduction onto the hyperbolic part of our equations, which has its origin in the elastic part of our problem.

For those data V^o for which the component in $J^{\perp}$ converges to zero not only weakly but strongly, we show the existence of a wave operator.

That is there is a map W^+ , $V^o \to W^+V^o \in D(A_{oo})$, where A_{oo} denotes A_o with $\gamma = 0$ (no coupling), with the property

(1.11) $\lim_{t\to\infty} \left\| KV(t) - V_{oo}(t) \right\| = \lim_{t\to\infty} \left\| KH(t)V^o - H_{oo}(t)W^+V^o \right\| = 0$.

Here K denotes the following operator: $K : (L_2(G))^m \to (L_2(\mathbb{R}^3))^m$, $(KW)(x) := W(x)$ for $x \in G$, $(KW)(x) := 0$ for $x \in \mathbb{R}^3 \setminus G$, $m = 9,10$.

$V_{oo}(t) = H_{oo}(t)W^+V^o$ and $\left\{H_{oo}(t)\right\}_{t\geq 0}$ denotes the semigroup generated by $-A_{oo}$.

Hence the solution in the exterior domain behaves asymptotically like a "free" solution, that is like a solution of the Cauchy problem with data W^+V^o .

Proof of (1.11):

Let A_J be the restriction of A onto $\mathcal{H}_J$ with domain $D(A_J) = J$, $\mathcal{H}_J$ being the closure of J in $\mathcal{H}$ (cf. Theorem 1.3).

For $W \in D(A_J)$ we have:

$$A_J W = \begin{pmatrix} E\,\pi W \\ 0 \end{pmatrix} \quad , \quad \text{where} \quad \pi W = \pi \begin{bmatrix} W^1 \\ W^2 \\ W^3 \end{bmatrix} := \begin{bmatrix} W^1 \\ W^2 \end{bmatrix} \quad \text{and}$$

$$E \equiv \begin{bmatrix} S & 0 \\ 0 & P^{-1} \end{bmatrix} \begin{bmatrix} 0 & -D \\ -D' & 0 \end{bmatrix} \quad , \quad E : D(E) \subset \pi\mathcal{H} \to \pi\mathcal{H} \ ,$$

$$D(E) := \left\{W \in \pi\mathcal{H} \mid W^2 \in W_o^{1,2}(G) \ , \ EW \in \pi\mathcal{H}\right\} \qquad \text{(E: elasticity operator).}$$

Under the assumptions made at the beginning of this section we know that the absolutely continuous subspace $\mathcal{H}_{ac}(iE)$ of the selfadjoint operator iE is the orthogonal complement of the null space of iE ; the usual wave operator $W_e^+ : \mathcal{H}_{ac}(iE) \to \mathcal{H}_{ac}(iE_o)$ exists, where E_o describes the corresponding Cauchy problem (cf. [Mo]).

$V^o \in D(A_J)$ implies $\pi V^o \in \mathcal{H}_{ac}(iE)$. If $\left\{H_e(t)\right\}_{t\geq 0}$ and $\left\{H_{e_o}(t)\right\}_{t\geq 0}$ denote the semigroups generated by E and E_o respectively, we obtain

$$\lim_{t\to\infty} \left\| K H_e(t)(\pi V^o) - H_{e_o}(t) W_e^+(\pi V^o) \right\|_{(L_2(\mathbb{R}^3))^9} = 0 \ .$$

Moreover $H_e(t)(\pi V^o) = \pi\big(H(t)V^o\big)$.

For $V^o \in D(A_J)$ we define W^+V^o by

$$W^+V^o := \begin{bmatrix} (W_e^+\pi)V^o \\ 0 \end{bmatrix} = (\pi^* W_e^+ \pi)V^o \ .$$

Since $H_{oo}(t)(W^+V^o) = \pi^*\big(H_{e_o}(t)(W^+V^o)\big)$ the asymptotic behaviour in J is given by:

$$(1.12) \qquad \lim_{t\to\infty} \left\| KH(t)V^o - H_{oo}(t)W^+V^o \right\| = 0 \,.$$

Since we assumed strong convergence to zero for the part in $J^\perp$ we obtain the claim (1.11) if we set

$$W^+V^o := (\pi^* W_e^+ \pi P_{J^\perp})V^o \,,$$

($P_{J^\perp}$: projection onto $J^\perp$ according to (1.6)).

The description of the asymptotic behaviour in J given by (1.12) always holds without assuming strong convergence of $V(t)$ to zero in $J^\perp$.

2. The nonlinear equations

For the fully nonlinear system (0.1), (0.2) there are until now only few results mostly on one-dimensional models. The questions arise whether one should look for weak or for strong (classical) solutions and if there are global solutions with regard to time t or if there will appear singularities after finite time.

Again the special hyperbolic-parabolic structure of the system is important and the natural question is if the dissipative influence - that is a possible damping effect - of the parabolic part

1. is strong enough to yield global smooth solutions $(t > 0)$ even for non-smooth data or
2. is sufficiently strong to yield global smooth solutions at least for smooth data or
3. is sufficiently strong to yield global smooth solutions at least for sufficiently small, smooth data or
4. is too weak to prevent solutions from blowing up after finite time even for small data.

This classification 1.-4. (for more general nonlinear systems) may also be found in [Da2] together with examples of nonlinear systems for each of the groups. The behaviour described in 4. is typical for nonlinear hyperbolic systems, which suggests to consider weak solutions to get global existence (cf. [K&M], [M&M], [Sm]).

The question to which group our system (0.1), (0.2) belongs is not yet finally answered. The investigations of special *one-dimensional* models classify this case into group 3.

For higher dimensions it is more likely that the hyperbolic part will predominate. Note that in the linear case in $\mathbb{R}^1$ the damping effect of the parabolic part is strong enough not to allow oscillating terms while these may appear in $\mathbb{R}^2$ and $\mathbb{R}^3$ (cf. Theorem 1.2). Possibly the geometry of the domain is important.

In the following we discuss the special one-dimensional models (for a homogeneous medium). For bounded domains and right-hand sides $b = 0$, $r = 0$ the following system is studied in [Sl] which arises from (0.1), (0.2) by assuming a purely longitudinal motion:

$$u_{tt} - \psi_{FF}(u_x+1,\theta+T_o)u_{xx} - \psi_{FT}(u_x+1,\theta+T_o)\theta_x = 0 , \tag{2.1}$$

$$\psi_{TT}(u_x+1,\theta+T_o)\theta_t + d(\theta,\theta_x)\theta_{xx} + \psi_{FT}(u_x+1,\theta+T_o)u_{xt} = 0 . \tag{2.2}$$

Here u and θ are functions of x and t, and $x \in G := (0,1)$.

Initial conditions are given by

$$u(\cdot,0) = u_o \quad , \quad u_t(\cdot,0) = u_1 \quad , \quad \theta(\cdot,0) = \theta_o \quad , \tag{2.3}$$

and the following boundary conditions are studied

$$\text{a)} \quad u_x\big|_{\partial G} = 0 \quad , \quad \theta\big|_{\partial G} = 0 , \tag{2.4}$$

$$\text{b)} \quad u\big|_{\partial G} = 0 \quad , \quad \theta_x\big|_{\partial G} = 0 .$$

Then Slemrod proves an existence and a uniqueness theorem, which we formulate for the case (2.4)a):

Theorem 2.1: *(Slemrod)*

(I) *(Local existence and uniqueness)*

Let $\psi \in C_4$, $d \in C_3$ *and*

$$0 < a_o \le \psi_{FF} \le a_1 < \infty ,$$
$$0 < b_o \le |\psi_{FT}| \le b_1 < \infty ,$$
$$0 < c_o \le -\psi_{TT} \le c_1 < \infty ,$$
$$0 < d_o \le d \le d_1 < \infty ,$$

with positive constants $a_o, a_1, b_o, b_1, c_o, c_1, d_o$ *and* d_1 .

For the data u_o, u_1, θ_o *we assume*

$$\begin{aligned} & u_{ox}, u_{oxx}, u_{oxxx} \in L_2(G) , \quad u_{ox} = 0 \quad \text{on } \partial G ; \\ (2.5)\quad & u_1, u_{1x}, u_{1xx}, \in L_2(G) , \quad u_{1x} = 0 \quad \text{on } \partial G ; \\ & \theta_o, \theta_{ox}, \theta_{oxx}, \theta_{oxxx} \in L_2(G) , \quad \theta_o = \theta_{oxx} = 0 \quad \text{on } \partial G ; \\ & -\psi_{FT}(u_{ox}+1, \theta_o+T_o)u_{1xxx} - d(\theta_o, \theta_{ox})\theta_{oxxxx} \in L_2(G) . \end{aligned}$$

Then (2.1), (2.2), (2.3), (2.4)a) *has a unique classical solution* u, θ *with*

$$u \in C_2(\overline{G} \times [0,t_o)) , \quad \theta \in C_1(\overline{G} \times [0,t_o)) ,$$
$$\theta_{xt}, \theta_{xx} \in C_o(\overline{G} \times [0,t_o)) , \quad \theta_{tt}(\cdot,t) \in C_o(\overline{G})$$
$$\text{for each } t , 0 \le t \le t_o ,$$

defined on a maximal interval of existence $[0,t_o)$, $0 \le t_o \le \infty$, *such that for* $t_1 \in [0,t_o)$

$$u_t, u_x, u_{tt}, u_{xt}, u_{xx}, u_{ttt}, u_{ttx}, u_{txx}, u_{xxx},$$
$$\theta, \theta_t, \theta_x, \theta_{tt}, \theta_{tx}, \theta_{xx}, \theta_{txx}, \theta_{xxx} \in L_\infty([0,t_1], L_2(G)) ,$$
$$\theta_{ttx} \in L_2([0,t_1], L_2(G)) .$$

(II) *(Global existence and uniqueness)*

If in addition to the assumptions from (I) *the* $L_2(G)$*-norms of*

$$u_{ox}, u_{oxx}, u_{oxxx}, u_1, u_{1x}, u_{1xx}, \theta_o, \theta_{ox}, \theta_{oxx}, \theta_{oxxx},$$
$$-\psi_{FT}(u_{ox}+1, \theta_o+T_o)u_{1xxx} - d(\theta_o, \theta_{ox})\theta_{oxxxx}$$

are sufficiently small, then (2.1), (2.2), (2.3), (2.4)a) *possess a unique global smooth solution.*

The arguments in [S1] strongly relied on the boundedness of the domain and the type of boundary conditions (2.4)a) resp. (2.4)b). These boundary conditions yield - using the differential equations - an additional boundary condition, which was crucial for obtaining the necessary energy estimates.
A similar result was obtained in [Zh1].

We have proved a local existence theorem also for the case of Dirichlet boundary conditions

$$(2.4) \qquad c) \quad u\big|_{\partial G} = 0 \quad , \quad \theta\big|_{\partial G} = 0 \, ,$$

also including unbounded domains (cf. [Ra4]).

For that purpose a modified approach and greater regularity of the data was necessary. At first weak solutions

$$u \in W_2^1([0,t_0), L_2(G)) \cap L_2([0,t_o), W_o^{1,2}(G)) \, ,$$
$$\theta \in L_2([0,t_o), W_o^{1,2}(G)) \quad , \quad 0 < t_o < \infty \, ,$$

of the linearized system (linearized from (2.1), (2.2)) were studied using Faedo-Galerkin methods (for the notation cf. [Ra4], [W1]), which has previously been used in [K&P] for the linear equations from section 1; the uniqueness result can be obtained as in [Ra3] for the linear equations.

Then regularity theorems for hyperbolic and for parabolic systems yielded the desired regularity of solutions of the linearized system. Finally a contraction argument as in [S1] yields the following theorem:

<u>Theorem 2.2:</u> *Let* ψ *and* d *be smooth and replace assumption* (2.5) *by*

$$(2.5') \qquad u_o, u_1, \theta_o \in C_o^\infty(G) \, .$$

Then Theorem 2.1(I) holds for the Dirichlet boundary conditions (2.4)c) for arbitrary bounded or unbounded domains $G \subset \mathbb{R}^1$ *. Moreover*

$$u,\theta \in C_\infty\left(\bar{G} \times [0,t_o)\right) .$$

For the Cauchy problem, $G = \mathbb{R}^1$, a global existence and uniqueness result was shown for small data in [Z&S]. There detailed L_p-estimates for the linearized problem are used. Also the decay at $t = \infty$ is analyzed - typically for the one-dimensional case. That the classification into group 3 is justified is shown in [D&H]. There it is proved that a C_2-solution of the Cauchy problem becomes singular in finite time if (u_{oxx}, u_{1x}) are sufficiently large.

In the future higher-dimensional models, possibly for anisotropic, inhomogeneous media, are to be studied; local existence theorems for bounded domains already exist (cf. [Zh2]), also for the Cauchy problem (from the results in [Ka]).
The possibility of obtaining global existence results or blow-up phenomena for the Cauchy problem in three dimensions is presently investigated.

References

(Papers marked with (*) have been supported by the SFB 72.)

[Ad] Adams, R.A.: Sobolev spaces. Academic Press; New York (1975)

[Ca] Carlson, D.E.: Linear thermoelasticity. Handbuch der Physik VIa/2. Springer-Verlag; Berlin, Heidelberg, New York (1972), 297-345.

[Da1] Dafermos, C.M.: On the existence and the asymptotic stability of solutions to the equations of linear thermoelasticity. Arch. Rat. Mech. Anal. 29 (1968), 241-271.

[Da2] Dafermos, C.M.: Dissipation, stabilization and the second law of thermodynamics. Lec. Notes Phys. 228 (1984),44-88.

[D&H] Dafermos, C.M. & Hsiao, L.: Development of singularities in solutions of the equations of nonlinear thermoelasticity. Quart. Appl. Math. 44 (1986), 463-474.

[D&G] Dassios, G. & Grillakis, M.: Dissipation rates and partition of energy in thermoelasticity. Arch. Rat. Mech. Anal. 87 (1984), 49-91.

[G&S] Gilliam, D.S. & Schulenberger, J.R.: Electromagnetic waves in a three-dimensional half-space with a dissipative boundary. J. Math. Anal. Appl. 89 (1982), 129-185.

[Ka] Kawashima, S.: Systems of a hyperbolic-parabolic composite type, with applications to the equations of magnetohydrodynamics. Thesis; Kyoto University (1983).

[K&M] Klainerman, S. & Majda, A.: Formation of singularities for wave equations including the nonlinear vibrating string. Comm. Pure Appl. Math. 33 (1980), 241-263.

[K&P] Kowalski, T. & Piskorek, A.: Existenz der Lösung einer Anfangsrandwertaufgabe in der linearen Thermoelastizitätstheorie. ZAMM 61 (1981), T250-T252.

[Ku] Kupradze, V.D.: Three-dimensional problems of the mathematical theory of elasticity and thermoelasticity. North-Holland; Amsterdam (1979).

[Le1] (*) Leis, R.: Außenraumaufgaben in der linearen Thermoelastizitätstheorie. Math. Meth. in the Appl. Sci. 2 (1980), 379-396.

[Le2] (*) Leis, R.:Über das asymptotische Verhalten thermoelastischer Wellen im $\mathbb{R}^3$. Math. Meth. in the Appl. Sci. 3 (1981), 312-317.

[Le3] (*) Leis, R.: Initial boundary value problems in mathematical physics. B.G. Teubner; Stuttgart. John Wiley & Sons; Chichester et al. (1986).

[M&M] MacCamy, R.C. & Mizel, V.J.: Existence and non-existence in the large of solutions of quasi-linear wave equations. Arch. Rat. Mech. Anal. 25 (1967), 299-320.

[Mo] Mochizuki, K.: Spectral and scattering theory for symmetric hyperbolic systems in an exterior domain. Publ. RIMS, Kyoto Univ., Vol. 5 (1969), 219-258.

[Ra1] (*) Racke, R.: On the time-asymptotic behaviour of solutions in thermoelasticity. Proc. Roy. Soc. Edinburgh 107A (1987), 289-298

[Ra2] (*) Racke, R.: Eigenfunction expansions in thermoelasticity. J. Math. Anal. Appl. 120 (1986), 596-609.

[Ra3] (*) Racke, R.: Uniqueness of weak solutions in linear thermoelasticity. Bull. Pol. Ac. Sci., Tech. Sci. 34, No. 11-12 (1986), 613-620.

[Ra4] (*) Racke, R.: Initial boundary value problems in one-dimensional non-linear thermoelasticity. Submitted to: Math. Meth. Appl. Sci..

[Sl] Slemrod, M.: Global existence, uniqueness, and asymptotic stability of classical smooth solutions in one-dimensional non-linear thermoelasticity. Arch. Rat. Mech. Anal. 76 (1981), 97-134.

[Sm] Smoller, J.: Shock waves and reaction-diffusion equations. Springer-Verlag New York, Heidelberg, Berlin (1983).

[Wl] Wloka, J.: Partielle Differentialgleichungen. B.G. Teubner; Stuttgart (1982).

[Zh1] Zheng, S.: Global solutions and applications to a class of quasilinear hyperbolic-parabolic coupled systems. Scienta Sinica Ser. A 27 (1984), 1274-1286.

[Zh2] Zheng, S.: Initial boundary value problems for quasilinear hyperbolic-parabolic coupled systems in higher dimensional spaces. Chin. Ann. of Math. 4B(4) (1983), 443-462.

[Z&S] Zheng, S. & Shen, W.: Global solutions to the Cauchy problem of a class of quasilinear hyperbolic parabolic coupled systems. To appear in: Scienta Sinica.

APPLICATIONS OF VARIATIONAL METHODS TO PROBLEMS IN THE GEOMETRY OF SURFACES

Michael Struwe

In this survey I would like to report on some variational problems arising in the study of surfaces. The problems are the following:

1. Morse theory for minimal surfaces,
2. Large solutions to the Plateau problem for surfaces of constant mean curvature ("Rellich's conjecture for H-surfaces"),
3. Harmonic mappings of Riemannian surfaces,
4. Minimal and H-surfaces with free boundaries.

As a common feature to these problems, "interesting" solutions do not (necessarily) correspond to relative minima of the associated functional in variation.

Problems 1)-4) moreover are related to variational problems which are on the border-line of standard variational methods; and indeed, in certain cases these methods <u>must fail</u> to produce solutions.

Thus, in order to solve these problems new variational techniques had to be found. Two developments, in particular, "<u>Ljusternik-Schnirelman and Morse theories on convex sets in Banach spaces</u>", and the characterization of "<u>loss of compactness</u>" <u>in a variational problem through</u> "<u>separation of spheres</u>" have (to a large extent) grown out of the analysis of the above problems; methods that considerably extend the applications of classical critical point theory.

Acknowledgement

A large part of the research which permitted me to contribute to these developments was funded by the Sonderforschungsbereich 72 of the Deutsche Forschungsgemeinschaft whose support I gratefully acknowledge.

Of all the members of the SFB 72 I am most indebted to Professors Frehse and Hildebrandt who have created the atmosphere of inspiration and challenge in a friendly and open scientific environment that has been a major stimulus for my work.

General references

Problems 1)-2) are covered by the lecture notes [42]. [5] and [24] will provide further background information also on problem 4). Harmonic maps will also be covered by the articles of Hildebrandt and Jost in this volume where general references will be given. Specific material will be quoted in the following as needed.

1. Plateau's problem for minimal surfaces

Given a Jordan curve $\Gamma \subset \mathbb{R}^N$, Plateau's problem asks for a surface of least area spanning Γ. Necessarily, such a surface will have vanishing mean curvature. If we specify the topological type of the surface to be that of the disc

$$B = \left\{ w = (u,v) \in \mathbb{R}^2 \;\middle|\; u^2 + v^2 < 1 \right\}$$

we may look for surfaces of least area spanning Γ which are conformally represented by a map $X : B \to \mathbb{R}^N$

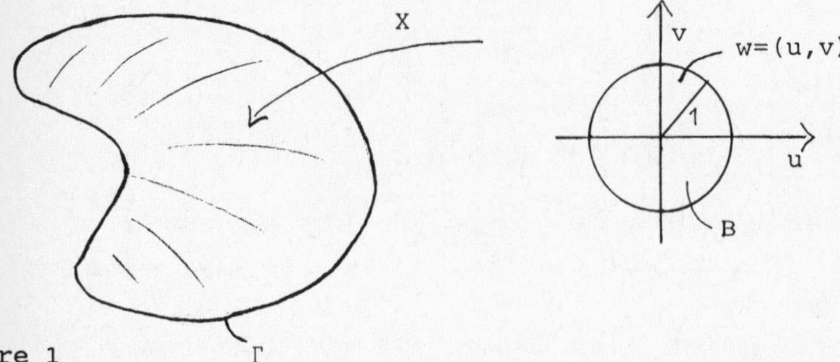

figure 1

In conformal coordinates a surface X will have vanishing mean curvature iff the coordinate functions $X = (X^1,\ldots,X^N)$ are harmonic. Thus as necessary conditions we obtain the following set of partial differential equations:

$$\Delta X = 0 \,, \tag{1.1}$$

$$|X_u|^2 - |X_v|^2 = 0 = X_u \cdot X_v \,, \tag{1.2}$$

$$X\big|_{\partial B} : \partial B \to \Gamma \quad \text{is an oriented homeomorphism .} \tag{1.3}$$

Here, e.g. $X_u = \frac{\partial}{\partial u} X$. (1.2) are the conformality relations; (1.3) is the Plateau boundary condititon. (1.1-3) defines the parametric Plateau problem for disc-type minimal surfaces. In the following, any solution $X \in C^2(B;\mathbb{R}^N) \cap C^0(\overline{B};\mathbb{R}^N)$ of (1.1-3) will be referred to as a minimal surface. Such a surface will be locally area-minimizing (away from branch points w_o where $\nabla X(w_o) = 0$); however, we no longer require the area of X to be globally minimal.

Although special cases of "Plateau's problem" already were studied by Lagrange (~1760) the general solution (for any prescribed boundary Γ) evaded mathematicians until progress in the direct methods in the calculus of variations ("Dirichlet's principle") led to the following result by Douglas [6] and Rado [27] in 1930/31:

<u>Theorem 1.1:</u> For any rectifiable Jordan curve Γ in $\mathbb{R}^N$ there is a disc-type minimal surface spanning Γ .

The proof of Theorem 1.1 rests on the fact that the solutions to (1.1-3) precisely correspond to the "critical points" of Dirichlet's integral

$$D(X) = \frac{1}{2} \int_B |\nabla X|^2 dw$$

on the space

$$C(\Gamma) = \left\{ X \in H^{1,2}(B;\mathbb{R}^N) \;\middle|\; X|_{\partial B} : \partial B \to \Gamma \text{ is a weakly monotone continuous map onto } \Gamma \right\}$$

of "surfaces" with finite "area" $D(X)$ and satisfying a weakened Plateau boundary condition. If Γ is rectifiable, $C(\Gamma) \neq \phi$, and

$$\inf\left\{ D(X) \;\middle|\; X \in C(\Gamma) \right\}$$

is attained at a generalized solution X to (1.1-3) in $C(\Gamma)$. Necessarily, X is harmonic, hence smooth in B ; the conformality relations arise as natural boundary conditions. Finally, (1.1-2) imply the strong form (1.3) of the Plateau boundary condition. Cp. [6], [27], [5], [42; Theorems I.4.10, I.5.3] for details.

Simple examples show that in general Plateau's problem may have more than one solution (of disc-type), cp. figure 2.

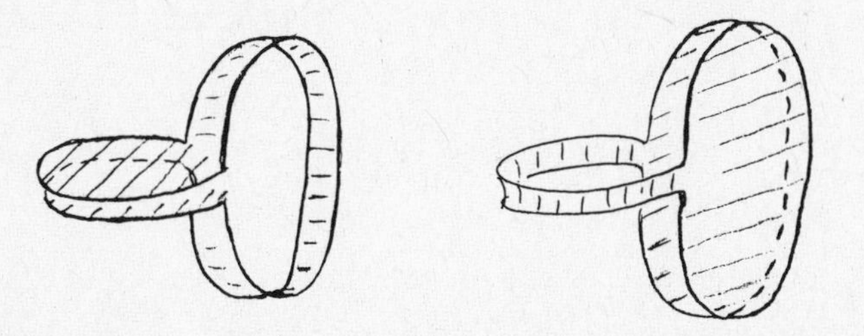

figure 2

Moreover, in general a Jordan curve may also span "unstable" minimal surfaces, i.e. (1.1-3) may possess solutions which do not correspond even to relative minima of D .

Theorem 1.2 (Morse-Tompkins [23], Shiffman [30]): Suppose a Jordan curve Γ of class C^2 spans two distinct minimal surfaces X_1, X_2 which do not belong to the same connected component of relative minima of D in $C(\Gamma)$. Then there exists an unstable minimal surface.

Remark 1.3: Originally, Theorem 1.2 was stated for rectifiable Γ with reference to the topology of uniform convergence on

$$C_o(\Gamma) = \left\{ X \in C(\Gamma) \;\middle|\; \Delta X = 0 \right\} .$$

This topology is weaker than the $H^{1,2}$-topology by the Courant-Lebesgue--Lemma, cp. [5, p. 103f].

Moreover, the respective connected components of relative minimizers of D ("blocks" of minimal surfaces) containing X_1, X_2 were assumed to be strictly D-minimizing in suitable neighborhoods. The above improvement (at the expense of a slightly stonger regularity condition on Γ) was obtained in [35], cp. also [42]. No non-degeneracy condition is required.

The method of Morse-Tompkins, Shiffman involves the Morse-inequalities for minimal surfaces which give a more detailed picture of the structure of the set of all minimal surfaces spanning a given curve Γ than the "mountain-pass-type" result stated in Theorem 1.2 itself. Let

$$m_k = \#\left\{ \text{minimal surfaces } X \text{ with "Morse Index" } k \right\}$$

be the number of geometrically district minimal surfaces spanning Γ corresponding to critical points of D with an "unstable manifold" of dimension k , measured by the "Morse index". The Morse inequalities for minimal surfaces consist of the following estimates:

$$(1.4) \qquad m_o \geq 1 \quad , \quad \sum_{k=0}^{\ell} (-1)^{\ell-k} m_k \geq (-1)^{\ell} \quad , \quad \forall \ell \in \mathbb{N}_o$$

$$(1.5) \qquad \sum_{k=0}^{\infty} (-1)^k m_k = 1 .$$

Remarks 1.4: i) The "last Morse-inequality" also may be interpreted as an equation for the "total degree of the differential of D on $C(\Gamma)$" with respect to 0 . However, this interpretation has only heuristic value since $C(\Gamma)$ does not have a differentiable structure.

ii) For the proof of Theorem 1.2 it now suffices to consider (1.4) for

$\ell = 1$: Under the hypotheses of Theorem 1.2 there holds $m_o \geq 2$, whence

$$m_1 \geq m_o - 1 \geq 1$$

corresponding to an unstable minimal surface.

Apparently, estimates (1.4-5) give much finer structural information than Theorem 1.2. However, working in a non-differentiable setting Morse-Tompkins and Shiffman resort to local definitions of the notions of critical point and Morse index. Although they are able to show that "critical points" of D (in their definition) correspond to minimal surfaces (solutions of (1.1-3)) it is not clear if the converse is true in general - even in the "non-degenerate" case. Moreover, the "Morse index" as defined by Morse-Tompkins and Shiffman may depend on the (coarse C^o-) topology used - instead of being an intrinsic property of the minimal surface itself.

For these reasons efforts were made to re-establish the Morse inequalities involving an intrinsic notion of index (and in a way which would make the theory accessible also for non-specialists, cf. [16, p. 324]).

The first partially successful attempt was made by Tromba, using the global structure properties of the set of minimal surfaces in $\mathbb{R}^N$ revealed by Böhme and Tromba [2]: Instead of fixing a curve Γ and studying the minimal surfaces it spans Böhme and Troba consider the bundle of all surfaces parametrized by harmonic vector functions and bounded by smooth Jordan curves in $\mathbb{R}^N$. As a subset this bundle contains the set of all minimal surfaces in $\mathbb{R}^N$ as a countable union (indexed by branching types) of differentiably immersed submanifolds M_λ^υ. Böhme and Tromba prove that the restriction of the bundle projection Π (which assigns to any surface X its boundary Γ) to any M_λ^υ is a Fredholm map and apply Smale's generalization of Sard's theorem to obtain generic finiteness and stability (with respect to variations of Γ) of minimal surfaces spanning a curve Γ in $\mathbb{R}^N$. Thus, looking at the Plateau problem from a global point of view, inspite of an apparent increase in complexity fine structure properties of the solution set also for (almost all) individual curves Γ are resolved.

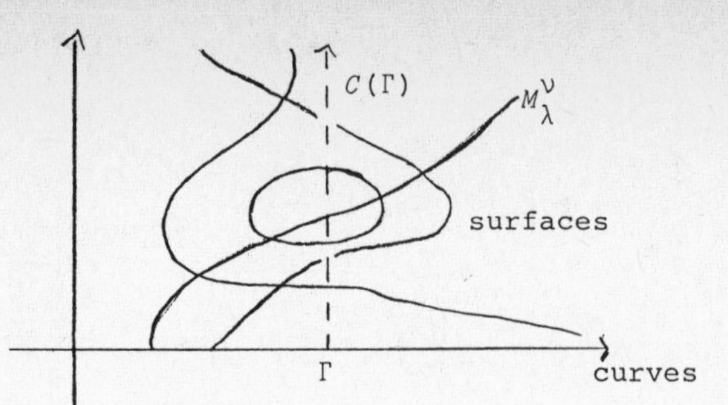

figure 3

Departing from the global stratification of the set of minimal surfaces obtained in [2] and using a variant of degree theory Tromba - while visiting the SFB 72 - in 1982 was able to prove the "last Morse inequality" (1.5) - and hence Theorem 1.2 - for a generic smooth boundary curve Γ in $\mathbb{R}^N$, $N \geq 3$; cp. [44]. Most important in this new derivation of identity (1.5) was the fact that Tromba was able to employ an intrinsic notion of Morse index rather than a local one.

Finally, inspired by Tromba's work but using a completely different approach, in my Habilitationsschrift [35] I was able to derive all Morse inequalities (1.4-5) for a C^5-curve $\Gamma \subset \mathbb{R}^N$, $N \geq 3$, spanning only non-degenerate minimal surfaces. By the work of Böhme-Tromba [2] this condition is satisfied for a generic curve $\Gamma \subset \mathbb{R}^N$, $N \geq 4$. The approach chosen in [35] was a natural and direct extension of the Morse theory of Palais and Smale [25], [26] to functionals defined on closed convex sets in Banach spaces. In particular, it could be shown that (with suitable modifications) the Palais-Smale condition (which is crucial for abstract variational methods) also held in the rather complex Plateau problem and a pseudo-gradient flow could be constructed.

In [40] the results of [35] were extended to doubly connected minimal surfaces bounded by two Jordan curves, and generalizations to minimal surfaces of arbitrary topological type will be given in [20].

Let us add that as a consequence of recent developments in abstract variational methods it now seems that the non-degeneracy condition appearing in [35], [40] can be completely dispensed with: Once the Palais-Smale condition has been established and the pseudo-gradient flow for D on $C(\Gamma)$ has been constructed Conley's Morse theory [4] and its generalization to infinite dimensional variational problems by Benci [1] may be immediately applied.

2. "Large solutions" to the Plateau problem for surfaces of prescribed constant mean curvature - "Rellich's conjecture"

Generalizing the Plateau problem for minimal surfaces, i.e. surfaces with vanishing mean curvature, one may also ask for (disc-type) surfaces of prescribed constant mean curvature $H \in \mathbb{R}$ spanning a given curve $\Gamma \subset \mathbb{R}^3$. In conformal representation such a surface will satisfy the following system of differential equations

(2.1) $$\Delta X = 2H\, X_u \wedge X_v$$

(2.2) $$|X_u|^2 - |X_v|^2 = 0 = X_u \cdot X_v$$

(2.3) $X|_{\partial B} : \partial B \to \Gamma$ is an oriented homeomorphism ,

"$\wedge$" denoting the exterior product in $\mathbb{R}^3$. Solutions to (2.1-3) will be called parametric surfaces of constant mean curvature $H \in \mathbb{R}$ or H-<u>surfaces</u> for short, corresponding to the fact that any solution X to (2.1-2) locally will parametrize a surface of mean curvature H (away from branch points).

Like (1.1-3) also (2.1-3) has a variational structure. Indeed, the solutions of (2.1-3) correspond to critical points of the functional

$$D_H(X) = D(X) + 2HV(X)$$

on $C(\Gamma)$, where D is the - already familiar - Dirichlet integral (measuring the "area" of X) and

$$V(X) = \frac{1}{3} \int_B X_u \wedge X_v \cdot X \, dw$$

denotes the (algebraic) "volume" enclosed by X. For smooth $X, X_o \in C(\Gamma)$ the quantity $V(X) - V(X_o)$ gives the Lebesgue measure of the oriented volume enclosed between X and X_o (counted with multiplicities).

Due to the cubic character of V for $H \neq 0$ the functional D_H will not be bounded from below unless certain geometric restrictions are imposed on admissible comparison surfaces.

<u>Theorem 2.1</u> (Hildebrandt [14]): Suppose $\Gamma \subset \overline{B_R(0)} \subset \mathbb{R}^3$ is a rectifiable Jordan curve. Then for all $H \in \mathbb{R}$ such that $|H|R \leq 1$ there exists a solution X_H to (2.1-3). Moreover, for $|H|R < 1$ the surface X_H furnishes a relative minimum for D_H on $C(\Gamma)$.

<u>Remarks 2.2:</u> i) If $|H|R < 1$ X_H is obtained as a minimizer of D_H in

$$C^{(H)}(\Gamma) = \left\{ X \in C(\Gamma) \;\middle|\; \|X\|_\infty < \frac{1}{|H|} \right\} ,$$

hence X_H is relatively D_H-minimal with respect to the $H^{1,2} \cap L^\infty$-topology.

By a result due to Brezis-Coron [3], however, relative D_H-minimality with respect to the $H^{1,2} \cap L^\infty$-topology implies relative D_H-minimality with respect to the $H^{1,2}$-topology, i.e. relative D_H-minimality in $C(\Gamma)$.

ii) Theorem 2.1 improves earlier results by Heinz [12], and Werner [47].

iii) By a result of Heinz [13] Theorem 2.1 is best possible if Γ is a circle. However, for lengthy curves the following result by Wente [45] and Steffen [33] will give better results:

<u>Theorem 2.3:</u> Suppose there exists $X_o \in C(\Gamma)$ such that $H^2 D(X_o) < \frac{2}{3}\pi$. Then there exists a solution X_H to (2.1-3) which yields a relative minimum for D_H on $C(\Gamma)$.

<u>Remark 2.4:</u> The bound stated in Theorem 2.3 is not optimal, cp. [42; Remark IV. 4.14]. It is conjectured that (as in the case of a circular curve) the condition $H^2 D(X_o) < \pi$ will suffice.

Theorems 2.1, 2.3 state existence results for stable (relatively D_H-minimal) H-surfaces. The following <u>thought experiment</u> proposed by Hildebrandt lends evidence that (2.1-3) may have still more solutions:

Dip a pipe into soap water so that a soap film forms spanning the (circular) opening of the pipe head of radius r. Blow slightly to create a small spherical bubble of radius $\rho > r$ - an H-surface with $H = \frac{1}{\rho}$. By blowing in more air we can create another H-surface (for the same parameter H) as the large spherical bubble which is the mirror image of the spherical complement of our small H-surface, reflected in the plane through the pipe opening. (Pipe smokers are invited to perform this experiment in a moment of leisure.)

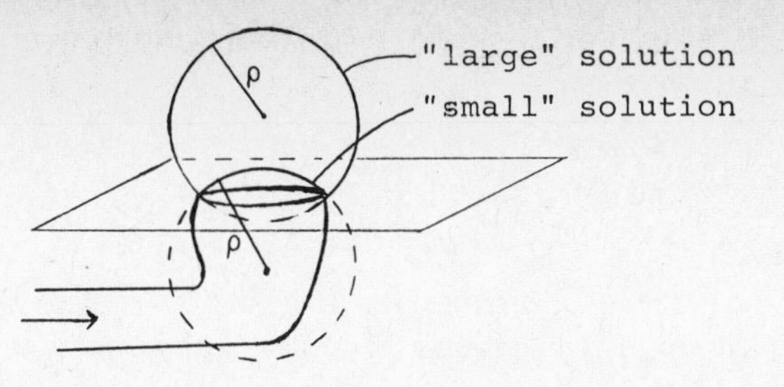

figure 4

More generally, we are led to conjecture that for <u>any</u> (smooth) Jordan curve $\Gamma \subset \mathbb{R}^3$ there is a range of curvatures $H \neq 0$ for which there exist at least two geometrically distinct ("small" and "large") solutions to (2.1-3). Sometimes this conjecture is attributed to Rellich, cf. [16; p. 325-326].

The fundamental tool for dealing with the functional D_H "in the large" is the isoperimetric inequality for closed surfaces in $\mathbb{R}^3$; cf. [46], [28]:

<u>Theorem 2.5:</u> For any $X,Y \in C(\Gamma)$ there holds the estimate:

$$36\pi \left| V(X) - V(Y) \right|^2 \leq \left(D(X) + D(Y) \right)^3 .$$

The constant 36π is best possible.

Like in our thought experiment, large H-surfaces can be constructed mathematically as solutions to the <u>volume-constrained</u> <u>Plateau</u> <u>problemm</u> for fixed $K \in \mathbb{R}$:

$$(2.4) \qquad D(X) \to \min \quad , \quad X \in C(\Gamma) : V(X) = K ;$$

the curvature parameter H arises as a Lagrange parameter corresponding to the (regular) volume constraint, cp. [46].

By using the isoperimetric inequality to compute asymptotic estimates for the set $H(K)$ of curvatures H arising as Lagrange multipliers in (2.4), Steffen was able to verify the existence of large solutions to (2.1-3) for a sequence $H_m \to 0$; cf. [32].

In 1982 non-uniqueness for (2.1-3) was established in general, independently by Brezis-Coron [3] and the author [39], with an important contribution by Steffen [34]. (Cf. [16; p. 327]).

In particular, Brezis-Coron were able to show non-uniqueness if $0 < |H|R < 1$, $\Gamma \subset B_R(0) \subset \mathbb{R}^3$. This result, again, is best possible for the circle.

However, the results of [3], [39], [34] may fail to give best results under the assumptions of Theorem 2.3. The following result from [37] also covers this case; in fact it may be best possible for <u>any</u> Jordan curve Γ of class C^2 :

<u>Theorem 2.6:</u> Suppose Γ is a C^2-Jordan curve in $\mathbb{R}^3$, $H \neq 0$. Assume there exists an H-surface X_H which yields a relative minimum for D_H on $C(\Gamma)$. Then there exists a solution X^H of (2.1-3) which is geometrically distinct from X_H .

<u>Remark 2.7:</u> X_H need not be a <u>strict</u> relative minimizer. Moreover X^H is unstable (in the sense that X^H is not a relative minimizer of D_H); cf. [42].

The proof given in [37] makes use of the variational methods presented in [35] for minimal surfaces. However, the Palais-Smale condition may fail "in the large" due to possible "separation of spheres". This loss of compactness must be overcome by suitable a-priori bounds for $D_H(X^H)$ <u>and</u> $D(X^H)$. The latter bounds are obtained by varying H - a technique which has proved useful also in other variational problems beyond the compactness range, cp. [41-43].

3. <u>Harmonic maps of Riemannian surfaces</u>

Another natural generalization of Plateau's problem is to study "energy-minimizing" maps of a (compact) Riemannian surface M into (compact) Riemannian manifolds N . For a (smooth) map $X : M \to N$ the energy of X is given by the intrinsic Dirichlet integral

$$E(X) = \int_M e(X)\,dM$$

with energy density given by

$$e(X) = \frac{1}{2}\,\gamma^{\alpha\beta}(w)\,g_{ij}(X)\,\frac{\partial}{\partial w^\alpha}X^i\,\frac{\partial}{\partial w^\beta}X^j$$

in local coordinates $w = (w^1, w^2)$ on M , $X = (X^1, \ldots, X^n)$ on N and in local representations $(\gamma_{\alpha\beta})_{1\leq\alpha,\beta\leq 2}$, $(g_{ij})_{1\leq i,j\leq n}$ of the metrics on M , N resp. $(\gamma^{\alpha\beta})$ denotes the inverse of the metric tensor $(\gamma_{\alpha\beta})$. A summation convention is used.

A smooth map X which is critical for E (with respect to smooth variations of X) is called harmonic. Necessarily, a harmonic map X satisfies the following elliptic system of Euler-Lagrange equations

$$\Delta_M X = \Gamma_N(X)(\nabla X, \nabla X)_M \tag{3.1}$$

in M , where in local coordinates

$$\Delta_M = \frac{1}{\sqrt{\gamma}} \frac{\partial}{\partial w^\alpha} \left(\gamma^{\alpha\beta}\sqrt{\gamma} \frac{\partial}{\partial w^\beta}\cdot\right)$$

denotes the Laplace-Beltrami operator on M , and the term appearing on the right of (3.1) with components

$$\Gamma_N(X)(\nabla X, \nabla X)^k = \gamma^{\alpha\beta}\Gamma^k_{ij}(X) \frac{\partial}{\partial w^\alpha} X^i \frac{\partial}{\partial w^\beta} X^j \quad , \quad 1 \leq k \leq n ,$$

involves the Christoffel-symbols Γ^k_{ij} of the metric g .

Harmonic maps - in particular smooth E-minimal maps $X: M \to N$ - are distinguished representants of maps $M \to N$. In order to understand how much of the topological structure of a space N is captured by harmonic maps $M \to N$ it is natural to study the following

Problem: Given a (smooth) map $X_o : M \to N$, is there a harmonic map $X : M \to N$ homotopic to X_o ?

Most likely, one would hope to find an E-minimizing harmonic map X in the homotopy class $[X_o]$ of X_o . In order to apply the "direct methods" we use the Nash embedding theorem to represent N as a submanifold of Euclidean $\mathbb{R}^N$ and consider the space

$$H^{1,2}(M;N) = \left\{X \in H^{1,2}(M;\mathbb{R}^N) \;\middle|\; X(M) \subset N \text{ a.e.}\right\} .$$

(The Sobolev space $H^{1,2}(M;\mathbb{R}^N)$ may be defined by covering M with local coordinate patches $\subset \mathbb{R}^2$.) The functional E is coercive and weakly lower semi-continuous on $H^{1,2}(M;N)$ with respect to the topology induced by $H^{1,2}(M;\mathbb{R}^N)$. However, since $H^{1,2}(M;\mathbb{R}^N)$ in two dimensions fails to be embedded into C^o, homotopy classes of maps $X_o: M \to N$ in general fail to be weakly closed in $H^{1,2}(M;N)$. The direct methods therefore in general will produce only trivial solutions to (3.1): constant maps $X(w) \equiv Q \in N$.

However, in 1964 Eells-Sampson [8] proved:

Theorem 3.1: Suppose the sectional curvature of N $\kappa_N \leq 0$. Then for any regular map $X_o : M \to N$ there is an E-minimizing harmonic map X homotopic to X_o .

Their method is based on an analysis of the evolution problem

$$(3.2) \qquad \partial_t X - \Delta_M X + \Gamma_N(X)(\nabla X, \nabla X)_M = 0 \quad , \quad X\big|_{t=0} = X_o$$

associated with (3.1). In view of the identity

$$\frac{d}{d\epsilon} E\left(X^i + \epsilon g^{ij}(X)\varphi^j\right)\Big|_{\epsilon=0} = \int_M \left\{-\Delta_M X + \Gamma_N(X)(\nabla X, \nabla X)_M\right\} \cdot \varphi dM$$

for smooth $X : M \to N$ and smooth (local) variations φ we may consider (3.2) as the L^2-gradient flow for E on $H^{1,2}(M;N)$ with respect to the metric $\left(g_{ij}(X)\right)$.

Eells and Sampson prove that for $\kappa_N \leq 0$ (3.2) has a unique regular solution $X(t)$ which as $t \to \infty$ converges to a regular harmonic map X_∞ homotopic to X_o .

If the curvature condition was weakened, similar results for (3.2) could only be established under a-priori restrictions of the size of the images $X(t)$ which are unnatural in our problem. Indeed, the result of Eells and Sampson ceases to hold true:

Example 3.2 (Eells-Wood [9]): For any map $X_o : T^2 \to S^2$ of degree ≥ 1 there is no harmonic representative $X \in [X_o]$.

T^2 is the standard two-dimensional torus.

In general, non-trivial harmonic maps of S^2 into a given target manifold N are the obstructions to the existence of harmonic representants of maps $M \to N$.

Theorem 3.3 (Lemaire [22], Sacks-Uhlenbeck [29]): Suppose the second fundamental group $\pi_2(N) = 0$. Then for any $X_o : M \to N$ there is an E-minimizing harmonic map X homotopic to X_o .

Sacks and Uhlenbeck approximate the functional E on the space $H^{1,2}(M;N)$ by a family of regularized functionals E_α , $\alpha > 1$, that admit regular E_α-minimal representants $X_\alpha \in [X_o]$. As $\alpha \to 1$, $E_\alpha \to E$ and a sequence $\{X_\alpha\}$ converges to a harmonic map $X_1 : M \to N$ away from finitely many points where non-constant harmonic maps $\overline{X} : S^2 \to N$ separate. If $\pi_2(N) = 0$, $[X_1] = [X_\alpha] = [X_o]$ is preserved. (Actually, $X_\alpha \to X_1$ globally.)

The following result makes it possible to prove Theorem 3.3 <u>directly</u> without using an approximation method. It extends the Eells-Sampson result to arbitrary compact target manifolds; of course, in view of Example 3.2 the concept of a solution to (3.2) has to be appropriately extended.

<u>Theorem 3.4</u> (Struwe [38]): For any $X_o \in H^{1,2}(M;N)$ there exists a global (distribution) solution X to (3.2) which is regular on $M \times]0,\infty[$ with exception of finitely many points $(w_k,t_k)_{1\leq k\leq K}$, $t_k \leq \infty$, and unique in this class. At a singularity $(\overline{w},\overline{t})$ a non-constant harmonic map $\overline{X} : \overline{\mathbb{R}^2} \cong S^2 \to N$ separates in the sense that for sequences $R_m \searrow 0$, $w_m \to \overline{w}$, $t_m \nearrow \overline{t}$ as $m \to \infty$

$$X_m(w) \equiv X(w_m+R_m w,t_m) \to \overline{X} \quad \text{in} \quad H^{1,2}_{loc}(\mathbb{R}^2;N) .$$

Moreover, if $\overline{t} = \infty$ is regular, $X(t)$ converges to a regular harmonic map $X_\infty : M \to N$ as $t \to \infty$.

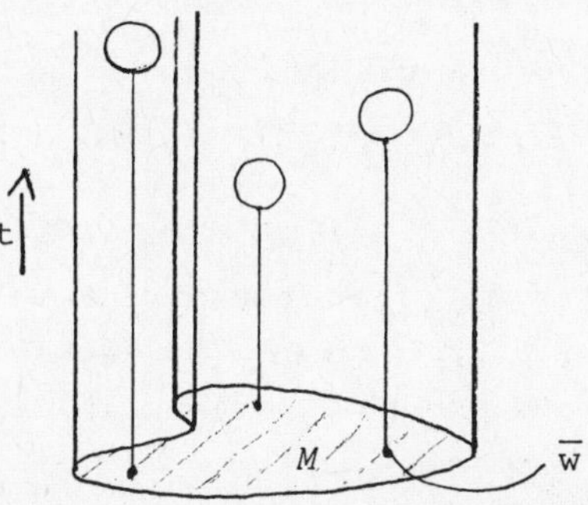

figure 5 "separation of harmonic spheres"

By Example 3.2 the flow (3.2) in general <u>cannot</u> be globally regular and asymtotically convergent. However, it is not known whether singularities may appear in finite time.

More generally, for higher dimensions: $\dim M = m \geq 2$, analogous to Theorem 3.4 we may conjecture that the (m-2)-dimensional Hausdorff measure (with respect to the parabolic metric $\delta\big((w,t),(w',t')\big) = |w-w'| + \sqrt{|t-t'|}$) of the set of singularities of the flow (3.2) is finite (and a-priori bounded in terms of $E(X_o)$).

4. Minimal and H-surfaces with free boundaries

Given a (smooth) compact surface $S \subset \mathbb{R}^3$ we now ask for disc-type surfaces X of prescribed constant mean curvature $H \in \mathbb{R}$ supported by S, in particular for minimal surfaces $(H = 0)$. The boundary $X(\partial B)$ of these surfaces may move freely on the supporting surface S; however, we shall prescribe the angle of contact to be vertical. For parametric surfaces $X \in C^2(B;\mathbb{R}^3) \cap C^1(\bar{B};\mathbb{R}^3)$ these conditions translate into the following non-linear system

$$\Delta X = 2H\, X_u \wedge X_v\,, \tag{4.1}$$

$$|X_u|^2 - |X_v|^2 = 0 = X_u \cdot X_v\,, \tag{4.2}$$

$$X(w) \in S\,,\ \forall w \in \partial B\,, \tag{4.3}$$

$$\frac{\partial}{\partial n} X(w) \perp T_{X(w)} S\,,\ \forall w \in \partial B\,, \tag{4.4}$$

where n denotes the (outer) unit normal along ∂B, "$\perp$" means vertical, and $T_p S$ designates the tangent space to S at p. For brevity, solutions to (4.1-4) will be called "H-surfaces on S", resp. "minimal surfaces on S", if $H = 0$.

<u>Example:</u> If S is an ellipsoid with half-axes $a > b > c > 0$ the three planes of symmetry (orthogonal to the axes of S) intersect S in minimal surfaces.

Minimal surfaces with (partially) free boundaries were considered by Gergonne (1816) and Schwarz (1872). Courant ([5]) observed that minimal surfaces on S may be obtained as (relative) minimizers of Dirichlet's integral on (essentially) the space

$$C(S) = \left\{ X \in H^{1,2}(B;\mathbb{R}^3) \;\middle|\; X(\partial B) \subset S \ \text{ a.e.} \right\}.$$

Applying the direct methods Courant obtained

<u>Theorem 4.1:</u> Let S be a compact surface of positive genus. Then there exists a non-constant minimal surface on S.

Courant reasons as follows: If $\text{genus}(S) > 0$ there exists a curve Γ in $\mathbb{R}^3$ which is not contractible in $\mathbb{R}^3 \setminus S$. Consider now surfaces $X \in C(S)$ whose parametric boundaries $X|_{\partial B} : \partial B \to S$ "link" with Γ, cp. figure 7. Then D attains its infimum in the class $C(S)$ at a minimal surface X intersecting Γ.

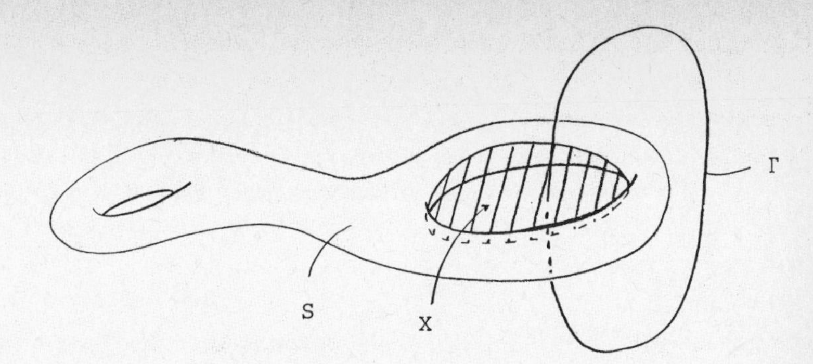

figure 6

Remarks 4.2: i) Courant's result requires minimal smoothness of the supporting surface, in particular polyhedral surfaces S are admissible.
ii) For regular supporting surfaces also the minimal surfaces obtained by Courant's method are smooth up to the boundary, cp. [21], [18], [17], [7], [10].
iii) Courant's result has been partially extended to H-surfaces by Hildebrandt [15]. For $S \subset B_r(0) \subset \mathbb{R}^3$ of positive genus and $|H|R \leq 1$ Hildebrandt obtains solutions to (4.1-3) with prescribed contact angle (in general $\neq 90^o$).

What happens if genus(S) = 0 , i.e. $S \cong S^2$, the standard sphere? In this case it is in general impossible to characterize a non-constant minimal or H-surface on S as a relative minimum in a suitably restricted class of comparison surfaces. Indeed, if S is strictly convex (i.e. if S bounds a strictly convex body in $\mathbb{R}^3$) the only solutions of (4.1-4) which are local minima of D in C(S) are the trivial (constant) solutions $X \equiv X_o \in S$, while any non-constant minimal surface on S will necessarily correspond to an unstable critical point of D .

The following result was obtained in [36]:

Theorem 4.3: Suppose S is C^4-diffeomorphic to S^2 . Then there exists a non-constant minimal surface supported by S .

Remarks 4.4: i) Simultaneously with [36], Smyth [31] proved the existence of 3 embedded minimal surfaces on a tetrahedron by an algebraic method. His approach, however, is limited to this particular configuration.
ii) For general convex surfaces S the existence of embedded minimal discs on S was later verified by Grüter and Jost [11], [19].

iii) The proof of Theorem 4.4 originally used the Sacks-Uhlenbeck approximation method; cp. §3. Indeed, there are certain relations between the free boundary problem (4.1-4) for minimal surfaces and harmonic maps of spheres which make this approach seem rather natural. However, there is also a direct method which extends to $H \neq 0$, as we shall see.

For general $H \in \mathbb{R}$ the Sacks-Uhlenbeck technique no longer seems to be applicable in (4.1-4). However, by studying "the" evolution problem associated with (4.1-4) one finds:

Theorem 4.5 (Struwe [41]): Suppose $S \subset \overline{B_R(0)} \subset \mathbb{R}^3$ is C^4-diffeomorphic to S^2. Then for a.e. $H \in \mathbb{R}$ such that $|H|R < 1$ there is a non-constant H-surface supported by S.

Actually, there are many ways to model the evolution of H-surfaces. By analogy with the harmonic map problem we try to set up the evolution problem as to generate the L^2-gradient flow for a suitable functional related to (4.1-4).

This functional is given by

$$D_H(X) = D(X) + 2H\big(V(X) - V(\eta(X))\big) ,$$

where η is a smooth extension operator, defined in a neighborhood of X, which associates to $X \in C(S)$ an extension $\tilde{X} = \eta(X)$ with the same boundary but lying completely on the surface S : $\tilde{X}(B) \subset S$. This determines $V(\eta(X))$ up to integral multiples of the volume enclosed by S. In particular, the "L^2-gradient" of D_H is uniquely determined and we are led to consider the evolution problem

$$\partial_t X - \Delta X + 2HX_u \wedge X_v = 0 \quad \text{in} \quad B \times]0,\infty[\tag{4.5}$$

with boundary and initial conditions

$$X(w,t) \subset S \quad , \quad \forall (w,t) \in \partial B \times]0,\infty[\ , \tag{4.6}$$

$$\frac{\partial}{\partial n} X(w,t) \perp T_{X(w,t)}S \quad , \quad \forall (w,t) \in \partial B \times]0,\infty[\ , \tag{4.7}$$

$$X\big|_{t=0} = X_o \in C(S) \ . \tag{4.8}$$

The boundary condition (4.7) allows to extend the flow X by reflection across S giving rise to a system similar to the evolution problem for harmonic maps.

Thus, under suitable conditions on H and the initial data one can now prove existence and regularity results for (4.5-8) analogous to Theorem 3.4. In particular, for $H = 0$ we obtain ([41])

Theorem 4.6: Let $H = 0$. Then for any $X_o \in C(S)$ there exists a unique global distribution solution to (4.5-8) which is regular on $\overline{B} \times]0,\infty[$ with exception of finitely many points $(w_k, t_k)_{1 \leq k \leq K}$, $w_k \in \partial B$, $t_k \leq \infty$. At a singularity $(\overline{w}, \overline{t})$ a non-constant minimal surface $\overline{X} : \overline{\mathbb{R}^2_+} \cong B \to \mathbb{R}^3$ supported by S separates in the sense that for sequences $R_m \searrow 0$, $w_m \to \overline{w}$, $t_m \nearrow \overline{t}$

$$X_m(w) \equiv X(w_m + R_m w, t_m) \to \overline{X} \quad \text{in} \quad H^{1,2}_{loc}(\overline{\mathbb{R}^2_+}) .$$

Finally, if $\overline{t} = \infty$ is regular, then as $t \to \infty$ $X(t)$ converges to a minimal surface supported by S.

Apart from possible extensions of Theorem 4.5 to arbitrary $H \neq 0$ one of the main open problems is the existence of multiple solutions. By analogy with the problem of closed geodesics on a surface S of genus 0 one may conjecture that for sufficiently small $|H|$ there will always be at least 3 geometrically distinct H-surfaces on S. However, so far only partial results are known, cp. [31], [19].

References

[1] Benci, V.: A new approach to the Morse-Conley theory, preprint, Pisa (1986)

[2] Böhme, R. - Tromba, A.J.: The index theorem for classical minimal surfaces, Ann. of Math. 113 (1981), 447-499

[3] Brezis, H. - Coron, J.-M.: Multiple solutions of H-systems and Rellich's conjecture, Comm. Pure Appl. Math. 37 (1984), 149-187

[4] Conley, C.: Isolated invariant sets and the Morse index, CBMS 38, AMS, Providence, 1978

[5] Courant, R.: Dirichlet's principle, conformal mapping and minimal surfaces, New York, Interscience, 1950

[6] Douglas, J.: Solution of the Problem of Plateau, Trans. AMS 33 (1931), 263-321

[7] Dziuk, G.: C^2-regularity for partially free minimal surfaces,Math. 7, 189 (1985), 71-79

[8] Eells, J. - Sampson, J.H.: Harmonic mappings of Riemannian manifolds,Am. J. Math. 86 (1964), 109-160

[9] Eells, J. - Wood, J.C.: Restrictions on harmonic maps of surfaces, Topology 15 (1976), 263-266

[10] Grüter, M. - Hildebrandt, S. - Nitsche, J.C.C: On the boundary behavior of minimal surfaces with a free boundary which are not minima of the area, Manusc. math. 35 (1981), 387-410

[11] Grüter, M. - Jost, J.: On embedded minimal discs in convex bodies, preprint, SFB 72, Bonn (1984)

[12] Heinz, E.: Über die Existenz einer Fläche konstanter mittlerer Krümmung bei vorgegebener Berandung, Math. Ann. 127 (1969), 258-287

[13] Heinz, E.: On the nonexistence of a surface of constant curvature with finite area and prescribed rectifiable boundary, Arch. Rat. Mech. Anal. 35 (1969), 249-252

[14] Hildebrandt, S.: On the Plateau problem for surfaces of constant mean curvature, Comm. Pure Appl. Math. 23 (1970), 97-114

[15] Hildebrandt, S.: Randwertprobleme für Flächen mit vorgeschriebener mittlerer Krümmung und Anwendungen auf die Kapillaritätstheorie II, Arch. Rat. Mech. Anal. 39 (1970), 275-293

[16] Hildebrandt, S.: The calulus of variations today, as reflected in the Oberwolfach meetings, Perspect. Math., Birkhäuser, 1985

[17] Hildebrandt, S. - Nitsche, J.C.C.: Minimal surfaces with free boundaries, Acta Math. 143 (1979), 251-272

[18] Jäger, W.: Behavior of minimal surfaces with free boundaries, Comm. Pure Appl. Math. 23 (1970), 803-818

[19] Jost, J.: Existence results for embedded minimal surfaces of controlled topological type, preprint, SFB 72, Bonn (1985)

[20] Jost, J. - Struwe, M.: in preparation

[21] Lewy, H.: On minimal surfaces with partially free boundary, Comm. Pure Appl. Math. 4 (1951), 1-13

[22] Lemaire, L.: Applications harmoniques de surfaces Riemanniennes, J. Diff. Geom. 13 (1978), 51-78

[23] Morse, M. - Tompkins, C.B.: The existence of minimal surfaces of general critical types, Ann. of Math. 40 (1939), 443-472

[24] Nitsche, J.C.C.: Vorlesungen über Minimalflächen, Grundlehren 199, Springer-Verlag, Berlin-Heidelberg-New York, 1975

[25] Palais, R.S.: Critical point theory and the minimax principle, Proc. Symp. Pure Appl. Math. 15 (1970), 185-212

[26] Palais, R.S. - Smale, S.: A generalized Morse theory, Bull. AMS 70 (1964), 165-172

[27] Rado, T.: On Plateau's problem, Ann. of Math. 31 (1930), 457-469

[28] Rado, T.: The isoperimetric inequality and the Lebesgue definition of surface area, Trans. AMS 61 (1947), 530-555

[29] Sacks, J. - Uhlenbeck, K.: The existence of minimal immersions of 2-spheres, Ann. Math. 113 (1981), 1-24

[30] Shiffman, M.: The Plateau problem for non-relative minima, Ann. of Math. 40 (1939), 834-854

[31] Smyth, B.: Stationary minimal surfaces with boundary on a simplex, Inv. Math. 76 (1984), 411-420

[32] Steffen, K.: Flächen konstanter mittlerer Krümmung mit vorgegebenem Volumen oder Flächeninhalt, Arch. Rat. Mech. Anal. 49 (1972), 99-128

[33] Steffen, K.: On the existence of surfaces with prescribed mean curvature and boundary, Math. Z. 146 (1976), 113-135

[34] Steffen, K.: On the nonuniqueness of surfaces with prescribed constant mean curvature spanning a given contour, Arch. Rat. Mech. Anal. 94 (1986), 101-122

[35] Struwe, M.: On a critical point theory for minimal surfaces spanning a wire in $\mathbb{R}^n$, J. Reine Angew. Math. 349 (1984), 1-23

[36] Struwe, M.: On a free boundary problem for minimal surfaces, Inv. Math. 75 (1984), 547-560

[37] Struwe, M.: Large H-surfaces via the mountain-pass-lemma, Math. Ann. 270 (1985), 441-459

[38] Struwe, M.: On the evolution of harmonic mappings of Riemannian surfaces, Comm. Math. Helv. 60 (1985), 558-581

[39] Struwe, M.: Nonuniqueness in the Plateau problem for surfaces of constant mean curvature, Arch. Rat. Mech. Anal. 93 (1986), 135-157

[40] Struwe, M.: A Morse theory for annulus-type minimal surfaces, J. Reine Angew. Math. 368 (1986), 1-27

[41] Struwe, M.: The existence of surfaces of constant mean curvature with free boundaries, Acta Math. (to appear)

[42] Struwe, M.: Plateau's problem and the calculus of variations, lect. notes SFB 72, Bonn (1986)

[43] Struwe, M.: Critical points of embeddings of $H^{1,2}$ into Orlicz spaces, preprint, Zürich (1986)

[44] Tromba, A.J.: Degree theory on oriented infinite dimensional varieties and the Morse number of minimal surfaces spanning a curve in $\mathbb{R}^n$,
Part I: $n \geq 4$, Trans. AMS 290 (1985), 385-413,
Part II: $n=3$, Manusc. Math. 48 (1984), 139-161

[45] Wente, H.C.: An existence theorem for surfaces of constant mean curvature, J. Math. Anal. Appl. 26, (1969), 318-344

[46] Wente, H.C.: A general existence theorem for surfaces of constant mean curvature, Math. Z. 120 (1971), 277-288

[47] Werner, H.: Das Problem von Douglas für Flächen konstanter mittlerer Krümmung, Math. Ann. 133 (1957), 303-319

OPEN PROBLEMS IN THE DEGREE THEORY FOR DISC MINIMAL SURFACES SPANNING A CURVE IN $\mathbb{R}^3$

A. J. Tromba

§0. Introduction.

It is now over fifty years since Leray and Schauder [8] introduced their degree theory on open sets of Banach spaces for non-linear mappings of the form (identity + compact). Leray used this degree to prove the existence of stationary solutions to the Navier Stokes equations on a bounded domain.

Since then many applications of this degree theory have been given in non-linear analysis [4], [9], [16], both to problems of existence, bifurcation and to counting the number of solutions to an equation. In addition there have been a very large number of papers generalizing this degree in many different directions. Unfortunately, unlike the original Leray-Schauder theory, several of these new theories have not been motivated by concrete analytical problems as was the original Leray-Schauder theory but rather by a search for generalization for its own sake.

In this paper we review a generalization of Leray-Schauder theory to a class of Fredholm maps on oriented infinite dimensional varieties and to a class of Fredholm vector fields on Banach manifolds. We show how these ideas enable one to compute the total Morse number of disc minimal surfaces spanning a contour in $\mathbb{R}^n$, $n \geq 3$. For $n > 3$ this total Morse number can be computed strictly by variational methods (Struwe [12]). We conclude with some open questions concerning this theory and its relation to minimal surfaces.

All of this work was supported by SFB 72, Bonn, and it is a great pleasure and honor to be able to present these ideas in a volume marking the close of this research group, and at the same time the beginning of another SFB.

1. A review of Elworthy-Tromba theory.

Let M be an infinite dimensional Banach manifold without boundary modelled on a Banach space E. Let $L_c(E)$ denote those linear operators of the form identity plus compact, $GL(E)$ the invertible linear operators on E, and $GL_c(E) = GL(E) \cap L_c(E)$. From [3] we have the following result.

Theorem 1.1: *The path component space* $\pi_0(GL_c(E))$ *of the Banach Lie group* $GL_c(E)$ *is isomorphic to* Z_2 *and consequently has two components. Let* $GL_c(E)$ *denote the component of the identity in* $GL_c(E)$ *and* $GL_c^-(E)$ *the other component.*

Definition 1.2: A smooth C^r Banach manifold M is a *Fredholm* manifold or admits a C^r Fredholm structure Φ if there is a collection of coordinate charts (φ_i, U_i) covering M so that the derivative of the transition maps $\varphi_i \circ \varphi_j^{-1}$ (when $U_i \cap U_j \neq \phi$) is in $GL_c(E)$. M is oriented with respect to Φ if a subcollection of charts of Φ can be found to have the property that the derivative of the transition maps is always in $GL_c^+(E)$.

Remark. Every open subset of a Banach space has a natural orientable Fredholm structure induced by $GL_c^+(E)$ itself. It is possible [3] that an infinite dimensional manifold can be orientable with respect to one Fredholm structure and not another.

From [15] we have results on the existence of Fredholm and orientable Fredholm structures. Before we describe these results we need the notion of a Fredholm map.

Defintion 1.3: A differentiable map between two Banach manifolds M and N is Fredholm if the derivative $df(x): T_xM \to T_{f(x)}N$ is linear Fredholm for each $x \in M$. This means that $\dim \operatorname{Ker} df(x) < \infty$, $\dim \operatorname{Coker} df(x) < \infty$ and the range of $df(x)$ is closed. The index of $df(x)$ at x is defined to be $\dim \operatorname{Ker} df(x) - \dim \operatorname{Coker} df(x)$. One can show that this index is constant on the components of M and thus if M is connected one can speak of the index of f. If M is not connected one requires the index to be the same on all components. Thus one speaks of Fredholm maps of index m.

We then have the following results from [3].

Theorem 1.4: *Let* M *be a Banach manifold modelled on* E *which admits* C^r *partitions of unity. Then* M *admits a* C^r *Fredholm structure if and only if there exists a Fredholm map* $f: M \to E$ *of index zero.*

Theorem 1.5: *Let* M *be a* C^r *Banach manifold admitting* C^r *partitions of unity and such that* $GL(E)$ *is contractible. Then* M *admits a Fredholm structure.*

Theorem 1.6: *A* C^r *Banach manifold* M *is orientable with respect to every possible Fredholm structure if* $H^1(M;Z_2) = 0$*; the first singular cohomology of* M *with* Z_2 *coefficients is zero. Conversely if* $GL(E)$ *is contractible and* M *admits* C^1 *partitions of unity, then the orientability of* M *with respect to every Fredholm structure implies* $H^1(M;Z_2) = 0$.

The existing application of this theorem which has arisen in analysis is where M is an open subset of a ball B in a Banach space E and the Fredholm structure on M is a restriction of the Fredholm structure on B [4]. Thus this Fredholm structure is orientable.

We shall desribe the Elworthy-Tromba degree theory for the case of Fredholm maps of index zero. from a Banach manifold M to an open subset $\mathcal{A}$ of a Banach space F. The case where $\mathcal{A}$ is a Banach manifold and the index is not necessarily zero is treated in [15]. In the case of nonzero index p the degree is no longer an integer but an element of a stable p-stem of the homotopy group of spheres.

Suppose $\pi: M \to \mathcal{A}$ if Fredholm of index zero and proper (by this we mean that the inverse image of compact sets is compact). We would like to assign an oriented degree to the map π. The two principal tools needed to accomplish this will be the generalization of the Sard theorem due to Smale and the pull back Theorem 1.9 which follows.

Definition 1.7: Let M be a C^r Banach manifold which admits a C^r-Fredholm structure Φ. A Fredholm map $\pi: M \to \mathcal{A}$ of index zero is said to be admissible with respect to Φ if for $(\varphi_i, U_i) \in \Phi$ the Frechet derivative $D(\pi \circ \varphi_i^{-1})(x) \in L_c(E)$; i. e. as a linear map is of the form identity plus compact.

Theorem 1.8 (Pull back Theorem): *Let* $\pi: M \to \mathcal{A}$ *be as above with* M *a Banach manifold,* $\mathcal{A} \subset F$ *an open subset and* π *Fredholm of index zero. The Fredholm structure* Φ *on* $\mathcal{A}$ *which* $\mathcal{A}$ *inherits from* F *(cf. Remark 1.3) and the map* π *induces a unique Fredholm structure* $\pi^*\Phi$ *on* M *with respect to which* π *is admissible. We call* $\pi^*\Phi$ *the pull back of* Φ *under* π.

In order to define a degree for such a map π we will need to know that π is proper and that $\pi^*\Phi$ is orientable.

The questions of properness is naturally a question of estimates, e.g. a priori elliptic estimates for solutions to nonlinear elliptic systems. The question of orientability is in a strong sense the heart of this paper. One could ask whether $H^1(M;Z_2) = 0$ and if so invoke Theorem 1.5 and this certainly works for the class of problems considered in [4]. However one may not be able to say anything whatsoever about $H^k(M;Z_2)$ for any k. This is the case with Plateau's problem. So we need some other condition to insure that $\pi^*\Phi$ is orientable and this will be given in §3. We shall therefore now assume outright that $\pi^*\Phi$ is orientable. Recall that this means that there is a subcollection of charts (φ_i, U_i) in $\pi^*\Phi$ covering M such that the derivative $D(\varphi_i \circ \varphi_j^{-1})$ when defined is in $GL_c^+(F)$.

In this case we define the oriented degree of π, $\deg \pi$, as follows. Let $y \in \mathcal{A}$ be a regular value of π (i. e. $x \in \pi^{-1}(y)$ implies that $D\pi(x): T_xM \to F$ is surjective). Since π is index zero this means that for each $x \in \pi^{-1}(y)$, $D\pi(x)$ is an isomorphism. The inverse function theorem guarantees that $\pi^{-1}(y)$ is a discrete set and properness implies that $\pi^{-1}(y)$ is a finite set of points $x_1,\dots,x_m$. For each x_i let (φ_i, U_i) be a coordinate chart in an orientation of M relative to $\pi^*\Phi$. Define

$$(1.9) \qquad \operatorname{sgn} D\pi(x_i) = \begin{cases} +1 & \text{if } D(\circ\varphi_i^{-1})(\varphi_i(x)) \in GL_c{}^+(F) \\ -1 & \text{if not} \end{cases},$$

and $\deg \pi = \sum_i \operatorname{sgn} D\pi(x_i)$.

If $\mathcal{A}$ is connected it can be shown (cf. [3]) that $\deg \pi$ is independent of the choice of regular value y and is also independent, under certain types of homotopies, of the choice of element in a restricted homotopy class of π. But most importantly if $\deg \pi \neq 0$, π is surjective.

2. Oriented varieties and degree theory.

We would now like to extend the notion of degree to maps $\pi: M \to \mathcal{A}$, where M is no longer a manifold, but some infinite dimensional oriented subvariety of some Banach space E.

We shall suppose as before that $\mathcal{A}$ is connected and open in a Banach space F. Moreover we shall also suppose M to be made up of pieces $\mathcal{M}_d$, $M = \cup_{d\in\mathcal{D}}\mathcal{M}_d$, these pieces being each C^2 Banach manifolds indexed by some linearly ordered set $\mathcal{D}$, with order relation $>$ in such a way that the following assumptions on stratification hold (cf. [13]):

(2.1) (i) $\partial\mathcal{M}_{d_i} \subset \cup_{d>d_i}\bar{\mathcal{M}}_d$,

(ii) $\pi: M \to \mathcal{A}$ is proper (the inverse image of a compact set is compact) and C^2 Fredholm of some nonpositive index on each $\mathcal{M}_d$,

(iii) that there is one "strata" or piece $\mathcal{M}_0$ such that π on $\mathcal{M}_0$ is C^2 Fredholm of index zero, and $d > 0$ for all d,

(iv) on all other $\mathcal{M}_d$, π is C^2 Fredholm of index less than or equal to negative 2, and finally

(v) $\mathcal{M}_0$ is orientable with respect to the pull back structure $\pi^*\Phi$ of the canonical Fredholm structure Φ on $\mathcal{A}$.

Then we claim that there is an oriented Elsworthy-Tromba integer degree for π. In order to see this we need a result due to Tomi and Tromba [13].

Theorem 2.2: *Under the stratification, index and properness assumptions on π, the image $\pi(\cup_{d>0}\mathcal{M}_d)$ has a closed nowhere dense image in $\mathcal{A}$ and does not disconnect $\mathcal{A}$.*

Remark 2.3: It will be very important to note that Theorem 2.2 holds under a weakening of the stratification hypotheses (2.1), namely we may modify (ii) to assume that each $\mathcal{M}_d$, $d \neq 0$, need not be a manifold but is locally contained in some manifold $\mathcal{N}_d$ such that $\pi|\mathcal{N}_d$ is Fredholm of index less than negative two (again see [13]).

Using 2.2 we can construct an oriented degree on $\mathcal{M}$ as follows. Let $\alpha \in \mathcal{A}$ be a regular value for π. By 2.2 we may assume that $\alpha \notin \pi(\cup_{d>0}\mathcal{M}_d)$. From the fact that π on $\mathcal{M}$ is proper and π is index zero on $\mathcal{M}_0$ we know that $\pi^{-1}(\alpha)$ consists of a finite number of points $x_1,\ldots,x_m$. Since $\mathcal{M}_0$ is orientable with respect to $\pi^*\Phi$ we can define

$$\deg(\pi,\alpha) = \sum_{x_i} \operatorname{sgn} D\pi(x_i) \quad ,$$

where $\operatorname{sgn} D\pi(x_i)$ is defined as in (1.10).

The main difficulty now, of course, is to show that $\deg(\pi,\alpha)$ is independent of the choice of $\alpha \in \mathcal{A}$. Therefore let us suppose that $\beta \in \mathcal{A}$ is another regular value of π, $\beta \notin \pi(\cup_{d>0}\mathcal{M}_d)$. By 2.2 we can find a path $\sigma\colon I \to \mathcal{A}$, I the unit interval such that $\sigma(0) = \alpha$, $\sigma(1) = \beta$, and with $\sigma(I) \subset \mathcal{A} - \pi(\cup_{d>0}\mathcal{M}_d)$. Since $\pi(\cup_{d>0}\mathcal{M}_d)$ is closed we may apply Smale's transversality theorem to perturb σ to a map $\tilde{\sigma}$ such that $\tilde{\sigma}(0) = \alpha$, $\tilde{\sigma}(1) = b$, π is transversal to $\tilde{\sigma}$, and $\tilde{\sigma}(I) \subset \mathcal{A} - \pi(\cup_{d>0}\mathcal{M}_d)$. Thus $\pi^{-1}(\sigma(I))$ effect a cobordism between $\pi^{-1}(\alpha)$ and $\pi^{-1}(\beta)$ and then the argument of Elworthy-Tromba [3] shows that $\deg(\pi,\beta) = \deg(\pi,\alpha)$ and we take this to be the degree of π.

3. <u>The existence of natural orientations on Banach manifolds.</u>

In §1 we described a result (Theorems 1.6 and 1.7) which give conditions when a Banach manifold admits a Fredholm structure which is orientable. In this section we show how oriented Banach manifolds may naturally arise in problems in nonlinear analysis. We begin with the first basic result on oriented infinite dimensional manifolds.

Let $\mathcal{A} \subset F$ be open and suppose that for each $a \in \mathcal{A}$ we have a Banach manifold $\mathcal{M}_a$ modelled on a Banach space E in such a way that $\cup_{a\in\mathcal{A}}\mathcal{M}_a = \mathcal{M}$ has the structure of a smooth fibre bundle with bundle projection map $\pi\colon \mathcal{M} \to \mathcal{A}$ (cf. [15]).

For each $a \in \mathcal{A}$ suppose that $X_a\colon \mathcal{M}_a \to T\mathcal{M}_a$ is a vector field on

$\mathcal{M}_a$ such that the induced vector field X on $\mathcal{M}$ defined by $X|\mathcal{M}_a = X_a$ is C^r, $r \geq 2$.

Definition 3.1: A vector field with the property that whenever $X_a(x) = 0$, the derivative $DX_a(x): T_x\mathcal{M}_a \circlearrowleft$ is in $L_c(T_x\mathcal{M}_a)$ is said to be *Palais-Smale*. For $a \in \mathcal{A}$ fixed, a zero x of X_a is nondegenerate whenever $DX_a(x) \in GL_c(T_x\mathcal{M}_a)$. Thus the sign of a zero is well defined depending on whether or not $DX_a(x)$ belongs to GL_c^+ or GL_c^-. Formally

$$\operatorname{sgn} DX_a(x) = \begin{cases} +1 & \text{if } DX_a(x) \in GL_c^+(T_x\mathcal{M}_a) , \\ -1 & \text{if } DX_a(x) \in GL_c^-(T_x\mathcal{M}_a) . \end{cases}$$

Warning. The derivative of a vector field is not a well-defined object as a morphism from a tangent space to itself away from a zero. Thus it makes no sense to speak of $DX_a(x) \in L_c(T_x\mathcal{M}_a)$ if x is not zero.

Theorem 3.2: *Let* $X: \mathcal{M} \to T\mathcal{M}$ *be a* C^r *vector field* $(r \geq 2)$ *on a trivial Banach fibre bundle* $\mathcal{M} = \mathcal{A} \times M$ *over an open subset* $\mathcal{A} \subset F$ *such that for each* $a \in \mathcal{A}$, $X_a = X|\mathcal{M}_a$ *is a Palais-Smale field on* $\mathcal{M}_a$. *Assume that* M *admits some orientable Fredholm structure* Φ. *Suppose further that whenever* $X(m) = 0$ *the total derivative* $DX(m): T_m\mathcal{M} \to T_m\mathcal{M}_a$ *is surjective. Then the zero set* Σ *of* X *is a smooth submanifold of* $\mathcal{M}$ *which has a natural orientable Fredholm structure* $\tilde{\Phi}$. *The projection map* $\pi: \mathcal{M} \to \mathcal{A}$ *restricts to a map* $\pi_\Sigma: \Sigma \to \mathcal{A}$ *which is Fredholm of index zero. If* Φ *denotes the Fredholm structure which* $\mathcal{A}$ *inherits as an open subset of* F, *then the pull back Fredholm structure* $\pi_\Sigma^*\Phi = \tilde{\Phi}$. *Moreover* $a \in \mathcal{A}$ *is a regular value for* π_Σ *iff* $DX_a(x): T_x\mathcal{M}_a \to T_x\mathcal{M}_a$ *is an isomorphism for any* x *for which* $X_a(x) = 0$.

Finally with respect to the natural orientation which X *induces on* Σ, *if* $x \in \pi_\Sigma^{-1}(a)$ *then*

$$\operatorname{sgn} D\pi_\Sigma(x) = \operatorname{sgn} DX_a(x) \quad . \tag{3.3}$$

For the proof see [15].

4. On the local winding number of a Rothe vector field.

Let $\mathcal{S}$ denote those linear operators $S: E \rightleftarrows$ such that for $0 \leq t \leq 1$, $tS + (1-t)I$ belongs to $GL(E)$, where I denotes the identity mapping. Let $R_C(E)$ denote all those linear operators of the form $S + K$ where $S \in \mathcal{S}$ and K is a compact linear operator. Denote by $G\mathcal{R}_C(E) = \mathcal{R}_C(E) \cap GL(E)$. Then from [15] we have

Theorem 4.1.: $\pi_0(GR_C(E)) = 2$. *Thus* $GR_C(E)$ *has two components.*

Denote by $GR_C^+(E)$ the component of the identity and $GR_C^-(E)$ the other component. These will be of importance momentarily.

Let U be an open subset of E with $Y: \bar{U} \to E$ a C^2-mapping.

Definition 4.2: Such a field Y is said to be a Rothe-field on U if for each $x \in U$, $DY(x) \in R_C(E)$. Thus a Palais-Smale field is a Rothe-field.

Since every element of $R_C(E)$ is a linear Fredholm operator of index zero we get immediately from the definition of non-linear Fredholm operator (one whose linearization is Fredholm)

Theorem 4.3: *A Rothe-field is a Fredholm map of index zero.*

The next result is due to Smale [11].

Theorem 4.4: *Fredholm maps are locally proper.*

Let x_0 be an isolated zero of a Rothe-field $Y: \bar{U} \to E$ $(Y(x_0) = 0)$. We wish to define the local winding number of degree of Y about x_0. Choose a neighborhood B of x_0 so that $Y|\bar{B}$ is proper and no other zero of Y is in $\bar{B}$. By properness $E - Y(\partial\bar{B})$ will be open and so let $\mathcal{O}$ be the component of $E - Y(\partial\bar{B})$ containing 0 and let $M = Y^{-1}(\mathcal{O})$ and $Y_M = Y|M$. The map $Y_M: M \to \mathcal{O}$ is a proper Fredholm map of index zero.

By the Smale-Sard theorem we can find a regular value $y \in \mathcal{O}$ for Y_M. Then $Y_M^{-1}(y)$ contains only a finite number of points $x_1, \ldots, x_m$, and since y is regular

$$DY_M(x_j) \in GR_C(E)$$

for all j. This permits us to define the signum of $DY_M(x_j)$ by

$$\operatorname{sgn} DY_M(x_j) = \begin{cases} +1 & \text{if } DY_M(x_j) \in GR_C^+(E) \\ -1 & \text{if } DY_M(x_j) \in GR_C^-(E) \end{cases} .$$

We now use this for the next basic

Definition 4.5: The local degree or winding number of Y about x_0, $\deg(Y,x_0)$ is defined by

$$\deg(Y,x_0) = \sum_{j=1}^{m} \operatorname{sgn} DY_M(x_j) .$$

Using the methods of Elworthy and the author it follows that this definition is independent of the choice of B and the choice of the regular value y.

More generally let U be any open subset of E with $X: \bar{U} \to E$ proper and Rothe on U. If $0 \notin Y(\partial U)$ we may then repeat the above construction to define a degree $\deg(Y,\bar{U})$. If no zeros of Y exist in U this degree will be zero. Moreover if Y has finitely many zeros $x_1,\ldots,x_m$ in U, then

$$\deg(Y,U) = \sum_{j} \deg(Y,x_j) . \tag{4.6}$$

Finally, we have the basic property of all degree theories, namely

Theorem 4.7: *If Y_0, Y_1 are two proper Rothe-fields on $\bar{U}$ such that there exists a homotopy Y_t, $0 \leq t \leq 1$ with the property that*

(i) each Y_t is Rothe on $\bar{U}$

(ii) $0 \notin \bar{Y}_t(\partial U)$ for all t

(iii) $(t,x) \to Y_t(x)$ is a proper map.

Then

$$\deg(Y_1,U) = \deg(Y_0,U).$$

5. The index theorem for classical minimal surfaces.

Our main reference for this section is the Index Theorem by Böhme and Tromba [2]. Let Γ be a smooth wire in R^n which is the image of a differentiable embedding $\alpha: S^1 \to R^n$, $\alpha(S^1) = \Gamma^\alpha$. A solution to the classical Pleateau problem for Γ is a mapping $x: \bar{D} \to R^n$, $\bar{D}$ the closed unit disc in R^2, $\partial D = S^1$ such that

(1) $\Delta x = 0$ (each component of x is harmonic),

(2) $x_u \cdot x_v = 0$ (the coordinates of R^2 being labeled by u,v),

(3) $\|x_u\|^2 = \|x_v\|^2$,

(4) $x: S^1 \to \Gamma^\alpha$ is a homeomorphism (or monotonic).

Conditions (2) and (3) imply that the surface is conformally parametrized. Moreover these are the Euler equations for critical points of Dirichlet's integral on the space of all disc surfaces spanning Γ.

A point $z_0 \in D$, where $F(z) = x_u + ix_v$. $i = \sqrt{-1}$. vanishes, is called an interior branch point of x. Since F is holomorphic p has a finite order λ $\{F(z) = (z-z_0)^\lambda G(z),\ G(z_0) \neq 0\}$. A point $\xi \in S^1$ where F vanishes is called a boundary branch point and by results of Heinz and Tomi [6] ξ has a well-defined order. Equations (2) and (3) also imply that all points where x fails to be an immersion are branch points. Moreover the monotonicity assumption (4) implies that boundary branch points will all be of even order.

For integers r and s, $r \geq 2s + 4$, define

$$\mathcal{D} = \mathcal{D}^s = \{u: S^1 \to S^1 \,|\, \deg u = 1 \text{ and } u \in H^s(S^1,\mathbb{C})\}\ ,$$

where H^s denotes the Sobolev space of s-times differentiable (in the distribution sense) functions with values in the complexes $\mathbb{C}$, set

(5.0) $\mathcal{A} = \{\alpha: S^1 \to R^n \,|\, \alpha \in H^r(S^1,R^n) = F,\ \alpha \text{ an embedding}\}$ [i. e. α is one-to-one and $\alpha'(\xi) \neq 0$ for all $\xi \in S^1$] and the total curvature of Γ^α is bounded by $\pi(s-2)$.

Let $\pi: \mathcal{A} \times \mathcal{D} \to \mathcal{A}$ denote the projection map onto the first factor. A minimal surface $x: \bar{D} \to R^n$ spanning $\alpha \in \mathcal{A}$ can be viewed as an element of $\mathcal{A} \times \mathcal{D}$, since x is harmonic and therefore determined by its boundary values

$$\alpha|\partial D = x|S^1 = \alpha \circ u \quad , \quad \text{where} \quad (\alpha,u) \in \mathcal{A} \times \mathcal{D} \ .$$

The classical approach to minimal surfaces was to understand the set of minimal surfaces spanning a given fixed wire α; that is the set of minimal surfaces in $\pi^{-1}(\alpha)$. Our approach is to first understand the structure of the set of minimal surfaces as a subset of the bundle $\mathcal{N} = \mathcal{A} \times \mathcal{D}$ as fibre bundle over $\mathcal{A}$ and then to approach the question of the set of minimal surfaces in the fibre $\pi^{-1}(\alpha)$ in terms of the singularities of the projection map π restricted to a suitable subvariety of $\mathcal{N}$. This is in the spirit of Thom's original approach to unfoldings of singularities.

Let us say that a minimal surface $x \in \mathcal{A} \times \mathcal{D}$ has branching type (λ,ν), $\lambda = (\lambda_1,\ldots,\lambda_p) \in Z^p$, $\nu = (\nu_1,\ldots,\nu_q) \in Z^q$, each $\lambda_i, \nu_i \geq 0$ if x has p distinct but arbitrarily located interior branch points $z_1,\ldots,z_p$ in D of integer orders $\lambda_1,\ldots,\lambda_p$ and q distinct boundary branch points $\xi_1,\ldots,\xi_q$ in S^1 of (even) integer orders $\nu_1,\ldots,\nu_q$. In a formal sense the subset $\mathcal{M}$ of minimal surfaces $\mathcal{N}$ is an algebraic subvariety of $\mathcal{N}$ [2] and is a stratified set, stratified by branching types. To be more precise let $\mathcal{M}_\nu^\lambda$ denote the minimal surfaces of branching type (λ,ν). We can now state the index result of Böhme and Tromba [2].

Theorem 5.1 (Index theorem for disc surfaces). *The set* $\mathcal{M}_0^\lambda$ *is a* C^{r-s-1} *submanifold of* $\mathcal{N}$ *and the restriction* π^λ *of* π *to* $\mathcal{M}_0^\lambda$ *is* C^{r-s-1}, *Fredholm of index* $I(\lambda) + 3 = 2(2-n)|\lambda| + 2p + 3$, *where* $|\lambda| = \Sigma\lambda_i$.

Moreover, locally, for $\nu \neq 0$, $\mathcal{M}_\nu^\lambda \subset W_\nu^\lambda$ *such that* W_ν^λ *is a submanifold of* $\mathcal{N}$ *and where the restriction* π_ν^λ *of* π *to* W_ν^λ *is Fredholm of index* $I(\lambda,\nu) + 3 = 2(2-n)|\lambda| + (2-n)|\nu| + 2p + q + 3$, *where* $|\nu| = \Sigma\nu_i$. *The number* 3 *comes from the equivariance of the problem under the action of the three dimensional conformal group of the disc.*

Ursula Thiel has shown [14] that if one uses weighted Sobolev spaces as a model the sets $\mathcal{M}_\nu^\lambda$ can indeed be given a manifold structure with the index of $\pi_\nu^\lambda = \pi|\mathcal{M}_\nu^\lambda = 3 + I(\lambda,\nu)$.

It is easy to see that if $n \geq 3$, then $I(\lambda,\nu) \leq 0$. This index measures (in some sense) the stability of minimal surfaces of branching type (λ,ν) in R^n and the likelihood of finding such surfaces; the more negative the index of π_ν^λ the less likely it is to find a wire admitting minimal surfaces of branching type (λ,ν) which span it.

These stratification and index results are the basis to prove the generic (open-dense) finiteness and stability of minimal surfaces of the type of the disc as discussed in [2]; i. e. there exists an open dense subset $\hat{\mathscr{A}} \subset \mathscr{A}$ such that if $\alpha \in \hat{\mathscr{A}}$, then there exists only a finite number of minimal surfaces bounded by α, and these minimal surfaces are stable under perturbations of α. This open set will be the set of regular values of the map π. Moreover we have

Remark 5.2: If $n > 3$ the minimal surfaces spanning $\alpha \in \hat{\mathscr{A}}$ are all immersed up to the boundary, and if $n = 3$ they are simply branched.

Finally there are some other surprising consequences of this index formula. For example minimal surfaces in R^3 are free of interior branch points if they minimize area [1, 5, 10], whereas most minimal surfaces with simple interior branch points are stable with respect to perturbations of the boundary.

We are now ready to proceed with the development of degree theory for minimal surfaces.

6. Degree theory for minimal surfaces of disc type in R^n, $n \geq 4$.

In this section we show how the degree theory developed in §2 applies to the Plateau problem discussed in the last section.

Let $\mathscr{N} = \mathscr{A} \times \mathscr{D}$ be the bundle over $\mathscr{A}$ introduced in the last section. Let $\alpha \in \mathscr{A}$ and let Γ^α be the image of such an embedding. Consider the manifold of maps $H^s(S^1,\Gamma^\alpha)$. In [2] it is shown that $H^s(S^1,\Gamma^\alpha)$ is a C^{r-s} submanifold of $H^s(S^1,R^n)$. Let $\mathscr{N}(\alpha)$ denote the component of $H^s(S^1,\Gamma^\alpha)$ determined by α. (In [2], the notation $\mathscr{N}_\alpha$ was used for $\mathscr{N}(\alpha)$.) Recall that the tangent space to $\mathscr{N}(\alpha)$ at the point $x \in \mathscr{N}(\alpha)$ can be identified with the H^s maps $h\colon S^1 \to R^n$ with $h(\theta) \in T_{x(\theta)}\Gamma^\alpha$ (the tangent space to Γ^α at $x(\theta)$). By harmonic extension we can identify elements of $\mathscr{N}(\alpha)$ with harmonic surfaces spanning Γ^α. We shall always assume this identification.

In [17] it was shown that there exists a smooth C^{r-s-1} vector field X_α on $\mathscr{N}(\alpha)$ whose zeros are precisely the minimal surfaces spanning Γ^α. We should note here that each of these zeros are minimal surfaces in a more general sense than classically defined, since a zero x of X_α viewed as a harmonic map $x\colon D \to R^n$ need not induce a homeomorphism of S^1 onto Γ^α. We have the following theorem which is of great importance to us.

Theorem 6.1: *If* $X_\alpha(x) = 0$, *then the Frechet derivative of* X_α *at* x *maps* $T_x\mathcal{N}(\alpha)$ *into itself and is of the form identity plus a compact linear operator. Thus each* X_α *is a Palais-Smale field.*

We can give a definition of the vector field X_α as follows. Let $\mathcal{T}_i: S^1 \to R^n$, $i = 1,\ldots,n$, be a smooth framing of Γ^α; i. e. for each $p \in \Gamma^\alpha$, $\{\mathcal{T}_i(p)\}_{i=1}^n$ forms an orthonormal basis for R^n. We shall assume that $\mathcal{T}_1$ is always tangential so that $\mathcal{T}_1(p) \in T_p\Gamma^\alpha$. Then the vector field is characterized by the following conditions. For each $x \in \mathcal{N}(\alpha)$, $X_\alpha(x): D \to R^n$ is harmonic so

$$\text{(i)} \quad \Delta X_\alpha(x) \equiv 0$$

satisfying the mixed Neumann-Dirichlet boundary conditions

$$\text{(6.2)} \qquad \text{(ii)} \quad \frac{\partial X_\alpha}{\partial r}(x) \cdot \mathcal{T}_1(x) = \frac{\partial x}{\partial r} \cdot \mathcal{T}_1(x) \quad ,$$

$$\text{(iii)} \quad X_\alpha(x) \cdot \mathcal{T}_j(x) = 0 \quad , \quad j = 2,\ldots,n \quad ,$$

where $\mathcal{T}_j(x)$ denotes the composition $\mathcal{T}_j(x(\theta))$ and $\partial/\partial r$ denotes the normal or radial derivative along S^1. We can paraphrase these boundary conditions as follows: Let $\Omega: \Gamma^\alpha \to OP(R^n)$, the orthogonal projections on R^n, be the C^{r-2} map such that $\Omega(p)$ is the projection of R^n onto $T_p\Gamma^\alpha$. Then (5.2) can be rewritten as

$$\text{(6.2')} \qquad \Omega(x)\frac{\partial X^\alpha}{\partial r}(x) = \Omega(x)\frac{\partial x}{\partial r} \quad ,$$

where $\Omega(x)$ again denotes the map $\theta \to \Omega(x(\theta))$.

Let $\mathcal{N}^* = \cup_\alpha \mathcal{N}(\alpha)$ and $\pi: \mathcal{N}^* \to \mathcal{A}$ be the natural projection map $\pi(\mathcal{N}(\alpha)) = \alpha$. The space $\mathcal{N}^*$ has the structure of a smooth fibre bundle over $\mathcal{A}$ which is bundle equivalent to the product bundle $\mathcal{N} = \mathcal{A} \times \mathcal{D}$ via the map $w: (\alpha,u) \to \alpha \circ u$ and hence $\mathcal{N}^*$ is globally trivial. We will identify $\mathcal{N}$ with $\mathcal{N}^*$ via this trivialization and hopefully no confusion should arise from this.

The family of vector fields X_α induces a C^{r-s-1} vector field X on $\mathcal{N}$ by the rule $X(x) = X_\alpha(x)$ if $x \in \mathcal{N}(\alpha)$. This vector field will be vertical in the sense that $X_\alpha(x) \in T_x\mathcal{N}(\alpha)$. If $x \in \mathcal{N}(\alpha)$ is a zero of X (and hence a zero of X_α) the Frechet derivative of X can be viewed as a map $DX(x)\colon T_x\mathcal{N} \to T_x\mathcal{N}(\alpha)$. If $x \in \mathcal{M}_\nu^\lambda$ we shall be interested in the corank of $DX(x)$; i. e. $\dim T_x\mathcal{N}/DX[T_x\mathcal{N}(\alpha)]$.

Before stating a theorem which gives the answer, we would like to discuss the action of the conformal group $\mathcal{G}$ of the disc on the space $\mathfrak{D}$ (and hence on $\mathcal{N}(\alpha)$ and the bundle $\mathcal{N}$). So let $\mathcal{G}$ be the group of conformal transformations of the disc onto itself. It is well known that every $g \in \mathcal{G}$ is of the form

$$g(z) = c \cdot \frac{z-a}{1-\bar{a}z} \quad , \qquad |z| \leq 1 \quad ,$$

where $|c| = 1$ and $|a| < 1$. $\mathcal{G}$ is a three dimensional Lie group which is not compact. $\mathcal{G}$ acts on $\mathfrak{D}$ in the following manner. Let $u\colon S^1 \to \mathbb{C}$. For $g \in \mathcal{G}$ define $g_\#(u) \in H^s(S^1,\mathbb{C})$ by $g_\#(u)(z) = u(g(z))$. For fixed g, $g_\#$ is a linear isomorphism of H^s to itself which fixes $\mathfrak{D}$. Thus $g_\#$ iduces a diffeomorphism of $\mathfrak{D}$ to itself and hence of $\mathcal{N}(\alpha)$ to itself. Again denote this diffeomorphism by g. The correspondence $\mathcal{G} \to \mathrm{Diff}(\mathfrak{D})$ given by $g \to g_\#$ defines the action of $\mathcal{G}$ on $\mathfrak{D}$.

This $\mathcal{G}$ action is not smooth and this of course creates many technical difficulties for the theory. However

<u>Remark 6.3:</u> If $u \in \mathfrak{D}$ and $u \in H^{s+1}(S^1,\mathbb{C})$, then $g \to g_\#(u)$ is a smooth map into $\mathfrak{D}$. Since the zeros of our vector field X_α are more than H^s smooth (in fact by Böhme's version of Hildebrandt's regularity result, H^r smooth) the orbits of $\mathcal{G}$ through zeros will be smooth submanifolds.

If $x = \alpha(u)$, then the $\mathcal{G}$ action on $\mathfrak{D}$ induces a $\mathcal{G}$ action on $\mathcal{N}(\alpha)$ via $g_\#(x) = \alpha(g_\#(u))$.

We have the following result from [17].

<u>Theorem 6.4:</u> *The vector field* X_α *(fixing* α*) is equivariant with respect to the* $\mathcal{G}$*-action on* $\mathcal{N}(\alpha)$*. This means that*

$$[Dg_\#(x)]^{-1}X_\alpha(g_\#(x)) = X_\alpha(x) \quad .$$

where D *denotes the Frechet derivative of the map* $x \to g_\#(x)$ *(this map being the restriction of a linear map is smooth).*

The fact that each X_α is equivariant implies that the corank of the total derivative $DX(x): T_x\mathcal{N} \to T_x\mathcal{N}(\alpha)$ must be at least three and hence the assumptions of Theorem 3.2 cannot be satisfied. One can ask if this is all that can happen and the negative answer is given in

Theorem 6.5: *If* $x \in \mathcal{M}_\nu^\lambda$, *then the corank of* $DX(x)$ *is equal to* $2|\lambda| + |\nu| + 3$. *The number three comes from the action of the conformal group, as was expected.*

For a proof see [17] or the appendix of [2]. Result 6.5 tells us that only at a nonbranched minimal surface x will the total derivative of X at x not be of maximal rank. We would like to apply "indirectly" the degree theory on varieties introduced in §2, and we begin as follows.

Let $\mathcal{G}(\alpha) = \{x \in \mathcal{N}(\alpha) | x(Q_j) = \alpha(Q_j)\}$, where Q_1, Q_2, Q_3 are three prescribed points on S^1. The codimension three subbundle

$$\mathcal{G} = \bigcup_\alpha \mathcal{G}(\alpha)$$

intersects each of the submanifolds $\mathcal{M}_0^\lambda$ (cf. Theorem 5.1) transversely and hence is also a submanifold which we denote by Σ_0^λ. The restriction of the projection map π, $\bar{\pi}_0^\lambda$, to Σ_0^λ is Fredholm of index $I(\lambda,0) = 2(2-n)|\lambda| + 2p$. Moreover if $\Sigma_\nu^\lambda = \mathcal{G} \cap \mathcal{M}_\nu^\lambda$ it follows again from the index theorem that locally Σ_ν^λ is contained in a C^{r-s-1} submanifold W_ν^λ such that $\bar{\pi}_\nu^\lambda = \pi | W_\nu^\lambda$ has index $I(\lambda,\nu) = 2(2-n)|\lambda|+(2-n)|\nu|+2p+q$.

The following calculation is elementary.

Theorem 6.6: *If* $n \geq 4$ *and* λ *or* ν *is not zero,* $I(\lambda,\nu) \leq -2$. *The index of* $\bar{\pi}_0^0$ *is zero.*

From the results of Tomi and Tromba [13] it follows that there is a partial ordering on these strata Σ_ν^λ and that the complement in $\mathcal{A}$ of the union of the images $\bigcup_{\lambda\neq 0, \nu\neq 0} \pi(\Sigma_\nu^\lambda)$ does not disconnect any component of $\mathcal{A}$.

Consequently, if $\mathcal{A}$ were connected and if the strata Σ_0^0 carries a natural orientation which is related to the vector field X^α as in 3.2

and such that π restricted to $\Sigma = \cup\Sigma_\nu^\lambda$, say, is proper we would have achieved a degree theory for minimal surfaces, by employing the degree theory for π_Σ developed in §2. The properness of π_Σ is a direct consequence of Hildebrandt's regularity theorem [7] that minimal surfaces are as regular as the curve they bound. Therefore we have

Theorem 6.7: *The map* $\pi_\Sigma: \Sigma \to \mathcal{A}$ *is a proper map.*

However in the beginning of §5 (cf. (5.0)) we imposed a curvature condition on the set $\mathcal{A}$ and so the set $\mathcal{A}$ is not "formally connected" even though the set of all H^r embeddings is. This is only a minor difficulty in constructing a degree theory for π_Σ.

Our next goal will be to put an orientation on Σ_0^0 related to the derivatives of the vector field X_α. To start we must introduce the weak Riemannian structure on the manifold $\mathcal{N}(\alpha)$.

Definition 6.8: Let $x \in \mathcal{N}(\alpha)$ and $h,k \in T_x\mathcal{N}(\alpha)$. Define $\langle\langle h,k\rangle\rangle_x$, the weak inner product of h and k over x, by the formula

$$\langle\langle h,k\rangle\rangle_x = \int_{S^1} \langle \frac{\partial h}{\partial r}, h\rangle \, d\theta = \sum_i \int_D \langle \nabla h^i, \nabla k^i\rangle \quad ,$$

where again h and k are identified with their harmonic extension.

Theorem 6.9: *The weak inner product is* $\mathcal{G}$*-invariant, and consequently the Dirichlet functional* $E_\alpha: \mathcal{N}(\alpha) \to R$ *defined by*

$$E_\alpha(x) = \sum_{i=1}^n \int_D \nabla x^i \cdot \nabla x^i$$

is also $\mathcal{G}$*-invariant.*

Moreover the vector field X_α *is the gradient of* E_α *with respect to* $\langle\langle\ ,\ \rangle\rangle$.

Proof: See [17].

Let $\mathcal{O}_x(\mathcal{G})$ be the orbit of the conformal group through x. By 6.3 for x a zero $\mathcal{O}_x(\mathcal{G})$ will be a smooth manifold. Let $T_x\mathcal{O}_x(\mathcal{G})$ denote the tangent space to this orbit at x. Then $T_x[\mathcal{O}_x(\mathcal{G})]$ has an orthogonal complement $T_x[\mathcal{O}_x(\mathcal{G})]^{\perp}$ w. r. t. the weak inner product $\langle\langle\ ,\ \rangle\rangle$. The equivariance of X_α, E_α and $\langle\langle\ ,\ \rangle\rangle$ under $\mathcal{G}$ implies the following

Theorem 6.10: *At a zero* x

$$DX_\alpha(x) : T_x[\mathcal{O}_x(\mathcal{G})]^{\perp} \to T_x[\mathcal{O}_x(\mathcal{G})]^{\perp} .$$

Definition 6.11: A minimal surface $x: D \to R^n$, $n \geq 4$, is said to be nondegenerate if $DX_\alpha(x)$ restricted to $T_x[\mathcal{O}_x(\mathcal{G})]^{\perp}$ is an isomorphism.

Remark 6.12: It is a consequence of the index formula that an open and dense set of wires $\hat{\mathcal{A}} \subset \mathcal{A}$ enjoy the property that if $\alpha \in \hat{\mathcal{A}}$, then there are only a finite number of nondegenerate minimal surfaces which span them. This nondegeneracy result is not true for surfaces in R^3 and in this case a new notation of nondegeneracy is required. This result allows us to define the $\operatorname{sgn} DX_\alpha(x)$ if x is nondegenerate. Since X_α is Palais-Smale, $DX_\alpha(x) \in GL_c[T_x(\mathcal{O}_x(\mathcal{G}))]^{\perp}$ and we define

$$\text{(6.13)} \qquad \operatorname{sgn} DX_\alpha(x) = \begin{cases} +1 & \text{if } DX_\alpha(x) \in GL_c^+[T_x(\mathcal{O}_x(\mathcal{G}))]^{\perp} , \\ -1 & \text{if } DX_\alpha(x) \in GL_c^- . \end{cases}$$

Using a modified version of theorem 3.2 one can prove the following

Theorem 6.14: *Let* $n \geq 4$. *Then there exists an orientable Fredholm structure* $\tilde{\Phi}$ *on the manifold* $\Sigma_0^0 \subset \mathcal{I}$ *such that (cf. 6.6)*

(1) *The projection map* $\bar{\pi}_0^0: \Sigma_0^0 \to \mathcal{A}$ *is admissible with respect to* $\tilde{\Phi}$.

(2) $(\bar{\pi}_0^0)^*\Phi = \tilde{\Phi}$, Φ *the natural Fredholm structure on* $\mathcal{A}$.

(3) *If* α *is a regular value for* $\bar{\pi}_0^0$ *(or* π_Σ*) there exists a finite number of nondegenerate minimal surfaces, say* $x_1,\dots,x_m$, *spanning* α *such that* $\operatorname{sgn} D\bar{\pi}_0^0(x_i) = \operatorname{sgn} DX_\alpha(x_i)$.

Although $\mathcal{A}$ (cf. remarks following 6.7) is not connected we may still construct a degree theory for $\pi_\Sigma: \Sigma \to \mathcal{A}$ and hence a degree for π_Σ, which we compute as follows.

Let α^* be the standard embedding of the unit circle S^1 into $\mathbb{R}^2 \subset \mathbb{R}^n$, $n \geq 2$. Then we know that $\pi_\Sigma^{-1}(\alpha^*)$ consists of only one point, namely the identity map id of the disc to itself. Moreover id is a nondegenerate critical point of E_{α^*} and one can easily show that

$$\operatorname{sgn} D\pi_{\Sigma_0^0}(\mathrm{id}) = 1 \quad .$$

Thus, combining what we know we obtain the following results.

Theorem 6.15: *The degree of the map* $\pi_\Sigma\colon \Sigma \to \mathscr{A}$ *is one; i. e.*

$$\deg \pi_\Sigma = 1 \quad .$$

Theorem 6.16: *There exists an open dense set of wires*
$\hat{\mathscr{A}} \subset \mathscr{A}$, $\mathscr{A} \subset H^r(S^1, R^n)$, $n \geq 4$,
such that if $\alpha \in \mathscr{A}$ *there are only a finite number of nondegenerate minimal surfaces* $x_1, \ldots, x_m$ *spanning* α. *Let* $D^2E_\alpha(x_i)\colon T_{x_i}\mathscr{N}(\alpha) \times T_{x_i}\mathscr{N}(\alpha) \to R$ *denote the Hessian of Dirichlet's function at* x_i. *Let* λ_i *be the maximal subspace in* $T_{x_i}\mathcal{O}_{x_i}(\mathscr{G})^\perp$ *on which* $D^2E_\alpha(x_i)$ *is negative definite. Then*

$$\sum_i (-1)^{\lambda_i} = 1 \quad .$$

This follows immediately from 6.15 and 6.14 and the fact that

$$\operatorname{sgn} DX_\alpha(x_i) | T_{x_i}\mathcal{O}_{x_i}(\mathscr{G}) = (-1)^{\lambda_i} \quad .$$

This equality is a consequence of the relationship between the Hessian of Dirichlet's functional and the derivative of our vector field, namely

$$D^2E_\alpha(x_i)(h,k) = \langle\langle DX_\alpha(x_i)h, k\rangle\rangle \quad .$$

Since $DX_\alpha(x_i)$ is of the form identity I plus a compact operator K, then

$$\operatorname{sgn} DX_\alpha(x_i) | T_{x_i}\mathcal{O}_{x_i}(\mathscr{G})^\perp$$

is the number, counted with multiplicities, of the eigenvalues of $K | T_x\mathcal{O}_{x_i}(\mathscr{G})^\perp$ strictly less than negative 1; but this is precisely λ_i.

Corollary 6.17: *Let* α *be any wire in* R^n, $n \geq 4$. *Then there exists a minimal surface spanning its image* Γ^α.

Proof: If not, then $\deg \pi_\Sigma = 0$.

Thus this theory also yields the existence of a minimal surface spanning an arbitrary embedded wire in R^n, $n \geq 4$. But we can recover the existence theorem in R^3.

Corollary 6.18: *Given any embedding* $\alpha \in H^r(S^1,R^3)$, *there exists a minimal surface spanning its image* Γ^α.

Proof: Find a sequence of wires $\alpha_n \in H^r(S^1,R^4)$ such that $\alpha_n \to \alpha$ in H^r. By 6.17 there exists a sequence of minimal surfaces x_n spanning Γ^{α_n}. By Hildebrandt's regularity result the x_n are uniformly bounded in $H^{s+1}(S^1,R^n)$ norm $(r >> s)$. Thus we can find a subsequence $x_{n_j} \to x_0$ in H^s. Clearly x_0 spans Γ^α and is a minimal surface.

7. The Morse number of minimal surfaces spanning a curve in $\mathbb{R}^3$

By the Index theorem we know that the generic curve $\alpha: S^1 \to \mathbb{R}^3$ spans at most finitely many minimal surfaces $x_1,\ldots,x_N$. As already mentioned these will not be nondegenerate in the sense of 6.11 (see remark 6.12). However they will be either immersed or simply branched. Thus each $x_i \in \Sigma_0^0$ or Σ_0^λ where $\lambda = (1,\ldots,1)$. Each x_i is nondegenerate in the sense that either $D\bar{\pi}_0^0: T_{x_i}\Sigma_0^0 \to T_a\mathscr{A}$ is an isomorphism if $x_i \in \Sigma_0^0$ or $D\bar{\pi}_0^\lambda: T_{x_i}\Sigma_0^\lambda \to T_a\mathscr{A}$ is an isomorphism if $x_i \in \Sigma_0^\lambda$.

Consider the minimal surface vector field X_α (theorem 6.1), which is Palais-Smale. In a local coordinate system about each x_i the local representative of X_α, in this coordinate system will be a Rothe-field. Unfortunately, although x_i will be isolated in Σ, by equivariance it will __not__ be an isolated zero of X_a. Nevertheless by an equivariant version of the ideas of §4 we can define the local degree, $\deg(X_\alpha,U_i)$, U_i some neighborhood about x_i containing non x_j, $j \neq i$. If $n \geq 4$ this local degree will be $\operatorname{sgn} DX_\alpha(x_i)$ and therefore ± 1. We define the total Morse number of minimal surfaces spanning α by

$$\text{(7.1)} \qquad \mathrm{Morse}(\Gamma^{\alpha}) = \sum_i \deg(X_{\alpha}, U_i) \quad .$$

We use the notation Γ^{α} because the number $\mathrm{Morse}(\Gamma^{\alpha})$ can be shown to depend solely on the geometric image Γ^{α} of α. The number $\mathrm{Morse}(\Gamma^{\alpha})$ agrees with the standard definition of the Morse number of critical points of E_{α}, in case $n \geq 4$.

We can now compute $\mathrm{Morse}(\Gamma^{\alpha})$ and the answer comes as no surprise.

Theorem 7.2: *For* $n \geq 3$

$$\mathrm{Morse}(\Gamma^{\alpha}) = 1 \quad .$$

Proof (a sketch): For the generic α in $\mathbb{R}^4$ this is the content of theorem 6.16. For α in $\mathbb{R}^3$ we slightly perturb α to α' in $\mathbb{R}^4$. Although, in principle X_{α} and $X_{\alpha'}$ are defined on different spaces ($\mathcal{N}(\alpha)$ and $\mathcal{N}(\alpha')$) by a fibre bundle trivialization we can assume they are defined on the same space. Moreover if α' is sufficiently close to α we can assume that (if the U_i are small enough) that all the zeros of $X_{\alpha'}$ are in $\bigcup_i U_i$ (for details see [15], [17]).

Now on each U_i, $X_{\alpha'}$ is homotopic to X_{α} with the homotopy introducing no zeros on $\bigcup_i \partial U_i$. By the homotopy invariance of the local degree of a Rothe-field, it follows that

$$\deg(X_{\alpha}, U_i) = \deg(X_{\alpha'}, U_i) \quad .$$

By 6.16

$$\mathrm{Morse}(\Gamma^{\alpha'}) = \sum_i \deg(X_{\alpha'}, U_i) = 1 \quad .$$

Thus

$$\mathrm{Morse}(\Gamma^{\alpha}) = 1$$

and this concludes our sketch of the proof of 7.2.

We are now ready to discuss some open problems and conjectures concerning this theory.

8. Open Problems.

Problem 1. Let x be a minimal surface spanning a curve α in $\mathbb{R}^3$ which is nondegenerate in the sense that $D\pi_{\Sigma}(x)$ is an isomorphism.

Let $\deg(X_\alpha, U)$ be the local degree of the minimal surface vector field about some U containing x, with U chosen so that x is the only zero of X_α in U. If x is immersed it follows from our discussions that

$$\deg(X_\alpha, U) = \pm 1 \quad .$$

Does this hold if x is <u>not</u> immersed?

<u>Problem 2.</u> Following upon problem 1, exactly what local data determine this degree. Does this local degree depend upon the coefficients of the Taylor expansion of x near its branch points?

<u>Problem 3</u> (A bifurcation problem). If x and α are in problem 1, and if α is slightly perturbed to a curve β in $\mathbb{R}^3$ it follows that if x is immersed there is a minimal surface $x(\beta)$ spanning the image curve Γ^β, such that $x(\alpha) = x$, $\beta \to x(\beta)$ is smooth, and $x(\beta)$ is locally the only minimal surface spanning Γ^β "near" x.

If x is <u>not</u> immersed it is formally a degenerate critical point of E_α. If one now slightly perturbs α to a curve β in $\mathbb{R}^4$ such that all minimal surfaces spanning β are nondegenerate

(a) does x follow along smoothly as a function of β?

(b) does x bifurcate into other minimal surfaces?

(c) does this bifurcation depend on the geometry of Γ^α or perhaps on the Taylor expansion of x about its branch points?

(d) more specifically describe this bifurcation if it occurs.

Consider the total derivative of the vector field $X: \mathcal{N} \to T\mathcal{N}$ (cf. §6) at a branched minimal surface x. The H^1 complement of the range of $DX(x)$ in $T_x\mathcal{N}(\alpha)$ is naturally spanned by tangent vectors in $T_x\mathcal{N}(\alpha)$ which Böhme and this author named <u>**forced Jacobi fields**</u> $\mathcal{J}(x)$ (cf. [2]) because they arise only from the branch points of x.

From 6.5 it follows that

$$\dim \mathcal{J}(x) = 2|\lambda| + |\nu| + 3 \quad .$$

These Jacobi fields are described as follows. Let (ν, θ) denote the polar coordinates for the disc. Then $\nu \in T_x\mathcal{N}(\alpha)$ if on ∂D, $\nu = \lambda x_\theta$, where λ extends to a meromorphic function on D, real on ∂D with poles at the branch points of x and where the orders of these poles do not exceed the orders of the respective branch points of x. These Jacobi fields are in both the kernel and cokernel of $DX_\alpha(x)$.

This brings us to our next question which we vaguely formulate as

Problem 4. Describe the geometric meaning of these forced Jacobi fields.

For problem 5 we go back to the relation of degree theory and the calculus of variations embodied in theorem 6.14. Let $n = 3$.

Problem 5. Does there exist a natural orientable Fredholm structure $\tilde{\Phi}$ on Σ_0^λ, $\lambda = (1,\dots,1)$ such that

(a) the projection map $\bar{\pi}_0^\lambda: \Sigma_0^\lambda \to \mathcal{A}$ is admissible

(b) $(\bar{\pi}^\lambda{}_0)^*\Phi = \tilde{\Phi}$, Φ the natural Fredholm structure on .

(c) if α is a regular value for $\bar{\pi}_0^\lambda$ there exists a finite number of minimal surfaces, say $x_1,\dots,x_m$ spanning α such that

$$\operatorname{sgn} D\bar{\pi}_0^\lambda(x_i) = \operatorname{sgn} DX_\alpha(x_i)$$

where $\operatorname{sgn} DX_\alpha(x_i)$ is defined as follows.

Let $\mathcal{J}(x_i)$ denote the forced Jacobi fields. Then the tangent space to the orbit of the conformal group at x_i, $T_{x_i}\mathcal{O}_{x_i}(\mathcal{G})$ is contained in $\mathcal{J}(x_i)$. Let $\mathcal{J}(x_i)^\perp$ denote the H^1 orthogonal complement of $\mathcal{J}(x_i)$ in $T_{x_i}\mathcal{N}(\alpha)$. $DX_\alpha(x_i): \mathcal{J}(x_i)^\perp$ and is an isomorphism. We define

$$\operatorname{sgn} DX_\alpha(x_i) = \operatorname{sgn} DX_\alpha(x_i)\,|\,\mathcal{J}(x_i)^\perp$$

(d) does $\deg(X_\alpha, U) = \operatorname{sgn} DX_\alpha(x_i)$, U some small neighborhood about x_i?

The local degree of a vector field X_α at a degenerate zero x surely depends, in a non-trivial way on the higher order jets of $DX_\alpha(x)$. We also know that whether x is a minimum or not depends on the higher order jets of the enery E_α. This brings us to

Problem 6. Examine the higher order jets of the enery E_α around a critical point x. Use this to give a proof and possible extension to $\mathbb{R}^n$ of the theorem of Alt-Gulliver-Osserman [1], [5] [10] that in $\mathbb{R}^3$ a branched critical point of E_α cannot be a minimum.

References

[1] Alt, H.W.: Verzweigungspunkte von H-Flächen I, Math. Z. 127, 333-362 (1972); II, Math. Ann. 201, 33-56 (1973).

[2] Böhme, R. and Tromba, A.J.: The index theorem for classical minimal surfaces; Annals of Mathematics 113 (1981), pp. 447-499

[3] Elworthy, K.D. and Tromba, A.J.: Differential structures and Fredholm maps on Banach manifolds; Proc. Pure Math. vol. 15, AMS (1970), pp. 45-94.

[4] Elworthy, K.D. and Tromba, A.J.: Degree theory on Banach manifolds; Proc. Symp. Pure Math. Vol 18, AMS (1970), pp. 86-94.

[5] Gulliver, R.: Regularity of minimizing surfaces of prescribed mean curvature; Ann. of Math. 97, 275-305 (1973).

[6] Heinz, E. and Tomi, F.: Zu einem Satz von Hildebrandt über das Randverhalten von Minimalflächen; Math. Z. 111 (1969), 372-386.

[7] Hildebrandt, S.: Boundary behavior of minimal surfaces; Arch. Rational Mech. Anal. 35 (1969), pp. 47-81.

[8] Leray, J. and Schauder, J.: Topologie et equations fonctionelles; Ann. Sci. Ecole Norm. Sup. 51 (1934), pp. 45-78.

[9] Nirenberg, L.: Variational and topological methods in non-linear problems; Bull. AMS 4 (1981), 267-302.

[10] Osserman, R.: A proof of the regularity everywhere of the classical solution of Plateau's problem; Ann. of Math. (2) 91 (1970), pp. 550-569.

[11] Smale, S.: An infinite dimensional version of Sard's theorem; Amer. J. Math. 87 (1965), pp. 861-866.

[12] Struwe, M.: On a critical point theory of minimal surfaces spanning a wire in $\mathbb{R}^n$; J. Reine Angew. Math. 349 (1984), pp. 1-23.

[13] Tomi, F. and Tromba, A.J.: On the structure of the set of curves bounding minimal surfaces of prescribed degeneracy; J. Reine Angew. Math. 316 (1980), 31-43.

[14] Thiel, U.: On the stratification of branched minimal surfaces; Analysis 5 (1985), 251-271.

[15] Tromba, A.J.: Degree theory on oriented infinite dimensional varieties and the Morse number spanning a curve in $\mathbb{R}^n$, Part II, n = 3; Manuscripta math. 48 (1984), pp. 139-161; Part I, Trans. AMS, vol. 290, Number 1 (July 1985), pp. 385-413.

[16] Tromba, A.J.: A general asymptotic fixed point theorem; J. Reine Angew. Math. 332 (1982), pp. 118-123.

[17] Tromba, A.J.: On the number of simply connected minimal surfaces spanning a curve; Memoirs of the AMS, 12 194, (1977).

ON A MODIFIED VERSION OF THE FREE GEODETIC BOUNDARY-VALUE PROBLEM

K.J. Witsch
Fachbereich 6 Mathematik der Universität Essen
Universitätsstraße 2, D-4300 Essen

§1. Introduction

By the free geodetic boundary-value problem one means the problem to determine from data for gravity and its potential at the surface of the earth its shape in a Euclidean coordinate system. In several papers [16,17,18,19], which partially have been supported by the SFB 72, I have considered this question and have proved existence, uniqueness, and regularity theorems. In this report the methods used in earlier papers shall be applied to a modified version of this problem: The data for the gravity-potential will be assumed to be known up to an additive constant only and instead, the distance between two points fixed in advance will be introduced as data of the problem as proposed in [5]. The results described here developed during my stay as a visitor of the SFB 72 in the summer semester 1986.

The earth is looked upon as being a rigid solid rotating around an axis fixed in $\mathbb{R}^3$ at constant angular velocity, having center of mass located on the axis of rotation, and having surface diffeomorphic to the unit sphere S^2 . The Euclidean coordinate system with respect to which the surface of the earth is to be determined has to be defined in such a manner that in it the coordinates of the gravity-vector (i.e. the vector of the attracting force) can be measured everywhere on the surface: It rotates together with the earth; the x_3-axis is the axis of rotation oriented towards the north; the direction of the x_1-axis is determined by the requirement that the gravity-vector observed in the Greenwich observatory has negative first and vanishing second component; finally, the origin is defined to be the center of mass.

The coordinates of the gravity-vectors are then determined by means of astronomical observations and weighing; the gravity-potential is gained up to an additive constant by combined measurement of height and weight which simulate an integration of the field.

The gravity and its potential are composed of a gravitational and a centrifugal part. The first part is assumed to be generated only by masses in the inside of the earth: the data thus is to be relieved from influence of the moon or other celestial bodies.

The mapping which maps each point of the surface of the earth into the direction of the force vector measured there is as usually [18] assumed to be a diffeomorphism onto S^2. Then the data for gravity--force and potential can be looked upon as being functions γ, U from S^2 into $\mathbb{R}^3, \mathbb{R}$, respectively. More strictly speaking we assume

$$(1) \qquad \gamma \in C^2_* \quad , \quad 0 \notin \bar{\Omega}_\gamma \quad , \quad \bigwedge_{\zeta \in S^2} \gamma(\zeta) \cdot n_\gamma(\zeta) \neq 0$$

to hold. Here C^k_* is the set of all k times continuously differentiable functions from S^2 into $\mathbb{R}^3$ $(C^k(S^2, \mathbb{R}^3))$, which are embeddings, Ω_γ is the unbounded component of $\mathbb{R}^3 \setminus \gamma(S^2)$, $n_\gamma(\zeta)$ is the unit normal to $\gamma(S^2)$ pointing towards Ω_γ at $\gamma(\zeta)$, and $\cdot$ is the inner product in $\mathbb{R}^3$.

Let now be given two fixed points $\zeta^1, \zeta^2 \in S^2$ as well as the data $\gamma \in C^2_*$, $U \in C^0(S^2)$ 1) with $\int_{S^2} U \, do = 0$ and $d \in \mathbb{R}^+ := (0, \infty)$, one then shall find an embedding $\lambda \in C^1_*$, a function $\hat{u} \in C^2(\bar{\Omega}_\lambda)$, a number $\beta \in \mathbb{R}$ as well as some vector $b \in \mathbb{R}^3$ having the following properties (2)-(6):

$$(2) \qquad |\lambda(\zeta^1) - \lambda(\zeta^2)| = d \,,$$

$$(3) \qquad \hat{u} = u + \rho \quad \text{with} \quad \rho(x) := \frac{\omega^2}{2}(x_1^2 + x_1^2) \quad \text{and} \quad \Delta u = 0 \quad \text{in} \quad \Omega_\lambda \,,$$

$$(4) \qquad \bigvee_{C>0} \; u = cr^{-1} + O(r^{-3}) \quad \text{as} \quad r(x) := |x| \longrightarrow \infty \,,$$

$$(5) \qquad (\nabla \hat{u}) \circ \lambda = \gamma \,,$$

$$(6) \qquad \hat{u} \circ \lambda = U + \beta + b_i B_i \;\; ^{2)} \,.$$

A potential function u regular in Ω_λ has the asympotics $cr^{-1} + c_i x_i r^{-3} + O(r^{-3})$ at ∞. In case of u being the potential of some mass distribution c_i/c are just the coordinates of the center of mass; the fact that these are vanishing is thus expressed by (4). Because of Fredholm's alternative one now expects that it is necessary to correct the data U by means of some linear combination of three

---

1) All functions are real-valued. For the function spaces we use the notations from [3].

2) Sum convenction: if an index appears twice in a product summation from 1 through 3 is implied.

suited linearly independent functions $B_1, B_2, B_3 \in C^\infty(S^2)$ in order to guarantee solvability (cf. [5,17,18]). This explains the requirement (6) which migth look strange. How to choose B_1, B_2, B_3 will be described later.

$(\lambda, \hat{u}, \beta, b) \in \mathcal{L} := \left\{(\lambda, \hat{u}, \beta, b) : \lambda \in C^1_*, \ \hat{u} \in C^2(\overline{\Omega_\lambda}), \ \beta \in \mathbb{R}, \ b \in \mathbb{R}^3\right\}$ satisfying (2)-(6), we say to be a solution of the free boundary-value problem (fbp) for $(U, \gamma, d) \in \mathcal{D} := \left\{(U, \gamma, d) \in C^0(S^2) \times C^2_* \times \mathbb{R}^+ : \int_{S^2} U \, do = 0\right\}$.

If we introduce knowledge of the shape of the earth into the problem, the fbp becomes a local problem: Near to a known solution $(\lambda_o, \hat{u}_o, \beta_o, 0) \in \mathcal{L}$ for data $(U_o, \gamma_o, d_o) \in \mathcal{D}$ one is looking for a solution $(\lambda, \hat{u}, \beta, b)$ for nearby data (U, λ, d). One calls λ_o telluroid and $\hat{u}_o$ (corresponding) normal potential.

Corresponding to a construction usual in geodesy we assume that the gravitational part $u_o := \hat{u}_o - \rho$ of the normal potential can be extended into a region Ω including $\overline{\Omega}_{\lambda_o}$ as a potential function, and that the following conditions (7)-(10) are satisfied:

(7) $$\gamma_o = \gamma \ ;$$

with

(8) $\psi \in C_o^\infty(\mathbb{R}^3)$, $\psi = \rho$ in a neighborhood of $\lambda_o(S^2)$, $\check{u}_o := u_o + \psi$

(9) $$\bigwedge_{x \in \Omega} \nabla \check{u}_o(x) \neq 0 \quad , \quad \det \nabla^2 \check{u}_o(x) \neq 0$$

holds, where $\nabla^2 w$ denotes the matrix of second derivatives of a function w (with $\hat{u}_o$ instead of $\check{u}_o$ (9) could not be satisfied); finally, every solution w of the homogeneous linear Molodenski problem $\mathcal{M}(\lambda_o, \hat{u}_o, 0, 0)$ at ∞ has the asymptotics

(10) $$w = cr^{-1} + (a \cdot x) r^{-3} + O(r^{-3}) \quad , \quad c \in \mathbb{R} \ , \ a \in \mathbb{R}^3 \setminus \{0\} \ .$$

The linear Molodenski problem, an exterior oblique derivative problem

of potential theory, occurs in case of linearization of the fbp "at the telluroid" [4,5]:
With the "isozenithal field"

$$\gamma := -(\nabla^2\hat{u}_o)^{-1} \nabla\hat{u}_o ,$$

$\mathcal{M}(\lambda_o,\hat{u}_o,g,G)$ denotes the boundary-value problem

$$\Delta w = g \text{ in } \Omega_{\lambda_o} \quad , \quad (\gamma\cdot\nabla w+w) \circ \lambda_o = G \quad , \quad w = o(1) \text{ as } r \longrightarrow \infty .$$

In [5] Hörmander states criteria showing when the above assumption on the linear Molodenski problem is fulfilled.

Because of the invertibility of $\nabla^2\check{u}_o$ the condition $n_\gamma\cdot\gamma \neq 0$ on S^2 (1) is equivalent to the transversality of γ. The linear Molodenski problem therefore is a regular boundary-value problem for which a Fredholm alternative holds. Eigensolutions surely are $\partial_3 u_o$ and in case of $\omega = 0$ also $\partial_1 u_o$, $\partial_2 u_o$. These functions still have the asymptotics (10).

The functions B_1,B_2,B_3 occurring in (6) are now chosen in such a way that for each $G \in H^{1/2}(S^2)$ a unique vector $b \in \mathbb{R}^3$ exists so that $\mathcal{M}(\lambda_o,\hat{u}_o,0,G + b_iB_i)$ has a solution w with the asymptotics

$$(11) \qquad \bigvee_{c\in\mathbb{R}} w = cr^{-1} + O(r^{-3}) \quad \text{as} \quad r \longrightarrow \infty$$

("Fredholm alternative", (10)); furthermore the solutions w_i , $i = 1,2,3$, of the Dirichlet problems

$$\Delta w_i = 0 \quad \text{in} \quad \Omega_{\lambda_o} , \; w_i \circ \lambda_o = B_i , \; w_i = o(1) \quad \text{as} \quad r \longrightarrow \infty$$

shall have the asympotics

$$w_i = c_i r^{-1} + x_i r^{-3} + O(r^{-3}) \quad , \quad c_i \in \mathbb{R} \text{ as } r \longrightarrow \infty .$$

Finally, one needs a relation between the points $\zeta^1,\zeta^2 \in S^2$ and the linear Molodenski problem: Let $x^1 := \lambda_o(\zeta^1)$, $x^2 := \lambda_o(\zeta^2)$, and let e be the uniquely determined solution of the Molodenski problem $\mathcal{M}(\lambda_o,\hat{u}_o,1 + b_i^eB_i)$, $b^e \in \mathbb{R}^3$, with the asympotics (11). Then

$$(12)\quad e_o := 2(x^1-x^2)\cdot[(\nabla^2\hat{u}_o(x^1))^{-1}\nabla e(x^1) - (\nabla^2\hat{u}_o(x^2))^{-1}\nabla e(x^2)] \neq 0$$

must hold.

Up to now in the literature the original problem has been treated where all of the potential is known at the free boundary and no distance function enters: In his fundamental paper [5] Hörmander examines the linear Molodenski problem and proves a hard implicit function theorem of the Nash-Moser type with whose help he derives local existence in the $C^{2,\epsilon}$-, and local uniqueness in the $C^{3,\epsilon}$-topology as well as regularity results for the free geodetic boundary-value problem. Sansó [10,11] introduces the Legendre transform method ("LT method") and obtains results similar to those of Hörmander in case of $\omega = 0$ [3].

The idea of this method is to transform the free boundary-value problem into a non-linear boundary-value problem with fixed boundary by means of a Legendre transform $l : \Omega_\lambda \longrightarrow \mathbb{R}^3$, $x \longrightarrow \nabla u(x)$ with respect to u . The inverse transformation then is a Legendre transform with respect to the solution of this non-linear boundary-value problem and yields the free boundary.

Motivated by a paper by Kinderlehrer and Nirenberg [7], in [17] in case of $\omega = 0$ the LT method is used for the proof of a local uniqueness theorem in the C^1-topology and in a special case the size of the uniqueness neighborhood is given a lower bound (see also [16]). Finally, in [18,19] also the case $\omega \neq 0$ is made amenable to the LT method and local uniqueness proved in the C^1-, local existence in the $C^{1,\epsilon}$-topology as well as regularity. In particular one can do without a hard implicit function theorem for the existence proof since one can solve the non-linear boundary-value problem resulting from the transformation by means of the usual implicit function theorem.

Analogous results can be derived for the modified version of the free boundary-value problem. In the following section the existence proof will be sketched and the uniquenes proof which needs the most modifications is carried out. As in [18] a Fredholm theorem must be proved for certain linear boundary-value problems in weighted Sobolev spaces, here in Sobolev-Banach instead of in Sobolev-Hilbert spaces though. This theory is provided in §3.

For a survey on the geodetic literature the reader is referred to [6].

---

3) That means that the centrifugal part of the data can be estimated sufficiently well.

§2 Existence, Uniqueness and Regularity for the Free Boundary Value Problem

In this section the uniqueness theorem for the modified geodetic boundary-value problem will be proved, the existence and regularity theorem formulated and shortly be talked about the modifications to the proof in [19].

Theorem 1: *Let there be given the data* $(U,\gamma,d) \in \mathcal{D}$ *as well as a telluroid* λ_o *with normal potential* $\hat{u}_o$ *, such that the conditions (7-10) and (12) are fulfilled. Then there is a constant* $\delta_o > 0$ *depending upon* $\lambda_o, \hat{u}_o$ *such that the fbp for* $(U,\gamma,d) \in \mathcal{D}$ *has at most one solution* $(\lambda,\hat{u},\beta,b) \in \mathcal{L}$ *with*

$$|\lambda_o - \lambda|_1 < \delta_o .$$

Here $|\cdot|_k$ *denotes the norm in* $C^k(S^2,\mathbb{R}^3)$.

Proof: For the proof let $(\lambda_j,\hat{u}_j,\beta_j,b^j) \in \mathcal{L}$, $j = 1,2$, denote solutions of the fbp for (γ,U,d) and analogously to (8)

$$\check{u}_j := u_j + \psi := \hat{u}_j - \rho + \psi \qquad (j = 0,1,2) .$$

According to [18, Lemma 4.1] $|\nabla\check{u}_j|$ and $\det \nabla^2\check{u}_j$ do not vanish in $\bar{\Omega}_{\lambda_j}$, if $|\lambda_j-\lambda_o|_1 < \delta_1$. Here and from now on $\delta_1,\delta_2,\ldots$, $\alpha_1,\alpha_2,\ldots,\rho_1,\rho_2,\ldots$ are positive constants depending on λ_o and $\hat{u}_o$. Because of [17, Theorem 2.1] the Legendre transforms $l_j : x \longrightarrow \nabla\hat{u}_j$ $(j = 0,1,2)$ are diffeomorphisms of the sets $\bar{\Omega}_{\lambda_j}$ onto $D \cup \Gamma$, where

$$\Gamma := \gamma(S^2) \quad , \quad D := (\mathbb{R}^3 \setminus \bar{\Omega}_\gamma) \setminus \{0\} .$$

Because of (1), $D \cup \{0\}$ is a bounded region in $\mathbb{R}^3$, starlike with respect to the origin. The inverse transforms l_j^{-1} $(j = 0,1,2)$ with functions $v_j \in C^2(D \cup \Gamma)$,

$$(13) \qquad v_j(z) := z\cdot l_j^{-1}(z) - \check{u}_j \circ l_j^{-1}(z) \quad , \quad z \in D \cup \Gamma ,$$

can be written as Legendre transforms $l_j^{-1} : z \longrightarrow \nabla v_j(z)$.

According to [17, Theorem 2.2] the functions v_j $(j = 0,1,2)$ have the asymptotics

$$v_j = -2c_j^{1/2} + O(r^{3/2}) \quad \text{as} \quad r(z) = |z| \longrightarrow 0 , \tag{14}$$

which can be differentiated twice, the derivative of an $O(r^s)$-term yielding an $O(r^{s-1})$-term; the constants c_j are those from the asymptotics (4). Because of $\Delta \check{u}_j = \Delta\psi =: f$ and $l_j^{-1} = \nabla v_j$ the v_j satisfy the differential equation

$$\text{Trace}(\nabla^2 v_j)^{-1} = f(\nabla v_j) \text{ in } D ,$$

and (6) and (13) yield the boundary condition

$$(\Re v_j)\circ\gamma = U + \beta_j + b_i^j B_i \quad , \quad \Re v_j(z) := z\cdot\nabla v_j(z) - v_j(z) .$$

For the differences $v := v_2 - v_1$, $\beta := \beta_2 - \beta_1$, $b := b^2 - b^1$ one then has the boundary condition

$$(\Re v)\circ\gamma = \beta + b_i B_i \tag{15}$$

and with $f_m := \partial_m f = \partial_m \Delta\psi$ the differential equation

$$\begin{aligned} &\left\{-c_o r^{-1}\left[\text{Trace}\left((\nabla^2 v_o)^{-1}\nabla^2 v(\nabla^2 v_o)^{-1}\right) + f_m(\nabla v_o)\partial_m v\right]\right\} + \\ &\left\{-c_o r^{-1}\left[\text{Trace}\left((\nabla^2 v_1)^{-1}\nabla^2 v(\nabla^2 v_2)^{-1} - (\nabla^2 v_o)^{-1}\nabla^2 v(\nabla^2 v_o)^{-1}\right)\right.\right. \\ &\left.\left. + f(\nabla v_2) - f(\nabla v_1) - f_m(\nabla v_o)\partial_m v\right]\right\} = 0 \quad \text{in} \quad D . \end{aligned} \tag{16}$$

The term between the first, second pair of braces in (16) we write as $L(\cdot,\partial)v$, $\tilde{L}(\cdot,\partial)v$, respectively, with linear differential operators

$$\begin{aligned} L(z,\partial) &:= -a_{mn}(z)\partial_m\partial_n + a_m(z)\partial_m , \\ \tilde{L}(z,\partial) &:= -\tilde{a}_{mn}(z)\partial_m\partial_n + \tilde{a}_m(z)\partial_m . \end{aligned} \tag{17}$$

With the notation v_j^{mn} for the component m,n of $(\nabla^2 v_j)^{-1}$, their coefficients are:

$$a_{mn}(z) := c_o|z|^{-1}v_o^{mk}(z)v_o^{kn}(z) \quad , \quad a_m(z) := -c_o|z|^{-1}f_m(\nabla v_o(z)) \ ,$$

$$\tilde{a}_{mn}(z) := c_o|z|^{-1}\left(v_1^{mk}(z)v_2^{kn}(z) - v_o^{mk}(z)v_o^{kn}(z)\right)$$

$$\tilde{a}_m(z) \ := -c_o|z|^{-1} \int_o^1 \left[f_m\left(\nabla v_1(z) + t\nabla v(z)\right) - f_m(\nabla v_o(z))\right]dt \ .$$

With

(18) $a_{mn}^o(z) := |z|^2\left(\delta_{mn} + 3z_m z_n|z|^{-2}\right)$ (δ_{mn} : Kronecker delta)

and

$$\alpha := c_o c_1^{-1/2} c_2^{-1/2} - 1$$

the coefficients satisfy the following conditions:

(19) $$a_{mn} = a_{nm} \ , \ \tilde{a}_{mn}, a_m, \tilde{a}_m \in C^o(D \cup \Gamma) \ ,$$

(20) $$\bigvee_{E>0} \ \bigwedge_{\xi\in\mathbb{R}^3} \ \bigwedge_{z\in D\cup\Gamma} a_{mn}(z)\xi_m\xi_n \geq E|z|^2|\xi|^2 \ ,$$

$$|a_{mn}-a_{mn}^o| = O(r^{-3}) \quad \text{as} \quad r(z) \longrightarrow 0 \ ,$$

$0 \notin \operatorname{supp} a_m$ (since the support $\operatorname{supp}\psi$ of ψ is compact),

(21) $$B(0,\rho_1) := \left\{z \in \mathbb{R}^3 : |z-0| < \rho_1\right\} \subset (D\setminus\operatorname{supp}\tilde{a}_m) \cup \{0\} \ ,$$

as well as

(22) $$|\alpha| + \sup_{D\cup\Gamma} r^{-3}\left(\sum_{m,n} |\tilde{a}_{mn}-\alpha a_{mn}^o|^2\right)^{1/2} + \sup_{D\cup\Gamma} r^{-2}\left(\sum_m \tilde{a}_m^2\right)^{1/2} < \alpha_1 \sum_{j=1}^{2} |\lambda_o-\lambda_j|_1 \ ,$$

if just $|\lambda_o-\lambda_j|_1 < \delta_2$. (21) and (22) are [18, Lemma 4.2].

In the following section boundary-value problems for L will be examined in appropriate weighted Sobolev spaces. With $k \in \mathbb{N}_o := \{0,1,2,\dots\}$, $s \in \mathbb{R}$, $p \in (1,\infty)$ and an open subset $\Omega \subset \mathbb{R}^3$ we define

$$W_s^{k,p}(\Omega) := \Big\{u \in L^p_{loc}(\Omega\setminus\{0\}) : r^{s/p+|\upsilon|}\,\partial^\upsilon u \in L^p(\Omega) \quad \text{for all}$$

$$\upsilon := (\upsilon_1,\upsilon_2,\upsilon_3) \in \mathbb{N}_o^3 \quad \text{with} \quad |\upsilon| := \upsilon_1 + \upsilon_2 + \upsilon_3 \le k\Big\} \tag{23}$$

and

$$\|u\|_{k,s,p,\Omega} := \Big(\sum_{0\le|\upsilon|\le k} \int_\Omega r^{s+p|\upsilon|}\,|\partial^\upsilon u|^p dx\Big)^{1/p}$$

as norm in $W_s^{k,p}(\Omega)$. For $s \in (-3-p/2,-3)$ let

$$X_s^p := \mathrm{span}(r^{1/2}) + W_{s-p}^{2,2}(D) := \Big\{ar^{1/2} + \tilde{u} : a \in \mathbb{R}\ ,\ \tilde{u} \in W_{s-p}^{2,p}(D)\Big\}.$$

With

$$\|\,ar^{1/2}+\tilde{u}\,\|_{(s,p)} := \Big(|a|^p + \|\tilde{u}\|^p_{2,s-p,p,D}\Big)^{1/p}$$

as norm X_s^p is a Banach space and with (14) $v \in X_s^p$ follows.

We now fix $p > 3$. Then according to the Sobolev embedding theorem the mappings

$$H' : X_s^p \longrightarrow \mathbb{R}\ ,\ w \longrightarrow 2\Big[\lambda_o(\zeta^1) - \lambda_o(\zeta^2)\Big]\cdot\Big[\nabla w(\gamma(\zeta^1)) - \nabla w(\gamma(\zeta^2))\Big]\ , \tag{24}$$

$$\tilde{H} : X_s^p \longrightarrow \mathbb{R}\ , \tag{25}$$

$$w \longrightarrow \Big[(\lambda_1+\lambda_2-2\lambda_o)(\zeta^1) - (\lambda_1+\lambda_2-2\lambda_o)(\zeta^2)\Big]\cdot\Big[\nabla w(\gamma(\zeta^1)) - \nabla w(\gamma(\zeta^2))\Big]$$

are continuous linear functionals and (2) yields

$$H'(v) + \tilde{H}(v) = \Big|\lambda_2(\zeta^1) - \lambda_2(\zeta^2)\Big|^2 - \Big|\lambda_1(\zeta^1) - \lambda_1(\zeta^2)\Big|^2 = 0\ . \tag{26}$$

With π as quotient mapping from $W^{1-1/p,p}(S^2)$ onto $W^{1-1/p,p}(S^2)/\mathrm{span}(1)$ we define the continuous linear operators

$$A,\tilde{A} : X_s^p \times \mathbb{R}^3 \longrightarrow W_{s-p}^{o,p}(D) \times \Big(W^{1-1/p,p}(S^2)/\mathrm{span}(1)\Big) \times \mathbb{R}$$

$$A(w,y) := \Big(L(\cdot,\partial)w\ ,\ \pi((\mathcal{R}w)\circ\gamma - y_iB_i)\ ,\ H'(w)\Big) \tag{27}$$

$$\tilde{A}(w,y) := \Big(\tilde{L}(\cdot,\partial)w\ ,\ 0\ ,\ \tilde{H}(w)\Big)\ .$$

Combining (14),(15),(16), and (26) now, we get

$$(v,b) \in N(A+\tilde{A}) := \left\{ w : (A+\tilde{A})w = 0 \right\} .$$

According to the corollary from §3 following below, A is a Fredholm operator of index 0, and the norm of $\tilde{A}$ because of (22) and (25) is estimated by

$$\|\tilde{A}\| \leq \alpha_2 \sum_{j=1}^{2} |\lambda_j - \lambda_0|_1$$

if $|\lambda_j-\lambda_o|_1 < \delta_2$. The stability theorem for Fredholm operators (see e.g. [13]) yields the existence of a constant δ_3 depending upon A and thereby upon λ_o, $\hat{u}_o$ only such that $A + \tilde{A}$ is a Fredholm operator of index 0 with

$$\dim N(A+\tilde{A}) \leq \dim N(A)$$

if $\|\tilde{A}\| < \delta_3$. Therefore with

$$\delta_o := \min \{\delta_2, \delta_3/(2\alpha_2)\}$$

the theorem is proved if

$$N(A) = \{0\}$$

can be shown.

For the proof of this one convinces oneself that for each $(w,y) \in N(A)$ the function $\tilde{w} := w \circ l_o$ solves the linear Molodenski problem $\mathcal{M}(\lambda_o,\hat{u}_o,0,\alpha + y_iB_i)$ with a constant function α and at ∞ has the asymptotics (11). Therefore $\tilde{w} = \alpha e$, and with (12) $\alpha = 0$ follows since $\alpha e_o = H'(w) = 0$. Now from (10) and the choice of the B_1,B_2,B_3, it follows that $\tilde{w}$ and thereby w vanishes.

One can prove a local existence theorem for the fbp as in [19] by determining by means of the implicit function theorem (v,β,b) near to $(v_o,\beta_o,0)$ as solution of the non-linear boundary-value problem

$$\begin{gathered} \operatorname{Trace}(\nabla^2 v)^{-1} - f(\nabla v) = 0 \quad \text{in } D \\ (\mathcal{R}v)\circ\gamma = U + \beta + b_iB_i \end{gathered}$$

$$H(v) := \left|\nabla v(\gamma(\zeta^1)) - \nabla v(\gamma(\zeta^2))\right|^2 = d^2$$

and then obtaining $\lambda = (\nabla v) \circ \gamma$ as free boundary.

In order to solve this non-linear boundary-value problem one introduces appropriately weighted Hölder spaces: with $\epsilon \in (0,1)$, $q \in \mathbb{R}$ we define

$$C_q^o := \Big\{ f \in C^{o,\epsilon}(D) : \bigvee_{f_o \in \mathbb{R}} \infty > \langle f\rangle := \sup_{z\in D} |z|^{-\epsilon/2} \Big| |z|^q f(z) - f_o \Big| + \sup_{\delta>0} \sup_{z,z'\in D\setminus B(0,\delta), z\neq z'} \delta^{\epsilon/2} |z-z'|^{-\epsilon} \Big| |z|^q f(z) - |z'|^q f(z') \Big| \Big\}$$

with the norm

$$|[f]| := |f_o| + \langle f\rangle$$

as well as

$$\tilde{C}_q^o := \Big\{ f \in C_q^o(D) : f_o = \lim_{z\to 0} |z|^q f(z) = 0 \Big\} .$$

Finally, let Y_o be the space of all $f \in C^{2,\epsilon}(D)$, which can be written in the form

$$f(z) = \alpha |z|^{1/2} + \alpha_{ij} z_i z_j |z|^{-1/2} + \tilde{f}(z)$$

with $\alpha, \alpha_{ij} = \alpha_{ji} \in \mathbb{R}$ and a function $\tilde{f} \in \tilde{C}^o_{-3/2}$, whose partial derivatives of first, second order are lying in $\tilde{C}^o_{-1/2}$, $\tilde{C}^o_{1/2}$, respectively. The spaces introduced are Banach spaces where the norm in Y_o is defined in an obvious way (cf. [19]).

If one now assumes $\gamma \in C^{2,\epsilon}(S^2;\mathbb{R}^3)$ to hold and if $\mathcal{U}$ denotes an appropriate Y_o-neighborhood of v_o (cf. (13)), then by means of

$$\phi : \mathcal{U} \times \mathbb{R} \times \mathbb{R}^3 \longrightarrow C^o_{-3/2} \times C^{1,\epsilon}(S^2) \times \mathbb{R}$$

$$(v,\beta,b) \longrightarrow \Big(c_o r^{-1} (\mathrm{Trace}(\nabla^2 v)^{-1} - f(\nabla v)), (\mathcal{R}v)\circ\gamma - \beta - b_i B_i, H(v) \Big)$$

a continuously differentiable function is defined whose derivative ϕ'_o in $(v_o, \beta_o, 0)$ with the notations from (17),(24) is given by

$$\phi_o'(w,\alpha,a) = \left(L(\cdot,\partial)w \ , \ (\mathcal{R}w)\circ\gamma - \alpha - a_i B_i \ , \ H'(w)\right) .$$

At $(v_o,\beta_o,0)$ ϕ meets the requirements of the theorem on the inverse function and thus to each (U,d^2) from a $C^{1,\epsilon}(S^2)\times\mathbb{R}$-neighborhood of (U_o,d_o^2) there is a solution (v,β,b) of

$$\phi(v,\beta,b) = (0,U,d^2)$$

in a $Y_o\times\mathbb{R}\times\mathbb{R}^3$-neighborhood of $(v_o,\beta_o,0)$. This solution can for instance be approximated by means of the Picard iteration scheme $(b^o := 0)$

$$\bigwedge_{n\in\mathbb{N}} (v_n,\beta_n,b^n) = (v_{n-1},\beta_{n-1},b^{n-1}) - \tag{28}$$

$$(\phi_o')^{-1}\left(\phi(v_{n-1},\beta_{n-1},b^{n-1}) - (0,U,d^2)\right) .$$

This way one obtains the following existence theorem

<u>Theorem 2:</u> *Let there be given the data* $(U,\gamma,d) \in \mathcal{D}$ *with* $U \in C^{1,\epsilon}(S^2)$, $\gamma \in C^{2,\epsilon}(S^2,\mathbb{R}^3)$ *as well as a telluroid* λ_o *with normal potential* $\hat{u}_o$ *, and the conditions (7-10) and (12) shall be fulfilled. Then there is a* $C^{1,\epsilon}(S^2)\times\mathbb{R}$*-neighborhood* $\mathcal{W}$ *of* (U_o,d_o) *as well as a* $C^{1,\epsilon}(S^2,\mathbb{R}^3)\times\mathbb{R}\times\mathbb{R}^3$*-neighborhood* $\mathcal{V}$ *of* $(\lambda_o,\beta_o,0)$ *such that for* (U,γ,d) *there is a solution* $(\lambda,\hat{u},\beta,b) \in \mathcal{L}$ *of the fbp with* $(\lambda,\beta,b) \in \mathcal{V}$ *if* $(U,d) \in \mathcal{W}$.

The sequence $(\lambda_n,\beta_n,b^n)_{n\in\mathbb{N}}$ *given by (29)-(33) converges.to* (λ,β,b) *in* $C^{1,\epsilon}(S^2,\mathbb{R}^3)\times\mathbb{R}\times\mathbb{R}^3$ *:*

$$\bigwedge_{n\in\mathbb{N}_o} \lambda_n := -\left[(\nabla^2\check{u}_o)^{-1}\nabla w_n\right] \circ \lambda_o \tag{29}$$

$$w_o(x) := -x\nabla\check{u}_o(x) + \check{u}_o(x) \quad \textit{for} \quad x \in \bar{\Omega}_{\lambda_o} \ , \ b^o := 0 \tag{30}$$

$$\bigwedge_{n\in\mathbb{N}} w_n := \tilde{w}_n + (\beta_n/e_o)e \ , \quad b^n := \tilde{b}^n + (\beta_n/e_o)b^e \quad \textit{(cf. (12))}; \tag{31}$$

here $(\tilde{w}_n,\tilde{b}^n)$ *is a solution of the linear Molodenski problem*

$\mathcal{M}(\lambda_o, \hat{u}_o, g_n, U + \tilde{b}_i^n B_i)$ *with the asymptotics* (11), *and*

$$
\begin{aligned}
g_n &:= \Delta w_{n-1} + \mathrm{Trace}\left\{(\nabla^2 \check{u}_o)^2 \left[\nabla^2 w_{n-1} - \check{u}_o^{ij}\partial_j w_{n-1}\nabla^2(\partial_i \check{u}_o)\right]^{-1}\right\} \\
&\quad + f\left(-(\nabla^2 \check{u}_o)^{-1}\nabla w_{n-1}\right) ,
\end{aligned} \tag{32}
$$

finally, with $x^1 := \lambda_o(\zeta^1)$, $x^2 := \lambda_o(\zeta^2)$:

$$
\begin{aligned}
\beta_n &:= d^2 - \left|\lambda_{n-1}(\zeta^1) - \lambda_{n-1}(\zeta^2)\right|^2 \\
&\quad - 2(x^1 - x^2)\cdot\left(\lambda_{n-1}(\zeta^1) - \lambda_{n-1}(\zeta^2)\right) \\
&\quad - 2(x^1 - x^2)\cdot\left((\nabla^2 \check{u}_o(x^1))^{-1}\nabla\tilde{w}_n(x^1) - (\nabla^2 \check{u}_o(x^2))^{-1}\nabla\tilde{w}_n(x^2)\right) .
\end{aligned} \tag{33}
$$

One obtains the sequence (w_n, β_n, b^n) defined recursively in (29)-(33) from the Picard iteration scheme (28) by the transformation

$$w_n := -v_n \circ l_o^{-1} ,$$

whereby boundary-value problems in D are becoming boundary-value problems in Ω_{λ_o} .

Finally, the regularity theorem [19, theorem 4.1] can be adopted without modification:

<u>Theorem 3:</u> *If* $(\lambda, \hat{u}, \beta, b) \in \mathcal{L}$ *is a solution for* $(U, \gamma, d) \in \mathfrak{D}$ *and* det $\nabla^2\hat{u}$ *does not vanish in a neighborhood of* $\lambda(S^2)$ *(which is true for the solutions from Theorem* 2), *then (i)-(iii) hold:*

(i) *If* $\gamma \in C^3(S^2, \mathbb{R}^3)$, $U \in C^2(S^2)$, *then* $\lambda \in C^{1,\epsilon}(S^2, \mathbb{R}^3)$.

(ii) *If* $\lambda \in C^{1,\epsilon}(S^2, \mathbb{R}^3)$, $\gamma \in C^{k,\epsilon}(S^2, \mathbb{R}^3)$, $U \in C^{k,\epsilon}(S^2, \mathbb{R}^3)$ *with an integer* $k \geq 2$, *then* $\lambda \in C^{k,\epsilon}(S^2, \mathbb{R}^3)$.

(iii) *If* γ, U *are analytic, then also* λ *is analytic.*

§3 <u>Linear Boundary-Value Problems in Weighted Sobolev Spaces</u>

Let $D \subset \mathbb{R}^3$ denote a bounded region as in §2: $0 \in \partial D$ and $D \cup \{0\}$ is a region of class C^2 with boundary $\Gamma := \partial D \setminus \{0\}$. In D regular boundary-value problems for the differential operator

$$L(z,\partial) := -a_{ij}(z)\partial_i\partial_j + a_i(z)\partial_i + a(z).,$$

elliptic in the sense of (20), will be examined. The coefficients $a_{ij} = a_{ji}, a_i, a$ are real-valued continuous functions on $D \cup \Gamma$ and at 0 have the asymptotics

$$r^{-2}|a_{ij}-a^o_{ij}| + r^{-1}|a_i| + |a| \longrightarrow 0 \ , \quad i,j \in \{1,2,3\} \ , \quad a^o_{ij} \text{ as in (18)}$$

with order of convergence still to be fixed.

For $\mu \in \mathbb{N}_o$, $\kappa \in \mathbb{Z} \cap [-\mu,\mu]$ let $S_{\mu,\kappa}$ be a spherical harmonic of degree μ so that $\{S_{\mu,\kappa}\}_{\kappa\in\mathbb{Z}\cap[-\mu,\mu],\mu\in\mathbb{N}_o}$ is a complete orthonormal system in $L^2(S^2)$. $S_{\mu,\kappa}$ we think of as being extended onto all of $\mathbb{R}^3 \setminus \{0\}$ such that it becomes a function homogeneous of degree 0. The functions

$$s_{\mu,\kappa} := \left(r^{(\mu+1)/2}\, S_{\mu,\kappa}\right)\Big|_D$$

in D satisfy the differential equation

$$L^o s_{\mu,\kappa} := -a^o_{ij}\partial_i\partial_j s_{\mu,\kappa} = 0 \ .$$

The boundary-value problems will be examined in the spaces

$$(34) \quad X^{p,\sigma}_s(D) := \text{span}\,\{s_{\mu,\kappa}\}_{\kappa\in\mathbb{Z}\cap[-\mu,\mu],\mu\in\mathbb{N}_o\cap[0,\sigma-1]} \oplus W^{2,p}_{s-\sigma p/2}(D)$$

with $\sigma \in \mathbb{N}_o$, $p \in (1,\infty)$ and

$$(35) \qquad s \in (-3-p/2,-3) \ ;$$

with regard to some properties of the weighted Sobolev spaces $W^{k,p}_t(D)$ defined in (23) it is referred to [2] and [15]. The aim is to prove normal solvability in $X^{p,\sigma}_s$ of the boundary-value problem for $Lu = f$ for right hand sides f from $W^{o,p}_{s-\sigma p/2}$.

Because of (35) the sum in (34) is direct, and for $\mu \geq \sigma$ $s_{\mu,\kappa} \in W^{2,p}_{s-\sigma p/2}(D)$. Finally, with $c_{\mu,\kappa} \in \mathbb{R}$, $\tilde{u} \in W^{2,p}_{s-\sigma p/2}(D)$ the norm in the Banach space $X^{p,\sigma}_s(D)$ shall be defined by

$$\Big\| \sum_{\mu=0}^{\sigma-1} \sum_{\kappa=-\mu}^{\mu} c_{\mu k} s_{\mu\kappa} + \tilde{u} \Big\|_{(\sigma,s,p)} := \Big(\sum_{\mu=0}^{\sigma-1} \sum_{\kappa=-\mu}^{\mu} |c_{\mu k}|^p + \|\tilde{u}\|^p_{2,s-\sigma p/2,p,D} \Big)^{1/p} .$$

By means of the transformation $T : x \longrightarrow |x|x$, L^o basically is transformed into $-\Delta$:

$$-\Delta \left((r^{-1/2}v) \circ T \right) = (r^{-3/2} L^o v) \circ T . \tag{36}$$

This connection is used to obtain similar results for the behavior of L^o in weighted spaces as for instance McOwen [8] for the Laplacian; contrary to that though here 0 is the singular point and not ∞.

The fundamental solution of the Laplacian

$$k_o(x,y) := \left(4\pi |x-y|\right)^{-1} , \quad x,y \in \mathbb{R}^3 , \ x \neq y$$

has the series expansion

$$k_o(x,y) = \sum_{\mu=0}^{\infty} \sum_{\kappa=-\mu}^{\mu} (2\mu+1)^{-1} S_{\mu\kappa}(x) S_{\mu\kappa}(y) |x|^{\mu} |y|^{-\mu-1} ,$$

absolutely and uniformly convergent with respect to $|x| < R$ and $|y| > R + \delta$ for each $R > 0$, $\delta > 0$. For $\sigma \in \mathbb{N}$ let

$$k_\sigma(x,y) := k_o(x,y) - \sum_{\mu=0}^{\sigma-1} \sum_{\kappa=-\mu}^{\mu} (2\mu+1)^{-1} S_{\mu\kappa}(x) S_{\mu\kappa}(y) |x|^{\mu} |y|^{-\mu-1} .$$

For $\sigma \in \mathbb{N}_o$

$$|k_\sigma(x,y)| \leq \left\{ \begin{array}{ll} c|x|^{\sigma-1}|y|^{-\sigma} & \text{for } |y| < |x|/2 \\ c|x-y|^{-1} & \text{for } |x|/2 \leq |y| \leq 2|x| \\ c|x|^{\sigma}|y|^{-\sigma-1} & \text{for } 2|x| < |y| \end{array} \right\} \leq c \left(\frac{|x|}{|y|}\right)^{\sigma} |x-y|^{-1}. \tag{37}$$

Here and from now on c stands for a positive constant whose numerical value can change during calculation and only depends upon specific quantities - σ in this case. The estimations of k_σ for the sets $|y| < |x|/2$ as well as $|x|/2 \leq |y| \leq 2|x|$ are trivial; one obtains it for the set $2|x| < |y|$ by using

$$k_\sigma(x,y) = |y|^{-1}k_\sigma\left(|y|^{-1}x, |y|^{-1}y\right)$$

from the fact that $k_\sigma(\cdot, |y|^{-1}y)$ is just the remainder of the Taylor expansion of order $(\sigma-1)$ at 0 for the function $k_o(\cdot, |y|^{-1}y)$.

Estimate (37) together with a theorem by Stein and Weiss [14, theorem B^*] (cf. also [9, Lemma 2.1]) shows that the integral operator given by

(38) $$(K_\sigma f)(x) := \int_{\mathbb{R}^3} k_\sigma(x,y)f(y)dy$$

maps the space $W^{o,p}_s(\mathbb{R}^3)$ continuously into $W^{o,p}_{s-2p}(\mathbb{R}^3)$ if $(2-\sigma)p-3 < s < (3-\sigma)p-3$. With inequality (2.7) from [9] one then obtains

<u>Lemma 1:</u> *For each* $\sigma \in \mathbb{N}_o$, $p \in (1,\infty)$, *and* $s \in \left((2-\sigma)p-3, (3-\sigma)p-3\right)$ *the integral operator given by (38) maps the space* $W^{o,p}_s(\mathbb{R}^3)$ *continuously into* $W^{2,p}_{s-2p}(\mathbb{R}^3)$.

From this follows:

<u>Lemma 2:</u> *Let* $p \in (1,\infty)$, $s \in (-p/2-3, -3)$, $\sigma \in \mathbb{N}_o$, *and* $R > 0$. *For each* $f \in W^{o,p}_s(B^R)$ 4) *there is a unique solution* $u \in W^{2,p}_s(B^R)$ *of the Dirichlet problem* "$L^o u = f$ *in* B^R , $u|_{\partial B(0,R)} = 0$ ", *and with a constant* $c > 0$ *independent of* f *and* R

(39) $$\|u\|_{2,s,p,B^R} \leq c\|f\|_{o,s,p,B^R}$$

holds.

---

4) $B(x,R)$: open ball with center $x \in \mathbb{R}^3$ and with radius R , $B^R := B(0,R)\setminus\{0\}$.

If $f \in W^{o,p}_{s-\sigma p/2}(B^R) \subset W^{o,p}_{s}(B^R)$ *, then* u *is in* $X^{p,\sigma}_{s}(B^R)$ *and with some constant independent of* f

$$\|u\|_{(\sigma,s,p)} \leq c\|f\|_{o,s-\sigma p/2,p,B^R} \tag{40}$$

holds.

Proof: It suffices to look at the case $R = 1$. In particular the independence of the constant c in (39) from R one obtains by coordinate transformation $x \longrightarrow Rx$. We shall write B instead of B^1.

The transformation $T : x \longrightarrow x|x|$ is a diffeomorphism from B onto itself, for each $t \in \mathbb{R}$, $k \in \mathbb{N}_o$, $p \in (1,\infty)$ induces a topological isomorphism

$$W^{k,p}_{t}(B) \longrightarrow W^{k,p}_{2t+3}(B) \quad , \quad w \longrightarrow w \circ T ,$$

and for L^o, $-\Delta$ and T (36) holds. From this follows the fact that $u \in W^{2,p}_{s}(B)$ solves the Dirichlet problem with right hand side $f \in W^{o,p}_{s}(B)$ if and only if for $v := (r^{-1/2}u)\circ T$ with $g := (r^{-3/2}f)\circ T \in W^{o,p}_{t}(B)$ and $t := 2s + 3p + 3 \in (2p-3,3p-3)$:

$$-\Delta v = g \text{ in } B \quad , \quad v \in W^{2,p}_{t-2p}(B) \quad , \quad v|_{\partial B(0,1)} = 0 \tag{41}$$

holds.

A solution for (41) one easily obtains in the form

$$v = K_o g + \hat{v}$$

with a potential function $\hat{v}$ in $B(0,1)$, which meets some appropriate boundary condition; in particular

$$\hat{u} := r^{1/2}(\hat{v}\circ T) \in X^{p,\sigma}_{s}(B)$$

for each $\sigma \in \mathbb{N}_o$.

If $f \in W^{o,p}_{s-\sigma p/2}(B)$, then $g \in W^{o,p}_{t-\sigma p}(B)$ and

$$K_o g = \sum_{\mu=0}^{\sigma-1} \sum_{\kappa=-\mu}^{\mu} (2\mu+1)^{-1} r^{\mu} S_{\mu\kappa} \cdot \int_B g(y)|y|^{-\mu-1} S_{\mu\kappa}(y)dy + K_\sigma g$$

with $K_\sigma g \in W^{2,p}_{t-\sigma p-2p}(B)$ because of Lemma 1. From this one obtains

$$r^{1/2}\big((K_o g)\circ T^{-1}\big) \in X^{p,\sigma}_s(B) .$$

To see that the solution of (41) is unique, one convinces oneself by means of Hölder's inequality that $W^{o,p}_{t-2p}(B) \subset L^1(B)$. Therefore solutions of (41) can be looked upon as being distributions in $B(0,1)$, and one can easily show that a solution of the homogeneous problem (41) solves Laplace's equation in $B(0,R)$ in the sense of distributions and hence vanishes by Weyl's lemma.

Finally one gets (39) and (40) from the Bounded-Inverse theorem.

From the Lemmas 1 and 2 one obtains the a priori esimates needed for the Fredholm theorem:

Lemma 3: *Let* $G := g_k\partial_k + g_o$ *be a differential operator on* Γ *with* $g_o, g_1, g_2, g_3 \in C^1(\Gamma)$ *such that* $(g_1,g_2,g_3)^t$ *nowhere is tangential to* Γ . *With* $\sigma \in \mathbb{N}_o$ *the coefficients of* L *at* 0 *shall have the asymptotics*

$$(42)\quad r^{-2}|a_{ij}-a^o_{ij}| + r^{-1}|a_i| + |a| = o(r^{\sigma/2}) \quad , \quad i,j \in \{1,2,3\} .$$

Moreover, let $1 < p < q \leq \infty$, $s \in (-p/2-3,-3)$, $t \in (-q/2-3,-3)$ *with* $qs - pt > 3p - 3q$. *Then for* $u \in W^{2,p}_s(D)$ *the following holds:*

(i) *If* $Lu \in W^{o,q}_t(D)$ *and* $Gu \in W^{1-1/q,q}(\Gamma)$, *then* u *lies in* $W^{2,q}_t(D)$.

(ii) *If* $Lu \in W^{o,p}_{s-\sigma p/2}(D)$, *then* $u \in X^{p,\sigma}_s(D)$.

(iii) *For all* $u \in X^{p,\sigma}_s(D)$

$$(43)\quad \|u\|_{(\sigma,s,p)} \leq c\left(\|Lu\|_{o,s-\sigma p/2,p,D} + \|Gu\|_{1-1/p,p,\Gamma} + \|u\|_{o,p,D\setminus B^\rho}\right)$$

with constants $c,\rho > 0$ *independent of* u .

Here $\|\cdot\|_{1-1/p,p,\Gamma}$, $\|\cdot\|_{k,p,\Omega}$ $(k \in \mathbb{N}_o$, $\Omega \subset \mathbb{R}^3$ *open) are the norms in* $W^{1-1/p,p}(\Gamma)$, $W^{k,p}(\Omega)$, *respectively.*

Proof: In case (i) known regularity results [1,12] yield that $u \in W^{2,q}(D\setminus B^\rho)$ holds for each $\rho > 0$. Because of (42) and Lemma 2 for sufficiently small $R > 0$ the Dirichlet problems "$Lu = f$ in B^R ,

$u|_{\partial B(0,R)} = 0$" for $f \in W_s^{o,p}(B^R)$, $f \in W_t^{o,q}(B^R)$ are uniquely solvable in $W_s^{2,p}(B^R)$, $W_t^{2,q}(B^R)$, respectively. If $\chi \in C_o^\infty(B(0,R))$ with $\chi \equiv 1$ in $B(0,R/2)$ and R so small that $B^R \subset D$, then χu is a solution in $W_s^{2,p}(B^R)$ of this Dirichlet problem corresponding to right hand side $f := L(\chi u) \in W_s^{o,p}(B^R)$. If v is the solution in $W_t^{2,q}(B^R)$ of this Dirichlet problem then it follows that $v = \chi u$; for the Hölder inequality yields $W_t^{k,q}(B^R) \subset W_s^{k,p}(B^R)$ for each $k \in \mathbb{N}_o$ if $qs-pt > 3p-3q$. With that (i) is proved.

With Lemma 2 and (42) one obtains (ii) from $L^o u = (Lu) + (L^o-L)u \in W^{o,p}_{s-\sigma p/2}(D)$.

For the proof of (iii) one chooses R and χ as above and decomposes u as $u = \chi u + (1-\chi)u$. The Bounded-Inverse theorem guarantees the validity of (40) also for solutions of the Dirichlet problems for the operator L :

$$\|\chi u\|_{(\sigma,s,p)} \le \|L(\chi u)\|_{o,s-\sigma p/2,p,D} \,.$$

One estimates $(1-\chi)u$ by means of the usual elliptic L^p-estimate:

$$\|(1-\chi)u\|_{2,p,D\setminus B^{R/2}} \le c\Big(\|L[(1-\chi)u]\|_{o,p,D\setminus B^{R/2}} + \|Gu\|_{1-1/p,p,\Gamma} + \|(1-\chi)u\|_{o,p,D\setminus B^{R/2}}\Big) \,.$$

Therefore it follows

$$\begin{aligned}\|u\|_{(\sigma,s,p)} &\le c\Big(\|\chi u\|_{(\sigma,s,p)} + \|(1-\chi)u\|_{2,p,D\setminus B^{R2}}\Big)\\ &\le c\Big(\|Lu\|_{o,s-\sigma p/2,p,D} + \|Gu\|_{1-1/p,p,\Gamma} + \|u\|_{o,p,D\setminus B^{R/2}}\\ &\qquad + \|u\|_{1,p,B^R\setminus B^{R/2}}\Big) \,.\end{aligned}$$

If one applies an inner elliptic L^p-estimate to the last term of the sum, the assertion follows with $\rho < R/2$.

In the following theorem the Fredholm theory is summarized:

<u>Theorem 4:</u> *With* L *and* G *as in Lemma* 3 *and for any* $p \in (1,\infty)$, $s \in (-p/2-3,-3)$, $\tau \in \mathbb{N}_o \cap [0,\sigma]$, *the operator*

$$F_s^{p,\tau} : X_s^{p,\tau}(D) \longrightarrow W_{s-\tau p/2}^{o,p}(D) \times W^{1-1/p,p}(\Gamma) \quad , \quad u \longrightarrow (Lu,Gu)$$

is a Fredholm operator of index 0 *and the kernels do not depend upon* p,s,τ .

Proof: Estimate (43) implies that $F_s^{p,\tau}$ is a semi-Fredholm operator with finite-dimensional kernel.[5] We first show that index and kernel do not depend upon p,s,τ and then determine the index of $F_{-7/2}^{2,0}$ as being equal to 0 .

For q_1,t_1 and q_2,t_2 with $t_i \in (-q_i/2-3,-3)$ and $1 < q_i < \infty$ there are $p > 1$ and $s \in (-q/2-3,-3)$ with $p \leq q_i$ and $q_i s - t_i p > 3p - 3q_i$ $(i \in \{1,2\})$. Therefore it suffices to show that kernel and index of $F_s^{p,o}$ and $F_t^{q,\tau}$ coincide if q and t are one of the numbers q_1,q_2 and t_1,t_2 , respectively, and $\tau \in \mathbb{N}_o \cap [o,\sigma]$.

Because of $X_t^{q,\tau}(D) \subset W_t^{2,q}(D) \subset W_s^{2,p}(D) = X_s^{p,o}(D)$ the kernel of $F_t^{q,\tau}$ is included in the kernel of $F_s^{p,o}$. On the other hand one obtains from Lemma 3, i,ii the reverse inclusion. It remains to show that the codimensions of the ranges $\operatorname{codim} R(F_t^{q,\tau})$ and $\operatorname{codim} R(F_s^{p,o})$ coincide.

If $d := \operatorname{codim} R(F_t^{q,\tau})$ then there is a d-dimensional subspace V of $W_{t-\tau q/2}^{o,q}(D)$ with $V \cap R(F_t^{q,\tau}) = \{0\}$. But then because of Lemma 3, i,ii also $V \cap R(F_s^{p,o}) = \{0\}$; therefore $\operatorname{codim} R(F_s^{p,o}) \geq d$.

If $\operatorname{codim} R(F_s^{p,o}) > d$ then there is a (d+1)-dimensional subspace V of $W_s^{o,p}(D)$ with $V \cap R(F_s^{p,o}) = \{0\}$. Since $C_o^\infty(D)$ and thereby also $W_{t-\tau q/2}^{o,q}(D)$ are dense in $W_s^{o,p}(D)$, V can even be chosen as subspace of $W_{t-\tau q/2}^{o,q}(D)$. But that yields the contradiction $\{0\} \neq V \cap R(F_t^{q,\tau}) \subset V \cap R(F_s^{p,o})$.

In order to determine the index of $F_{-7/2}^{2,0}$ let $\psi \in C^o(D \cup \Gamma)$ such that $\psi \geq 0$, $\psi \equiv 1$ near Γ and $\psi \equiv 0$ in a neighborhood of 0 . With $\lambda > 0$ we define

5) The results for Fredholm and semi-Fredholm operators used here can be found in [13].

$$F_\lambda : W^{2,2}_{-7/2}(D) \longrightarrow W^{0,2}_{-7/2}(D) \times W^{1/2,2}(\Gamma)$$

$$u \longrightarrow (L^o u + \lambda\psi u, Gu) \; .$$

For sufficiently large λ F_λ is a continuous bijection, therefore in particular Fredholm operator of index 0 : Namely, for sufficiently large λ the form corresponding to the boundary-value problem characterized by F_λ is strictly coercive with respect to $W^{1,2}_{-7/2}(D)$, and the Lax-Milgram theorem as well as a regularity theorem yield the unique solvability of $F_\lambda u = (f,g)$ for all right hand sides (cf. [16]).

For each $t \in [0,1]$ the operator $t\,F^{2,0}_{-7/2} + (1-t)F_\lambda$ is a continuous semi-Fredholm operator with index independent of t , therefore $\operatorname{ind} F^{2,0}_{-7/2} = \operatorname{ind} F_\lambda = 0$.

<u>Corollary:</u> A *in* (27) *is a Fredholm operator with index* 0 .

For $G := \mathfrak{R}$ and L from (27) Theorem 4 holds with $\sigma = 2$, and $X^{p,2}_s(D) = X^p_s \oplus \operatorname{span}\{s_{1,-1}, s_{1,0}, s_{1,1}\}$.
A is finite-dimensional perturbation of $A_2 \circ F^{p,2}_s \circ A_1$ with

$$A_1 = X^p_s \times \mathbb{R}^3 \longrightarrow X^{p,2}_s(D) \quad , \quad (u,b) \longrightarrow u \; ,$$

$$A_2 : W^{o,p}_{s-p}(D) \times W^{1-1/p,p}(\Gamma) \longrightarrow W^{o,p}_{s-p}(D) \times \left(W^{1-1/p}(S^2)/\operatorname{span}\{1\}\right) \times \mathbb{R}$$
$$(f,g) \longrightarrow \left(f, \pi(g\circ\gamma), 0\right) ,$$

and its index is 0 , the sum of the indices of $A_1, F^{p,2}_s$, and A_2 .

<u>References</u>

[1] Agmon, S.; Douglis, A.; Nirenberg, L.: Estimates near the boundary for elliptic partial differential equations satisfying general boundary conditions I. Comm. Pure Appl. Math. 12 (1959), 623-727.

[2] Edmunds, D.E.; Kufner, A.; Rákosnik, J.: Embeddings of Sobolev spaces with weights of power type. Z. Anal. Anw. 4 (1985), 25-34.

[3] Gilbarg, D.; Trudinger, N.S.: Elliptic partial differential equations of second order. Berlin, Heidelberg, New York: Springer 1977.

[4] Heiskanen, W.A.; Moritz, H.: Physical geodesy. San Francisco, London: Freeman 1967.

[5] Hörmander, L.: The boundary problems of physical geodesy. Arch. for Rational Mech. Anal. **63** (1976), 1-52.

[6] Holota, P.: Boundary value problems in physical geodesy: present state, boundary perturbation and the Green-Stokes representation. Proc. of the 1st Hotine-Marussi Symposium on Mathematical Geodesy at Rome, 1985; ed..: F. Sansó, Istituto di Topografia, Fotogrammetria e Geofesica, Politecnico di Milano, Piazza Leonardo da Vinci 32: Milano 1986.

[7] Kinderlehrer, D.; Nirenberg, L.: Regularity in free boundary problems. Ann. Scuola Norm. Sup. Pisa, Ser. IV, Vol. IV (1977), 373-391.

[8] McOwen, R.C.: The behaviour of the Laplacian on weighted Sobolev Spaces. Comm. Pure Appl. Math. 32 (1979), 783-795.

[9] Nirenberg, L.; Walker, H.F.: The null spaces of elliptic partial differential operators in $\mathbb{R}^n$. J. Math. Anal. Appl. **42** (1973), 271-301.

[10] Sansó, F.: The geodetic boundary value problem in gravity space. Atti Accad. Naz. Lincei Mem. Cl. Sci. Math. Natur. Sez. I^a, Ser. VIII, Vol. **XIV** (1977).

[11] Sansó, F.: The local solvability of Molodensky's problem in gravity space. Manuscripta Geodetica **3** (1978), 157-227.

[12] Schechter, M.: On L^p estimates and regularity, I. Amer. J. Math. **85** (1963), 1-13.

[13] Schechter, M.: Principles of functional analysis. New York: Academic Press 1971.

[14] Stein, E.M.; Weiss, G.: Fractional integrals on n-dimensional Euclidean space. J. Math. Mech. **7** (4) (1958), 503-514.

[15] Triebel, H.: A remark on a paper by D.F. Edmunds, A. Kufner and J. Rákosnik "Embedding of Sobolev spaces with weights of power type". Z. Anal. Anw. **4** (1985), 35-38.

[16] Witsch, K.J.: Ein schiefes Randwertproblem zu einer elliptischen Differentialgleichung zweiter Ordnung mit einer expliziten L_2-Abschätzung für die zweiten Ableitungen der Lösung. Math. Meth. in the Appl. Sci. **1** (1979), 214-240.

[17] Witsch, K.J.: A uniqueness result for the geodetic boundary problem. J. Diff. Equ. **38** (1980), 104-125.

[18] Witsch, K.J.: On a free boundary value problem of physical geodesy, I (Uniqueness). Math. Meth. in the Appl. Sci. **7** (1985), 269-289.

[19] Witsch, K.J.: On a free boundary value problem of physical geodesy, II (Existence). Math. Meth. in the Appl. Sci. **8** (1986), 1-22.